Methods in
Olfactory Research

Methods in Olfactory Research

edited by

D. G. MOULTON

Monell Chemical Senses Center and Department of Physiology, University of Pennsylvania, and Veteran's Administration Hospital, Philadelphia, Pennsylvania

AMOS TURK

Department of Chemistry, The City College of the City University of New York, N.Y.

J. W. JOHNSTON JR.

8940 Jones Mill Road, Chevy Chase, Maryland

1975

Academic Press

London . New York . San Francisco

A Subsidiary of Harcourt Brace Jovanovich, Publishers

ACADEMIC PRESS INC. (LONDON) LTD.
24-28 Oval Road,
London NW1

United States Edition published by
ACADEMIC PRESS INC.
111 Fifth Avenue
New York, New York 10003

Library of Congress Catalog Card Number: 73–19016
ISBN: 0-12-508950-3

Text set in 11/12 pt. IBM Press Roman, printed by photolithography, and bound in Great Britain at The Pitman Press, Bath

CONTRIBUTORS

H. ALTNER, *Facbereich Biologie, Universitat Regensburg, 84 Regensburg Universitatsstrasse 31, West Germany.*

M. G. M. BEETS, *Vice President, International Flavors and Fragrances, Liebergerweg 72–98, Post box 309, Hilversum, Netherlands.*

R. N. L. BROOKE, *Department of Anatomy, The University of Oxford, South Parks Road, Oxford, England.*

R. L. DOTY, *Monell Chemical Senses Center, University of Pennsylvania, 3500 Market Street, Philadelphia, Pennsylvania 19104.*

K. B. DØVING, *Institute of Zoophysiology, University of Oslo, P.O. Box 1051, Blindern, Oslo 3, Norway.*

A. DRAVNIEKS, *Odor Sciences Center, ITT Research Institute, 10 West 35th St., Chicago, Illinois 60616.*

R. C. GESTELAND, *Department of Biological Sciences, Northwestern University, Evanston, Illinois 60201.*

P. P. C. GRAZIADEI, *Department of Biological Sciences, Florida State University, Tallahassee, Florida 32306.*

I. KOLNBERGER, *Facbereich Biologie, Universitat Regensburg, 83 Regensburg Universitatsstrasse 31, West Germany.*

E. P. KÖSTER, *Psychologisch Laboratorium, Varkenmark 2, Utrecht, Netherlands.*

P. MACLEOD, *Laboratoire de Physiologie Sensorielle, College de France, 11 Place Marcelin Bertholet, Paris 53, France.*

D. G. MOULTON, *Monell Chemical Senses Center and Department of Physiology, University of Pennsylvania and Veterans Administration Hospital, Philadelphia, Pennsylvania 19104.*

A. J. PINCHING, *Department of Anatomy, The University of Oxford, South Parks Road, Oxford, England.*

R. M. SILVERSTEIN, *Department of Chemistry, SUNY College of Environmental Sciences and Forestry, Syracuse, New York 13210.*

D. A. STEVENS, *Department of Psychology, Clark University, Worcester, Massachusetts 01610.*

AMOS TURK, *Department of Chemistry, The City College of The City University of New York, New York, NY 19104.*

JONATHAN TURK, *Department of Chemistry, University of Colorado, Boulder, Colorado.*

J. C. YOUNG, *Agriculture Canada, Chemistry and Biology Research Institute, Ottawa, Ontario K1A OC6, Canada.*

PREFACE

This volume owes its conception to a NATO sponsored Summer School on 'Odour Perception: multidisciplinary research methods' held at the State University of Utrecht in the Netherlands in 1970 and organized by Dr. E. P. Köster. The success of the meeting underlined the need for a text that would not only summarize the increasingly specialized methods evolving in olfactory research but also guide in matching technique to problem, provide theoretical orientation where appropriate and quicken interest in new approaches and ideas. The chapters in this – the resulting book – range from those covering techniques peculiar to olfaction, such as olfactometry, to those of more general applicability, such as scanning electron microscopy and psychophysical techniques.

D.G.M.
A.T.
J.W.J.

CONTENTS

1 Instrumental Aspects of Olfactometry

ANDREW DRAVNIEKS

Odor Sciences Center
IIT Research Institute
Chicago, Illinois 60616

I. Introduction

Olfactometry deals with measurements of odours, and an olfactometer would be an instrument for such measurements. However, by usage, the term "olfactometer" has come to describe an instrument for the preparation and delivery of an odorant—an odour stimulus—to a chemoreceptor system The responses to the stimuli are measured outside of the olfactometer, on the chemosensing system, either physiologically or electrophysiologically, or psychophysically. This chapter is limited to the description of odour stimuli consisting of various vapours in air, which is the most frequently used form of stimuli. Another form—delivering odorants to the chemoreceptor by direct contact with aqueous solutions—is used in studies on aquatic animals, specialized electrophysiological studies, and chemotaxis of micro-organisms, but this will not be discussed here.

Olfactometers prepare stimuli of various odorants, concentrations, durations, and flow rates. The needs of researchers vary with the experimental design and with the subjects of their works. Olfactometric requirements in studies of the electrophysiological response of single receptor cells in insects, the behavioural response of rats, and the psychophysical response of humans are different. Consequently, numerous olfactometers have been described by dozens of researchers, and several new ones are added every year.

Despite the large number of different designs, similar principles in various combinations are utilized for stimulus preparation and control, and the same effects—adsorption, diffusion, etc.—influence the stimulus parameters. The objective of this chapter is to provide an insight into the processes used in stimulus preparation and control. Specific olfactometers are referred to as illustrations of some design features. However, the bibliography contains a more complete listing of references to other olfactometers.

II. Principal physical dimensions of stimuli

The stimuli may differ in duration, flow rates, and concentrations of odorants and other substances, e.g. water vapour, in air carrying these species.

Duration refers to the length of exposure of the chemoreceptive system to the stimulus. Since odorants are propagated through air by air flow, convection, and diffusion, it is practically impossible to switch instantaneously from a non-stimulus to a stimulus condition as would be possible with a visual stimulus. The stimulus concentration builds up from a zero value to its final preselected value over a certain period of time which can be short, fractions of a second, but not virtually instantaneous in relation to the response time of chemosensing organs. The time course of this build up period depends very much on the design of the olfactometer. Similarly, a stimulus does not disappear from air instantaneously when its delivery is interrupted. Instead, its concentration decays fast or gradually, again depending on the design of the olfactometric system. In the so-called pulsed delivery, when an air flow supplies the stimulus in constant preselected amounts, the stimulus strength can be varied either by changing the duration of the pulse or by holding the duration of the pulse constant but changing the amount of the odorant per pulse. Thus, the pulses C and D, Fig. 1, deliver the same amount of odorant but differ in their duration and the maximum concentration exhibited during the concentration pulse.

Again, the pulse by necessity consists of an ascending concentration phase, an optional constant concentration phase, and a concentration decay phase; a "square wave" form of the concentration-time relationship is virtually impossible. The principal forms of the time-stimulus relationships are depicted in Fig. 1.

In reality, there is no difference between the so-called constant stimuli and pulsed stimuli, except in the relative durations of the temporal phases. A pulsed delivery is best for measurements of the sensitivities in terms of the total number of molecules of odorant delivered to the chemosensing system, since obviously the maximum number of molecules which can possibly reach the chemoreceptors is then limited. In the longer-duration stimulus delivery, A in Fig. 1, the ascending concentration phase can be made short and the stimulus quantified in terms of the steady-phase concentration, but an estimate of the number of molecules of odorant actually reaching the chemoreceptor requires many assumptions about odorant transport to the receptor.

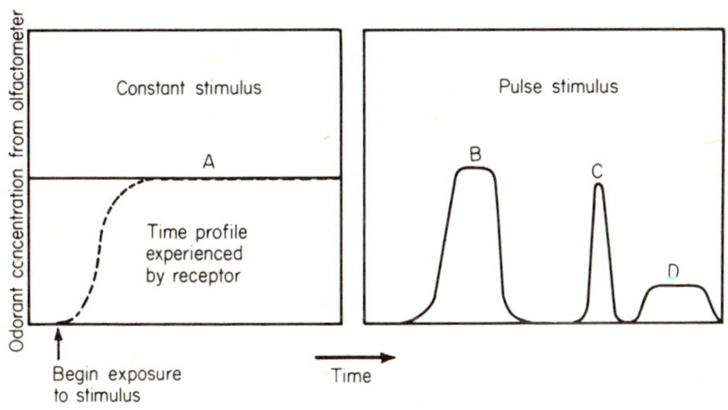

FIG. 1. Time profiles of odorant concentrations in stimulus delivery from olfactometers.
Even when a constant concentration stimulus (solid line in A) continues to emerge from the olfactometer, a receptor, during the initial phase of its exposure to the stimulus, experiences a concentration-time profile as indicated by the dashed line.

Even if the odorant concentration at the exit of a stimulus-delivering port has reached its constant level, the stimulus receptor, e.g. human nose, cannot immerse into such a stimulus instantaneously: the receptor unavoidably has to traverse the intermediate concentration zones between the ambient air and the constant stimulus.

Examples of instrumentally controlled pulsed deliveries of a stimulus are found in work by Stuiver (1960) and Köster (1967), cf. the section on stimulus delivery techniques. In most work, stimuli are prepared at constant rates and concentrations. They are made available to the chemoreceptive system with the shortest feasible concentration build up phase for a duration prescribed by the experimenter's needs, and are regulated by controls, or by moving the chemoreceptive system, e.g. human nose, in and out of the constant concentration stimulus. Thus, most modern olfactometric systems are "constant dynamic stimulus" olfactometers.

In practical comparisons of odour intensity by the human nose the reproducibility of simple "blast" procedures, in which the stimulus is forcibly injected into the nose, may be sufficient, as evidenced by usage of Turk's (1967) kit for the measurement of diesel exhaust intensities. This kit consists of polyethylene flasks containing a small pool of odorous materials to supply a headspace with odorant vapours. When squeezed, the flask delivers a pulse of odorous headspace to the human nose. Even such simple methods are preferable to even simpler techniques where the odorant is smelled from a glass flask so that the concentration in air reaching the nose depends on the convection between the static headspace in the flask and the room air. For more exact work, Jones (1955), Wenzel (1955) and Köster (1971) consider the reproducibility of forced injection method to be inadequate.

Systems where constant flows of a stimulus are not supplied, but where surges of vapours are delivered from a closed and odorized gas phase space, are frequently termed "static stimulus", olfactometers. This is somewhat of a misnomer, because the stimulus must anyway be delivered to the chemosensing system dynamically.

Stimulus delivery rate refers to the flow of the gas, e.g. air, carrying the vapours of the stimulus substance towards the chemosensing system. The rate may be expressed either in volumetric units–l/s–or linear units–cm/s. If the cross-section of the opening from which the stimulus is delivered is known, conversion from the volumetric to the linear units is possible. However, such value of the linear rate is only approximate: the actual linear rate of air movement is not constant across the cross-section of the opening, and furthermore decreases rapidly after entering open air (Hama, 1958). Only the volumetric rate can be accurately defined.

The linear rate of air movement needs to be considered because it relates to the mechanical impact of the stimulus-carrying air. Humans begin to feel the air flow from cylindrical ports at estimated flow rates of 10 cm/s; little is known about the air flow sensitivity of other chemoreceptive systems. At high flow rates, the stimulus becomes both mechanical and chemical. Accounting for the possible role of the mechanical component, as well as attempting to hold it constant while changing the odorant or its concentration, become important considerations in the experimental design.

The volumetric flow imposed by the olfactometer can range from zero—static stimulus from a flask—to many litres per second. The highest flow rates are needed where rapid changes in odorant concentration in large spaces are required, as in behavioural research on animals and in odour detection work with humans where an entire room may serve as an olfactometer.

In much of the olfactometric work, the simplest direct measurement to characterize the odorant vapour dilution is volumetric. If w volumes of air saturated with an odorant vapour are diluted with V volumes of odorant-free air, the resulting stimulus is at dilution Z:

$$Z = \frac{w}{V + w} \qquad \text{(I)}$$

The letter Z and the term "dilution quotient" were proposed for this dilution factor at the Symposium on Odorous Air Pollution (Karolinska Symposium, Stockholm 1970), to eliminate confusion arising from the coexistence of many methods of expressing the dilution factors. Correspondingly, at the same symposium, a logarithmic form pZ in analogy to pH, has been proposed as an alternative, to handle the problem of large numbers:

$$pZ = -\log_{10} \frac{w}{V + w} = \log_{10} \frac{V + w}{w} \qquad \text{(II)}$$

Thus, for a dilution by a factor of 10^6, $pZ = 6$. Chemical stoichiometry permits calculations of other concentration units from pZ if the vapour pressure of the odorant and its molecular weight are known.

Odorant concentration in the stimulus is the third physical dimension of an olfactometric stimulus. In many cases studied, the response of the chemosensing systems to an odorant appears to be proportional to a fractional power of this odorant; e.g. a tenfold concentration increase may cause only a twofold or threefold odour intensity change, in the judgement of human subjects. The detectability of an odorant also can vary among individuals by one or two orders of magnitude (Amoore, 1967). Hence, to study a sufficiently broad range of responses to the *same* odorant, an olfactometer should be able to change the concentration of this odorant in the stimulus over several orders of magnitude. For *different odorants* with respect to the same chemoreceptive system, or the same odorant with respect to *different chemoreceptive systems*, this working range of several orders of magnitude has to be selected at different absolute levels. For instance, a dilution of the air saturated with an odorant vapour at room temperature by an additional 10^6 volumes of air will bring all but a few odorants below their level of detectability for humans. Therefore, a three-orders-of-magnitude range set at different levels within the range between dilutions of 1:1 and $1:10^6$ will suffice for most work with humans. On the other hand, with animals who exhibit an extreme sensitivity to certain compounds, e.g. sex attractants, the range may need to be set at much lower levels. Stimuli preparation methods may vary correspondingly.

A trivial but perpetual problem is the choice of units to express the concentration of the odorant in the stimulus. Parts per million, per billion, etc., usually taken in terms of volume of odorant vapour per volume of air, represent one usage. For the molecular considerations of the receptor mechanisms, a usage based on the number of molecules of odorant per ml of air appears to be preferable. Since one ml of air at 25°C and atmospheric pressure contains a total of $2.47 \cdot 10^{19}$ molecules (all molecules considered), n molecules/ml of an odorant in air correspond approximately to

$$\frac{n}{2.5 \cdot 10^{19}} \qquad \text{(III)}$$

volume parts of odorant vapour in air, e.g. if $n = 2.5 \cdot 10^9$, within the range of the lowest odour thresholds for humans, air will contain 10^{-10} or 0.1 parts per billion, by volume, of this odorant.

The other usage is based on mass units of odorant per unit volume of air; this usage exists since in calculating the odorant concentration in the stimulus one frequently starts either with data obtained by chemical analysis of the stimulus, or from the gravimetrically measured consumption of the odorants in the olfactometer by the odorant-carrying air. The air pollution literature in the U.S.A., in work with gaseous contaminants in air, tends toward the use of the units: micrograms (10^{-6}g) per litre of air. For some odorants down to their lowest thresholds, small fractions of microgram per litre would have to be quoted, and perhaps nanograms/litre (1 ng = 10^{-9}g) or picograms/litre (1 pg = 10^{-12}g), or all these—whichever results in a more convenient number for the particular odorant—would be more suitable.

Because of the extremely large range of concentrations that occur in the olfactometric studies, suggestions have been made to characterize the odorant concentrations in logarithmic units, in analogy to pH in chemistry where pH is the negative logarithm of the hydrogen ion concentration in mol per litre. Thus, Laffort (1963) proposed reporting the concentrations in mol of odorant per litre of air and adapting a symbol of p . ol for its logarithmic transform:

$$p\,.\,ol = -\log\,(\text{mol/l}) = \log \frac{1}{(\text{mol/l})} \qquad \text{(IV)}$$

III. Stimulus generation

Single odorants for olfactometric work are usually obtained in liquid or solid form; only a few odorants are gases at ambient conditions. Stimulus preparation from gaseous substances normally takes the form of dilution in one or several stages, and differs little from preparation of various concentrations of vapourized liquids or solids. Liquid or solid odorants, however, must be transformed into vapours. To obtain the desired concentrations of such odorants in air, the odorants can be either diluted by liquid non-odorous solvents, or the odorants can be vapourized and their vapours subsequently diluted.

Solvent dilution method

The preparation of diluted vapours by using solutions of odorants in solvents has many inherent difficulties and is best reserved for simple preliminary experiments.

First, there is a *sensory purity* problem. It is difficult to obtain perfectly odourless solvents suitable for the dilution of all odorants. For the water-soluble odorants, water can be used as a diluent, but many odorants are not chemically stable in water, e.g. esters slowly hydrolyse to acids and alcohols. For odorants with a low solubility in water, which is the predominant case, highly refined medicinal paraffin oil, propylene glycol U.S.P., and some low-vapour-pressure esters such as benzyl benzoate have been used. Usually, these solvents exhibit weak odors, most probably due to residual impurities.

Second, a *thermodynamic difficulty* exists in defining the stimulus strength. The concentration of the odorant in the stimulus—air headspace *above the solution*—is proportional to the concentration of the odorant *in the solution* by a factor called the "activity coefficient", γ. This factor differs from different odorants in the same solvent, changes with the solvent; and sometimes changes with concentration, even for the same odorant/solvent combination. The corresponding physico-chemical relation is

$$p_i = (\gamma_i)\,(x_i)\,(p_i^0) \qquad \text{(V)}$$

p_i = the vapour pressure of odorant above the solution;
p_i^0 = the vapour pressure of pure odorant at the temperature of solution;
x_i = the concentration, in molfraction units, of the odorant i in the solution;
γ_i = the activity coefficient which reflects the inter-molecular interactions between the odorant and the solvent.

Thus for a hydrocarbon such as hexane in water, $\gamma >> 1$, while for ethanol in water $\gamma < 1$; for a solution in paraffin oil, the magnitudes of γ are in opposite directions.

An illustration of the solute/solvent interaction effect is a solution of 10 parts/10^6 of *n*-hexane and *n*-propanol in water. Both solutes have vapour pressures of the same order of magnitude at 25°C and are in the solution at comparable molar concentrations. However, the ratio of concentrations of (hexane): (propanol) in air above the water is approximately 10^4:1, since γ in water is large for hexane and small for propanol.

Esters as solvents tend to exhibit a narrower range of γ values. Consequently, there is a better proportionality between p_i and p_i^0 across a broad selection above the solutions of odorants in esters.

The consequences of the above relations is that the strength of the stimulus above the solution is ill-defined unless the activity coefficient is known or the actual concentration in the headspace above the solution is analytically measured. When several odorants are dissolved. together in the same solvent, their relative concentrations in air may be radically different from those in the solution.

The third is a *kinetic difficulty* with a continuous generation of the vapour stimulus from a solution of the odorant in a solvent. For the presentation to the chemoreceptive system, either repetitive samples of headspace above the solution, or a continuous slow flushing of the headspace with a non-odorous carrier gas, usually air, is required. The odorant is gradually depleted from the solution, so that the stimulus strength drops with time and, except at the beginning, the actual concentration of the stimulus remains uncertain without additional measurements. Also, the odorant evaporates from the surface of the solution, and must be continuously replenished by a diffusion in the solvent phase from the bulk of the solution toward the surface. This process must keep pace with the removal of the vapourized stimulus substance from above the solution. An energetic stirring, or air bubbling through the solution becomes necessary to reduce this short-term depletion effect. Bubbling leads to aerosol formation and associated problems (see below).

In work with animals, separate experiments are needed to establish that the "non-odorous" solvent is really non-odorous to the animals.

On the whole, preparation of diluted stimuli by dissolution in solvents invokes many problems which complicate data interpretation, especially with respect to the uncertainty about the actual stimulus strength. In olfactometric work with humans, where odour quality or hedonic tone is studied, the actual odorant concentration is of secondary importance. The convenience of having many odorants available for smelling from static systems, such as an assembly of solutions in sniffing bottles, can then outweigh the drawback of the uncertainty about the actual concentration.

Vapourization from pure odorant phase

Here, two techniques are possible, Fig. 2. In a complete direct vapourization, the odorant in a pure form, or in a solution in a non-odorous solvent, is fed at a constant rate, usually by a motorized syringe, to a vapourizer, and the vapour is mixed with a constant flow of volume of non-odorous air (Scherberger *et al.*, 1958). In the saturated vapour method, air or another non-odorous gas is brought in contact with the pure odorant and saturated with the odorant's

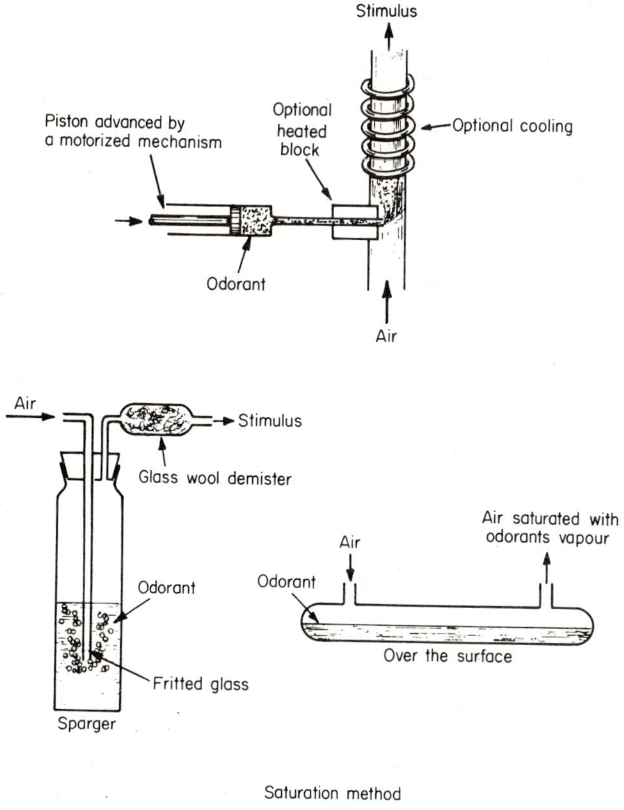

FIG. 2. Odorant vapourization techniques for stimulus preparation.

vapour; the subsequent mixing of the vapour-saturated air with additional non-odorous air permits the production of any desired dilution of the saturated vapour with air.

Before elaborating on these methods, the role of impurities in the odorant needs a few comments. The usual concept of chemical purity, e.g. 99 + %, satisfactory in much synthetic work, is not adequate in the chemoreception studies, since some impurities can exhibit odour at extremely low concentrations. Purification by gas chromatography or zone refining may be necessary to remove odour-relevant impurities. However, the impurities influence the complete vapourization and the vapour saturation methods in different ways.

When, for instance, 1 mg of odorant is totally vapourized every second, all impurities are co-vapourized; if the vapourizer is properly designed, there is no change in the impurity level of the remaining supply of the odorant. The relative concentration of an impurity with respect to the odorant remains exactly the same as in the supply of the odorant; thus, if the odorant contained 0.1% of impurity, the ratio of the impurity to the odorant concentrations in the prepared stimulus will be 1:1000.

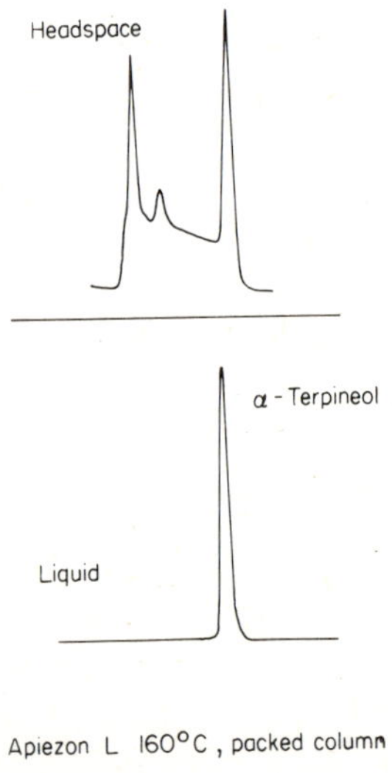

FIG. 3. Gas chromatograms of a liquid α-terpineol and of its vapour. Time advances from left to right.

On the other hand, if the stimulus is prepared by saturating an air stream with the odorant's vapour, the concentration of the impurity in air will be determined by eqn. V, with the odorant playing the role of a solvent and the impurity being the substance i. Thus, if the impurity has a higher vapour pressure than the odorant and/or large activity coefficient γ_i, its vapour pressure in air in contact with odorant will be out of proportion to its concentration x_i in the odorant.

Two effects then result. The impurity will be initially over-represented in the saturated vapour of the odorant, in comparison with the complete vapourization method, but gradually will be depleted, and unless the supply of the odorant is frequently changed, finally will drop below the level obtained in the complete vapourization method. The situation is reverse if the impurity is less volatile than the odorant and its γ_i value is not excessively large. As an illustration, Fig. 3 represents two gas chromatograms. Below is the chromatogram of α-terpineol injected directly. If an impurity is present, its concentration is well below one %. Above is the chromatogram of the vapour above the same substance. At least two impurities are present, one at a concentration, judging from the sizes of the peaks, almost as high as the α-terpineol itself.

Complete vapourization method

This method is useful when large volumes of air must be odorized (Scherberger *et al.*, 1958). Limitations arise if very low concentrations of the odorant in air must be generated, since sufficiently slow continuous injection rates of odorant into the total vapourizer are experimentally difficult. Thermal expansion in the capillary tubing leading from a motorized syringe to a heated zone cannot be maintained absolutely constant in time: small changes in the heat distribution produce fluctuations in the liquid flow, and these reach substantial proportions of the total syringe-imposed flow at very slow delivery rates. The actual performance of such systems at slow flows has not been reported.

If large volumes of air are to be supplied with the stimulus, and a large holdup volume is inserted after the stimulus preparation to smooth out the short-direction fluctuations, this method performs satisfactorily to lower stimulus strength levels. For many odorants with not-too-low odour thresholds, the total vapourization method is quite adequate. For substances with low thresholds, the odorant may be dissolved in a larger volume of inert solvent, but then a study of the possible effect of the solvent itself also is needed.

The problem with the preparation of a highly diluted stimulus by this method is best illustrated by an example. Suppose that the olfactometer room is to be supplied with a stimulus in air flowing at a rate as high as 1000 l/s. If the substance such as butyric acid must be maintained in this air at a concentration of 10^{-10} g/litre—still within the detectability for some people and certainly for dogs—butyric acid must be evaporated at a constant rate of 10^{-7} g/s. This is an extremely difficult technical task, possible only by a force delivery to the evaporator of highly dilute solution of the butyric acid in some inactive solvent, if the complete vapourization method is used without subsequent splitting of the generated vapour-gas mixture and successive dilutions of the resultant streams.

Vapour saturation methods

In the vapour saturation method, air or another non-odorous carrier gas is brought into contact with the condensed odorant phase under conditions where the odorant's vapour saturates the carrier gas. This first step prepares the strongest stimulus possible at the saturator's temperature. Weaker stimuli are prepared by further dilution in the gas phase. The role of impurities in such a process has been already discussed in another section.

The actual concentration of the odorant in the stimuli prepared by such a method can be easily calculated if the vapour pressure is known and the conditions for reaching the vapour/condensed phase equilibrium are met. If the vapour pressure of the substance is not known, the stimulus strength for many purposes, e.g. for studies of the response/concentration relationships, can still be adequately characterized in terms of a relative pressure:

$$p_r = \frac{p_i}{p_i^0} \qquad \text{(VI)}$$

p_i = the vapour pressure of the odorant in the diluted stimulus;
p_i^0 = the saturated vapour pressure of the odorant in the saturator;
p_r = the relative vapour pressure, equivalent to the fraction of saturation and can be calculated from the dilution in the gas phase assuming ideal gas law to hold, which is essentially correct at low concentration of vapours.

Since p_i^0 for an odorant is constant as long as the temperature of the saturator is constant, p_r values are proportional to the concentrations of the same odorant in the stimuli.

A comment on p_i^0 is needed here. Some substances, e.g. aliphatic acids, dimerize in the gas phase, so that two types of species will be present in the vapour: single molecules, and paired molecules. The concentration ratio of the two molecular species is concentration-dependent and only the total formed concentration can be calculated from the total vapour pressure tables for such substances. If the dimerization coefficient is known, the concentration of each species at different dilutions of vapour can be calculated. Usually, at high dilutions, the monomer will dominate, but at higher concentrations both species may coexist at substantial concentrations.

The saturation of a gas by vapours from liquids is conducted either by sparging–bubbling of non-odorous gas through the liquid–or by passing the gas over as extended surface of the odorant, Fig. 2. With solids, these methods are feasible but not trouble-free since small dust particles from the solid odorant can become entrapped in the gas and carried into the dilution system. Aerosol filtration elements can reduce this problem, but are not entirely dependable. as discussed in the subsequent section. In this author's experience, the dust problem can be eliminated by making a pasty mixture of the solid odorant powders with an odourless medium in which they are not too soluble, e.g. water, and passing a carrier gas over an extended surface of such a paste. This appears to provide sufficient contact to saturate the carrier gas with the vapour of the solid odorant. Since upon exceeding the solubility of an odorant in water the vapour pressure of the odorant above the two-phase mixture is equal to that of pure odorant, the only limitation is a kinetic one: diffusion of the odorant's vapour through the thin films of water covering the odorant crystals. With enough surface exposed, and with a sufficiently slow flow of the carrier gas, this hindrance is insignificant.

Since the vapour pressure increases with temperature, the odorant in the saturator must be maintained at a constant temperature. This temperature should be less than the temperature of the subsequent ductwork, especially before dilution with additional gas: otherwise the excess odorant vapour will condense at the cooler spots, and the lowest temperature in the system will determine the concentration. Furthermore, when a relatively volatile odorant is rapidly vapourized, the odorant may cool in order to supply the heat of vapourization. In such cases, efficient thermostatting of the odorant in the saturator is necessary.

The required constancy of temperature can be estimated from vapour pressure tables (Handbook of Chemistry). For l-butanol, the vapour pressure at ambient temperature increases by 6%/°C.

In practice, two modes of temperature fluctuation are of concern, each for a different reason. Long-term fluctuations simply produce fluctuation in vapour pressure; at least in humans the minimum concentration difference (differential threshold) that can be discriminated by the olfactory sense is in excess of 15%, a temperature control must be constant within 0.2 to 1°C to be adequate for the olfactometric work.

Short-term fluctuations can influence the stimulus delivery rate by a temporary expansion or contraction of the vapour volume; if the stimulus is further diluted, pulses of lower and higher concentrations will occur. For instance, if the vapour-saturated headspace is 300 ml, and the rate of its delivery is 1 ml/min. an increase in its temperature by 1°C/min will expel into the dilution system an additional 1 ml/min, doubling the stimulus strength. The reverse, a temperature drop by 1°C/min, can temporarily very significantly weaken the stimulus.

Thus, rapid temperature fluctuations, such as may be caused by too active thermostatting, are particularly undesirable non-fluctuating temperatures, supplied by room-temperature non-thermostatted aluminium or copper blocks, and good thermal insulation, may be more desirable than an overdesigned thermostatic system.

Saturation by sparging

The passage of a carrier gas through a porous glass filter into a liquid, Fig. 2, is a very efficient way of producing a good contact between the gas and liquid. However, it generates aerosols when the gas bubbles burst at the liquid surface. Such methods are actually used to generate aerosols (Blanchard, 1954; Day, 1963). For instance, in the case of single small bubbles bursting through the water, typically five to six airborne droplets per bubble are formed, with sizes ranging down to 1 micron in diameter. The resulting stream then consists of the odorant-vapour-saturated gas, plus excess odorant in the form of entrapped liquid odorant particles. As little as one droplet, one cubic micron in size, contributes approximately 10^{-12} g; if converted, during subsequent dilution, to a vapour, 10^{-12} g per cm^3 of air would be a significant amount of some fatty acids for the human nose, and an enormous amount of a pheromone for an insect.

An aerosol filter can reduce the number of droplets entrapped in the gas stream. However, none of the aerosol removal devices are absolute filters for particles of all sizes, so that a complete elimination of the aerosol, once it has formed in the sparger, is not possible.

If an aerosol filter is used, a tightly packed glass wool element is the most suitable filter for olfactometric work. As an example, 5 g of Pyrex glass wool, 6 micron diameter fibre, packed in a cylindrical tube 2.5 cm diameter and 12.5 cm long is used in air pollution measurements and is considerably better than 95% efficient in the removal of aerosols, encompassing a broad range of the particle sizes. Such a filter does not clog up easily because of the large volume fraction of the void volume. It can operate at air flows on the order of 20 l/min with a pressure drop of 50 mm of Hg. Porous plate or sheet filters are not too suitable in olfactometric work: the increase in the flow resistance, as the aerosol accumulates on the filter, complicates regulation of the flow system.

If a sparger/filter system is used, it is necessary to maintain the filter at the same temperature as the liquid in the sparger. If the filter is warmer, the aerosol will evaporate there supplying additional odorant. If the filter is cooler, the excess vapour will condense there and the saturator vapour pressure will correspond to the temperature of the filter rather than that of the sparger.

On the whole, sparger systems are of questionable value for accurate work with low-volatility low-threshold odorants, unless many precautions are taken, and the aerosol content, or its absence in the stimulus, are ascertained.

Over-the-surface saturators

The static surface of a pure liquid odorant emits vapours into the gaseous space above it ("headspace") in accordance with Ficks diffusion law

$$J = A \, . \, D_i \, . \, \frac{n_i^0 - n_i}{d} \qquad \text{(VII)}$$

J = the emission rate, molecules/second, from the total exposed surface;
A = the surface area, cm^2;
D = diffusion coefficient of the odorant in air at the saturator temperature, $cm^2\ s^{-1}$, with numerical values on the order of 0.1 at ambient temperature for typical odorants;
n_i^0 = the concentration of odorant, molecules/cm^3, in the saturated vapour of the odorant at the saturator temperature;
d = a parameter, cm, characterizing the formal thickness of the non-turbulent gas film adjoining the odorant surface. The odorant molecules diffuse through this film into the bulk air by a process of molecular non-turbulent diffusion.

A typical over-the-surface saturator was shown in Fig. 2 (Cheesman and Kirkby, 1959). Air enters at end I, flows over the odorant's surface and accumulates the odorant's vapour. Vapourization is fastest at inlet I where the vapours diffuse into the vapour-free air, and slows down towards the exit end E where the vapours must diffuse into air which already contains accumulated odorant vapour.

At each surface location, eqn. VII applies, but the value of n_i changes. A simple mass transfer balance calculation, ignoring, for simplicity, the effects of turbulence in the gas phase and assuming a constant boundary layer thickness d, gives the following equation for the final concentration, c_f, in the exit air from the saturator (Wnek, 1971):

$$C_f = C_o\left[1 - e - \left(\frac{L}{lh}\ \frac{D}{d}\right)\right] \tag{VIII}$$

and

$$\frac{C_f}{C_o} = 1 - e\ -\left(\frac{L}{lh}\ \frac{D}{d}\right) \tag{IX}$$

C_o = the concentration of odorant in air when air is saturated with odorant's vapour;
$\frac{C_f}{C_o}$ = the ratio indicating the degree of saturation at the exit from the saturator;
e = the base of natural logarithms and equal to 2.178;
L = the length, cm, of the surface over which air flows;
l = the linear flow rate cm/s of air along the surface;
h = the height, cm, of the air space above the odorant surface;
D = cm^2/sec, the diffusion coefficient of the odourant in air;
d = cm, the boundary layer thickness, approximately 0.1–0.2 cm.

For a fixed temperature, geometry and flow rate, the degree of saturation achieved in the saturator, C_f/C_o is independent of the vapour pressure of the odorant but dependent only on its diffusion coefficient in air. Thus, the low volatile odorants will reach practically the same degree of saturation as the more volatile ones for any reasonable length of tube L. A numerical example illustrates the degree of saturation for a specific case:

L = 15 cm
h = 1 cm
v = 1 cm/s (60 ml/min for air space, 1 cm^2 in cross-section)
D = 0.1 cm^2/s
d = 0.1 cm
$\frac{C_f}{C_o} = 1 - e^{-15} \approx 1.00$ (practically full saturation)

If $D = 0.02$ as for a very large molecule probably not sufficiently volatile to exhibit an odour, the degree of saturation is still:

$$\frac{C_f}{C_o} = 1 - e^{-3} = 1 - 0.05 = 0.95 \tag{X}$$

so that the air exists 95% saturated. By slowing the volumetric flow rate to 30 cm^3/min, saturation in the second case will reach 99.8%. Since a 5% difference in the odorant concentration is of little physiological or psychophysical consequence, the approximate physical dimensions of the saturators and the flow rates in the above example are adequate for most conditions. For faster flow rates of saturated vapours, longer surfaces would be necessary. Usually, the saturated vapour is diluted considerably after the removal from the saturator, so that much more dilute vapour flow rates can be generated from a single saturator.

In the author's experience, nitrogen or air flowing at rates of up to 100 ml/min along a 15 cm long and 1.25 cm wide surface of an odourant reaches 95 to 98% saturation at the exit from such a saturator.

Squeeze-flask technique

In the squeeze-flask technique, the odorant liquid is at the bottom of a plastic flask; the subject opens the flask, squeezes it, and sniffs the rapidly expelled air. Fresh air replaces the expelled air. A question arises: what is the recovery time needed to restore the headspace in the flask above the liquid to the original—pre-squeeze—vapour concentration? Calculations by Wnek (1971) indicate that for a 15 cm high headspace above the liquid, in the absence of mixing, the re-equilibration is slow. If an equivalent of 75% of the vapour-saturated headspace were expelled in a few rapidly repeated squeezes, the headspace would temporarily be 25% saturated. The vapour concentration will slowly rise by an evaporation of the liquid, but 90% of saturation will be reached only in about 30 minutes. Much faster—fraction of a minute—re-equilibration occurs if the liquid is shaken in a circular pattern. This illustrates the need for mixing after the squeeze-sniff exercise. A too-vigorous shaking should be avoided to prevent an aerosol formation.

Diffusion method

The diffusion tube principle is illustrated by Fig. 4 (Altshuller and Cohen, 1966). The odorant is in the closed bottom end of the tube, while the rest of the channel contains a stagnant column of a gas phase. The device operates as a constant-rate odorant vapour emitter, as long as the free gas phase at the open end of the tube does not accumulate the odorant's vapour. If the open end is in a duct, as shown, and a gas, e.g. air, flows past the channel's opening, this condition is met.

The rate of emission of vapour in molecules per second into the duct, carrying a flowing gas, e.g. air, is determined by a modified eq. VII

$$J = AD_i \frac{n_i^0}{d} \tag{XI}$$

J = the emission rate, molecules/s, from the open end of the tube;
A = the cross-sectional area, cm^2, of the diffusion tube;
d = the distance, cm, between the surface of the odorant and the open end of the tube.

If the odorant vapour is rapidly carried away by the flow of air in the duct, $n_i << n_i^0$, and one can assume that in eqn. VII $(n_i^0 - n_i) \approx n_i^0$. The values of the diffusion coefficient D can be

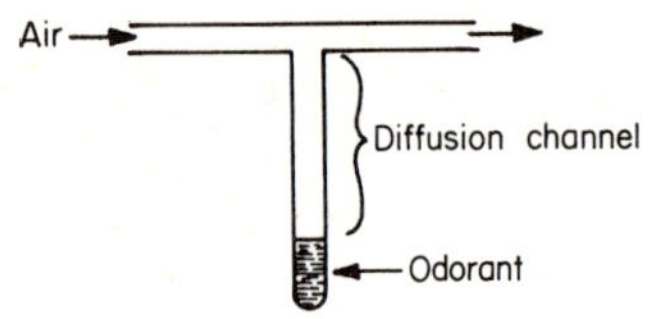

(a) Gaseous diffusion

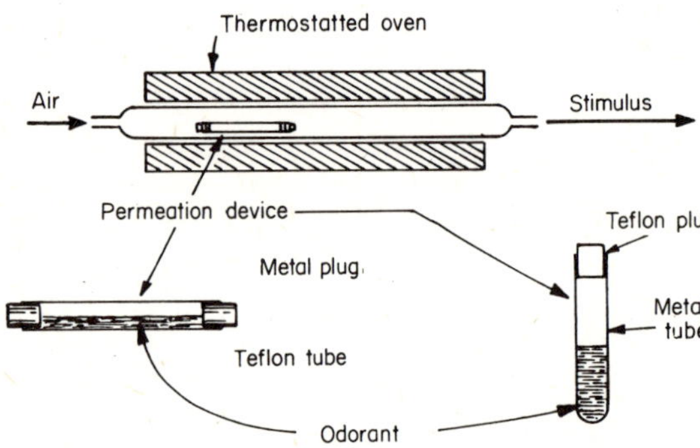

(b) Permeation through Teflon

FIG. 4. Diffusion and permeation techniques in odorant stimulus preparation.

estimated (Fuller *et al.*, 1966). It is instructive to insert numbers into the equation. Suppose

A = 0.01 cm^2 (one square mm)
d = 1 cm
D = 0.1 cm^2/s, a typical value for a reasonably large vapour molecule in air at ambient temperature
w = flow of air in duct = 10 ml/s

Let n_s be the concentration, molecules/ml, of the odorant in the stimulus after mixing of the odorant vapour with the air flow in the duct,

$$n_s = \frac{J}{w} = 0.01 \times 0.1 \, \frac{n_i^0}{10} = 10^{-4} n_i^0$$

The stimulus contains the saturated vapour of the odorant diluted by a factor of 10^4. Thus, this method easily produces stimuli equivalent to the saturated vapour diluted by several order of magnitude with air. Dilution can be increased by narrowing the cross-sectional area or increasing the length of the channel. The resulting dilution can be estimated from the dimensions of the channel.

If the vapour pressure of the odorant is relatively high–hundreds of mm of Hg–the equation still applies, but the emission rate calculation needs to be corrected for the self-diffusion of the odorant in its own vapour at the bottom of the tube (cf. Bird *et al.*, 1960). Similarly, a mole complex dependence on the diffusion channel length occurs when a dimerization or other equilibrium process affects the concentrations of the odorant molecules in the gas phase.

The method has limitations. First, it is sensitive to the contamination of the tube walls by a liquid odorant above the liquid surface: obviously, the diffusion path would then be shorter than from the open end to the surface of the odorant on the bottom. Second, any pressure change in the duct would interfere; for instance, a sudden decrease would pull out of the tube a small amount of odorant vapour-rich gas, temporarily enriching the stimulus. In reverse, a pressure increase would temporarily reduce the stimulus strength. This effect disappears after the steady state of diffusion is again established. However, the half-time to re-establishing the steady state is proportional to d^2

$$t_{1/2} = \frac{d^2}{2D} \tag{XII}$$

For $D = 0.1$ cm²/s and $d = 1$ cm, the steady state re-establishes within a few seconds. For $d = 10$ cm, $t_{1/2}$ may extend to 10 or more minutes. Channels with $d > 5$ cm become impractically slow in restoring themselves to a steady diffusion rate after the pressure disturbances which are unavoidable in those flow systems where valve manipulations produce pressure pulses.

The diffusion tube must not widen in any place. As an example, suppose that there is additional vapour space above the odorant. This extra volume will be essentially saturated with the odorant's vapour and will dispense a large surge of odorant vapour if for some reason the pressure in the duct suddenly drops.

A similar "vapour storage" effect arises if the walls of the tube adsorb the odorant. In our work, glass capillaries were found to cause considerable additional odorant emission when a negative vapour surge occurred and a polar odorant was used. Teflon capillary channels were satisfactory provided that some skill was used in placing a drop of the odorant with a syringe on the bottom of the channel. Other filling methods may be applicable to smaller capillary channels, at the cost of increasing complexity.

For dilutions of vapours by a fixed factor as large as 10^5 in one single simple device, the method is useful in spite of the above shortcomings. Channel sizes for various dilutions can be pre-calculated, and small adjustments in the emission rates made by changing the level of the odorant in the channel, thus changing d in the equation. The method is less suitable for solid odorants in powder form, since the spaces between particles will serve as a saturated vapour "bulge space", as explained above.

Permeation methods

In the permeation methods, the odorant is enclosed in a sealed container consisting of at least in part of an organic polymer. The emission-controlling process is the permeation of the odorant vapour through the polymer, usually of FEP Teflon (O'Keefe and Ortman, 1966; Martin and Debbrecht, 1970). Fig. 4b illustrates two typical forms. In one, the container is a Teflon tube; in the other designed to obtain lower permeation rates, the container walls are either glass or metal, and the open end is sealed with a Teflon plug or disc. Teflon has a high thermal expansion coefficient and metal containers are desirable if the plug form is used to avoid wall cracking.

The rate of emission from the permeation devices depends on the vapour pressure of the odorant and the solubility and diffusion coefficient of the odorant in the Teflon. To maintain a constant stimulus concentration, the permeation tube is maintained at a constant temperature in a constant flow rate of non-odorous carrier gas. The stimulus concentration at a constant flow rate can be changed by changing the temperature of the permeation tube; typically, the emission rate increases tenfold with a 30° to 50°C temperature increase. The emission rates are measured empirically, by determining the weight loss of a permeation tube containing the odorant and maintained in a carrier gas flow over a long period of time. From the emission rate measurements at two temperatures, the rates at other temperatures can be easily estimated from a plot of the log of (emission rate) versus the reciprocal of absolute temperature.

The permeation devices arrive at steady-state emission conditions after several days. They are most suitable for obtaining and maintaining very low concentrations of stimuli substances with relatively high vapour pressures. Gradual changes in the permeation rates may occur if the polymeric wall material undergoes internal changes—further polymerization, escape of lighter polymeric fractions present in most polymers, or swelling. Re-checking of the emission rates may be necessary at high temperatures or with odorants which may change the wall properties. Permeation tubes with low-molecular weight odorants are in use as calibration elements in instruments for monitoring low concentrations of these odorants in air.

Exponential dilution or concentration methods

In the exponential dilution or concentration methods, the stimulus strength in the stimulus preparation vessel changes with time at a predictable rate, dependent upon the size of the vessel and the flow rates. The instrumentation is very simple and can yield a timed series of stimuli of decreasing or increasing concentration. The method is much less suitable if the stimuli strengths must be changed up and down rapidly, with the timing determined dependent on some other experimental requirements.

Figure 5 illustrates the principle of exponential dilution. The principal part is a vessel with a volume v ml which initially contains the

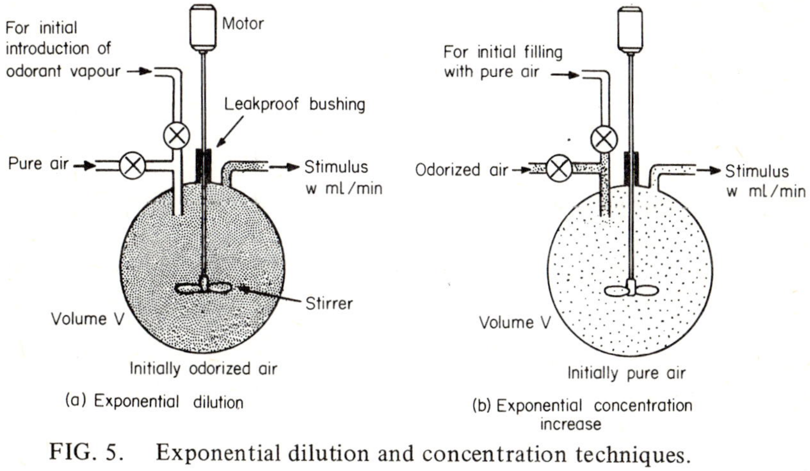

FIG. 5. Exponential dilution and concentration techniques.

non-odorous air kept in rapid turbulence by an efficient mixing. At a time zero, a known concentration of the odorant's vapour is generated in the vessel from some stimulus preparation system. There should be no excess condensed odorant in the vessel. The initial known concentration of the odorant in the vessel is then C_o. During the exponential dilution process, pure air at a constant flow rate, w ml/min, is supplied to the vessel, and the odorant-air mixture flows out of the vessel at the same rate. Because of the efficient stirring, the clean entering air rapidly mixes with the pre-existing mixture in the vessel. The concentration of the odorant in the vessel, and consequently the concentration of the odorant in the w ml/min flow of air leaving the vessel, decrease exponentially. The concentration C_t, at t minutes after the beginning of the dilution is

$$C_t = C_o e^{-t/\tau} \qquad \text{(XIII)}$$

where τ is the time constant, here in minutes, depending only on the volumetric rate of flow and the volume of the vessel:

$$\tau = \frac{V}{w} \qquad \text{(XIV)}$$

Thus, the concentration of the odorant in the stimulus emerging from the vessel at any given moment is easily predicted.

In a reverse procedure, the vessel initially contains pure air, Fig. 5b. The odorant-containing air, with a constant concentration, C_o, of the odorant, is fed at a constant rate, w ml/min, into the vessel, and the mixture leaves the vessel at the same rate. The concentration of the odorant in the vessel gradually increases, and at any given time t min after the start can be calculated:

$$C_t = C_o\,(1 - e^{-t/\tau}) \qquad \text{(XV)}$$

where τ has the same meaning as above.

The exponential method has been utilized in some industrial odour threshold level studies and in analytical chemistry to prepare low concentration mixtures of gases for instrument calibration. The exponential concentration method may be suitable in odour threshold studies.

Delivery from gas chromatograph

The gas chromatograph, schematically shown in Fig. 6, is used in the study of the composition of complex mixtures of substances, say odorants in this case. The sample is injected by a needle syringe through a silicone rubber or similar septum into a heated port A. The substances contained in the sample are transported in vapour form by an inert carrier gas, usually helium or nitrogen, through a partition tube or "column" B which contains a liquid or a grease-like "stationary phase" dispersed as a thin film on the walls of the tube, or on a powder of a ceramic or an organic polymer "support", with which the tube is filled. Because of the differences in the vapour pressures and in the solubilities of the different odorants in the stationary phase, these migrate along the partition tube at different

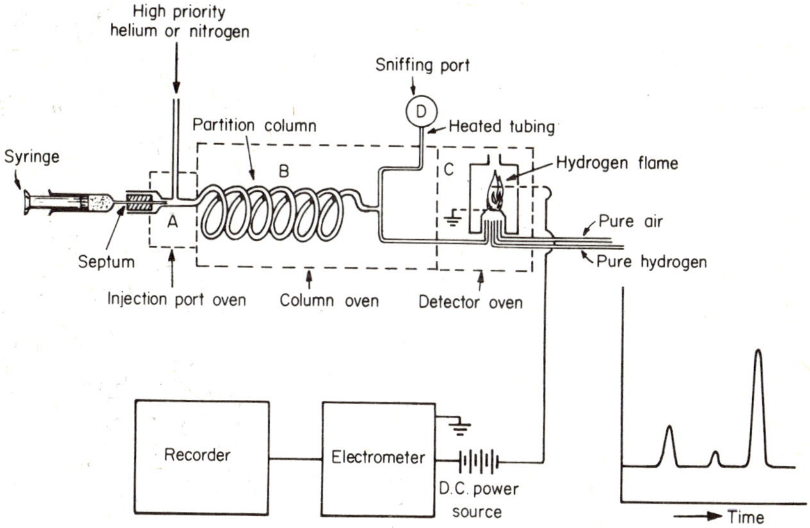

FIG. 6. Principal parts of a gas chromatograph equipped for sniffing the odours of gas-chromatographically separated odorants.

rates and emerge from the other end of the partition tube separated as vapour packages in the carrier gas, and in a certain sequence determined by the properties of the odorant and the stationary phase. The degree of the separation of the odorants depends significantly on the length of the partition tube. The elution of the organic odorants is customarily monitored by a hydrogen flame ionization detector, C.

Such detectors respond by an increase in the electrical conductivity of the flame when an organic substance enters the flame. The response is approximately proportional to the mass (g) of the respective odorant sample. Typical chromatograms were shown in Fig. 3. The time course of the concentration of an odorant during its elution tends to correspond to a normal distribution curve, so that the temporal shape of the odour stimulus which is delivered from sniffing port D, parallel to the detector, is well known.

The gas chromatographic process separates many odorants, so that a work with less pure odorants is possible, delivering for sniffing only that "peak" which corresponds to the desired odorant itself. Optical isomers and some positional isomers separate only with great difficulty on some special stationary phases. Impurities which belong to the same homologous class but differ by one or more carbon atoms are separated very easily. The stimulus can also acquire impurities consisting of vapours of the stationary phase.

The criteria for selection of the stationary phase for the partition column are many. In the work with odorants in general, the stationary phase should be odourless, as temperature-stable as possible, and produce well defined elution peaks with any odorant. The last requirement is difficult, since especially organic acids do not elute as sharp peaks from most stationary phases. At present, the most suitable compromise phase is a polar phase SP1000, which is a modified form of a widely used polyether phase Carbowax 20M.

The gas chromatographic separation as a means for purer stimuli preparation so far can be applied only to the pulse-type stimuli. It is in much use in work with odorant mixtures from natural and industrial sources, since it permits a fast estimate of the olfactory properties of various components of the mixture (Guadagni, 1968; Dravnieks and O'Donnell, 1971). It also has been used in electrophysiological studies on frogs (Drake *et al.*, 1969) and may assist in the search for pheromones.

The rate of stimulus flow from various types of the gas chromatographic partition tubes is from a fraction of ml/min to 100 ml/min. Further dilution is a possibility. The duration of a pulse of an odorant ranges from a few seconds to a sizeable fraction of minutes and can be changed for the same odorant by changes of the temperature and the flow rate.

In one study (Dravnieks and O'Donnell, 1971) a cursory comparison was made of odour thresholds of several compounds smelled from olfactometers with those thresholds obtained through the gas chromatographic stimulus delivery. The size of the injected sample was increased until the substance began to exhibit odour at the exit from the chromatograph. Reasonable agreements in the threshold values were attained when it was assumed that the odorant was equally distributed in 30 ml of effluent which corresponded to the peak width.

Carrier gas purity

Obviously, the gas—usually air—used to prepare different vapour dilutions and carry the stimulus must be clean and non-odorous. In critical cases, an analysis for trace odorants, cf. stimulus analysis, may be necessary to establish the absence of possible artefacts in the dilution air.

When the air consumption by the olfactometer is low, "zero grade" air, such as used in gas chromatography for hydrogen flame ionization detectors is the best choice. Odorous air occasionally occurs in the usual grade of compressed air from cylinders since odour absence is not a technical specification. Sometimes the pressure control gauge is contaminated or contains some odorous gasket. When a new cylinder and/or gauge are installed, it is always useful to insure that the air obtained is non-odorous.

Cylinder air is dry. If the olfactory experiment is conducted in a room with normal moisture, and the stimulus is delivered with dry air, even a clean dry air blank may appear to generate a vague sensation of difference. This effect should be taken into account, and a humidification in the final dilution stage may be desired.

Gases in cylinders invariably contain aerosols which separate by settling-out very slowly. The aerosols can be removed at the exit from the cylinder, after the first flow control valve, where the pressure is less than in the cylinder, by commercially available porous Teflon filters.

When the use of cylinder air is not desirable, air is pumped to the olfactometer from free atmospheric or room air. Obviously, oil-sealed compressors and similar devices are not suitable. Some membrane and other pumps can also be unsuitable because of the presence of odorous materials in membranes, gaskets, seals, etc. A sensitive nose is one of the most critical devices to check these points. Graphite vane pumps, and some small piston pumps such as used in aquaria aeration are among those that can supply essentially non-odorous air if the intake air is non-odorous.

The air intake must be away from any possible sources of odour: hood exhausts, chemical storage areas, even from the motor of the pump; and would usually be equipped with a particle filter to remove dust before the air reaches the pump.

The air purification is usually carried out after the compression and is intended to remove residual odorants and impurities. If air is taken from the room, and the room is supplied with deodorized air or is in a clean area with plenty of ventilation, a purification step may sometimes be omitted, provided that the compressor really does not contribute odorants. There is little point in cleaning up air to the olfactometer much beyond air purity prevalent in the room where human panelists smell stimuli from the olfactometer–such purification would simply produce more difference with respect to the environment of the panelist.

There is no "best simple" procedure for air purification. Even freshly activated carbon at ambient temperatures produces a faint odour of unknown origin (Stuiver, 1960), and trace organic vapours can be collected from air treated in this way (author's experience). Activated carbon does efficiently remove gross odours. When a carbon bed filter is chilled in a temperature gradient so that only the downstream end is at dry ice temperature, –80°C, it purifies air flow very well: each vapour is adsorbed before it reaches its dew point, so that an aerosol formation by a shock-cooling is avoided. However, it also cryogenically removes air moisture. To avoid clogging with ice, it needs a desiccant upstream. Dry air is usually not desirable in olfactometric work. A carbon bed needs also a glass fibre or other similar high-capacity particle filter to collect carbon dust, which is highly undesirable in the olfactometer.

Air can be purified also by passage through an alumina chip bed at 700°–800°C. This oxidizes all organic matter to carbon dioxide (Valenta, 1970), with only a negligible increase over the ambient carbon dioxide levels and without materially influencing the water content of air. Cooling and particle removal by a filter may be necessary in such a process.

IV. Stimulus handling in the olfactometer

Olfactometry requires stimuli of various odorants and various concentrations of odorant vapours in the carrier gas. The arrangements for changing from one concentration to another, or from one odorant to another can be loosely classified in two groups: one-channel, and multi-channel.

In some olfactometers, flow pathway is used to connect the stimulus generation device to the port where the stimulus emerges in its final form; to obtain different concentrations, either the setting of the stimulus generator is modified, or an adjustable system of dilutor and splitter passages is used to proportion the stimulus/carrier gas flows. In all such systems, the last segments of the flow pathway are common to all stimuli. When a change in the stimulus generation or dilution is made, the new stimulus does not reach its final steady state instantaneously: a certain purge time is required to replace the prior stimulus with the new one.

In the more recent olfactometers, attempts have been made to reduce the needed purge time as much as possible. This is usually accomplished by multichannel systems, where all desired different stimuli are prepared continuously and then directed by the shortest possible route toward the delivery port for any desired interval of time in any desired sequence. An arrangement where each of the prepared stimuli exits through a separate port and the stimuli are changed by changing the ports, eliminates the purge problem but may not be suitable for all olfactometric needs; it is quite convenient in work with human subjects.

The actual arrangements for stimulus modification vary widely. Capillaries, two- and three-way valves, manometers, and flow meters are among the design components used. Problems of common significance are the duration of the transition time when stimuli are changed and the selection of materials for the tubing and ducting (Mateson, 1955).

Transition time in stimulus change

Figure 7 illustrates most of the possible effects that influence the transition time. At A, the entrance point, a stimulus enters from a generator device. Even if its concentration there is instantaneously changed from the initial steady-state value n_1 molecules/ml to n_2, as shown in an idealized set of sketches at the left, this change does not propagate through the rest of the system to the exit port E as a sharp front, with a velocity equivalent to the volumetric flow velocity of the carrier gas.

Suppose, for simplicity, that $n_1 = 0$; then it is abruptly changed to some finite value n_2 at time = 0. In a smooth, straight, clean tube with non-adsorbing walls, odorant molecules and the carrier gas diffuse with respect to each other, and the initially sharp concentration front separating the $n_1 = 0$ and n_2 regions changes into a sigmoid shape curve approximating that of cumulative normal distribution. Such a diffusion effect, well known in the gas chromatography, depends on the residence time of the front in the tube, nature of the flow (usually laminar for smaller tubes, turbulent for larger tubes at higher flows) and the diffusion coefficient of the odorant in the carrier gas. One of the important corollaries of the diffusion effect is that if the odorant adsorption on walls is low, the odorant molecules may appear in the effluent at E well in advance of the time calculated simply from the flow rate of the carrier gas. Thus, after a residence time of t s x cm ahead of the front one half of the final concentration would be reached.

$$x_{1/2} = 2Dt \qquad \text{(XVI)}$$

For a residence time of 20 s and $D = 0.1$, $x = 2$ cm; if flow was 2 cm/s, one-half of n_2 would be reached at 19 s, etc. while 90% of the final concentration n_2 would be attained only at 22 to 23 seconds. These effects need to be considered in work with suprathreshold concentrations, since in some olfactometers the threshold concentration may occur well before the full stimulus is calculated to reach the

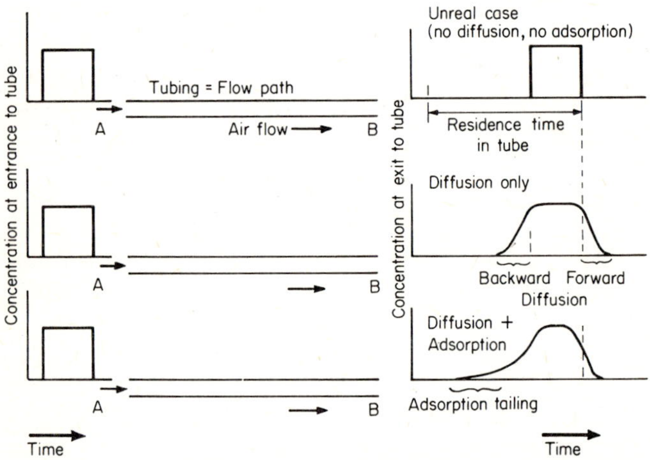

FIG. 7. An idealized representation of the degeneration of concentration-time profiles during changes of the stimulus in a stimulus flow line.

effluent port. A reduction in the length of the flow trains and an increase in the linear flow rates minimize the diffusion effect. It becomes negligible above linear flow rates of 10 to 20 cm/s and residence times of less than a few seconds.

The above was applied to smooth clean non-adsorbent tubes. In real systems, one encounters turns, flow cross-section changes, valves and holdup volumes; walls adsorb odorant, and impurities on walls, e.g. grease, serve as solvents for the odorants. The effect of the adsorbtion (or absorption) is illustrated by the bottom sketch in Fig. 7.

Turns, cross-sectional changes, and valves cause temporary turbulence which subsides after a distance of 10 to 20 diameters of the tube length. When two flows are to be mixed, turbulence is desirable. Otherwise, its role in the usual olfactometric systems, except in cases of larger holdup volumes, is of little further importance.

The holdup volume problem arises when a chamber, a box, or a larger cross-section sniffing port, or similar element is used in the olfactometer. The situation is somewhat analogous to the exponential concentration or dilution process described in an earlier section, cf. eqns., XIII and XV. Suppose that one tries to purge with clean air a box in which a stimulus at concentration C_0 molecules/cm^3 was previously maintained. The progress of purging depends on the extent of mixing of the clean air and pre-existing mixture in the box. Such mixing occurs under the spontaneous turbulence caused by the clean air injection and normal small temperature differentials; if the mixing is very efficient, a longer purging is needed; however, an even longer purging is needed if air pockets exist, conducive to stagnation, such as in the corners of a box. As a rule of thumb, an amount of air equivalent to at least 20 volumes of the box volume would normally reduce the initial odorant concentration in a non-adsorbing box by a factor of 1000 or better.

Box purging is accelerated if several entrances are dispersed on one side of the box, and exits correspondingly on the other end, so that the air flow in the box is reasonably streamlined. Also, corners are preferably avoided by using rounder shapes. In chambers with intricate shapes, air mixing may however, be more effective than streamlining if stimuli need changing.

Wall materials

The most significant requirement is that the walls are clean from grease and oils and are preferably constructed of inorganic materials. Plastics always contain more volatile lower molecular monomers of the polymeric materials, and frequently contain plasticizers, which are slightly volatile and can contribute complications in more exact work. Furthermore, plastics dissolve organic vapours and are much more difficult to decontaminate. Probably the only plastics worth considering as permanent parts of an olfactometric apparatus are various Teflons, which can be decontaminated, if needed, by prolonged flushing assisted by heating, if feasible.

Polyethylene tubing is reasonably odourless, at least for humans, but cannot be heated without major deformations much beyond 100°C. Teflon can be decontaminated at higher temperatures, up to 200°C. Greased joints, valves, etc. invariably dissolve odorants and need cleansing and regreasing when the odorants are changed in critical work (Cheesman and Kirkby, 1959). Their action in reducing the odorant concentration by the removal of small amounts of odorants is usually inconsequential, since on properly greased stopcocks and joints, the grease should not extend to the rims of the inner passages. At reasonable rates of odorant/carrier flow, diffusion of the odorant to the grease is slow and would not materially influence the contents of the odorant in the stimulus. However, after exposure to higher concentrations, such elements can dispense low-level concentrations of the odorant into the cleaner carrier gas. for long periods of time.

The same applies to rubber, Tygon and other plastic tubing. These are best used in short lengths, if at all, simply to hold together tubes of metal or glass end to end, and discarded after experiments with higher levels of odorants or when returning to lower levels or another odorant.

Tygon (vinyl) tubing, specially formulated for use with food products (Formulation B-44-4x) is lower in odour than the usual laboratory variety.

One must realize that as always the final test for non-odorosity is simply to smell the tubing, since even the same nominal formulations appear to vary in their odour level and a catalogue designation "odourless" cannot be accepted blindly.

Inorganic materials such as glass or metal do not dissolve odorants but do adsorb them. The physically adsorbed amount increases with the molecular size of the odorant, and depends upon the relative vapour pressure of the odorant, cf. eqn. VI. the adsorbed amount, per cm^2 of a wall surface, corresponds to less than a monomolecular layer of odorant in any reasonably diluted (by a factor in excess of 10) vapour. Therefore, a physical limit exists on the maximum amount of odorant that can be held by physical adsorption on glass or metal.

The ultimate amount of adsorbed material in the monolayer, per actual wall surface, including a correction for roughness, is on the order of 10^{15} molecules/cm^2. From dilute vapours, much less than that will be adsorbed. However, when one considers that concentrations at the threshold need to be only 10^2 molecules/ml for some insect pheromones (Kaissling, 1971) and 10^9 to 10^{10} for some substances in humans, one cm^2 of surface exposed to a higher concentration of an odorant can store up relatively substantial amounts of an odorant. They might interfere with subsequent work unless the system is purified by prolonged purging assisted, if possible, by heating, prior to critical threshold experiments. It is generally desirable to reduce the wall effect by decreasing the dimensions of the tubular passages in the olfactometers.

Stainless steel tubing is probably the best choice in materials for olfactometric purposes. It withstands considerable abuse, can be used in sizes much smaller than glass, and can be conveniently joined either by overlapping disposable neoprene or Tygon tubing or by various systems of fittings, such as Swagelok; or easily silver brazed by a hand torch, into overlapping larger tubing or into holes drilled in larger tubing. With the advent of the gas chromatography, stainless steel tubing in 1.6 mm (1/16 in.) o.d. size has become available with a wide range of internal diameters from 0.1 mm (0.004 in.) to 0.75 mm (0.030 in.), making it easy to construct small-volume splitters. Glass tubing in larger sizes is the traditional choice. Metals other than stainless steel may create problems with certain odorants. Thus copper-containing metals interact with nitrogen bases and with some sulphur compounds. A recent advance is the availability of thin wall internally Teflon coated aluminium and stainless steel tubing down to 0.3 cm (1/8 in.) o.d. Teflon reduces the adsorption on metal; however, it can dissolve odorants and the use of such tubing at ambient temperature has not yet been sufficiently established. Bare aluminium exhibits a strong adsorption since it naturally contains a film of hydrated aluminium oxide. This leads to a high initial adsorption when an odorant flow is begun, but also to a low but long lasting desorption rate when a pure gas is passed through a contaminated tube.

Small size thin wall Teflon tubing appears, in the author's experience, satisfactory for channelling prepared stimuli. Teflon walls do dissolve small amounts of odorants. At ambient temperatures this does not lead to any significantly fast loss of the odorant from the flow line if the tubing is not too long (20–40 cm is safe). However, the dissolved odorant is released into the stream if the concentration of the odorant is reduced. Thus, concentration can be increased from experiment to experiment with little difficulty, but when returning to much lower concentrations, or changing the odorant, the Teflon tubing needs decontamination. In such situations, when time is of the essence, new Teflon tubing is substituted. Contaminated tubing can be cleaned by storage at 100°C for some hours, preferably with a clean gas flowing through the tube, or by heating at 150°C–

200°C with a heat gun under rapid clean gas flow in a few minutes. Thin wall Teflon tubing dissipates dissolved odorant vapours also when stored for several days at ambient temperature.

Splitters, flow controls

Splitters serve to reduce the odorant concentration in the carrier gas by taking a fraction of the more concentrated odorant stream and diluting it to a lower concentration by adding some non-odorous carrier gas. The splitter systems usually utilize capillaries as flow restrictors.

The odorant vapour/carrier gas mixture to the splitters is normally supplied at a constant pressure. The flow rates when a constant pressure difference is maintained between the upstream and the downstream end of the capillary is approximately:

$$w = \frac{\Delta P\,\eta}{8\pi r^4 L} \qquad \text{(XVII)}$$

w = the flow rate, ml/s;
ΔP = the difference in pressure between the ends of the capillary, dynes/cm^2;
π = 3.14;
η = the viscosity of the gas, poises; and is assumed to be independent of pressure, over small pressure (η = 186 micropoises for air at 25°C) pressure ranges;
r = the radius, cm, of the capillary;
L = the length, cm, of the capillary.

In olfactometry, the influence of the odorant vapour on the viscosity of air can usually be ignored, since the odorant concentrations are low. Thus, for a constant ΔP, the flow rate decreases by a factor of 2 if the capillary length is increased by a factor of 2, but decreases by a factor of $2^4 = 16$ if the radius is halved. Changing the radius of the capillary becomes the primary flow resistance adjustment factor; changing the length can be considered as a finer mode of this adjustment. Table I gives equivalent lengths and approximate empirical flow rates at 10 cm water column head pressure through 10 cm long capillaries of several nominal sizes, for air at ambient pressure. If everything else is kept constant, the flow rate would decrease by approximately 2% for each 1°C increase in the temperature.

By using splitters, one can split off a slow stream of the odorant vapour carrier and dilute it with a large amount of pure carrier. It is desirable to select a ΔP large enough so that small changes in upstream or downstream pressure will not influence ΔP significantly. When the flow of the carrier gas used for dilution at the downstream end of a capillary is fast, the exit from the capillary into this flow must be designed so that the capillary stream and the dilutor stream meet at a 90° angle, to avoid a pushing or pulling effect by the velocity components of the main flow on the capillary exit pressure.

Capillary exit flows much below 0.1 ml/min are impractical even with small capillaries, especially since such slow flows are difficult to measure exactly. If the dilution gas flow is on the order of 10 l/min, one dilution stage can produce a dilution of not much more than a factor of 10^5. Two or more dilution stages become needed to reach the lowest physiologically active concentrations, especially if each stage produces less than a 10^5 dilution.

TABLE I Examples of air flow rates through stainless steel capillaries at room temperature

*Approximate (nominal) internal diameter of the capillary**		*Air flow rate, ml/min through 10 cm long piece of tubing at 10 cm (4 in.) water head pressure†*
mm	in.	
1.1	0.045	1500
0.74	0.030	300
0.50	0.020	59
0.25	0.010	3.7
0.15	0.006	0.5
0.10	0.004	0.1

* The batches of capillaries with the same nominal i.d. may differ in the actual i.d.
† Equivalent to 7.3 mm mercury.

Capillaries produce a constant flow when the pressure is held constant. For larger head pressures, 1 atm (15 PSI) and over, the usual gas cylinder two-stage pressure regulators are usually adequate. For smaller pressures, several manostats may be needed. By far the simplest and most dependable is a combination of a usual gas pressure regulator gauge or even a pump with a water-column manostat, shown in a later Fig. 13, in connection with a description of several olfactometers. The flow is adjusted to maintain a slow bubbling of excess air through the water-immersed end, so that the entire downstream system is at a constant pressure, approximately equal to the water head pressure.

Another system of flow rate regulation is found with constant flow valves, e.g. Brooks valves, widely used in gas chromatography. These adjustable valves supply, at some predetermined setting, a constant volumetric flow, ml/min. at ambient pressure, as long as the pressure differential between the upstream and downstream ends of the valve is not less than 0.2 atm (3 lb/in.^{2}). With an upstream pressure of 7 atm (90 lb/in^2), the flow rate is constant for any downstream pressure between the ambient and 6.8 atm; this assures constant flow rates of a carrier gas regardless of wide changes of the backpressure of the rest of the system.* The passage of odorous gases through such valves is undesirable, and aerosol particles must be removed from the carrier gases before entering the valves.

Other valving is used in controlling stimulus flows through parts of the olfactometers. The valves can be used to direct the flows of pure carrier gases, and here the selection depends on the pressure range, desirability of manual or electrically actuated solenoid operation, and dependability of the valve action. Other uses of valves may involve the passage of odorous streams. In such cases the valves should not contain gaskets and seats of materials which can adsorb or absorb odorants—a feature that can best be checked by an experiment, as the valve designs and construction materials change.

In the author's experience, the use of valves in the odorous gas carrying streams at lower air velocities is not desirable because of residual contaminations and the tendency to retain odorants, especially the less volatile odorants. This may be less of a difficulty when a long-term stability of a stream of a definite composition is to be maintained and the gas flows are very fast.

In the solenoid valves, the coil heats up when the valve is actuated, and a surge of desorbed odorant may result. Systems with a minimum of valves or even without valves in the odorant stream are possible and may be preferred.

Flow rate measurements

The dilution of a stimulus-carrying stream with additional carrier gas is measured with various flowmeters, Fig. 8.

The most accurate device for flow rate measurements is a wet gasometer used for the calibration of various flowmeters and described in most laboratory supply catalogues. It measures the total flow over any given period of time, and after various correction factors, gives data with an accuracy to 0.5%. It operates with the gas flow at essentially atmospheric pressure and contains a sealant liquid, usually

* Another method for maintaining constant flow is based on a "critical orifice" effect (Corn and Bell, 1963).

water. It can not be used with gases which are considerably soluble in the sealing liquid, nor with odorous gases which would contaminate the meter for a long time to come. It is more suitable for checking and calibrating parts of the flow system with air or other non-odorous gases.

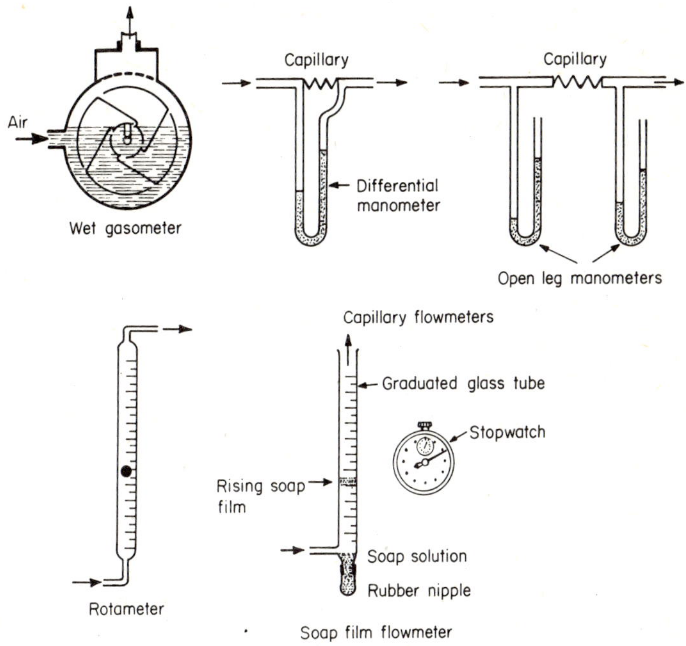

FIG. 8. Devices for measuring gas flow rates. Descriptions of other flow rate measuring devices have been given by Considine and Ross (1964) and in Gas Engineers' Handbook (1965).

For very large flow rates, liquidless expansion chamber gasometers, such as those used in metering household heating gaps, are useful.

Capillary or orifice flowmeters measure the gas pressure drop which results when the flowing gas passes through a flow obstacle such as a capillary or an orifice. The pressure drop is measured differentially either with a U-tube manometer, which contains a manometric liquid, such as mercury, water, or some oil, or with a pressure gauge. Two open-ended manometers (upper left, Fig. 8) can be used instead of a differential flowmeter; then the actual pressure of the flowing gases can be measured in the various parts of the system, but the measurement of the capillary pressure drop becomes less accurate, since it requires calculating the differences of the readings from two manometers, each by itself involving a subtraction of the readings of two meniscus levels.

The capillary flowmeter is very easily constructed from laboratory glassware. It must be calibrated. It is not suitable for monitoring those portions of the olfactometric flow which carry odorant vapours. when the odorant concentrations are frequently changed; odorant vapours drift into the legs of the flowmeters, dissolve or adsorb on the manometric liquids, linger in the manometer, and are difficult to remove. The differential diaphragm or Bourdon tube manometers do not use sealing liquids. However, they also present passages with blind ends into which the odorant vapours may drift and linger. The capillaries in the flowmeters are sensitive to dust contamination and need rechecking.

Rotameters are slightly tapered calibrated glass or plastic tubes in which a metal or glass float or ball floats at an elevation dependent on the mass flow rate of the gas. Plastic material flowmeters are suitable only for non-odorous flows, since plastics dissolve the odorants. Glass flowmeters are preferable, and can be cleaned if needed, by a sufficient flow of non-odorous gas and even gentle heating. Rotameters are usually supplied with calibration charts, and guides exist to correct for the pressures and densities of different gases. The rotameters have an accuracy of 5–10%, depending upon their quality, and such an accuracy is generally reasonably sufficient for most olfactometric work. It is important to check if the float or ball moves freely if the flow rate is modified; if not, some particle may have stuck in the flowmeter and cleaning with a solvent would be necessary.

For checking, but not for a continuous monitoring of slow flow rates, soap film flowmeters are popular. They are applicable to flow rates of a fraction of a ml/min up to 100–500 ml/min, and measure the flow at the exit into the atmosphere. A soap film is set across a calibrated vertical burette and the rate of the film movement along the burette due to the flowing gas is timed.

Clue avoidance

Clues which result from improper statistical designs, specimen designations, and operator's errors and attitudes, are amply discussed in experimental psychophysics texts. Another group of clues can arise from the mechanical design features, and may influence human as well as animal experiments. Effects such as the noises of the flowing gases, the clicking of valves, a glance at some indicating gauge, the positions of the switches, can generate either correct or false clues and might influence the results; acoustic effects certainly can influence animals. Generally, the olfactometers that have several stand-by stimuli, or require none or very little adjustment and change when the stimulus is changed are more likely to be free of such unnecessary clues.

V. Stimulus composition measurement

Olfactometers prepare and deliver the odorant-containing stimuli. The composition of a stimulus is characterized by the chemical species present and their concentrations; these represent the physicochemical dimensions of the stimulus.

We consider the stimulus as it emerges from the olfactometer. One can either estimate these stimuli properties from simplified assumptions based on the method of the stimulus preparation, or else proceed to analyse the stimulus to obtain more exact values on its composition. The first course, an estimate, is the one most commonly used in olfactometric work and may be supplemented by simple measurements. The second course, an analysis, in reality is a very complex procedure, wrought with pitfalls because of the extremely low concentration levels which sometimes need to be measured.

Estimates

Estimates involve assumptions on the purity of the stimulus source chemicals; on the rate of its consumption, or the rate of the process used to vapourize the chemical; on the dilution ratios which require the knowledge of the various flow rates; losses of the stimulus substance, if any, in the olfactometric system; and on the possible degree contamination of the stimulus in the olfactometric system.

The purification and testing of the purity of the stimulus chemical is outside the scope of this discussion. However, a few comments are in order. First, a "chemically pure" odorant can emit, in vapour form, substantial amounts of potentially olfactorily active impurities. A direct analysis of the odorant is of little utility, and a vapour space analysis with a pre-concentration is a necessity; thus, "estimates" on the purity of the odorant itself are almost impossible. Second, the desired level of purity cannot be confirmed without a laborious research effort. The normal procedure is to obtain the odorant at the best available purity, and to rationalize the need for further purification on the basis of the objective of the particular olfactometric study. This aspect will be discussed in more detail in the following section on the stimulus analysis.

The rate of the odorant's consumption in the stimulus generator can be estimated gravimetrically or, for gases, volumetrically. Such an estimate in practice may become quite complex and give erroneous results. If the gas which serves as the odorant vapour carrier is not dry, or contains carbon dioxide or other impurities, the odorant source can dissolve moisture and other contaminants. For the low-volatility odorants, a weight gain may actually result. For low-volatility odorants with residual and not very odorous solvents, the weight loss may be primarily due to the loss of impurities. Only with dry pure carrier gas and reasonably volatile odorants can reliable weight losses be obtained.

When the stimulus is generated by saturating a clean carrier gas with the odorant's vapour, the concentration calculations are based on the vapour pressure of the odorant at the temperature of the odorant. For dilute vapours, which is the usual case with odorants, the vapour pressure is independent of the total pressure (vapour + carrier gas above the odorant source), and the resulting molecular concentration at reaching the equilibrium vapour pressure is:

$$n_i = \frac{(6 \,.\, 10^{23}) \,.\, 273\, p_i}{(22\,400)\, T\, 760} \approx 0.96 \,.\, 10^{19} \frac{p_i}{T} \qquad \text{(XVIII)}$$

or, approximately,

$n_i = 3 \,.\, 2 \,.\, 10^{16}$ at 25°C.

n_i	= molecules/ml
p_i	= vapour pressure of odorant, mm Hg at T
T	= temperature of the odorant in the stimulus generator, °Kelvin
$6 \,.\, 10^{23}$	= Avogadro number (number of molecules per mol)
22 400	= molvolume of gas, ml, at 0°C and 760 mm Hg.

Since the rest of the vapour space consists of the carrier gas, the total number of molecules of the odorant and carrier gas per ml of vapour space at a reasonable pressure is:

$$n_t = \frac{(6 \,.\, 10^{23}) \,.\, 273\, p_t}{(22\,400) \,.\, 760\, T} \qquad \text{(XIX)}$$

or, approximately $3 \,.\, 2 \,.\, 10^{16}\, p_t$ at 25°C;

n_t = the total number of molecules per ml;

p_t = total pressure, mm, odorant and carrier gas, in the vapour space above the odorant.

The mixture of the odorant vapour and the carrier gas leaves the vapour saturator vessel with n_i molecules of odorant per n_t molecules of the mixture. This ratio n_i/n_t remains constant, as long as the pressure in the saturator remains constant, irrespective of any further changes in the pressures in the flow system, although the number of molecules per ml obviously will decrease if the pressure decreases. The ratio remains constant until the next dilution–introduction of more molecules of the carrier gas–is encountered, unless some odorant is lost by adsorption or solution in the flow duct construction materials. The important point is that the ratio will remain the same even if the mixture is prepared at a higher pressure and then depressurized to atmospheric pressure.

Consequently, the molecular concentration of the odorant in a stimulus leaving the olfactometer's port for the atmosphere is

$$n_a = \frac{n_i\,(6\,.\,10^{23})\,.\,273\,p_a}{n_t\,(22\,400)\,T\,760} \approx \frac{n_i}{n_t}\,0.96\,.\,10^{15}\,.\,\frac{p_a}{T} = 10^{15}\,\frac{p_a}{T}\left(\frac{n_i}{n_t}\right) \qquad \text{(XX)}$$

n_a = number of the odorant molecules per ml of stimulus;
p_a = barometric pressure in the test room;
T = temperature, °Kelvin, of the stimulus on leaving the olfactometer.

For rapidly vapourizing odorants, their temperature may drop to supply the heat of vapourization, and then temperature should be either measured or controlled at a known constant temperature by an efficient thermostat. A change in temperature by one degree Centigrade may cause a several percent change in the vapour pressure.

The above estimates are possible if the vapour pressure of the odorant is known. The data for many substances have been listed (Handbook of Chemistry and Physics, 1970). Engineering equations for estimating vapour pressures from various physico-chemical approximations (Perry, 1963) are also available. For some compounds, vapour pressures can be estimated if those of its homologues are known, since the vapour pressures vary systematically within a homologous series.

When the vapour pressure at two temperatures is known, its value for the intermediate temperatures (if the melting point is not within this interval) can be obtained from a plot of logarithm of the vapour pressure versus $1/T$ where T is the temperature in °Kelvin.

If the vapour pressure of an odorant is not accessible through any of these approaches, two courses remain: measure the vapour pressure, which is not a simple task, or report all of the concentrations in terms of dilution with respect to the saturated vapour of the odorant. Of course, an analysis of the stimulus is also possible. although difficult, because of the extremely high sensitivity that may be needed.

Tracer methods

When one odourless gas is used to generate a certain concentration of an odorant vapour in this gas, for instance by the over-the-surface saturation method, and some other gas is used to produce further dilutions of such a stream, the dilution ratio can be measured by analysing the resultant gas mixture for the first gas. This author has used methane, which is odourless to humans, to produce an odorant-vapour-saturated stream; pure nitrogen to dilute this stream; and a hydrogen flame ionization detector to continuously monitor the resulting methane concentration in nitrogen in a branch of the flow line parallel to the sniffing port. Dilutions of up to $1:10^6$ can be monitored. Because of the flammability of methane, air cannot be used as a diluent, and fast flows that may build up methane in the room are unwise. One assumption is that neither of the two carrier gases have any olfactory activity and that the dilution and migration of the odorant vapour closely follows that of the carrier gas. The latter is certainly not true if the mixture passes through complicated flow channels where adsorption can occur and when not enough time is given for a steady-state adsorption to establish after each change in the dilution.

Choices of non-toxic non-odorous and non-inflammable gases for which sufficiently sensitive and continuous monitoring devices are available are very limited. Their principal uses are probably only in tracing the paths and dilutions of odorous effluents in surroundings. Here, two principal choices exist. One is odourless sulphur hexafluoride. Since it is an electro-negative gas, a very sensitive electron capture detector can be used for its detection; however, this detector cannot be used directly in air because of its sensitivity of oxygen and some other electron-capturing compounds, mostly organic halides. Air must be monitored by periodic analyses (Turk *et al.,* 1965). Electrical power transformers use SF_6 as an isolator, and leaks from such equipment may interfere with tracing experiments.

The other choice is non-toxic low-odorosity gas dichloro-difluoromethane (Freon 12) which can be detected by a portable commercial detector (General Electric, 1969) used for finding refrigerant leaks from refrigeration systems, here interferences may result from the leaks from nearby refrigerators and air conditioners.

In regular olfactometry, tracer methods to follow the dilutions of the stimulus in the olfactometer have not found much use because of the complexity and doubts as to the effect of the tracer gas on the olfactory perception. A radio-isotope method has been described (Smith and Hochstetter, 1969).

Isotopic and radio-isotopic methods have definite application in studying the fate of the odorant once it has reached the organism, for instance in attempts to establish what amount of the traced odorant really reaches the receptors (Schneider *et al.*, 1968). This subject is outside the scope of the present chapter.

Stimulus analysis

A continuous analysis of low-concentration stimuli present is possible only in rare cases, for instance a flame-photometric sulphur detector (Stevens and O'Keefe, 1970) may monitor for most sulphur compounds down to human detection levels. A batch analysis, taking an aliquot of the stimulus, is possible and usually would utilize gas chromatographic techniques (Johnston and Sandoval, 1962; Schneider *et al.*, 1963; Moulton *et al.*, 1970).

Such an analysis only rarely can be performed by taking a stimulus sample with a gas syringe and injecting it into the gas chromatograph. First, the maximum gas sample that can be injected is on the order of only a few ml; even for humans the odour thresholds for many substances are well below 10^{-10} g/ml of air, and with a typical routine GC sensitivity at least 10^{-9}g of an odorant is needed to produce a reasonably dependable measurable response. Second, the heavier odorants adsorb in the syringe and cannot be fully injected for analysis. A heated sampling loop is sometimes used to avoid the latter difficulty, but does not solve the inadequate concentration problem. Thus, direct analysis methods have severe limitations.

Techniques have been developed for the preconcentration of odorous vapours from air, and samples extracted from many litres of air can be reliably analysed (Dravnieks and O'Donnell, 1971). However, such cumbersome methods would rarely be suitable in routine olfactometric work where one starts with substances of reasonably known purity.

Gas chromatographic analysis with sample preconcentration is a useful checking tool for cases where the olfactory purity of the stimulus is a crucial point in proving some point. As explained in connection with Fig. 3, a direct analysis of an odorant is not sufficient to establish its purity. It is necessary then to analyse the stimulus itself, after it has been prepared and diluted, and had a chance to pick up additional impurities in the olfactometer. In such cases, stimulus recollection, e.g. by the resin adsorption method (Dravnieks *et al.*, 1971) for a gas chromatographic analysis is desired. The analysis indicates if impurities are present. If the gas chromatograph is equipped with an effluent sniffing port, it is possible, even without knowing the chemical identity of the impurities, to crudely estimate if the impurities are odorous and to what extent they may modify the odour of the stimulus.

Of course, similar analytical methods can be applied to test the purity of gases used as vapour carriers in olfactometry.

VI. Delivery of stimulus to receptor

The stimulus receptor may be an electrophysiological preparation, an immobilized animal, a mobile animal or insect in an olfactory chamber, or human in an odour presentation room. The method of stimulus delivery is very much a function of the experimental design and the preference of the experimenter.

In the best arranged electrophysiological and in immobilized animal experiments, the stimulus delivery port or tube remains in a constant position with respect to the actual chemoreceptor area of the organism. Stimulus changes are produced, without changing the delivery port, internally in the olfactometer. Here the preferable arrangement consists of the shortest possible passage between the point on the olfactometer where the stimulus concentration is changed and the exit opening of the stimulus. Up to the point where the stimulus enters this passage, the stimulus flow rate should be high, so that the time for removal of adsorbed odorants and the volumetric change-over time for the stimulus, are as short as possible. Lingering of the stimulus in the final delivery tube also should be minimized, by avoiding high-absorbancy materials and by maintaining the internal surfaces of this passage free of grease and condensed moisture. Experiments with all other mechanical and electrical operations must be conducted to ascertain that some other effects do not interact with the animal.

In the mobile animal experiments two basic situations can be distinguished. In one, the olfactory chamber is well-ventilated, and the location of the stimulus entrance into the chamber is searched and compared by the animal. In another type of animal experiment, the entire olfactory chamber is permeated with the stimulus when desired. Here careful aerodynamic design must be considered, to effect a fast and uniform change of air. This requires many properly distributed stimulus injection ports, or one end of the chamber operated from the same stimulus preparation device, and with a similarly distributed air outflow registers on the other end of the chamber. Relatively non-turbulent air exchange can be achieved by such design. The next best design has one or a few stimulus introduction ports, but a rapid mixing of air in the chambers; here high turbulence is unavoidable and should be accounted for as a sensory noise.

By far the largest variety of stimulus delivery systems is encountered in work with humans. Figure 9 depicts the principal variations.

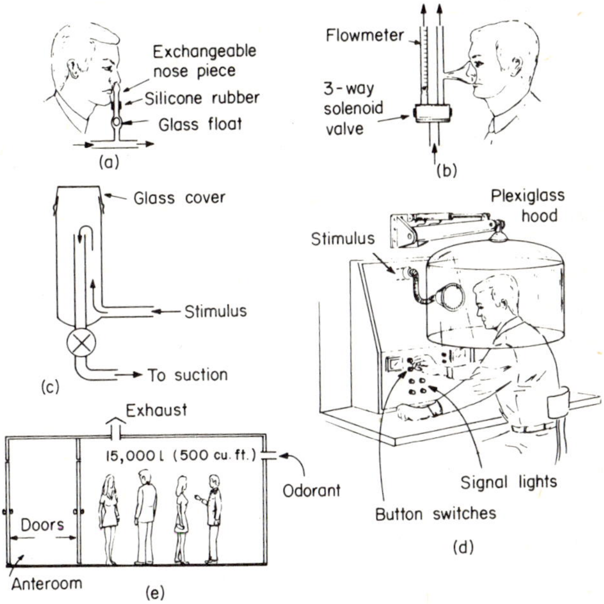

FIG. 9 Some techniques for odour stimulus delivery to humans. A, after Köster (1967, 1971); B, after Johnston and Sandova (1962); C, after Cheesman (1959); D, after Stone (1968) modified; Stone used a fixed rectangular hood; E, after Deininger and McKinley (1954).

Considerable disagreement exists among researchers on the most proper method for the delivery of the stimulus to the human nose. On one extreme, a direct non-leaky connection is made between the olfactometer and the nostrils, inserting tubes into the nostrils (Köster, 1967), or by placing the nose in a glass mask (Johnston and Sandoval, 1962). In Köster's arrangement, Fig. 9a, a small glass float opens the passage between the odour stimulus duct and the nose when a negative pressure is generated by inhaling through the nose; this admits a small controlled volume to the nostril since shortly afterward the float hits the upper glass orifice and interrupts any further inhalation. Such systems define well the concentration of the odorants in the inhaled air and the volume inhaled. In Johnston's olfactometer, a glass nose cone is used, Fig. 9b.

On the other side, some researchers (Jones, 1955; Wenzel, 1955) argue that each individual's olfactory apparatus, including the nasal passages and the inhalation mechanism, operate best when least interfered with. Sniffing from a freely operating port or a flask, or simply entering a room in which the entire air serves as the stimulus, represent more natural modes of smelling. Normalizing sniffing volumes to the same value for everyone may optimize odour evaluation to some individuals, but make it less effective for others; masks, tubes, etc., also interfere with the skin, insides of nostrils, and so on, and may have an effect on the judgements.

When sniffing from a flask, or a smaller port, the actual concentration of the odorant in air is less well defined than when smelling a stimulus delivered by a positive flow from a mask or a nostril tube, as some extraneous air enters. Since most odour studies deal with relative effects, this factor must be considered in comparing data obtained by different stimulus delivery systems. Mullins (1955) reported that the odour thresholds become stable when the stimulus flow rate reached 4 litre/minute. However, much useful data has been obtained at much slower flow rates, not excluding natural sniffing from glass or squeeze flasks. There appears to be no reason why simpler and logistically more feasible techniques cannot usefully serve in much of the olfactometric work.

Figure 9 depicts also some arrangements for stimulus delivery in a more open fashion. Sketch C is Cheesman's (cf. 1959) modification of the port method. The port is covered and the stimulus is pulled by a pump into the centre tube within the port; this prevents an unnecessary contamination of the room air by the odour. For sniffing, the cover is removed and the suction from the centre tube is stopped.

Sketch D represents a hood device. The subject's head is in a hood. The odourous or blank stimuli are supplied to the hood; a signal light calls for a judgement, and the judgement is signalled by depressing the corresponding pushbutton. The hood method dates back to Zwaardemaker (1921), and is used in some contemporary olfactometers. Thus, Ough and Stone (1961; Stone, 1968) used a rectangular Plexiglas hood, with the signal system in the backpanel of the hood, and with 260 l/min (9.5 cu fl/min) stimulus delivery rate. (Others:

Bozza *et al.,* 1960; Cederlöf *et al.,* 1966; Lindvall, 1970, annotated references.)

Sketch E shows a method in which air in a room serves as a stimulus. The individual enters a room in which the odorant vapour is dispersed, and makes an odour judgement upon entrance (Deininger and McKinley, 1954; Kerka and Humphreys, 1956). The room is lined with aluminium sheets to facilitate cleaning and to reduce residual odours, and is preceded by an anteroom for the conditioning of the subjects.

The mode of the stimulus delivery to the nose should be selected to functionally meet the desired objective. If one is concerned with the smallest detectable absolute amount of some substance, the port inserted or attached to the nose is probably the most reliable technique to quantify the delivered absolute amount. If one is concerned with the odorant concentrations that may be noticeable in a room, entering an odour-stimulus containing room is the most suitable procedure.

In some procedures, e.g. if the subject has to select, from a set of three stimuli, the one which is different from the two others (triangle test), several ports supplying separate stimuli are a necessity.

Delivery of stimulus by a blast injection technique, where a rapid pulse of odorous air is forcibly injected into the nostril, while the subject holds his breath, has been used by Elsberg and Levy (1936). It is also used, in connection with the syringe dilution method for establishing the odour thresholds of atmospheric pollutants, in an ASTM (1967; Mills *et al.,* 1963) test. Jones (1955), Wenzel (1955), and Köster (1971, p. 66) reported that the blast injection technique tends to give less reproducible results than the other techniques, based on a natural inhalation.

VII. Several typical olfactometers

A separate list of references, at the end of the text of this chapter, enumerates a variety of olfactometers that are described in the literature. Each represents a somewhat different combination of the olfactometric instrumentation elements discussed in this chapter.

Four contemporary olfactometers, as examples, are somewhat arbitrarily selected for a more complete description. Moulton *et al.* (1970) olfactometer is typical for work with animals. Köster's (1967, 1971) is suitable for work with humans and as a source of several exchangeable stimuli. Sanders *et al.* (1970) represents a portable device, equipped with a standardizing odorant, for odour studies in various environments. Finally, a simple continuous flow olfactometric system used by the author is described; it serves as a convenient substitute for the sniffing flask method, but avoids the need for diluents other than room air for preparing the diluted stimuli.

Simpler olfactometric devices have been given a variety of names: osmoscopes, osmometers, scentometers (Barnebey-Cheney, 1963), stinkometers, etc. In the design of many of these, expediency has dominated. When their limitations are understood, they find some partial utility in practical industrial work, mostly for relative and gross comparisons.

Moulton's olfactometer

Moulton (1972) has described an olfactometer for animal studies. Stimulus preparation follows a traditional approach utilizing gas dispersion bottles to prepare a saturated odorant vapour, and flowmeters and needle valves to split and dilute the gas flows, Fig. 10. A gas chromatograph to monitor the stimulus concentration is an added feature. Since the concentrations indicated by the hydrogen flame ionization detector reasonably approached those calculated from the flowmeter readings, Moulton implied that the chromatographic monitoring was not an absolute necessity. It would appear that such monitoring conducted with sufficient sensitivity is a useful tool for checking the purity of the stimulus in critical experiments. The highly instrumented monitoring of the responses of the animal is a particularly interesting feature of Moulton's olfactometer (Moulton *et al.*, 1970).

The animal is in a glass test chamber with a Teflon-lined floor. One end of the chamber is equipped with two compartments. Either pure air or an odorous stimulus can be supplied to the base of the compartment by switching valves in the flow loop L. The compartments are separated from the test chamber by upward sliding glass doors. The rabbit initiates each odour experiment by pressing a glass treadle, not shown, on the other end (bottom of the figure) of the chamber. The compartment doors then open, the animal smells the air in the compartments, and if trained with a specific odour, signals his choice by interrupting a photocell-monitored horizontal light beam for 7 s in the selected compartment. If the choice was correct, water is delivered to a cup in this compartment; and in 7 s, the hinged floor in the compartments moves up and expels the rabbit. If the choice was incorrect, no water appears, the floor moves up to expel the rabbit, and the glass doors to the compartments close.

Aside from this monitoring of the animal's choices, his electrophysiological responses are monitored by chronically implanted electrodes in the olfactory bulb, and transmitted by a miniature radio transmitter to an omnidirectional antenna, not shown, but located on the top of the test chamber, for the further processing and recording.

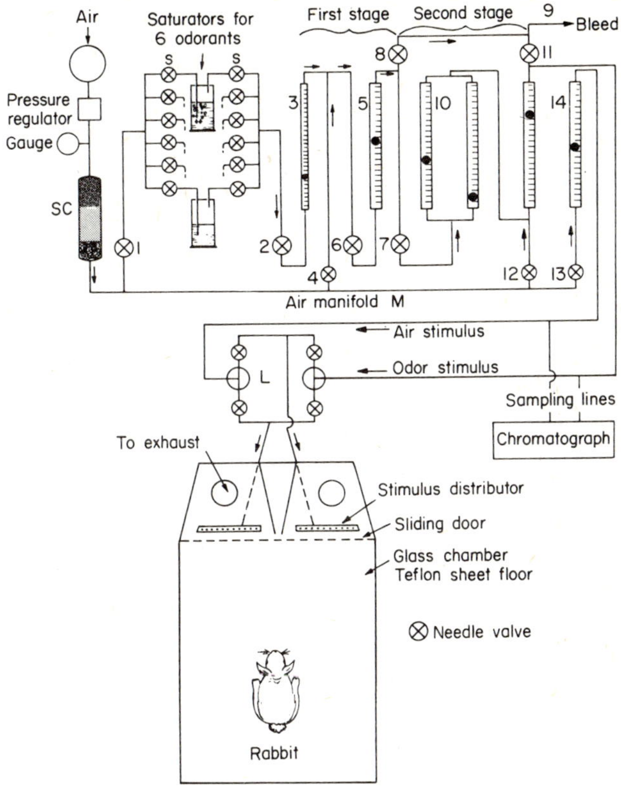

FIG. 10. An olfactometer for animal studies, after Moulton *et al.* (1970) and Moulton (1972).

The stimulus preparation includes a two-step dilution. Pump P delivers air through a purifier SC which contains silica gel to remove moisture, and activated carbon to significantly reduce the content of the organic vapours. The air flow is split. One part goes through a manifolded dispersion flask system with odorants. Stopcocks S permit selection of the dispersion flasks. The flow of air saturated with the odorant vapour is controlled with needle valves 1 and 2 and measured with the flowmeter 3. The other portion of the purified air flows through the manifold M. A certain fraction of it, controlled by the needle valve 4, is used to dilute the saturated odorant vapour flow; the joint flow is measured with the flowmeter 5, and the dilution achieved in this first step is calculated from the readings of 3 and 5; the needle valve 6 provides additional flow control. In the second stage, the diluted stimulus is split, with the split ratio adjusted by needle valves 7 and 8. The fraction passing through 8 is discarded through bleed line 9. The fraction passing through 7 is measured with the flowmeter 10 (several flowmeters may be needed to cover a broader range of flows) and further diluted with purified air, admitted by valve 12, to the second dilution stage. Dilution by a factor of 10^7 can be reached with this system. The "blank"—non-odorous air stimulus—is supplied from the manifold via 13 and 14. If only the first dilution stage is used, valves 8, 11, and 12 are closed.

The dispersion flask and the manometer system are enclosed in a chamber thermostatted at 20°C.

Further lines permit analysis of the prepared stimulus in the gas chromatograph and the delivery of either the diluted stimulus or the purified air to either of the two compartments of the test chamber.

Ducts in the roof of the chamber permit air to exhaust from the chamber to the outside of the building.

Köster's olfactometer

Köster (1967, 1971) has described a modified and partially automated Stuiver's (1960) olfactometer for use in human olfactory studies. The stimulus is prepared at a rate of 600 ml/s and thus its preparation system should be suitable for many tasks in behavioural and electrophysiological olfactory studies in many animals. A special sniffing valve is used in the work with humans to control the stimulus volume sniffed.

The olfactometer is typical of a design in which valves, capillaries, and flowmeters are used on a massive scale. The principal features are shown in Fig. 11.

An air compressor pumps outside air through an air conditioner where the air is washed by water, cooled to 0°C, and reheated to 20°C before entering the system. An air flow control valve and a capillary flowmeter with two U-tube manometers permit adjusting the air flow rate, typically at 600 ml/s. This air flows through a manifold, picks up the odorant-carrying side streams when they are delivered from the odorant vapour preparation system, and is split into three streams to supply three branches each equipped with one sniffing port. Valves 2 are used to equalize the flow rates in the branches.

The stimulus can be simultaneously evaluated by three subjects each using a separate port. They are alerted by a signal light, and reminded by a sound alarm if they do not respond with a judgement within a brief time interval.

The subjects express their judgements by pressing buttons. These judgements are automatically recorded on a punch tape.

The odorant vapour preparation and dosage system utilizes cylinder nitrogen as the vapour carrier. Valve 3 permits adjusting the nitrogen flow to 30-60 ml/sec. An expansion volume helps in smoothing the nitrogen flow.

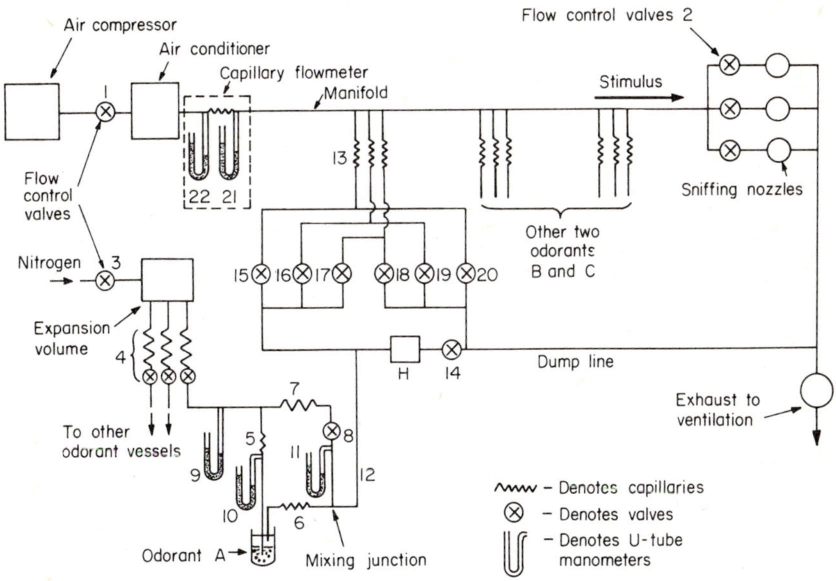

FIG. 11. An olfactometer for psychophysical studies on humans, after Köster (1971).

Capillary and valve combinations 4 subdivide the nitrogen flow into three equal parts, each fed to one of the three vapour preparation systems. Thus vapours of three different odorants singly or in any desired mixture can be supplied to the sniffing port manifold.

Only one of the odorant vapour preparation systems is shown in the figure. The nitrogen flow is split: approximately 1 ml/sec flows via capillary 5, picks up the odorant vapour in the thermostatted odorant container, and continues through capillary 6 toward the mixing junction; approximately 10 ml/s flow through capillary 7 and valve 8, again toward the mixing junction. Both streams mix in the junction. Pressure drop along the capillary 5 is measured by U-tube manometers 9 and 10 and permits calculating the actual rate of the nitrogen flow through the odorant vessel. Similarly, the pressure drop along the capillary 7 is measured by the manometers 9 and 11, and is used to calculate the exact flow rate through the bypass. Thus, the extent of the dilution of the saturated vapour of the odorant in the mixture leaving the mixing junction through line 12 can be calculated.

The prepared vapour/nitrogen mixture then enters a stimulus proportioning and switching system. This system communicates with the air manifold through a series of capillaries 13, and through a dump line to the ventilation system. When the electro-magnetically operated valve H is open, the odorant vapour is short-circuited to the dump line and does not enter the manual valving system leading to the capillaries 13. This would be a "no stimulus" position for the odorant shown. Valve 14 serves to adjust flows in such a way that the dump line which generates a slightly negative pressure (because of its connection to the exhaust) does not cause a large influx of air from the manifold through capillaries 13 when H is open.

The three H valves (only one is shown), one for each of the three odorants, are operated by a punch tape which contains instructions on the order and the duration of the stimulus to be delivered to the sniffing ports. The coded information on the actually presented stimuli is automatically recorded on the same punch tape which records the subjects' judgements; both types of information are then sent to a computer for tabulations and counts.

When the valve H is closed, the odorant vapour/nitrogen mixture is manipulated by the manual valves 15 through 20. Valve 20 leads to the dump line. If for instance valve 15 is open, valves 16 to 19 closed, and valve 20 is gradually restricted, the pressure in the manual valving system rises. When the manometer 11 shows a higher pressure than the manometer 21, nitrogen and odourant vapour mixture through valve 15 and the left one of the capillaries 13 will enter the manifold producing an odour stimulus at the sniffing ports. The other valve pairs, 16 and 19, or 17 and 18, can operate jointly in the same manner, by sending the stimulus through the middle or the right capillary 13, respectively. The flow rates of the odorant vapour/nitrogen mixture into the manifold are calculated from the pressures indicated by the manometers 11 and 21 and the calibration data on the capillaries 13.

The calculations of the dilution of the odorant vapour which reaches the sniffing ports require recording the indications of the manometers 21, 22, 9, 10, and 11 and having calibrations (flow rate versus pressure drop) of capillary 23, of the three capillaries denoted 13, and of capillaries 5 and 7. The indications of the manometers 21 and 22 and the calibration of 23 apply to all three odorants.

Other manometers and capillaries occur respectively in the two other odorant systems, not shown in Fig. 11, but identical to the system for controlling the stimulus of the odorant A.

The design of the sniffing ports used by Köster has been shown in Fig. 9a.

Sanders' olfactometer

Sanders *et al.* (1970) have described a transportable dynamic olfactometer designed with provisions for a continuous adjustment of the odour intensity by the subjects themselves.

The principal use of this device was in establishing the odour thresholds on air pollution odours: the subject begins by sniffing a continuously flowing purified air stream delivered to a face mask, and by the gradual turning of a needle value continuously increases the addition of odorous air to the purified air until an odour is noticed. At this point, the concentration of the odorous air in the purified air is calculated from the flowmeter readings.

The same principle can obviously be used for measuring odour thresholds of prepared odorous vapours or for maintaining the psychophysical odour intensity at some desired level.

The olfactometer is provided also with a facility for calibrating each subject's sensitivity with a known odorant. Sanders was concerned with odorous emissions from paper and pulp mills, and therefore used methylmercaptan as the standard odorant; this substance is considered one of the odour carriers in such emissions.

One part of his olfactometer which was arranged for atmospheric sampling is shown in Fig. 12. A carbon vane pump P sucks air from the atmosphere. Such pumps do not contribute oil or plastic odours to the moved air. Excess air is discharged through a needle valve 1. The rest of the air passes into the olfactometer through a volume V, which smoothens out the pressure surges created by the pump, and through another needle valve 2. The air flow is then split. The branches B and S are closed, so that one part of the air flows through a dilution chamber D (which has no specific function in this mode of operation, but is needed in the calibration of subjects, see lower), and line 3 into the air deodorizer I, then, through line 4 and rotameter (flowmeter) 5 into a mixing chamber C and a breathing mask. The flow rate into the mask is maintained at 12 1/m.

The deodorizer column utilizes silica gel to absorb moisture from the air and active carbon to remove odorants, with glass wool plugs to hold the absorbents in place and to trap dust resulting from carbon.

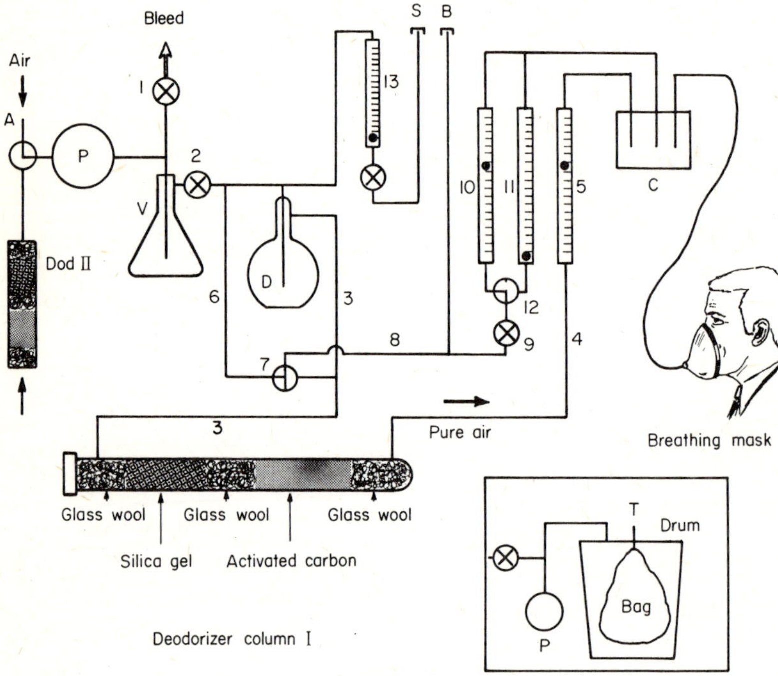

FIG. 12. A transportable olfactometer, after Sanders *et al.* (1970).

The other branch, 6, leads pumped odorous air through three-way stopcock 7 into line 8. By a gradual opening of the needle valve 9, the odorous air can be admitted in a continuous increasing stream to the mixing chamber C, where it mixes with the deodorized air before reaching the breathing mask. The odorous air is metered with the rotameters 10 and 11, one for slower, and the other for faster air flows. A three-way stopcock 12 permits selection of either rotameter. The needle valve 9 does not carry positional markings. This is necessary to avoid a bias in repetitive measurements on the same stimulus.

The subjects are standardized in each session with an odorant. A compressed gas cylinder containing known concentrations of an odorant, for instance methylmercaptan, in nitrogen is attached at S. The flow is metered by rotameter 13, and dilution chamber D. Outside air is again delivered by the pump P, but in this mode of operation air enters the pump through a second deodorizing column, DOD II, instead of inlet A.

The valve 7 is set so as to connect line 3 to line 8; in this valve position, there is no flow in line 6. The line 3 sends part of the odorized mixture through the column I, where it is deodorized and sent via line 4 to the breathing mask at 12 litre/minute.

The odorous mixture from D via 7 and 8 is gradually admitted through the valve 9 to the mixing chamber, C.

When the desired odour level, e.g. threshold, is reached, the rotameters are read. The flow in 13 divided by the sum of flows through 5 and 10 (or 5 and 11, if 11 was used) gives the dilution of the cylinder gas entering at the valve 9. The flow in 10 (or 11) divided by the sum of flows in 10 and 5 gives the dilution which the flow through 9 experiences in C. The product of these two dilution factors gives the final dilution of cylinder gas in the breathing mask.

An air sampling mode frequently used in the odorous air pollution studies, especially in sampling odorous effluents, utilizes plastic bags (Schuette, 1967), Fig. 12, lower right inset. The bags, typically 20 to 100 litres in size, are placed in a drum, with the sampling tube T outside of the drum. By applying pump suction to the drum, the bag is made to expand and draw in the gaseous sample through T.

In the reverse, by pumping air into the drum, the bag is made to collapse and expel the sample into an olfactometer.

In Sanders' olfactometer, the bag sample is introduced through line B, while the valve 7 is turned to disconnect line 8 from lines 3 and 6. Non-odorous air prepared from outside air and driven by P is again supplied by line 4 into C; the concentration of the bagged sample in C is again gradually increased by opening the needle valve 9 until an odour is noticeable.

Flexible simple valveless olfactometers

Because of its simplicity, the flask method in which odorants are presented for sniffing in pure form or diluted by odourless solvents is still in wide use. As explained earlier, eqn. V, the vapour concentration of the odorant cannot be calculated from the concentration in the liquid unless additional information on the activity coefficient of the odorant in the solvent is available. Furthermore, truly odourless solvents (except water in which many odorants dissolve poorly) at present are not available. Also, sniffing from a stagnant vapour space in the flask is not a well-defined method of vapour delivery to the nose.

The author of this chapter has developed a flexible olfactometric technique in which the lower odorant concentrations are prepared by the dilution of the vapours of pure substances by air and the resulting stimuli are delivered for sniffing with a positive flow from glass sniffing ports. Valves are avoided except those which may be in the pumps, and the simple olfactometric system can be put together or dismounted and discarded when the odorants are changed. The technique serves as a substitute for the flask method. Figure 13 shows the principal elements of the technique. Air is pumped by an aquarium

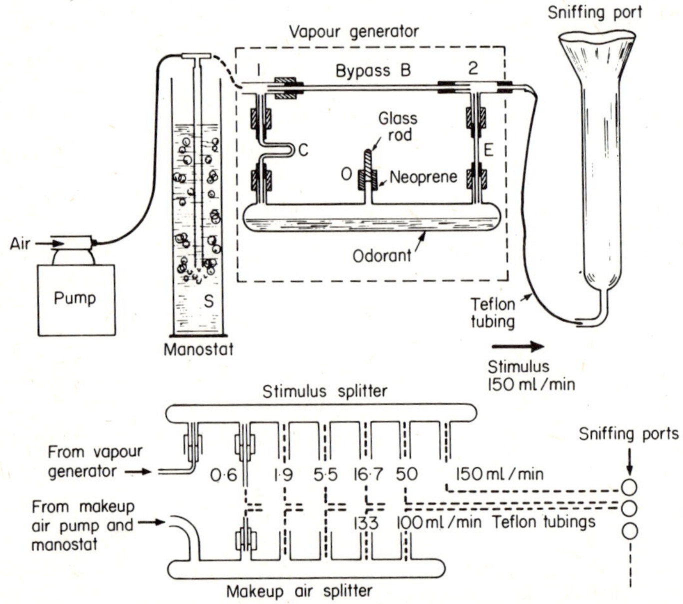

FIG. 13. A simple valveless olfactometry (Dravnieks, this text).

type of pump 7; an odourless pump is selected, thereby avoiding plastic membranes and plastic disc valves, since these materials would impart odours to the pumped air. The air pressure at the inlet to the vapour generator is controlled by a water column manostat. The pressure in the tee above the manostat, when an excess of air continues to bubble through the water, remains constant and is dependent merely upon the depth of immersion of the standpipe S in the water.

The pumped air is delivered, via nominally odourless Tygon tubing

(Formulation B-44-4X, used for pumping milk products) to a brass tee 1, such as those used in aquaria for air flow splitting.

The manostatted air enters the vapour generator. The odorant is placed with a medical disposable syringe into a 15 cm long 1.25 cm o.d. glass vessel (shown oversized in the figure) through the middle opening, O, which is then closed with a short piece of neoprene tubing and a glass rod; the rod is pushed against the glass tube stub to minimize the contact between the odorant vapour and the neoprene. One portion of the air bypasses the odorant vessel through an approximately 15 cm long, 0.75 mm (0.030 in.) i.d., 1.5 mm (1/16 in.) o.d. stainless steel tubing B. The other portion enters the vapour space of the odorant vessel through a stainless capillary C selected so as to produce the desired flow rate. The bypass tubing size is maintained constant, only C is changed as needed.

The odorant is maintained at a measured ambient temperature, from which the vapour pressure can be calculated. Air saturated with the odorant vapour exists from the vessel through a 5 cm long 0.75 mm i.d. stainless steel exit capillary E and joins the by-pass air by another brass tee, 2.

The connections between the capillaries, brass tees, and glass stubs are formed by two sizes of neoprene tubing. A short length of ~ 5 mm (3/16 in.) o.d., ~ 0.7 mm (1/16 in.) i.d. neoprene tubing is slipped over the stainless tubing, and a short length of a larger neoprene tubing, 3 mm (1/8 in.) i.d. and 6 mm (1/4 in.) o.d. is used to hold the capillary in the brass tees or glass stubs. The end of the capillary protrudes significantly into the connected pieces of the tubings. Thus, air and odorant vapours contact the neoprene only through a dead-ended annular space between the stainless tubing and the larger glass or brass tubing. Vapours can exchange between the flowing gas and the neoprene only by a static diffusion through the annular air space. This process is so slow that neoprene odours, loss of the odorant by diffusion into the neoprene, and even contamination of the air flow with odorants absorbed by neoprene do not interfere with the prepared stimuli. In the case of severe contamination, neoprene parts and capillaries can be discarded and easily replaced by new sets.

The flow rates are manipulated by changing the capillaries. The 1.5 mm (1.16 in.) o.d. stainless steel capillaries are available with internal diameters from 0.08 mm (0.004 in.) to 1.1 mm (0.045 in.). Since at the same head pressure and length the flow rates, ml/min, are proportional to the 4th power of the internal diameter, such a selection provides an extremely wide range of flow resistances. Thus,

10 cm of 0.08 mm tubing provides ~ 7500 times slower flow than 0.75 mm tubing of the same length at the same pressure. Further flow adjustments are obtained by changing the lengths of the tubing.

In practice, an inventory of several tubing sizes is maintained. These are obtainable from gas-chromatographic supply houses. When necessary, tubing is easily cut by holding it in pliers, filing a modest notch on one side of the tubing with the sharp edge of a small metal file, and breaking the tubing by bending it at the notch a few times. Tubing ends do not close when this procedure is followed. Edge burrs do not interfere with tubing attachment and need no deburring, except for craftsmanship. Table I lists typical flows for some capillaries.

To measure the flow split ratio in the vapour generator shown, the brass tee 2 is pulled off to expose the end of the bypass capillary B and of the exit tubing E. A soap film flowmeter is used to measure the flowrates from the bypass and from E. The ratio, but not the absolute flow rate, remains the same when the both ends are in the brass tee. The prepared diluted stimulus is then directed through a thin-wall Teflon spaghetti tubing to a glass sniffing port equipped with an elliptical opening to permit a simultaneous sniffing with both nostrils. The port is 12 cm (~ 5 in.) long and the opening is 2 by 3.5 cm (~ 3/4 by 1 1/2 in.). A stainless steel capillary serves as a connector F and is selected in such a way that at a manostat's water column pressure of 20 to 40 cm the stimulus flow from the sniffing port is on order of 150 ml/min. The port with the Teflon tubing can be held by the subjects and handled about as easily as a sniffing flask, and stored in a ventilated hood when not in use by the subject.

For further dilutions, the stimulus from the device shown can be directed to a splitter manifold. This is a glass tube with side tubes which can carry stainless steel capillaries of various sizes. A six-way splitter is shown at the bottom of the figure. Here the first capillary delivers 150 ml/min of the prepared stimulus, the second 50 ml/min, etc. These fractional flows are also delivered to the sniffing ports. In addition, makeup air is pumped, for the simplicity of flow adjustment, by another pump/manostat system, and is fed to the ports through a separate splitter and Teflon tubes; this is needed to maintain the same total flow, 150 ml/min in all six ports.

Similar systems have been used to make mixtures of odorant vapours. One pump drives several vapour generators in parallel, each with a different odorant. Mixtures are obtained by simply sticking the corresponding Teflon outlet tubes into the same sniffing port.

Annotated Bibliography

Eyferth, K. and Krueger, K. (1970). The construction of a programmable olfactometer. *Behav. Res. Meth. Instr.* 2, 19–22.

Provides 10 binary dilution steps by a system of 1:1 splitters at each step. Stores 24 different odorants. Electromagnetically controlled values provide stimuli of each of these, or pairs of different stimuli for comparisons, or mixtures of two odorants. Order of presentation controlled by punched tape. The volume of stimulus flow is greatly minimized in comparison with other olfactometers. Provisions for GC analysis of the stimulus.

Foster, D., Scofield, E. H. and Dallenbach, K. M. (1950). An olfactorium. *Am. J. Psychol.* **63**, 431–440.

Subjects, after thorough bathing and coating with a plastic film, enter a 7500 litre (250 cu. ft.) glass chamber to evaluate odours. (Note by Dravnieks: plastics contain volatile and frequently odorous components.)

Gee, A. H. (1954). Organoleptic appraisal of three-component mixtures. *ASTM Spec. Techn. Publ.* **164**, 33–55.

Adjustable mixtures of odorous vapours are continuously delivered to a sectioned glass manifold, 7.5 cm (3 in.) i.d., for simultaneous evaluations by 4 judges at separate sniffing ports.

Hemeon, C. L. (1968). Technique and apparatus for quantitative measurement of odour emissions. *J. Air Poll. Control Assoc.* **18**, 166–170.

Collects gaseous emission samples in a plastic bag for a later expulsion into an olfactometer equipped with 3 ports for simultaneous evaluation by 3 judges. In some industrial use now for odour intensity and threshold measurements.

Katz, S. H. and Talbert, E. J. (1930). Intensities of odours and irritating effects of warning agents for inflammable and poisonous gases, *U.S. Dept. Commerce Techn. Paper* **480**, 37 pages.

One of the classical papers in dynamic olfactometry, describing five methods of continuously adding odorant vapours to air stream, to deal with substances of different volatilities and to encompass the dilution ratios from 1:10 to $1:10^{13}$. Stimulus flow rate 8 litre/min.

Lindvall, T. (1970). On sensory evaluation of odorous air pollutant intensities, *Nord. Hyg. Tidsk., Suppl. 2.*

Several olfactometers described: Portable olfactometer (pages 22 to 23) with a system of exchangeable capillaries to produce dilutions in geometric progression. Subject smells by inserting face through a hole in a hood where stimulus flows at 100 litre/min. Extensive use of Teflon and Teflon-coated surfaces. Mobile olfactometric laboratory (pages 23 to 26) consists of two connected vans—waiting room and laboratory—with climate control. Pneumatically operated stimulus control system to provide 16 different dilution levels. Two odour exposure hoods similar to the hood in the portable unit. Low odour absorption materials used throughout.

Nader, J. (1958). An odor evaluation apparatus for field and laboratory use. *Am. Ind. Hyg. Assoc. J.* **19**, 1–7.

Technically, a precursor of Sanders *et al.* (1970) portable olfactometer. Special features: diffusion cell for supplying vapours of liquid reference odorants; a portable Teflon hood to fit on the judge's head and isolate him from odorous environment.

Randebrock, R. E. (1970). Über den Einsatz eines Olfaktometers, *J. Soc. Cosm. Chem.* **21**, 289–297.

A relatively compact 3-stage dilution olfactometer. Presents one stimulus at a time. No valving in the flow system between a tampoon-type saturator (no bubbling) and the sniffing port. Prepares dilutions up to $1{:}10^9$.

Skramlik, E. von (1936). A device for testing the efficiency of the sense of smell. *Münch. med. Wochensch.* **38**, 1564.

The odorant is in a dish in a long vertical tube. The dish is lifted until odour is perceived at the upper end of the tube. (Dravnieks: lifting shortens the diffusion path for the vapours of the odorant; very interesting simple device, subject to problems discussed in text under "Diffusion Method.")

Earlier olfactometers:

Moncrieff (1967) gives an excellent illustrated review of earlier olfactometers.

References

Altshuller, A. P. and Cohen, I. R. (1960). Application of diffusion cells to the production of known concentrations of gaseous hydrocarbons. *Anal. Chem.*, **32**, 802–810.

Amoore, J. E. (1967). Specific anosmia: a clue to olfactory code. *Nature (Lond.)* **214**, 1095–1098.

ASTM (Amer. Soc. Test Materials) (1967). Measurement of Odor in Atmosphere (Dilution Method), Standard Test D-1391-57. Philadelphia, Pa.

Barnebey-Cheney Company (1963). Odor control with activated charcoal. *J. Air Poll. Control Assoc.* **13**, 150–153.

Bird, R. B., Stewart, W. E. and Lightfoot, E. N. (1960). "Transport Phenomena", 522–526, John Wiley and Sons, Inc., New York.

Blanchard, D. C. (1954). A simple method for the production of homogeneous water drops down to one micron radius. *J. Coll. Sci.* **9**, 321–328.

Bozza, G., Calearo, C. and Teatini, G. P. (1960). On the making of a rational olfactometer. *Acta Otolaryng. (Stockholm)* **52**, 189–209.

Cederlof, R., Edfors, M. L., Friberg, L. and Lindvall, T. (1966). On the determination of odor thresholds in air pollution control–an experimental field study on flue gases from sulfate cellulose plants. *J. Air Poll. Control Assoc.* **16**, 92–94.

Cheesman, G. H. and Kirkby, H. M. (1959). An air dilution olfactometer suitable for group threshold. *Quart. J. Exp. Psychol.* **11**, 115–123.

Considine, M. D. and Ross, S. D. (1964). "Handbook of Applied Instrumentation", 5–12 to 5–21. McGraw-Hill Book Co., New York.

Corn, M. and Bell, W. (1963). A technique for construction of predictable low-capacity critical orifices. *Industr. Hyg. J.* **24**, 502–504.

Day, J. A. (1963). Small droplets from rapturing air bubble films. *J. Rech. Atmos.* **1**, 191–197.

Deininger, N. and McKinley, R. W. (1954). The design, construction, and use of an odor test room. ASTM *Spec. Techn. Publ.* **164**, 23–32.

Drake, B., Johansson, B., von Sydow, E. and Döving, K. B. (1969). Quantitative psychophysical and electrophysiological data on some odorous compounds. *Scand. J. Psychol.* **10**, 89–96.

Dravnieks, A. and O'Donnell, A. (1971a). Principles and some techniques of high-resolution headspace analysis. *J. Agr. Food Chem.* **19**, 1049–1056.

Dravnieks, A., Krotoszynski, B. K., Whitfield, J., O'Donnell, A. and Burgwald, T. (1971b). High-speed collection of organic vapors from atmosphere. *Envir. Sci. Techn.*, **5**, 1220–1222.

Elsberg, C. A. and Levy, I. (1936). The sense of smell. A new and simple method of quantitative olfactometry. *Bull. Neur. Inst., N.Y.* **4**, 5–19.

Fuller, E. N., Schettler, P. D. and Giddings, J. C. (1966). A new method for prediction of binary gas-phase diffusion coefficients. *Ind. Eng. Chem.* **58** (5), 19–27.

Gas Engineers Handbook (1965). Section 7. "Gas Measurements". 7/1–7/62. The Industrial Press, New York, N.Y.

Köster, E. P. (1967). Olfactomètre modifié et automatisé. *Parf. Cosm. Sav.* **10**, 509–513.

Köster, E. P. (1971). Adaptation and Cross-Adaptation in Humans, p. 63–69. Thesis, University of Utrecht, Netherlands.

Laffort, P. (1963). Essai de standardization des sevils olfactifs humains pour 192 corps purs. *Arch. Sci. Physiol.* **17**, 75–105.

Martin, A. J. and Debbrecht, F. J. (1970). "Devices for preparing low-level gas mixtures". Bulletin distributed by Anal. Instr. Devel., Inc., 250 So. Franklin St., West Chester, Pa. 19380.

Mateson, J. P. (1955). Olfactometry: its techniques and apparatus. *J. Air Poll. Contr. Assoc.* **5**, 167–170.

Mills, J. L., Walsh, R. T., Luedtke, K. D. and Smith, L. K. 1963. Quantitative odour measurements. *J. Air Poll. Control Assoc.* **13**, 467–475.

Moncrieff, R. W. (1967). "The Chemical Senses", 208–226. CRC Press, Cleveland, Ohio.

Moulton, D. G., Celebi, G. and Fink, R. P. (1970). Olfaction in mammals-two aspects: proliferation of cells in the olfactory epithelium and sensitivity to odours. *In* "Taste and Smell in Vertebrates" (Wolstenholme, G. E. W. and Knight, J., eds), 227–250. J. and A. Churchill, London.

Moulton, D. G. (1972). Monell Chemical Senses Center, U. Pennsylvania, Philadelphia, Pa. Personal Communication.

General Electric (1969). General Electric H-2 and H-5P Industrial Halogen Leak Detectors and Accessories, Sales bulletin GEA 8672, West Lynn, Mass.

Guadagni, D. G. (1968). Requirements for coordination of instrumental and sensory techniques. In ASTM STP 440, Correlation of Subjective-Objective Methods in the Study of Odors and Taste: 36–48, Philadelphia, Pa.

Hama, G. M. (1958). Calibration of Alnor Velometers. *Amer. Ind. Hyg. Assoc. J.* **19**, 477–480.

Johnston, J. W. and Sandoval, A. (1962). The sterochemical theory of olfaction 4. The validity of muskiness as a primary odor. *Proc. Sci. Sect. Toilet Goods Assoc.* **37**, 34–45.

Jones, F. N. (1955). A comparison of the methods of olfactory stimulation: blasting *vs.* sniffing. *Amer. J. Psychol.* **68**, 486.

Kaissling, K. E. (1971). Insect Olfaction. *In* "Handbook of Sensory Physiology", Vol. IV Chemical Senses, Part 1, Olfaction, 351–431, Springer-Verlag, Berlin.

Karoliska Symposium (3rd) on Environmental Health (1970). Methods for measuring and evaluation odorous air pollutants at the source and in the ambient air. *Nord. Hyg. Tidskr.* **51**, No. 2, 78 pages, Stockholm, Sweden.

Kerka, W. F. and Humphreys, C. M. (1956). Temperature and humidity effect on odour perception. *Heat. Piping Aircond. J.* **28** (4), 129–136.

Mullins, L. J. (1955). Olfaction. *Ann. N. Y. Acad. Sci.* **62**, 246.

O'Keefe, A. E. and Ortman, G. C. (1966). Primary standards for trace gas analysis. *Anal. Chem.* **38**, 760–763.

Ough, C. S. and Stone, H. (1961). An olfactometer for rapid and critical odor measurement. *J. Food Sci.* **27**, 452–456.

Perry, J. H. (ed) (1963). Prediction and correlation of physical properties. *In* "Chemical Engineers' Handbook", 3–213 to 3–231. McGraw-Hill, New York.

Sanders, G. R., Umbraco, R. A., Twiss, S. A. and Mueller, P. K. (1970). The measurement of malodor in a community by dynamic olfactometry. Presented Conf. Methods for Measuring and Evaluating Odorous Air Pollutants at the Source and in the Ambient Air, Stockholm.

Scherberger, R. F., Happ, G. P. Miller, F. A. and Fasset, D. W. (1958). A dynamic apparatus for preparing air-vapor mixtures of known concentrations. *Ind. Hyg. J.* **19**, 494–498.

Schneider, R. A., Costiloe, J. P., Vega, A. and Wolf, S. (1963). Olfactory threshold technique with nitrogen dilution of *n*-butane and gas chromatography. *J. Appl. Physiol.* **18**, 414–417.

Schneider, D., Kasang, G. and Kaissling, K.-E. (1968). Determination of odour threshold of *Bombix mori* with tritium-marked bombykol. *Naturwiss.* **55**, 395–396.

Schuette, F. J. (1967). Plastic bags for collection of gas samples. *Atm. Environm.* **1**, 515–519.

Stevens, R. K. and O'Keefe, A. E. (1970). Modern aspects of air pollution monitoring. *Anal. Chem.* **42**, 143A–148A.

Smith, H. O. and Hochstetter, A. D. (1969). Determination of odor thresholds in air using C^{14} labeled compounds to monitor concentrations. *Envir. Sci. Techn.* **3**, 169–170.

Stone, H. (1968). Physiological and psychological aspects of the olfactory response in man. *In* "Theories of Odour and Odour Measurement" (Tanyolac, N. N., ed.), 337–346. Istanbul, Turkey, Robert College (presently Bosphorus University).

Stuiver, M. (1960). An olfactometer with a wide range of possibilities. *Acta Oto-Larynol.* **51**, 135–142.

Turk, A., Edmonds, S. M., Mark, H., Collins, G. F. and Bartlett, F. (1965). Use of gas tracers for micro-meteorological studies. *Am. Chem. Soc., Div. Water Waste Chem.* Preprints **5** (1), 178–179, also *Chem. Abstr.* **65**, 18363.

Turk, A. (1967). Selection and Training of Judges for Sensory Evaluation of the Intensity and Character of Diesel Exhaust Odors. PHS Publ. No. 999-AP-32, Cincinnati, Ohio, U. S. Department of Health, Education, and Welfare, Natl. Center for Air Pollution Control.

Valenta, J. (1970). Rockhurst College, Kansas City, Mo., Personal Communication.

Weast, R. C. (ed), (1970–1971). "Handbook of Chemistry and Physics", p. D-146 to D-172. The Chem. Rubber Co., Cleveland, Ohio.

Wenzel, B. M. (1948). Techniques in olfactometry: a critical review of the last one hundred years. *Psychol. Bull.*, **45**, 231–247.

Wenzel, B. M. (1955). Olfactometric method utilizing natural breathing in odor-free "environment". *Science, N.Y.* **121**, 802–803.

Wnek, W. (1971). Personal communication, calculated per author's inquiry. IIT Research Institute, Chicago, Illinois.

Zwaardemaker, H. (1921). A camera inodorata. *Perf. Ess. Oil Rec.* **12**, 243–244.

2 The Purity of Odorant Substances

AMOS TURK

Department of Chemistry
The City College of
The City University of New York

JONATHAN TURK

Department of Chemistry
University of Colorado
Boulder, Colorado

I. Introduction

The idea of "ultimate purity", usually considered to be the condition of a substance composed entirely of like molecules, has no operational meaning in the laboratory. We cannot examine each molecule in a sample, and even if we could, we might not be able to describe the purity of the sample because we would not necessarily know which differences between molecules were sufficiently permanent to persist after separation. All the chemist can do is to look for evidence of impurity. A "pure substance" is then taken to be one for which no evidence of impurity is found. The criterion of purity is thus always conditional and temporary.

II. Purification of odorants

A substance is impure if it can be separated into components, which is to say, if it can be further purified. Purification involves physical separation, which occurs as a consequence of the differences in properties among the components of the original sample. Therefore purification involves (1) a motive (moving) process because components must move if they are to separate, (2) a differentiating action that distinguishes among the components, and (3) a sensing method. An idea of the degree of variety possible in purification procedures may be gleaned from the listings in Table I. Examples of methods frequently used by chemists are in Table II.

TABLE I Requirements for the purification of materials

A. *Motive processes* (*processes that involve motion*)
- (a) Gravitational or centrifugal settling
- (b) Motion in response to hydraulic or other mechanical force
- (c) Convection (motion in a moving stream of gas or liquid which acts as a carrier)
- (d) Acceleration in a magnetic field
- (e) Acceleration in an electric field
- (f) Diffusion or effusion
- (g) Motion associated with changes in states of aggregation (examples: precipitation of a solid; rise of bubbles of a vapour)

B. *Properties in which components may differ*
- (a) Mass
- (b) Density
- (c) Electron affinity
- (d) Ionization energy
- (e) Thermal conductivity
- (f) Magnetic properties
- (g) Temperatures at which changes of state occur
- (h) Solubility characteristics
- (i) Particle size

C. *Sensing or detection methods*
- (a) Direct sensations: vision, touch, taste, smell
- (b) Determination of physical properties of the type used by chemists since the last century for characterization of substances; melting point, boiling point, refractive index, etc.
- (c) Electrical methods: detection of charge, ability to form ions on being heated or otherwise excited, acceleration in an electric field, electrical conductivity
- (d) Magnetic methods: deflection in magnetic fields
- (e) Methods that use electromagnetic radiation: light microscopy, photography, X-ray diffraction, infrared or ultraviolet photometry, other spectrometric methods
- (f) Thermal methods: measurement of conduction of heat through a substance

TABLE II Examples of purification methods

Method	Motive process	Differential action	Sensing or detection
In filtration	material is moved by gravity or pressure through a sieve, paper, sand bed, or other medium.	The smaller particles move through the medium more easily than the larger ones, and	the separated components can be seen and their amounts can be determined by weighing.
In distillation	a liquid is converted to a vapour by boiling. The vapour is lighter than the liquid and rises from it.	Components which vapourize more easily leave the liquid earlier, and	are sensed by noting the temperatures at which they are produced, or by their indices of refraction (ability to bend light).
In column chromatography	material is carried by a liquid stream through a column of powder or porous granules of large surface area.	The components which adhere to the surface least emerge from the column earliest.	The progress of separation of coloured materials can be followed visually. Colourless components may be detected spectrophotometrically.
In electrophoresis	charged particles in a liquid medium migrate in an electric field to a positive or negative electrode.	The components are separated according to their electric charge and mass and	can be rendered visible by appropriate dyeing or staining methods.
In gas-liquid partition chromatography	a mixture of vapours is carried in a stream of helium through a tube containing a non-volatile liquid.	The components least soluble in the stationary liquid emerge earliest.	They are detected by their ability to produce ions when they are burned in a hydrogen flame or by other methods.

In recent decades, several highly effective purification methods have been developed. What is characteristic about these methods is that a given separation step occurs repeatedly, and the cumulative effect of a large number of such steps yields cleanly separated components. Furthermore, when the process proceeds automatically, purification is rapid and requires little effort. Thus, the development of fractional distillation in the early part of the twentieth century was a major advance over the simple distillation that had been practised for centuries, although the fundamental process of vapourization remained the same.

For the purification of odorous materials, the critical requirement is that the odour of the sample should be that of the major component. Thus, 0.1% of nonane in a sample of nonyl mercaptan is insignificant, but the reverse situation (a mercaptan impurity in nonane) would make the sample worthless for use in any investigation of paraffinic odours. Since the effectiveness of separation methods bears no relationship to odour, and since odorous impurities may be detected in very small concentrations, odour research demands the use of the most powerful methods of purification available.

The most generally useful methods for ultrapurification of odorous (and therefore volatile) substances are preparative gas-liquid chromotography, and zone refining (Zief and Speights, 1972). The various chromatographic methods have been extensively reviewed and are widely applied to the separation and purification of odorants, and so will not be surveyed here. Zone refining, on the other hand, is less frequently used, though it is uniquely applicable to the removal of trace quantities of highly odorous matter from a weaker odorant that the investigator wishes to characterize. It will therefore be helpful to review this technique and to describe an experimental example of its utility.

III. Zone refining

(Pfann, 1958; Herrington, 1963)

When part of a solution freezes, the solute distributes itself between the liquid and solid phases. Let us designate the concentration of solute, B, in the solid or crystalline phase as $[B]_{(c)}$ and in the liquid phase as $[B]_{(l)}$. The ratio of these concentrations, K, is the distribution Coefficient.

$$K = \frac{[B]_{(c)}}{[B]_{(l)}}$$

In most solutions, the solute is more soluble in the liquid phase of the solvent than in the solid phase, and the value of K is therefore less than 1. When an aqueous solution of salt, NaCl, is partially frozen, practically all the salt remains in the water and almost none in the ice; the value of K for this system is therefore close to zero. The technique of zone refining makes use of this distribution phenomenon by melting only one region of a solid material, and passing this molten zone slowly through the entire sample. The solid is usually a cylindrical bar of which a narrow segment is melted by a ring-shaped heater (Fig. 1). The heater and the melted zone which it produces move

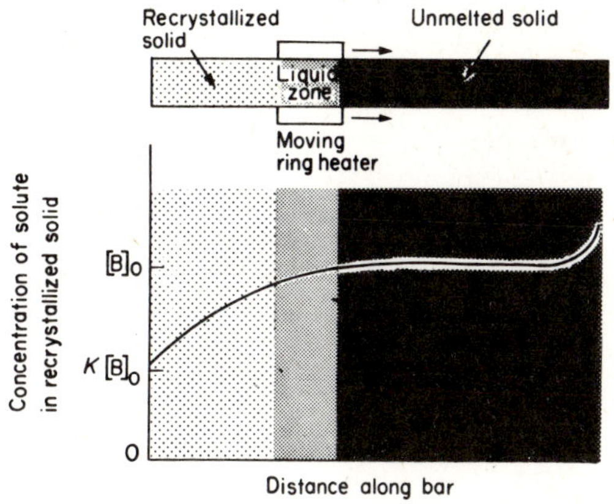

FIG. 1. Zone melting.

along the bar from one end to the other. Let us say that the initial concentration of the solute, $[B]_0$, is 0.2, and that the value of K is 0.5. When the first portion of solid melts, no change in composition occurs, and the concentration in the liquid zone is therefore also $[B]_0$, or 0.2. As the heater moves along the bar, the trailing edge of the molten zone refreezes. The composition of the first recrystallized material thus produced is:

$$[B]_{(c)} = K[B]_{(l)} = 0.5 \times 0.2 = 0.1$$

The recrystallized material thus contains only half of the original concentration of impurity. This condition does not last, however, because the rejected solute accumulates in the molten zone, where the concentration $[B]_{(l)}$ continuously increases. The solute concentration in the trailing recrystallized solid therefore also increases, as shown in the graph of Fig. 1. When $[B]_{(l)}$ reaches a value of $[B]_0/K$, or 0.2/0.5 = 0.4, zone purification ceases. The concentration of impurity (solute) in the trailing solid will now be 0.5 x 0.4 = 0.2, which is the same as the initial concentration, $[B]_0$, in the unmelted solid. This steady condition lasts until the molten zone reaches the end of the bar; when the last liquid crystallizes, all its solute appears in the solid phase, and $[B]_{(c)}$ then rises sharply.

Zone refining becomes highly effective when molten zones are passed through the solid repeatedly. Each successive pass transfers more impurity in the direction of the moving molten zone, leaving more of the recrystallized material in a higher condition of purity. Ultimate levels of purification obtainable by this method are very high. Germanium and silicon, for example, which are used in semiconductors, are routinely purified by zone refining to reduce the concentration of their impurities to less than one part in ten billion.

IV. Analysis of impurities

A bottle of brown liquid that smells of caramel and is labelled "ethanol" is obviously impure; it is not necessary to show that the colour and odour change upon distillation. Of the various kinds of evidence of impurity that do not necessarily involve physical separation of components, thermal analysis is probably the most classical, and mass spectrometry the most versatile.

V. Thermal analysis

The degree of impurity in a sample may be determined by an analysis of its time *v* temperature cooling curve (Rossini and Taylor 1944). This classical method has been extended by the use of modern automated thermal analysis equipment (Levy, 1970), and by applications of zone melting (Van Wijk *et al.*, 1967). However, applications to analysis of odorous impurities have been meagre.

VI. Mass spectral analysis (Perry *et al.*, 1971)

In brief, mass spectral analysis involves the ionization of a sample and the determination of the masses of all resulting ions. Micrograms of sample are vapourized into an ionization chamber. The pressure of the vapour is kept very low (below 10^{-7} torr) in order to prevent reactions between particles. A beam of energetic electrons is generated, and on impact with one of these electrons, the vapour molecule is ionized to give the "*parent ion*".

$$\mathrm{M{:}} + e^- \longrightarrow \underset{\text{parent ion}}{\mathrm{M}\cdot^+} + 2e^-$$

Many parent ions are formed in highly excited vibrational states and therefore undergo fragmentation. The *parent ions* and all *fragment ions* are separated and *detected* by their mass/charge (m/e) value.

For most ions the charge is 1, hence *m/e* is simply the mass of the ion. Neutral molecules or fragments are not detected. The mass spectrum is recorded as a series of sharp peaks, the position of each peak representing the mass of an ion. The height of each peak is a measure of the relative abundance of the corresponding ion. The mass spectrum for methanol, CH_3OH, is shown in Fig. 2.

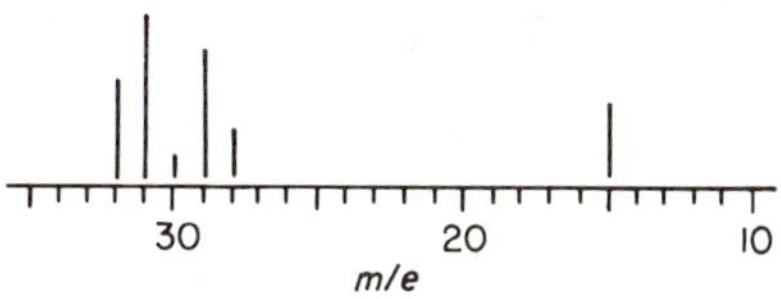

FIG. 2. Mass spectrum of methanol, CH_3OH.

The chemist assigns a structure to each ion, as we now illustrate for the four most prominent peaks from methanol, those of masses 32, 31, 29, and 15:

$$H_3C\ddot{O}H + e^- \longrightarrow [H_3C\dot{O}H]^+ \text{ (mass 32)} + 2e^-$$
parent ion

$$[H_2C\text{–}\dot{O}H]^+ \longrightarrow [H_2C\text{=}OH]^+ \text{ (mass 31)} + H\cdot$$

$$[HC\text{=}O\ H]^+ \longrightarrow [HC\equiv O]^+ \text{ (mass 29)} + H_2$$

$$[H_3C\text{:}OH]^+ \longrightarrow H_3C^+ \text{ (mass 15)} + \text{:}\dot{O}H$$

Notice that two of the ions appear to come directly from fragmentation of the parent ion, while one of them can only come from further dissociation of a fragment ion.

The following types of situations will illustrate the use of mass spectrometry as a means of examining a sample for the presence of impurities.

Mixtures of known dissimilar compounds

The mass spectra of the separate components of a mixture are additive. Therefore a composite spectrum can be analysed into its *n* separate component spectra by a computer program that solves *n* simultaneous equations. This methods is ideally suited to the examination of mixtures of compounds of similar structure but different molecular weights, such as the series of normal primary alcohols.

Mixtures of similar compounds

In general, the quantitative estimations of peak magnitude may vary up to about ± 5%. It is therefore difficult to identify an impurity of lower molecular weight because its parent or other significant ion may be obscured by fragment ions from the major component. Thus, a 1% impurity of benzene (mol. wt. 78) in toluene (mol. wt. 99) is "lost" because the fragment ions from toluene include peaks at m/e 77 and 78. The opposite situation, in which the impurity has a higher molecular weight, is obviously favourable for mass spectral detection. An example would be butyric acid (parent peak at m/e 88) in butyraldehyde (m/e 72).

Mixtures of isomers

The cracking patterns of many isomers, for example, those of the *cis* and *trans* bicyclodecanes, are sufficiently similar to each other that the identification of a trace quantity of one in a larger amount of the other may be very difficult.

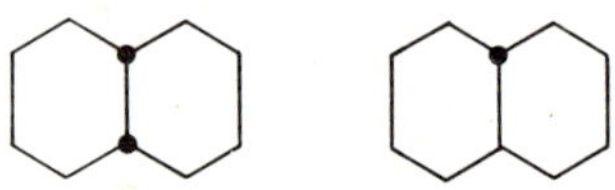

cis *trans*

Isotope patterns

Chlorine, atomic weight 35.453, consists of ^{35}Cl (~ 75%) and ^{37}Cl (~ 25%). These proportions persist in chlorine compounds, and therefore a CH_3Cl peak will consist of mass numbers 50 and 52 in the same ratio (3:1). The recurrence of such ratios in a cracking pattern of a supposedly non-chlorinated compound therefore signals the presence of a chlorinated impurity. Another such example is bromine, atomic weight 79.909, which consists of approximately equal proportions of ^{79}Br and ^{81}Br. It would be helpful if such isomer ratios could point to sulphidic or nitrogenous impurities in an odorant, but here the abundance ratios are less favourable. Thus, sulphur, atomic weight 32.064, is 95% ^{32}S, 4.2% ^{34}S, and smaller proportions of ^{33}S and ^{36}S. Since errors in estimations of magnitude can approach ± 5%, it is easier to detect sulphidic impurities by the nose than by the mass spectrometer. A parallel situation obtains for nitrogen compounds.

VII Experimental: the purification of acetophenone

Two samples of acetophenone were selected for instrumental and sensory evaluation for impurities, and for attempts at purification by zone refining. Sample (I) was acetophenone obtained from a chemical laboratory supply company. Sample (II) was a mixture of 99.5% (I) plus 0.5% eugenol. Both samples were examined, then purified by zone refining, and re-evaluated. The following results were obtained.

Zone refining was carried out in a cold room (acetophenone melts at 19.6°C) by the method of Bollen *et al.* (1967), using the Enraf Nonius Model X-521 (Delft, Holland) instrument. The special features of this device are (a) a high-speed rotation of the tube with periodic reversals to exert shearing forces at the boundary layer and promote turbulence for improved mixing, and (b) an upward movement of the molten zones to guarantee a good contact of liquid and solid at the freezing interface.

GC analysis. All samples were analysed on an Aerograph Model 200 instrument with flame ionization detector, using a 45 m stainless steel capillary column; 0.03 inch i.d., packed with silicone o.v. 17 at 125°C., with nitrogen carrier gas flow at 7.5 ml/min. The sample size was 0.1 μl. Figures 3a, 3b, 4a, and 4b refer respectively, to sample (I) before and after zone refining and sample (II) before and after zone refining. Figures 3a and 3b show that the commercial sample of acetophenone contained a variety of impurities, and that the zone refining eliminated almost all that could be detected. Figure

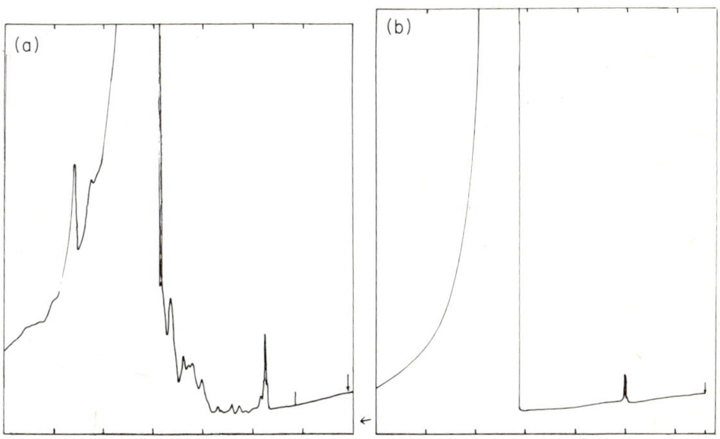

FIG. 3. Gas chromatograms of acetophenone (I), attenuation x 8, (a) before purification; (b) after zone refining.

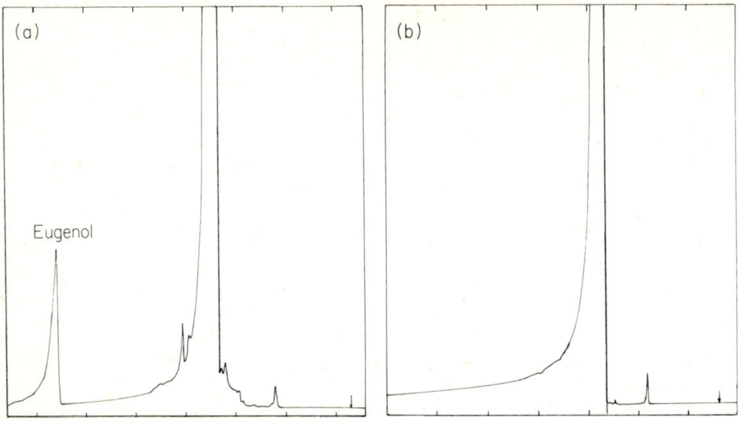

FIG. 4. Gas chromatograms of acetophenone plus 0.5% eugenol, (II), attenuation x 16,
(a) before purification; (b) after zone refining.

4a clearly shows the peak corresponding to the added eugenol, and Fig. 4b again shows that zone refining was practically complete.

Sensory evaluation. The original acetophenone (I) was pale yellow and had a gassy, styrene-like odour note superimposed on its fundamental ketonic odour. When eugenol was added to it (sample II), a slight characteristic eugenol odour was added to the other qualities. After zone refining, both samples were colourless, and the gassy and eugenol odours were absent.

Mass spectral analysis. The fact that acetophenone is one of two major byproducts in the manufacture of phenol from cumene, the other byproduct being α-methylstyrene, led us to look for the latter impurity.

The mass spectrum of sample (II) was complex and indicated the presence of at least four components. The major component, acetophenone, was indicated by peaks at m/e 120, 105, 77, 51, and 39:

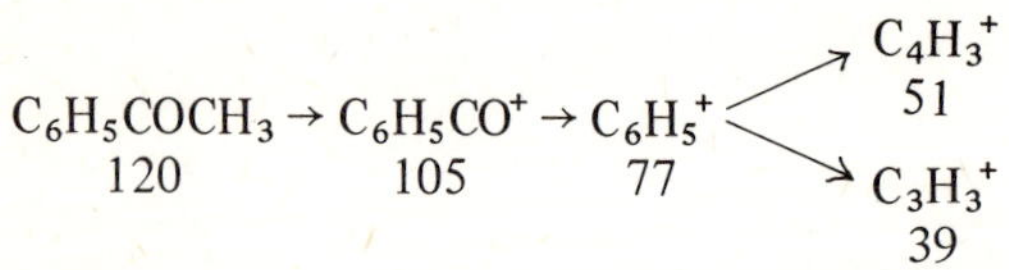

The intensity order (Cornu and Massot, 1966) is:

105	77	51	120	43	50	39
1000	826	305	247	177	128	52

The spectrum of (II) showed m/e peaks at 121, 120, and 105 in approximately equal ratios. These findings are consistent with the presence of the eugenol, which has significant peaks at m/e 120 and 121.

The spectrum of (II) also showed significant peaks at m/e 118, 117, 104, 103, and 91. Four of these peaks are also exhibited by α-methylstyrene (Cornu and Massot, 1966):

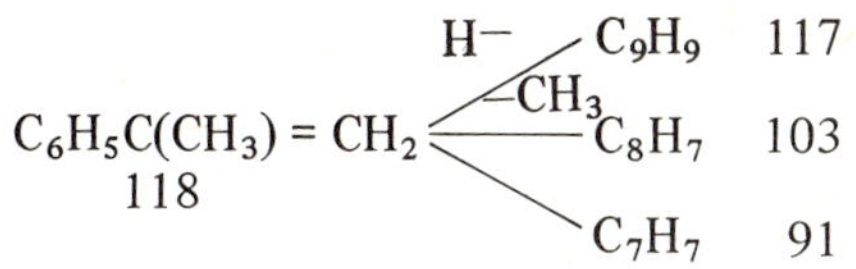

Thus, the styrene-like gassy odour, the circumstances of the chemical manufacturing process, and the mass spectral data all reinforce each other to point to α-methylstyrene as one of the impurities that affects the odour of acetophenone.

Acknowledgement

The technical staff at the Institute for Physical Chemistry, TNO, Utrecht (The Netherlands) provided facilities, equipment, and guidance in the purification of acetophenone.

References

Bollen, N. J. G., Van Essen, M. J. and Smit, W. M. (1967). A fast method of zone melting as an aid in analytical chemistry. *Anal. Chem. Acta* **38**, 279–284.

Cornu, A. and Massot, R. (1966). "Compilation of Mass Spectral Data". Heyden and Sons, London.

Herrington, E. F. C. (1963). "Zone Melting of Organic Compounds". Blackwell, London.

Levy, P. F. J. (1970). Thermal analysis, an overview. *Amer. Laboratory* pp. 46–58.

Perry, R., Long, R. and Majer, J. R. (1971). The use of mass spectrometry in the analysis of air pollutants. *In* "Proceedings of the Second International Clean Air Congress" (Englund, H. M. and Beery, W. T. eds.), 557–562. Academic Press, New York, and London.

Pfann, W. C. (1958). "Zone Melting", John Wiley, New York.

Rossini, F. D. and Taylor, W. J. (1944). *J. Res. Natn Bur. Stds.* **32**, 197.

Van Wijk, H. F., Van der Most, P. F. J. and Smit, W. M. (1967). Zone melting as an aid to impurity determinations by thermal analysis. *Anal. Chem. Acta.* **38**, 285–290.

Zief, M. and Speights, R. M. (1972). "Ultrapurity: Methods and Techniques". Marcel Dekker, New York.

3

Biological and Chemical Methodology in the Study of Insect Communication

J. CHRISTOPHER YOUNG

Agriculture Canada
Chemistry and Biology Research Institute
Ottawa, Ontario K1A OC6, Canada

ROBERT M. SILVERSTEIN

Department of Chemistry
SUNY College of Environmental Sciences and Forestry
Syracuse, New York 13210, USA

I. Introduction

A. Scope of the review

It is well established that many organisms use chemical signals for communicating between individuals. The mode of transmission varies: the message may come through the air, via solution, or simply by contact with another organism. This review will be limited to airborne chemical messages between insects and some other arthropods such as centipedes and millipedes.

In this chapter, we discuss the experimental techniques employed in isolation and identification of the chemicals used for intra- and interspecific communication. Critical attention is paid to the difficulties likely to be encountered since an awareness of these problems is often necessary to avoid arriving at erroneous conclusions.

B. Definition of terms

Many terms have been coined to designate the various types of chemical interaction between organisms, and those in common use are defined below.

The chemical substances delivering the message have been recently termed *semiochemicals* (from the Greek *semeion*, a mark or signal) (Law and Regnier, 1971).

Semiochemicals used for interspecific communication are called *allelochemicals* (Whittaker and Feeny, 1971), These in turn are divided into *allomones* (Brown, W. L. *et al.*, 1970), which give adaptive advantage to the emitter (for example the spray of a skunk, venom of wasps and snakes, and repellent secretions of many arthropods, and *kairomones* (Brown, W. L. *et al.*, 1970), which give adaptive advantage to the receiver (for example substances that enable predators to locate their prey).

The term *pheromone* (Karlson and Butenandt, 1959) (from the Greek *pherein*, to carry, and *horman*, to excite or stimulate) is applied to semiochemicals used for intraspecific communication between individuals. Pheromones can be classified according to the response they elicit. Chemical stimuli that trigger an immediate and reversible change in the behaviour of the recipient are called *releasers*, whereas those inducing delayed, lasting responses are referred to as *primers* (Wilson, 1965). A response may be due to a single compound or a mixture of compounds.

It should be noted that the major categories of semiochemicals are not mutually exclusive. For example the sex pheromones of the bark beetle *Ips paraconfusus* also serve as a kairomone to attract other beetles that prey on *Ips* (Wood *et al.*, 1968). Depending upon concentration, the mandibular secretion discharged by the ant *Lasius alienus* may function as an alarm pheromone or a defensive allomone (Regnier and Wilson, 1969).

In some semiochemical mixtures, all components must be present to trigger a maximum response (e.g. Wood *et al.*, 1967b). Such a phenomenon is termed *synergism* when the total effect from the combined action of individual compounds is greater than the sum of the several effects taken independently. Synergism need not be limited to those substances emitted by the organism, but may include substances contributed by the host (e.g. Bedard *et al.*, 1969).

C. Kinds of behaviour

Insects communicate for a variety of fundamental reasons associated with their need for food and protection, and to reproduce. It has long been known that auditory and visual systems are used as perception mechanisms to communicate these needs, but only recently have we become aware of the widespread use of chemical signals to transfer information. The complex behavioural patterns of insect societies very often have a chemical basis. We propose to summarize the various kinds of insect behaviour that depend upon olfactory communication.

Insect defense mechanisms against predators involve concealment, flight, or resistance. To protect themselves from predators against which physical defense is futile, arthropods have evolved diverse chemical defense systems involving various methods of dispersing chemical agents. Some glands allow the secretion to ooze onto the arthropod's surface, some release the allomone as a foam, and others spectacularly discharge their contents as a directed spray. The arthropod generally discharges only in response to tactile contact by a predator. Although the outcome is usually not fatal, the predator is left in no doubt as to the message communicated to it. Individuals may release milligram amounts of material; nearly 20% of the body weight of the ant *Formica rufra* L. is formic acid. Insects show great chemical diversity in their defensive substances, and may secrete very complex mixtures. The types of compounds isolated thus far include terpenes (**1–6**),

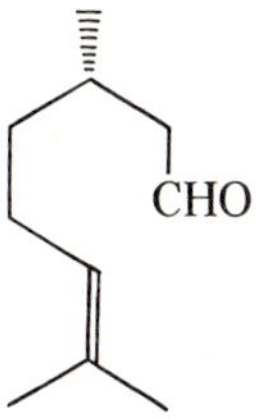

1
L-Citronellal

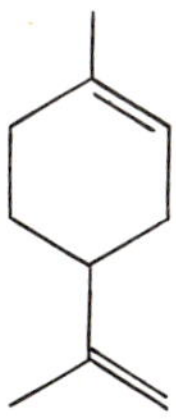

2
Limonene

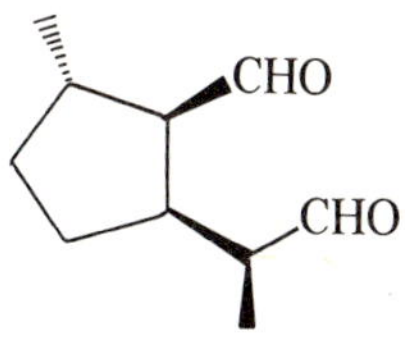

3
Iridodial

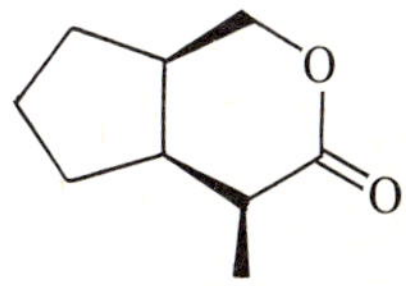

4
Iridomyrmecin

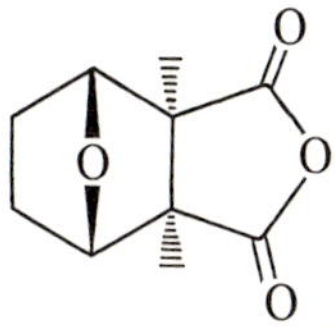

5
Cantharidin

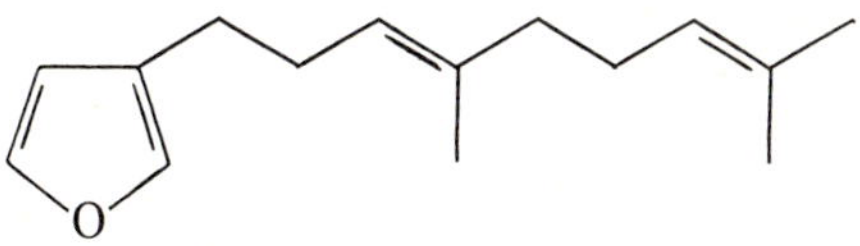

6
Dendrolasin

benzoquinones (**7**–**9**),

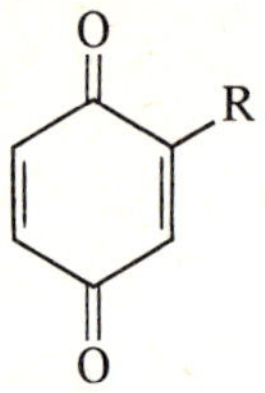

7

R = H, CH_3, C_2H_5, OCH_3

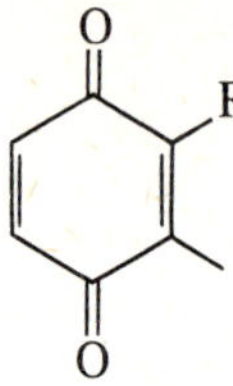

8

R = CH_3, OCH_3

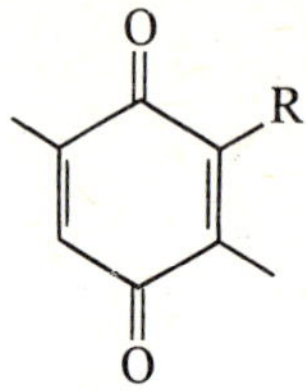

9

R = H, CH_3

hydrocarbons, aliphatic acids, aliphatic aldehydes of the type **10**,

R = nC_5–C_{11}

10

aromatic compounds, hydrogen cyanide, several other miscellaneous complex compounds (**11**–**14**), and even the halogen (**15**) (Eisner, T., 1970).

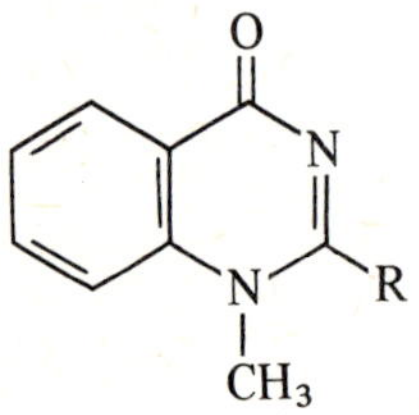

11

R = CH_3, C_2H_5

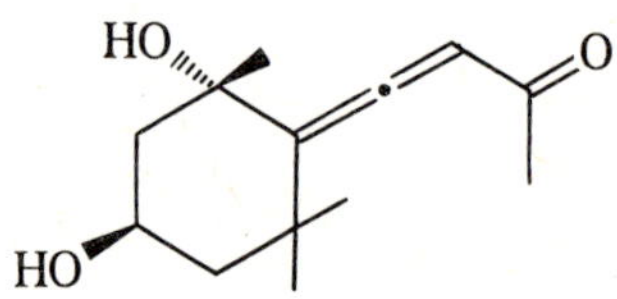

12

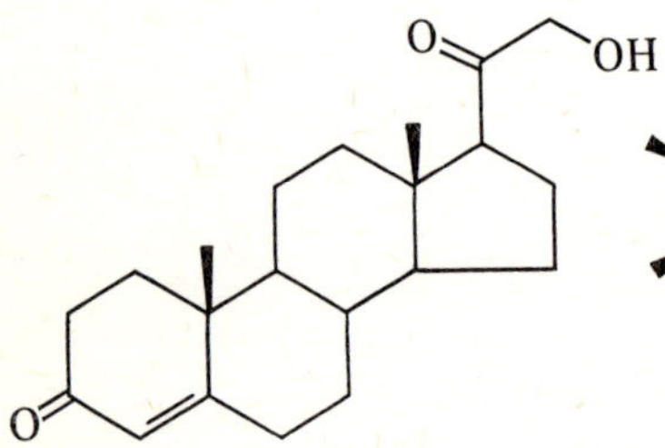

13

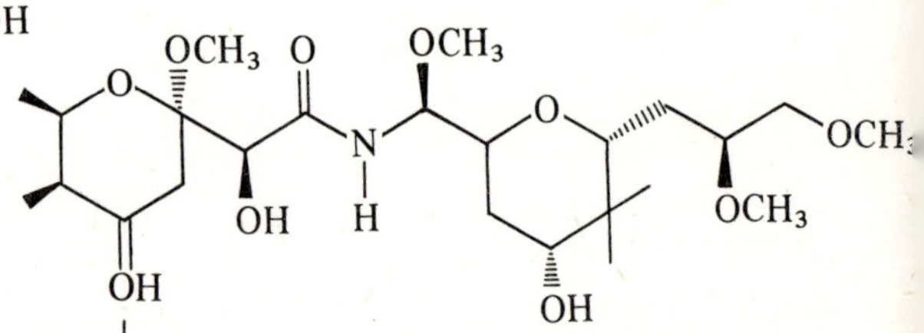

14
Pederin

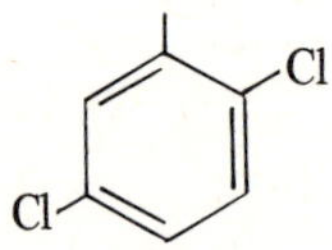

15

In addition to secreting chemicals from a distance, insects can "defend" themselves by being toxic or distasteful once eaten to their predators, which quickly learn by experience. The Monarch butterfly, *Danaus plexippus*, is protected from insectivorous birds by this method (Reichstein *et al.*, 1968). Chemical defence mechanisms of arthropods have been extensively reviewed (Maloeuf, 1938; Pavan, 1958; Eisner, T. and Blumberg, 1959; Pavan, 1959; Cavill, 1960; Cavill and Hinterberger, 1960b; Gupta, 1961; Schildknecht, 1961; Chopard, 1962; Roth and Eisner, T. 1962; Beard, 1963; Brower, L. P. and Brower, J. V., 1964; Cavill and Robertson, 1965; Eisner, T. and Meinwald, J., 1966; Jacobson, 1966b; Weatherston, 1967; Herbert, 1968; Moore, 1969; Taylor, 1969; Eisner, T., 1970; Schildknecht, 1970; Weatherston and Percy, 1970; Blum, 1971; Meinwald, J. and Muetterties, 1971; Pain, 1971; Pavan and Dazzini, 1971; Schildknecht, 1971). Packard (1896) lists many papers from the early literature (from 1638) on defensive glands of insects. For a discussion of other interspecific interactions see Whittaker and Feeny, 1971.

Scent marking by mammals has long been recognized as an important form of communication, although very little is known about the chemical nature of these signals. For reviews of mammalian behaviour that appears to have a chemical basis see: Lederer (1950); Bruce (1970); Meinwald, J. and Muetterties (1971); and Ralls (1971).

One major function of mammalian pheromones is to mark a territory, which the marking individual will defend. Another function is to establish the relative status of the individual in a group. Scent marking is also used for laying trails, for alarm signals, for individual, group or species recognition, for sexual attraction or stimulation, and as a primer to influence reproductive processes. The roles of substances produced by skin glands in mammalian communication (Mykytowycz, 1970) and of pheromones in mammalian reproduction (Whitten and Bronson, 1970) have been discussed.

The likelihood of human pheromones has been recently reasserted (Comfort, 1971), and of particular interest in this regard is the isolation of male sex attractants from female rhesus monkeys (Curtis *et al.*, 1971; Michael *et al.*, 1971).

Of all the forms of chemical communication, the use of pheromones by insects has been the most extensively studied. (For general reviews see Butenandt, 1959; Wilson, 1963b; 1965; Jacobson, 1966b; Novak, 1966; Butler, 1967; Karlson, 1967; Shorey and Gaston, 1967b; Regnier and Law, 1968; Wilson, 1968; Magnetti and Tibaldi, 1969; Moore, 1969; Blum, 1970; Butler, 1970; Gary, 1970; Moser, 1970; Stuart, 1970; Sudha and Nayar, 1970; Wilson, 1970; Blum, 1971a; Law and

Regnier, 1971; Meinwald, J. and Muetterties, 1971; Regnier, 1971; Vadasz, 1971; Blum and Brand, 1972; Jacobson, 1972; Levinson, 1972; Regnier, 1972; MacConnell and Silverstein, 1973; Pain, 1973; Shorey, 1973; Birch, M. C., 1974).

The most thoroughly documented cases of long-range communication are those of the sex pheromones used in signalling sexual receptivity. Females of the emperor moth have been reported to attract males from up to 11 km (Butenandt, 1963). It is likely however that most insects are attracted from much shorter distances. Usually the female releases the pheromone, which the males perceive through specialized sensory receptors on their antennae, although in some species the males are the emitters. (For reviews of antennae see Schneider, 1964, 1969). Detection is followed by a simple or complex rigidly ordered series of events leading to copulation. After the male and female have been brought together by the attractant pheromone, other pheromones acting as aphrodisiacs may also be released. As will be shown, it is important to recognize the total behavioural pattern when designing bioassays for active material. Sex pheromones are among the most biologically active substances known. Threshold concentrations required to induce 50% of male silkworm moths, *Bombyx mori*, and cabbage looper moths, *Trichoplusia ni*, to respond within 30 seconds have been calculated to be 2×10^2 (Boeckh *et al.*, 1965) and 5×10^5 molecules per ml of air (Shorey *et al.*, 1967) respectively.

In addition to being highly sensitive to sex pheromones, many insects may be highly selective. For example the cabbage looper responds only to **16** and not at all to the *trans* isomer **17** (Jacobson *et al.*, 1968b), the sex pheromone of the false codling moth, *Argyroploce leucotreta* (Read *et al.*, 1968).

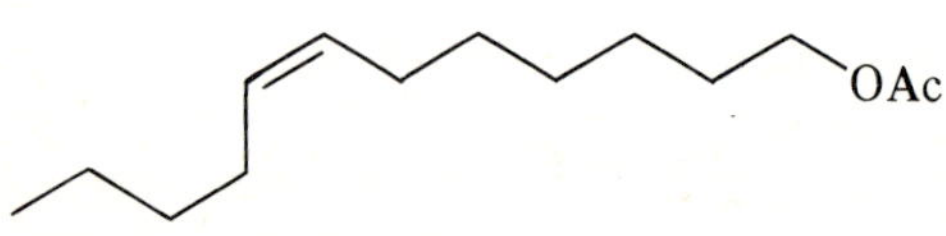

16

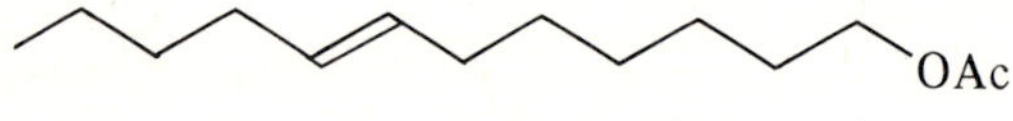

17

On the other hand, examples of cross attraction of several species in the same genus have been reported (Rodin *et al.*, 1969; Sanders, C. J., 1971b; Nielson and Balderston, 1973); and even of species in different genera (Berger and Canerday, 1968).

Many of the moth sex pheromones are unsaturated, even-numbered, straight-chain acetates or alcohols. Whereas some pheromones have been identified as a single compound, others contain several components, such as **18**–**21** secreted by the boll weevil, *Anthonomus grandis* (Tumlinson *et al.*, 1969).

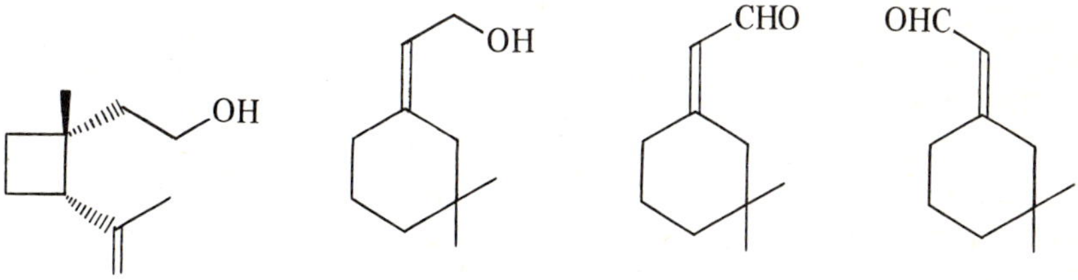

18 **19** **20** **21**

(For reviews of sex pheromones see Richards, 1927; Hardy, 1947; Karlson and Butenandt, 1959; Beroza, 1960; Green *et al.*, 1960; Hatanaka, 1963; Jacobson, 1963; Jacobson and Beroza, 1963; Wilson, 1963b; Jacobson, 1965; 1966a, b; Butler, 1967; Carr and Levin, 1967; Karlson, 1967; Shorey and Gaston, 1967b; Herbert, 1968; Lebedeva, 1968; Muto, 1968; Regnier and Law, 1968; Shorey *et al.*, 1968a; Silverstein, 1968; Williams, C. M. and Bowers, 1968; Pyatnova *et al.*, 1969; Blum, 1970; Burkholder, 1970; Callow, 1970a, b; Eiter, 1970; Jacobson, 1970; Jacobson *et al.*, 1970a; Moser, 1970; Nakajima, 1970; Pain, 1970; Renwick, 1970; Shorey, 1970; Silverstein, 1970a, b; Skirkevicius and Vaitkeviciene, 1970; Sudha and Nayar, 1970; Tumlinson *et al.*, 1970; Wood, 1970; Beroza, 1971b; Borden and Stokkink, 1971; Law and Regnier, 1971; Silverstein, 1971; Beroza, 1972a, b; Fukami, 1972; Hardee, 1972; Jacobson, 1972; Myers, 1972; Regnier, 1972; Trier and Metzger, 1972; Warren, 1972; Karlson and Schneider, 1973; Pain, 1973; Shorey, 1973).

Bark beetles (Scolytidae) produce attractants that have more than one function. Males or females boring in trees produce a frass that is highly attractive, usually to both sexes, and triggers mass infestation of the host tree. Thus the pheromones produced serve as sex and recruitment (or aggregation) pheromones. (For reviews of recruitment pheromones see: Franke-Grossman, 1963; Jacobson, 1966b; Shorey and Gaston, 1967b; Regnier and Law, 1968; Ishii, 1970; Silverstein,

1970a, b; Wood, 1970; Borden and Stokkink, 1971; Law and Regnier, 1971; Wood, 1971).

The social development of some insects (e.g. bees, wasps, ants and termites) into colonies with large populations where the individuals have specific roles appears to be chemically regulated. Many of these insects lay aerial or terrestrial trails leading to food sources or nesting sites, and possibly for other purposes. Most of the trails laid are of short duration (minutes to an hour) but some may persist for a week or more. (For reviews of trail pheromones see: Wilson, 1963a; Blum *et al.*, 1964; Wilson, 1965; Jacobson, 1966b; Butler, 1967; Blum, 1970; Gabba and Pavan, 1970; Moser, 1970; Pain, 1970; Sudha and Nayar, 1970; Pavan and Dazzini, 1971; Blum and Brand, 1972; Pain, 1973).

When alarmed, insects may release secretions to alert other members of their own species to the danger. Some species respond by retreating into the interior of the nest whereas others will vigorously attack the intruder. Some of these secretions also act to repel the enemy (i.e. function as defensive allomones). Whereas sex pheromones are highly active and specific, alarm pheromones threshold behaviour concentrations are quite high, ranging from 10^7 to 10^{13} molecules per ml of air. Some species may use the same chemical; for example 4-methyl-3-heptanone is secreted by at least a dozen species. There is some species specificity, however, for often the secretion contains a mixture of compounds. Alarm pheromones are frequently relatively simple ketones in the C_6 to C_{12} range. (For reviews of alarm pheromones, see: Cavill, 1960; Cavill and Hinterberger, 1960b; Wilson, 1963a; Boch and Shearer, 1965; Wilson, 1965; Maschwitz, 1966; Butler, 1967; Carr and Levin, 1967; Regnier and Law, 1968; Blum, 1969, 1970; Cavill, 1970; Gabba and Pavan, 1970; Moser, 1970; Law and Regnier, 1971; Pain, 1971; Pavan and Dazzini, 1971; Blum and Brand, 1972; Regnier, 1972; Pain, 1973).

Thus far, pheromones eliciting a more or less immediate response have been considered. Much more difficult to evaluate are those continually produced pheromones which induce a long-term response. Termites produce some pheromones that keep the colony together and others that determine caste and enable recognition within the colony. (For reviews of termite pheromones, see: Karlson and Butenandt, 1959; Wilson, 1965; Moore, 1969; Stuart, 1970). Honeybee queens secrete a pheromone complex whose function depends upon the context in which it is released. The so-called queen substance 22 functions as a sex attractant and as an aphrodisiac of a virgin queen outside the hive, but inside the hive inhibits the development of ovaries in worker bees.

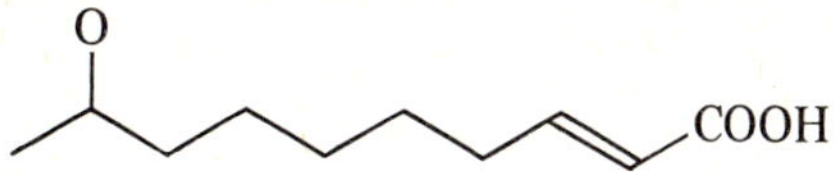

22

(For reviews of honeybee pheromones, see: Karlson and Butenandt, 1959; Butler, 1967; Carr and Levin, 1967; Regnier and Law, 1968; Butler, 1969; Blum, 1970; Callow, 1970a, b; Gary, 1970; Pain, 1971, 1973; Butler, 1973). Some social insects are able to distinguish between intra-specific colonies, possibly by perception of differing proportions of chemicals emitted or a colony odour adsorbed on the body.

One important question to be answered is: How do insects orient to an odour source? The simplest explanation is that an insect senses a concentration gradient of odour molecules and follows the gradient from an area of low molecular density through increasing density to the highest density at the source. This phenomenon (*chemotaxis*) is probably quite rare except within a few centimetres of the pheromone source, because over greater distances air turbulence will disrupt an odour gradient. The most widely accepted explanation is that after sensing a pheromone, an insect is stimulated to move upwind (*positive anemotaxis*) toward the air source rather than the chemical source (Schwinck, 1954, 1955, 1958). When leaving odour-bearing air, the insect turns frequently, across and downwind until encountering the pheromone which once again stimulates positive anemotaxis. Recent studies (Farkas and Shorey, 1972, 1973; Grubb, 1973) have cast doubt on the use of anemotaxis by at least one species for orientation to an odour source. The mechanisms and effective distances of orientation to pheromone sources have been reviewed (Dethier, 1957; Karlson and Butenandt, 1959; Jacobson, 1965; Shorey and Gaston, 1967b; Shorey *et al.*, 1968a; Shorey, 1970, 1973). See also Wright (1958) and for a discussion of odour meteorology, Chapman, (1967). Some workers have used smoke plumes to illustrate diffusion of odour molecules. In reality, the rates of diffusion of solid particles and gaseous molecules are likely much different.

Another important question is: How can only a few chemicals communicate the information needed to control the behaviour of an insect species? The answer is that an individual compound is capable of communicating more information than that implicit in a simple on-or-off signal. For Wilson and Bossert's treatment of this problem see Bossert and Wilson (1963), Wilson and Bossert (1963), and Wilson, (1965,

1970). The amount of information transmitted by a single compound may be increased by:

(1) Adjustment of the time between release of the pheromone and final disappearance of the threshold concentration required for a behavioural response. This may be done by altering the emission rate and/or threshold concentration. The life time of a pheromone can also be shortened by deactivation.

(2) Different responses to different concentrations. At different concentrations of a given compound, one species may respond in different ways (for example, Bartell and Shorey, 1969a; Regnier and Wilson, 1969) or various species may respond only to certain concentrations of that compound (Kaae *et al.*, 1973).

(3) Change of meaning through change of context, for example the honeybee, *Apis mellifera*, "Queen substance."

The amount of information transmitted can also be increased by secretion of a mixture of compounds. A 9:1 mixture of *cis*-9- and *cis*-11-tetradecenyl acetate will attract males of only the moth *Adoxophyes orana*, whereas a 1:4 mixture will attract males of only the moth *Clepsis spectrana* (Minks *et al.*, 1973). At least 32 volatile compounds have been isolated from the heads of honeybee, *Apis mellifera,* queens (Callow *et al.*, 1964), and 47 compounds from some *Formica* species ants (Bergström and Löfquist, 1973). An insect can use multiple glands, each secreting its own set of chemicals; up to five glands may be employed by individual ants and bees. Glandular secretions may acquire additional or different meanings when presented together. Wright (1964) discusses the importance of pheromone mixtures.

D. General methodology of isolation and identification

A rigorous approach to the study of chemicals involved in insect communication involves the following steps (Silverstein *et al.*, 1967b):

(1) Development of a sensitive bioassay that requires only a small amount of active material. Usually this means a laboratory bioassay, which should be based on sound biological principles derived from the behaviour of natural populations. Ideally, field bioassays should be performed at every major step in the isolation but the small quantity of material available and difficulty in controlling conditions seldom permits the luxury of such experiments. Discrepancies between laboratory and field results would necessitate field evaluations at an earlier step in the isolation.

(2) Production of large amounts of material either by mass rearing or by large-scale field collecting.

(3) A series of progressively refined isolation steps, each monitored by bioassay, until individual components are obtained in a very high state of purity. The presence of even very small amounts of highly active or inhibitory impurities is an ever-present hazard.

(4) Identification of the individual components. Usually these are obtained in milligram or microgram quantities so that identification rests heavily on spectrometric techniques. The basic instruments required are mass, infrared, nuclear magnetic resonance, and ultraviolet spectrometers. These instruments give mutually complementary information on chemical structures, and with the exception of mass spectrometers, are non-destructive of the sample.

(5) Confirmation of postulated structures by comparison with unambiguously synthesized compounds.

(6) Confirmation of biological activity of synthesized compounds in both the laboratory and field.

During the past few years, numerous studies on semiochemicals have been reported. Most of these reports describe little more than the response of the insect to a partially purified active fraction. Characteristically, failure to achieve the ultimate objective, chemical identification of the semiochemical(s), can be attributed to deficiencies in performing steps 3 and 4.

Most of the successful identifications of semiochemicals have involved teamwork between entomologists and chemists. Initially, the role of the entomologist is to recognize the behavioural patterns of the insect that are controlled chemically. He then directs the production and collection of active material and develops the important bioassay. The chemist then takes the crude material and proceeds to purify, identify and synthesize the semiochemical(s), all the while monitoring each stage of the process with bioassays usually conducted by the entomologist. In these days of specialization, one rarely finds a scientist fully capable of conducting all stages of this work by himself.

II. Bioassay

A bioassay is an assay for the presence (qualitative) and/or quantity (quantitative) of a biologically active substance, with the detector being a living organism or some portion of that organism (Shorey, 1970).

A. *Development of a bioassay*

When semiochemical molecules stimulate the chemosensory organs, a response is elicited from the insect receiving the stimulus. The observed response thus triggered may range from an apparently simple to an obviously complex sequence. In bioassaying fractions from a chemical separation or samples of synthetic compound, the tacit assumption is that the physical presence of the emitting insect is not required.

Ideally, a bioassay should be the response under natural conditions, the role change being the substitution of a sample for the emitting insect. However this is usually impractical, and compromises become necessary; therein lies the art.

Although it is helpful to recognize each step in a series of responses to a single stimulus, it may not be necessary to base the bioassay on the total sequence, because different pheromone concentration thresholds are probably required for the initiation of the individual steps. Bartell and Shorey (1969a) have shown quantitatively that the initial step in the mating sequence of the male moth *Epiphyas postvittana* has the lowest pheromone concentration threshold and that each successive step has a higher threshold than the preceding one. Thus the assay may be based on only one or a few steps, and to detect the lowest possible pheromone concentration, an early step should be selected.

Probably not many semiochemical-producing insects continually produce and/or release their semiochemicals from birth until death. Similarly a potential receptive insect is not capable of being stimulated at all times. The factors controlling pheromone-stimulated sexual behaviour in insects have been reviewed by Shorey *et al.*, (1968a).

Physiological states of insects that need to be considered when developing a bioassay include:

(1) Age of the organism. The release of sex pheromones is likely associated with maturation of the reproductive system.

(2) Effect of circadian rhythms, time of day, or time of year. Most insects are sexually active only during particular times of the day or night.

(3) Effect of mating. Some insects mate only once and cease sex pheromone production, whereas others continue production of sex pheromones and will mate several times more.

(4) Effect of hormones. These influence (1), (2) and (3).

(5) Effect of adaptation or responsiveness to semiochemicals. The response to a persistent chemical stimulus may fade with time (Ignoffo, *et al.*, 1963; Shorey *et al.*, 1964; Borden, 1967; Vick *et al.*, 1973) and if the pheromonal concentration is much above the threshold concentration, the response may be completely inhibited (Shorey *et al.*, 1967). If a pheromone stimulus is removed and presented again at a later time, a higher concentration may be needed to cause the same degree of response (Shorey and Gaston, 1964).

Semiochemical-stimulated behaviour may be affected by environmental factors such as:

(1) Air velocity. Many insects require air movement before being stimulated to orient toward the source, whereas others are capable of locating the source in quiet air.

(2) Temperature. This will affect both the rate of dispersal of the semiochemical and the behaviour of the insect.

(3) Light intensity. Insects may orient either to or away from a light source instead of the chemical stimulus or may not respond at all.

(4) Relative humidity.

Qualitative bioassays have been widely used in pheromone research to detect the presence of biologically active material and to determine responses. A major disadvantage with qualitative bioassays is that losses of up to 99% of pheromone may go undetected. For example Butenandt *et al.*, (1961a, b) isolated 0.024 μg of sex pheromone per female silkworm, *Bombyx mori*, whereas it was later shown (Steinbrecht, 1964) that each female contains about 1.5 μg.

Of greater value are the quantitative bioassays, which give some measure of the amounts of active material present in a sample. Some workers report minimum concentrations required to elicit a behavioural response, the degree of which is often not specified. A better method of expressing quantitative bioassay measurements is to plot the percentage of organisms responding (on a logit (Berkson, 1953) or probit (Finney, 1964) scale) *versus* the logarithm of the concentrations tested (usually a dilution series) and thus obtain relatively straight lines that can be readily compared. (For examples see: Wharton *et al.*, 1954b; Shorey and Gaston, 1964. 1965a; Shorey *et al.*, 1964; 1965; Henneberry *et al.*, 1966; Shorey and Gaston, 1967a; Brady and Smithwick, 1968; Bartell and Shorey, 1969a, b; Howland *et al.*, 1969; August, 1971). Although one often does not know absolute concentrations of pheromones, especially during extractions, arbitrary units can be chosen which still yield useful information. For example, by using an equivalency unit representing the mass or number of insects,extracted per unit volume, relative activities of different samples can be determined.

Quantitative measurements on insects can also be made by use of the electroantennogram (EAG), which measures the change in voltage potential occurring along the antenna in response to an olfactory stimulus (Reviews: Boeckh *et al.*, 1965; Schneider, 1965, 1969; Yamada, 1970; Roelofs, 1971). The EAG is useful for investigating physiological and environmental factors, molecular specificity, and other variables on antennal olfactory responsiveness, and for comparison of these data with behavioural responses (Schneider and Hecker, 1956; Schneider, 1957a, b, 1962; Boeckh, 1962; Boeckh *et al.*, 1963; Lacher, 1964; Schneider *et al.,* 1964; Stürckow, 1965a,b; Boeckh, 1967; Lacher, 1967; Schneider *et al.*, 1967; Priesner, 1968; Ruttner and Kaissling, 1968; Moorhouse *et al.*, 1969; Schneider and Seibt, 1969; Boeckh *et al.*, 1970; Jacobson *et al.*, 1970c; Payne, 1970, 1971; Payne *et al.*, 1970; Grant, 1971; Roelofs *et al.*, 1971; Roelofs and Comeau, 1971b; Adler *et al.*, 1972; Mayer, 1973).

The uses to which quantiative bioassays of pheromones have been put include determination of:

(1) The efficiency of extraction by various solvents (Gaston *et al.*, 1966; Shorey and Gaston, 1967a).

(2) The absolute amounts produced by an organism (Jacobson *et al.*, 1963; Steinbrecht, 1964; Shorey and Gaston, 1965a; Berger, 1966; Gaston *et al.*, 1966) and the relative amounts produced by related species (Shorey *et al.*, 1968a).

(3) Relative amounts produced by different parts of the organism (Roth and Dateo, 1966; Zenther-Møller and Rudinsky, 1967), with changes in the physiological state of an organism such as age (Shorey and Gaston, 1964, 1965a; Shorey *et al.*, 1968c; Struble and Jacobson, 1970; Lawrence and Bartell, 1972; Negata *et al.*, 1972) and time of day (Shorey and Gaston, 1964), or changes in environment (Shorey and Gaston, 1964; Bartell and Shorey, 1969b).

(4) The rate of pheromone release (Sower *et al.*, 1971).

(5) Responsiveness with age (Sekul and Sparks, 1967; Payne *et al.*, 1970), time of day (Bartell and Shorey, 1969b; Payne *et al.*, 1970), or adaptation to the presence of a pheromone (Shorey and Gaston, 1964; Borden, 1967; Payne *et al.*, 1970; Traynier, 1970; Vick *et al.*, 1973).

(6) The concentration of extracted material required to elicit various degrees of response (Wharton *et al.*, 1954b; Shorey *et al.,* 1964; Roelofs and Feng, 1967; Sekul and Cox, 1967; Brady and Smithwick, 1968; Guerra, 1968; Onsager *et al.*, 1968; Bartell and Shorey, 1969b; Tashiro *et al.*, 1969b; August, 1971; Takahashi, F., *et al.*, 1972).

(7) Absolute amounts of pheromone required to elicit a specific degree of response (Boeckh *et al.*, 1965; Shorey *et al.*, 1967; Moser *et al.*, 1968).

(8) The concentration thresholds for each step in a sequence of responses (Bartell and Shorey, 1969a; Regnier and Wilson, 1969; Takahashi, F., *et al.*, 1972).

(9) Concentrations needed to inhibit responses to the pheromone (Shorey *et al.*, 1967; Klun and Robinson, 1970).

(10) Activity of synthetic material (Sturckow, 1965; Toba *et al.*, 1968; Howland *et al.*, 1969).

(11) The effect of change in pheromonal structure with changes in be behavioural response (Butenandt *et al.*, 1962; Boeckh *et al.*, 1965; Berger and Canerday, 1968; Jacobson *et al.*, 1968b; Boch and Shearer, 1971).

(12) The effect of treating an organism with non-pheromonal chemicals (such as sterilants or insecticides) on its behaviour (Henneberry *et al.*, 1966).

(13) The response of an organism to pheromones of another species (Shorey *et al.*, 1965; Priesner, 1968).

B. Problems encountered with bioassays

Unless the investigator pays careful attention to experimental detail he will likely encounter difficulties when conducting bioassays. It is advantageous to anticipate and thus avoid these difficulties. The following list enumerates many of the problems frequently met, such as:

(1) Choice of response for bioassay. In the search for sex attractants of moths, many workers have unfortunately ignored the distinction between stimulatory responses (e.g. the first six responses for insect sex pheromones on page 92) and attraction. This had led to the isolation of excitants, many of which fail to attract effectively in the field. In many cases, it appears that the long-range attractant pheromone may consist of several components, one of which may be the compound responsible for the stimulatory response. Use of a walking response in the laboratory for insects showing a flying response in the field has sparked controversy over interpretation of results (Vité, 1970, and Wood *et al.*, 1967a). It should be noted that the bark beetles in question do walk to the source once on the tree trunk.

(2) Specificity of response. The response may be non-specific and triggered by chemicals other than the pheromone. This is a common problem with bioassays based on excitation (Block, 1960) or attraction. Even the solvent may have attractive powers (Block, 1960).

(3) Change in response pattern. At a certain concentration one type or pattern of response may be elicited whereas at higher concentrations a completely different pattern may manifest itself (Moser *et al.*, 1968; Bartell and Shorey, 1969a; Regnier and Wilson, 1969).

(4) Physiological state of the insect. The responsiveness of an insect is subject to changes in age, circadian rhythms (daily, seasonal, or of other periods), and effects of mating. Circadian rhythms can be manipulated by conditioning with artificial day-night cycles. Laboratory

reared insects may be affected by their diet (Pliske and Eisner, T., 1969). When studying sex pheromones, care must be taken to isolate the sexes while they are still virgins.

(5) Environmental factors. Insects exhibiting positive anemotaxis may respond only when there is airflow above a certain velocity (Casida *et al.*, 1963; Butler and Fairey, 1964; Soo Hoo and Roberts, 1965). Nocturnal insects usually respond only under dim light or complete darkness. It will then be necessary to screen out most or all light, otherwise the insect may orient either toward or away from the source or not respond at all. Day-time insects also are sometimes affected by light sources. Temperature and relative humidity ranges should also be studied for effects on responsiveness.

(6) Choice of individuals for response to test samples. Large numbers should be used to obtain statistically significant results, although this may present a new problem: conformity of behaviour. The response of one individual may affect others in that vicinity. The size of the insect may restrict the number which can be accommodated at one time in the bioassay apparatus. In the study of sex pheromones, 10 to 25 individuals are commonly bioassayed at one time and each substance tested on 5 to 10 groups. If only limited numbers are available, measurements can be repeated using the same individuals. A major disadvantage of reusing insects for bioassay is that adaptation or sensitization (Borden, 1967) to the pheromone may occur and higher or lower concentrations may later be required to elicit the same degree of response. The time between initial exposure to complete loss of adaptation may be only minutes or longer than a day (Wharton *et al.*, 1954b). Preselection of responding individuals has been used (Pitman *et al.*, 1965). The use of only those individuals that exhibit a positive response does not allow for the inherent variability within and between populations, and the results may be biased by such variables as preconditioning, learning or fatigue. In-bred laboratory stocks may have lower pheromone production (Minks, 1971).

(7) Inhibition. If the pheromonal concentration is much above the threshold concentration, the response may be completely inhibited (Shorey *et al.*, 1967).

(8) Air contamination. Pheromones are highly active and may cause habituation (see for example Vick *et al.*, 1973). Thus great care must be exercised to keep stray molecules, either from live insects or extract solutions, from entering the assay system. Whenever possible, bioassay facilities should be isolated from rearing areas, and exhaust air from the bioassay site should be vented outside. The air may also have to be "scrubbed" to remove impurities that may affect the

response. Before removing water from the air stream, it must be established that this change in relative humidity will not affect the response.

(9) Re-use of bioassay apparatus. Pheromone molecules may remain in the apparatus after use and affect the outcome of the next bioassay. The apparatus should be thoroughly cleaned between bioassays.

(10) Masking or inhibition. Some workers have reported that some crude pheromonal extracts are inactive, but after purification become active (Casida *et al.*, 1963; Berger *et al.*, 1965; Jacobson and Smalls, 1966; Beroza, 1967). Masking by impurities from the insect or in the solvent or by the solvent itself has been suggested by some as the reason, but this is doubted by others (Shorey and Gaston, 1967a; Shorey *et al.*, 1968a). Jacobson (1968) observed masking by 15% of the geometric isomer of the pink bollworm moth, *Pectinophora gossypiella*. Tumlinson *et al.*, (1972) observed that field response to *cis*-7-dodecenyl acetate, the sex pheromone of the cabbage looper, is significantly reduced by as little as 0.1% of the corresponding alcohol. It is therefore necessary to ensure that synthetic pheromones are of high purity.

(11) Syngerism. If a pheromone consists of several semiochemicals (Wood *et al.*, 1967b; Tumlinson *et al.*, 1969), all of which must be present to trigger a maximum response, then bioassay of individual fractions will show greatly reduced or no response. Activity will be regained on recombination. The degree of response may be markedly affected by the relative amounts of the components, (Beroza *et al.*, 1973; Klun *et al.*, 1973; Minks *et al.*, 1973). It is also possible that there there may be synergism between pheromones released by the insect and compounds emanating from its host (Bedard *et al.*, 1969; Hardee *et al.*, 1971).

(12) Release of non-uniform amounts of pheromone in quantitative bioassays. Neither the "contaminated pipet" method nor the "wetted glass rod" procedure allows for a high degree of reproducibility. Placement of uniform volumes of solution on an absorbent is not difficult; however delivery of uniform concentrations in the air stream is subject to such variables as the nature of the absorbent and rate of evaporation from the absorbent surface, and maintenance of a uniform and steady air flow. (See the section on methods of bioassay for details of these methods).

(13) Acclimatization to bioassay apparatus. When an insect is introduced to the bioassay apparatus, time must be allowed for adjustment to the new surroundings. For the insect this may be an alarm situation.

(14) Contamination of test sample with other pheromones. In the initial stages of isolation of trail pheromones from the ant, *Atta texana*, Silverstein *et al.*, (unpublished results) observed confusing responses

due to contamination by alarm pheromones. This difficulty was overcome by fractionation, and in later studies the heads (source of the alarm pheromones) were removed and only the bodies were extracted.

(15) Insect communication by means other than chemical. For example some Scolytid beetles use chemical and acoustical communication (Rudinsky and Michael, 1972).

(16) Field bioassays. Bioassays that use traps and lures in the field are affected by environmental factors (Saario *et al.*, 1970; Mitchell, E. B., *et al.*, 1972), trap design (Guerra, 1969; Debolt, 1970; Sharma *et al.*, 1971; Kaae and Shorey, 1972; Sharma *et al.*, 1973), trap color (Hendricks *et al.*, 1972; Roach *et al.*, 1972), trap position (Gary, 1962; Saario *et al.*, 1970; Lilly and Shorthouse, 1971; Klein *et al.*, 1972; Roach *et al.*, 1972), the amount of lure used (Roelofs *et al.*, 1972), and the solvent used to apply the lures (Lange and Weissinger, 1972). Traps might even be robbed by predaceous insects (Howell and Davis, 1972). Replication is necessary and reports of a small catch on a single trap are highly suspect.

C. Methods of bioassay

Bearing in mind all the pitfalls to be avoided while conducting a bioassay, we now review the various methods which have been used by workers in this field.

When developing a bioassay, one must first observe the response pattern to a semiochemical, then choose one or more steps of the sequence to serve as a basis for the bioassay, and finally devise a method for presenting the semiochemical to the organism in such a way that, hopefully, quantitative measurements of the response can be made.

a. Sex pheromones, laboratory

Observation of behaviour under natural conditions is desirable but usually impractical and one must witness responses within the confines of the laboratory. One notable exception was the the erection of a large flight cage in the field for study of the queen butterfly, *Danaus gilippus berenice* (Pliske and Eisner, T., 1969). In preliminary laboratory behavioural studies, one usually delivers pheromones from either live organisms (George, 1965; Burkholder and Dicke, 1966; Adams and Mulla, 1968; Hammond and Hensley, 1970; Fatzinger and Asher, 1971; Percy *et al.*, 1971; Birch, M. C., 1970) or extracts (Shorey, 1964; Jacobson and Smalls, 1967; Weatherston and Percy, 1968; Daterman, 1970; Hammond and Hensley, 1970; Shorey and Bartell, 1970; Sower *et al.*, 1970) to test the organism and observes the overall pattern of response. Although most organisms will respond in normal or dim light, some require complete darkness before they will react. To overcome this difficulty, time-lapse photography with strobe lights can be used (Baldwin *et al.*, 1971; Fatzinger and Asher, 1971). Very small insects have been observed by use of a microscope (Gravitz and Willson, 1968; Cone *et al.*, 1971).

Responses serving as a basis for the laboratory bioassay of the insect sex pheromones include:

(1) antennal waving,

(2) wing vibration (Shorey *et al.*, 1965),

(3) clasper extension (Perez and Long, 1964),

(4) extension of genitalia,

(5) abdominal curving,

(6) attempted copulation (Tashiro *et al.*, 1969a),

(7) "dance of love",

(8) a combination of several of (1) to (7) (Wharton *et al.*, 1954a; Block, 1960; Shorey, 1964; Klun, 1968; Bodenstein, 1970; McDonough, *et al.*, 1970; Redfern *et al.*, 1970),

(9) "striking" or sexual leaping of one sex onto the other (Rogoff *et al.*, 1964; Fletcher *et al.*, 1966),

(10) rapid spinning flight (Adeesan *et al.*, 1969), or "buzzing" (Sanders, C. J., 1971a, b),

(11) licking of a glass rod (Onsager *et al.*, 1968),

(12) positive anemotaxis (Mayer and Thaggard, 1966; Wood *et al.*, 1966; Happ *et al.*, 1970),

(13) simple attraction to a pheromone source on filter paper (Yinon and Shulov, 1967; Gravitz and Willson, 1968), or in an olfactometer (Adams and Mulla, 1968; Henzell *et al.*, 1969; Happ *et al.*, 1970; Rejesus and Reynolds, 1970; August, 1971),

(14) arrested movement toward a light source (Jantz and Rudinsky, 1965; Borden *et al.*, 1968),

(15) "hairpencilling" (Brower, L. P. *et al.*, 1965; Pliske and Eisner, T., 1969; Birch, M. C., 1970; Grant, 1970).

Substances being bioassayed in the laboratory for insect sex pheromonal activity have been presented to the insect by use of various substrates such as:

(1) The "contaminated pipet". This method involves filling and then emptying a pipet or medicine dropper with an extract solution followed by passage of air through the pipet into a cage or closed container (Sekul and Cox, 1965; Jones, W. A. *et al.*, 1966; Berger, 1968; Klun, 1968; Shorey *et al.*, 1968b; McDonough *et al.*, 1970; Redfern *et al.*, 1970; Sanders, C. J., 1971b).

(2) The "wetted glass rod." A glass rod is momentarily dipped into an extract solution, the solvent allowed to evaporate, and the rod is placed in the presence of the insect (Brady and Smithwick, 1968; Onsager *et al.*, 1968; McDonough *et al.*, 1969).

(3) An absorbent substrate. A drop of extract is placed on a suitable absorbent such as filter paper (Wharton *et al.*, 1954b; Ouye and Butt, 1962; Lilly and McGinnis, 1965; Yinon and Shulov, 1967; Gravitz and Willson, 1968; Fletcher, 1969; Tashiro *et al.*, 1969a; Bodenstein, 1970; Vick *et al.*, 1970; Roelofs and Cardé, 1971b), dental roll (Bosman and Brand, 1971), glass wool (Daterman, 1972), paraffin (Tashiro and Chambers, 1967), plastic (Cone *et al.*, 1971), or metal pin (Fatzinger, 1972), and the substrate is presented to the insect after evaporation of the solvent.

(4) Surrogates. A surrogate insect is impregnated with a pheromonal extract (Rogoff *et al.*, 1964; Roth and Dateo, 1966; Henzell and Lowe, 1970).

(5) Gas chromatographs. The effluent stream of vapours from a gas chromatograph (GC) has been passed over insects mounted on a rack or allowed movement within a cage (Gaston *et al.*, 1966; Berger, 1966, 1968; Beroza, 1971a, b), or onto an antenna prepared for recording EAGs (Moorhouse *et al.*, 1969). Detection of one picogram of pheromone has been reported for the gypsy moth, *Porthetria dispar* (Beroza, 1971a, b). This method allows rapid bioassay of GC fractionations, provided that the insect gives a readily measurable response, is relatively insensitive to environmental factors, and there is only one compound in the pheromone.

To facilitate observation of their behavioural responses to substances, insects have been restricted by use of mounting racks (Block, 1960; Waters and Jacobson, 1968; Tardif and Stevens, 1972), tethers (Hughes and Pitman, 1970; Bennett and Borden, 1971; Hsiao, 1972), and flight mills (Atkins, 1961; Roeder, 1966; Smith and Furniss, 1966; Rowley *et al.*, 1968; Borden and Bennett, 1969).

The olfactometer has been widely used for bioassays. This is an enclosure designed to guide or limit the movement of the insect. Individual creativity has come to the fore and there appear to be as many olfactometer designs as there are experimenters. A common type involves passing air at a controlled rate over an absorbed or adsorbed sample, then into an enclosure containing the insects where the degree of behavioural response can be observed, and finally venting, preferably into a hood (Gaston and Shorey, 1964; Shorey *et al.*, 1965; Gaston *et al.*, 1966; Roelofs and Feng, 1967; Bartell and Shorey, 1969b; Struble, 1970; Takahashi, F., *et al.*, 1972; Vinson, 1972). A variation of this type is the so-called "choice olfactometer" in which an organism chooses between an air stream laden with molecules to be tested or an air stream devoid of a scent. Movement toward the pheromone source is considered a positive response. Choice olfactometers have been constructed in Y-(Adams and Mulla, 1968; Keys and Mills, 1968; Toba *et al.*, 1968; Henzell *et al.*, 1969; Howland *et al.*, 1969; Tashiro *et al.*, 1969b; Rejesus and Reynolds, 1970; Mitchell and Mau, 1971; Chang and Curtis, 1972; Chaudhury *et al.*, 1972; Takahashi, S. and Kitamura, 1972), T- (Guerra, 1968), swastika (Rejesus and Reynolds, 1970) and other complex shapes (Hardee *et al.*, 1967; Ali Naizee and Stafford, 1971; August, 1971) and on a board with up to five air streams (four controls) converging on the test insects (Wood and Bushing, 1963; Wood *et al.*, 1966; Browne *et al.*, 1974). Miscellaneous olfactometer designs include: containers having only one air stream inlet and outlet with orientation to the inlet taken as a positive response (Mayer and Thaggard, 1966; Happ and Wheeler, 1969; Happ, 1970; Daterman, 1972; Nation, 1972; Takahashi, F. *et al.*, 1972); a walkway over a perforated area with test substances beneath and a positive response being arrestment of motion toward (Jantz and Rudinsky, 1965) or turning away from (Borden *et al.*, 1968a, b) a light source; a walkway to observe the responses of individual walking insects to attractants and light (Francia and Graham, 1967; Moeck, 1970); a flight box to determine the effect of light on orientation to a pheromone source (Shorey and Gaston, 1965b); a pegboard in an enclosure to simulate galleries for bark beetles (Klieforth *et al.*, 1964); timed turntables (Tashiro *et al.*, 1969b; Rejesus and Reynolds, 1970); and a design to study orientation to simultaneously attenuated odour streams (Wright, 1966).

Sex pheromones have also been bioassayed by use of the electroantennogram. Capillary electrodes are implanted between the base and tip of an intact or excised antenna. Air containing test molecules is passed over the antenna, and the change in voltage potential is electronically measured, displayed on an oscilloscope, and photographed (Schneider and Hecker, 1956; Schneider, 1957b; Payne *et al.*, 1970; Roelofs and Comeau, 1971). Although there is often good correlation between EAG and behavioural data, sometimes there is not (Payne *et al.*, 1970; Stürckow, 1970; Birch, M. C., 1971; Grant, 1971; Grant *et al.*, 1972; Mayer, 1973). Chemicals other than the pheromone may elicit an EAG response (Boeckh *et al.*, 1963; Stürckow, 1965; Birch, M. C., 1971). There is evidence that the techniques used are very important: some investigators have elicited typical "EAGs" from saline-saturated pipe cleaners. The convenience of using an EAG on a single insect must be balanced against the fact that such responses are even further removed than laboratory bioassay from a field response. Current research may help to bridge this gap.

b. Sex pheromones, field

Field bioassay of insect pheromones generally involves placing a trap containing live insects (Coppel *et al.*, 1960; Solomon and Morris, 1966; Davis and Henderson, 1967), extracts (Keys and Mills, 1968; Struble, 1970), or synthetic pheromones (Kishaba *et al.*, 1970; Pitman and Vité, 1970) outdoors and at a later time counting the number of insects captured. Field assays have also been conducted in a greenhouse (Gentry *et al.*, 1964). Although natural populations are generally used, some investigators have released insects in the field. This is a valid procedure provided that the investigator reports the distinction. Marked individuals are sometimes released at various positions relative to the trap to measure environmental factors on response to the pheromone (Kishaba *et al.*, 1970; Patrick and Hensley, 1970; McGovern *et al.*, 1973). For studies on effect of trap design see: Snow and Copeland (1969); Debolt (1970); Sharma *et al.* (1971); Howell (1972); Granett (1973); Sharma *et al.* (1973).

Trap designs include:

(1) Adhesive coated cups (Holbrook *et al.*, 1960; Ouye and Butt, 1962), boards (Coppel *et al.*, 1960; Davis and Henderson, 1967), paper plates (Batiste and Joos, 1972), cylinders (Henneberry *et al.*, 1967; Pitman, 1971), folding cardboard (Roelofs and Comeau, 1970; Howell, 1972), cardboard boxes (Granett, 1973), wooden boxes (Sharma *et al.*, 1973), hardware cloth (Bedard *et al.*, 1969), and wire mesh (Bedard and Browne, 1969);

(2) Maze (Ladd, 1970; Killinen and Ost, 1971; Sharma *et al.*, 1971; Kaae and Shorey, 1972), and glass barrier traps (Chapman, 1966; Nijholt and Chapman, 1968);

(3) Traps used in conjunction with lights (Henneberry and Howland, 1966; Henneberry *et al.*, 1966; Wolf *et al.*, 1967, 1969; Manley and Ferrier, 1969; Debolt, 1970).

(4) Traps simulating tree trunks (Gara, 1967a, b; Pitman and Vité, 1969; Vité and Pitman, 1969);

(5) A device that indicates the time of day capture occurred (Batiste, 1970);

(6) A lure suspended from a helium filled balloon elevated 7 to 30 m (Gary, 1962; Shearer *et al.*, 1970; Strang, 1970; Boch *et al.*, 1975) for the honey bee, *Apis mellifera*;

(7) Electric grids (Mitchell, E. R., *et al.*, 1972); and

(8) Miscellaneous traps (Vité and Gara, 1961).

Substrates used for dispersal of sex pheromones in the field include:

(1) Absorbents such as corrugated paper (Holbrook *et al.*, 1960),

filter paper (Holbrook *et al.*, 1960; Cuthbert and Reid, 1964; Berger *et al.*, 1964; Graham *et al.*, 1966; Wong *et al.*, 1971), cotton wads (Holbrook *et al.*, 1960; Daterman and McComb, 1970; Lilly and Shorthouse, 1971), cellucotton pads (Cross, W. H. *et al.*, 1969), nylon carpet (Debolt, 1970), cotton thread wicks (Debolt, 1970), rubber bands (Madsen and Vakenti, 1972), fire brick (Hardee *et al.*, 1972), cigarette filters (Hardee *et al.*, 1972; McKibben, 1972; Moody *et al.*, 1972), and cork granules (Klun *et al.*, 1973).

(2) Reservoirs such as glass dishes (Saario *et al.*, 1970), bottles (Pitman, 1971), glass vials inside perforated 35mm film containers (Rudinsky *et al.*, 1972), polyethylene caps (Glass *et al.*, 1970), polyethylene tubing (Pitman, 1971), glass capillaries (Bedard *et al.*, 1971), and stainless steel planchets (Sharma *et al.*, 1971).

(3) Sand in glass jars (Debolt, 1970), cheese cloth bags (Debolt, 1970) and polyethylene bags (Toba *et al.*, 1969).

(4) Gas chromatographic liquid phases and supports that are treated with pheromones, placed in tubes and then desorbed by aeration (Bedard *et al.*, 1969).

(5) Microencapsulation (Beroza *et al.*, 1974).

c. Alarm pheromones

Alarm pheromones of social insects are usually bioassayed by placing solutions of test samples on filter paper (Blum *et al.*, 1966; Bowers *et al.*, 1972), index cards (Ayre and Blum, 1971), paper at the apex of a plastic cone (Blum *et al.*, 1971) or other substrates (Boch *et al.*, 1970) at the entrance of the nest or hive and observing the behaviour of the insects. Olfactometers have also been employed (Moser *et al.*, 1968). Placement of test substances along trails in the field have been used to determine alarm-causing activity (Moser *et al.*, 1968).

d. Trail pheromones

To bioassy for trail-following activity, a solution of the sample is streaked over a surface of paper (Moser and Blum, 1963; Karlson *et al.*, 1968; Birch, A. J. *et al.*, 1970; Pasteels, 1972; Tschinkel and Close, 1973), slick cardboard (Tumlinson *et al.*, 1972), or glass (Smythe *et al.*, 1967) and the degree of response is noted. Olfactometers have also been used (Ritter *et al.*, 1973). In a preliminary study, Wilson (1959) laid artificial trails near natural trails and observed the response. Macgregor (1948), Carthy (1951), and Hölldobler and Wilson (1970) designed special arenas to observe orientation of ants to trail odours.

e. Miscellaneous pheromones

Robertson and Orton (1971) describe techniques and apparatus developed for pheromone bioassay studies of ants. Boch *et al.* (1975) describe the field bioassay of the pheromone released by queen honey bees, *Apis mellifera*, to stabilize swarms of workers.

III. Isolation

A. Source of the semiochemicals

The number of insects required in the isolation stages of the investigation will depend, in large measure, upon the amount of semiochemical(s) given off by each individual, and to a lesser degree upon the structural complexity. Obviously, the more material one can obtain, the easier becomes the technical process of identification. At the present state of the art, structural elucidations can be made quite readily at the milligram level, and with some difficulty at the submilligram level.

Insect sex pheromones are usually produced in extremely small amounts (for example: 0.5 to 1.5 ng per female cabbage looper, *Trichoplusia ni*, (Shorey and Gaston, 1965a); 3.2 ng per female Indian meal moth, *Plodia interpunctella,* (Kuwahara and Casida, 1973)) and one therefore, may require large numbers of individuals to isolate enough material for identification. If the semiochemical has a simple structure, then only a small amount of material may be needed; for example the hydrocarbon structure of a tiger moth sex pheromone was determined by using material obtained from only 50 females (Roelofs and Cardé, 1971c). Three hundred male Danaid butterflies, *Lycorea ceres ceres*, provided enough material for identification because the semiochemicals could be isolated directly from extruded glands without extensive purification (Meinwald, J. *et al.*, 1966). At the other extreme, material from 4.5 million beetles was used to identify the sex pheromones (**18** to **21**) of the cotton boll weevil, *Anthonomus grandis*, (Tumlinson *et al.*, 1969). Structural elucidation of the majority of sex pheromones has required material from 10,000 to 800,000 individuals. The recently developed GC-EAG technique (described in the section on structure determination) of Roelofs and Comeau (1971a, b) for identifying sex attractants, especially of Lepidoptera, requires only 50 to 200 individuals.

Trail following pheromones are also produced in minute quantities and are highly active. Tumlinson *et al.*, (1971) were able to identify a pheromone from 3.7 kg of leaf-cutting ants, *Atta texana*. In the laboratory bioassay, worker ants responded at a concentration of 0.08 pg per cm of trail; at this level, only 330 μg of this compound would be needed to draw a detectable trail around the world. Aided by the discovery of Smythe *et al.* (1967) that a trail-following substance employed by the termite *Reticulitermes flavipes* is also produced by a fungus-infested wood, Matsumura *et al.* (1968) identified the pheromone using the wood as the major source of active material.

Alarm pheromones are frequently simple ketones in the C_6 to C_{12} range and are quite easy to identify, often by comparison with known compounds. Only 200 honey bees, *Apis mellifera* (Shearer and Boch, 1965) and 500 *Atta texana* ants (Moser *et al.*, 1968) were needed to identify their respective alarm pheromones.

The defensive allomones of arthropods are sometimes secreted in milligram amounts, are often easily isolated from their reservoirs, and with few exceptions have fairly simple molecular structures. Material from only one (Eisner, T. *et al.*, 1961), or several (Roth *et al.*, 1956), individuals may be enough to identify simple allomones, although for identification of the complex molecule 12, isolated from the grasshopper, *Romalea microptera*, 1000 insects were used (Meinwald, J. *et al.*, 1968b). Throughout much of the defensive allomone studies, the secretion from several hundred arthropods has been sufficient for identification.

After choosing an insect to study for semiochemicals, and estimating how many individuals will be required, one must decide between field collection or mass rearing. Field collection, when feasible, usually presents fewer complications. Mass rearing often requires considerable expertise to raise healthy colonies. In addition to simply keeping the insects healthy, it is necessary to keep environmental factors, such as the day-night cycle, as close to natural conditions as possible. The diet used to feed the insects may affect semiochemical production. Thus, male queen butterflies, *Danaus gilippus berenice*, raised indoors on a normal food plant (fresh and refrigerated cuttings) were deficient in a particular semiochemical, whereas males raised outdoors on the same plant (living) were not (Pliske and Eisner, T., 1969). Minks (1971) observed decreased sex pheromone production in an in-bred stock of laboratory reared moths.

B. Collection of the semiochemical(s)

The most important concern is to ensure collection of the maximum amount of active material. One first chooses between a single collection or a collection over a period of time, and then from among the various methods of collection. Selection of the procedure will depend on a number of factors.

Suppose one discovers that the semiochemical is produced and stored in a single large accessible gland. One may:

(1) remove the gland contents from the intact insect (Defensive allomones: Blum, *et al.*, 1960; Calam and Scott, 1969) or excised gland (Defensive allomones: Blum, 1961; Blum *et al.*, 1961; McCullough, 1966); or

(2) excise and extract the gland (Alarm pheromones: Butenandt *et al.*, 1959; McGurk *et al.*, 1966; Morse *et al.*, 1967; Blum *et al.*, 1968; Moser *et al.*, 1968; Crewe *et al.*, 1970. Defensive allomones: Roth *et al.*, 1956; Chadha *et al.*, 1962; Blum, 1966; Cavill and Williams, 1967; Eisner, T. *et al.*, 1968; Baker, J. T. and Jones, P. A., 1969; Blum *et al.*, 1969).

One may also rinse or extract:

(1) segments (Sex pheromones: Acree, 1953; Berger, 1966; Sekul and Sparks, 1967; Henzell and Lowe, 1970; Jacobson *et al.*, 1970b; Dahm *et al.*, 1971; Roelofs and Cardé, 1971; Sower *et al.*, 1973);

(2) hairs (Sex pheromones: Meinwald, J. *et al.*, 1966, 1968a, 1969; Alpin and Birch, M. C., 1968);

(3) hindguts (Recruitment pheromones: Pitman *et al.*, 1968); or

(4) whole insects (Alarm pheromones: Blum *et al.*, 1963; Blum and Wharter, 1966; Crewe *et al.*, 1970. Recruitment pheromones: Ikan *et al.*, 1969. Sex pheromones: Jones *et al.*, 1966; Silverstein *et al.*, 1967a; Jacobson *et al.*, 1968a; Rodin *et al.*, 1969; Tumlinson *et al.*, 1969; Kuwahara *et al.*, 1971. Trail pheromones: Moore, 1966. Defensive allomones: Cavill and Hinterberger, 1960a; Blum and Woodring, 1962; McGurk *et al.*, 1968).

Ställberg-Stenhagen (1972) and Bergström (1973) have devised techniques for collecting volatile material from very small samples without prior extraction by solvent and proceeding directly with the purification by gas chromatography.

On the other hand, the semiochemicals may be produced over a period of time, and a continuing collection process is used. This may involve:

(1) passing air over live insects (aeration) into a cold trap (Sex pheromones: Jacobson *et al.*, 1963; Yamamoto, 1963; Jones, P. A. *et al.*, 1965; Silverstein and Rodin, 1966; Tumlinson *et al.*, 1968; Moreno *et al.*, 1972; Browne *et al.*, 1974. The techniques of Browne *et al.* (1974) is unique in that all of the volatiles and the air itself are condensed. Defensive allomones: Roth and Howland, 1941; Ladisch and McQue, 1953; Loconti and Roth, 1953; Roth and Stay, 1958; Blum and Crain, 1961; Englehardt *et al.*, 1965);

(2) aeration into a solution (Defensive allomones: Eisner, T. *et al.*, 1963b. Sex pheromones: George and McDonough, 1972);

(3) aeration into an absorbent (Sex pheromones: Yamamoto, 1963; Silverstein and Rodin, 1966; Solomon *et al.*, 1972. Silverstein *et al.* (unpublished) have efficiently trapped bark beetle pheromones on Poropak Q. Defensive allomones: Regnier and Wilson, 1968).

(4) keeping live insects in contact with, and then extracting, an absorbent surface (Sex pheromones: Roth and Willis, 1952; Coppel *et al.*, 1960; Wharton *et al.*, 1962; Tumlinson *et al.*, 1968; Brady *et al.*, 1971; Weatherston *et al.*, 1971; Solomon *et al.*, 1972); or

(5) collecting over a period of time and then extracting frass or fecal material produced by the insect (Sex pheromones: Wood *et al.*, 1966; Borden *et al.*, 1968; Silverstein *et al.*, 1968; Tumlinson *et al.*, 1968, 1969).

Insects have been "milked" by being forced to discharge their defensive allomones:

(1) into a solution (Blum, 1964; Schildneckt *et al.*, 1967);

(2) into a centrifuge tube (Dateo and Roth, 1967) or other container (Chadha *et al.*, 1961; Hurst *et al.*, 1964);

(3) onto filter paper (Meinwald, J. *et al.*, 1962; Monro *et al.*, 1962; Weatherston and Percy, 1969; Eisner, T. *et al.*, 1971); or

(4) onto the surface of the insect's body where the secretion is picked up by capillary tubes (Meinwald, Y. C. *et al.*, 1966; Staddon and Weatherston, 1967; Meinwald, J. *et al.*, 1968b).

One must consider the physiological state of the insect and the environmental factors affecting it. For example, the female cabbage looper does not begin maximum production of sex pheromone until the fourth night after emergence from the pupa (Shorey and Gaston, 1965a), and then only releases the pheromone between several hours after sunset and dawn (Shorey, 1964). If the semiochemical is stored mainly as an inactive precursor, the investigation may lead to the identification of the precursor; such is the case for some of the defensive allomones (Schildkneckt and Holoubek, 1961; Eisner, T. *et al.*, 1963a). In the case of the gypsy moth, *Porthetria dispar*, only minute amounts of the sex pheromone, an epoxide, were available. The presence of an olefin precursor was postulated, and a relatively large amount was isolated, identified, and chemically transformed to the active material (Bierl *et al.*, 1970).

The properties of the semiochemical must also be considered. Lack of stability to air, light, or chemicals has stymied many an investigation. Chemical interaction with solvents (e.g. transesterification with alcohols (Jacobson *et al.*, 1970b)) or with impurities or decomposition products in solvents (e.g. peroxides in ether, hydrochloric acid or phosgene in chlorinated solvents) may occur. Volatility of the semiochemical will also determine the collection procedure and solvent used. Schlunegger and Leuthold (1972) used high boiling 1,3-dimethoxybenzene to extract volatile compounds in ants and then proceeded directly to combined gas-chromatography – mass spectrometry for identification.

C. Purification

The choice of purification method will depend upon the physical state, stability, and amount of semiochemical available. To ensure that the semiochemical does not undergo rearrangement or transformation, one should use the mildest possible conditions for purification.

a. Distillation

One of the simplest methods of purification is distillation. Useful discussions of distillation may be found in the texts by Vogel (1956), Wiberg (1960a) and Perry and Weissberger (1965). Although vacuum distillation (Moore, 1966; Bernardi *et al.*, 1967; Cavill and Williams, 1967; Cavill *et al.*, 1968), steam distillation (Moore, 1964; McGurk *et al.*, 1966; Tumlinson *et al.*, 1969; Moreno *et al.*, 1972), sweep co-distillation (Jacobson *et al.*, 1970b) or co-distillation with mineral oil (Matsumura *et al.*, 1968) have been occasionally used, the amounts of material usually extracted do not permit use of the classical distillation techniques. Wood *et al.* (1966) were able to separate volatile material from polymeric and other non-volatile material extracted from frass by short path vacuum distillation

b. Sublimation and crystallization

In a few of the instances where the semiochemicals are solids, sublimation has been employed (Loconti and Roth, 1953; Blum and Crain, 1961; Meinwald, Y. C. *et al.*, 1966; Meinwald, J. *et al.*, 1969). Recrystallization, a classical method for purification of solids, has been used only infrequently, largely because of the minute amount of material available.

c. Solution chilling

Polymeric, other high molecular weight, and lipid impurities can sometimes be removed by dissolving the extract in a suitable solvent such as methanol or acetone, cooling the solution to −20° to −40°C, and separating the resulting precipitate by filtration (Jones *et al.*, 1966; Sekul and Sparks, 1967; Jacobson *et al.*, 1970b; Kuwahara *et al.*, 1971).

d. Extraction and derivative formation

Carboxylic acids can be removed by extraction with dilute aqueous alkali (Meinwald, Y. C. and Eisner, T., 1964; Silverstein *et al.*, 1967a; Cavill *et al.*, 1968; Meinwald, J. *et al.*, 1968c) or precipitated by bubbling dry ammonia through their ether or hexane solutions for 30 seconds. (Silverstein and Yarger, unpublished). Benzoquinones have been removed from other semiochemicals by adsorption on an alumina slurry (Hurst *et al.*, 1964). Esters have been hydrolysed in alkaline solution to their corresponding carboxylic acids and alcohols that can be separated by extraction (Moore, 1966; Jacobson *et al.*, 1970b; Beroza, 1971a) and then reconverted to the starting esters. Butenandt *et al.*, (1959, 1961a) purified the sex pheromone of the silkworm moth, *Bombyx mori*, *via* its 4′-nitroazobenzenecarboxylic acid ester. Some ketonic semiochemicals have been converted to their 2,4-dinitrophenylhydrazone (Staddon and Weatherston, 1967; Calam and Scott, 1969) or semicarbazone derivatives (Chadha *et al.*, 1962) which were then purified. A danger with these purification procedures is the possibility of acid or base catalysed rearrangements.

e. Chromatography

By far the most extensively exploited method of purification is that of chromatography. All types of chromatography are based on the partitioning of compounds between a stationary phase (solid or liquid) and a moving phase (liquid or gas). The major types of chromatography are column chromatography, thin-layer chromatography (TLC), paper chromatography, and gas chromatography (GC).

i. Column chromatography. In column chromatography, a vertical cylindrical tube, often made of glass is packed with a slurry of a solid adsorbent and a non-polar liquid. Excess liquid is drained off, the sample, in a minimum volume of solvent, is applied to the top of the column, and fresh solvent is allowed to flow through the column. This process separates the sample into bands containing the individual components. These bands are then eluted from the bottom of the column in sequence. Techniques of column chromatography are discussed in Cassidy (1951), Cassidy (1957), Wiberg (1960b), and Bobbitt *et al.* (1968b). Common adsorbents for column chromatography include alumina (aluminium oxide), Florisil (a magnesium silicate), silica gel, and silicic acid. Silica gel is the most versatile and can be used in most cases. Florisil is the least active, making it a very useful adsorbent for the chromatography of sensitive compounds. Since silver ions will complex with carbon-carbon double bonds, silver nitrate treated silica gel can be used to separate isomeric unsaturated compounds (Sekul and Sparks, 1967; Roelofs and Arn, 1968; Tumlinson *et al.*, 1969; Kuwahara *et al.*, 1971). The composition of the solvent may remain unchanged, or be varied either stepwise or gradually to increased polarity when separating the components of complex mixtures. Column chromatography is capable of high resolution but can be fairly laborious. One new technique for obtaining high resolution from column chromatography involves use of small particles size adsorbents, long (15 to 100 cm) and narrow (0.5 to 20 mm i.d.) columns, and high pressure to get high eluate flow rates; use of sensitive detectors and recorders has sparked a new era in column chromatography. Techniques of liquid chromatography are discussed by Hadden *et al.*, (1971); Kirkland, (1971); Perry *et al.*, (1972).

Ionizable substances, such as acids, can be separated on a support consisting of fine heads of an organic or inorganic resin that has an affinity for anions or cations. This type of column chromatography is known as ion-exchange chromatography. For discussion of ion-exchange chromatography, see: Calmon and Kressman (1957); Marinsky (1966, 1969).

ii. Thin-layer chromatography. Thin-layer chromatography employs an active adsorbent, such as alumina or silica gel, spread as a layer of 0.05 to 2 mm thickness (usually 0.25 mm) onto a solid plate of glass, plastic or aluminium. Samples are applied to the base of the plate in volatile solvents and the chromatogram is then developed by being placed into a closed chamber with a shallow pool of solvent. The solvent ascends on the coated layer by capillary action from just below the point of application to near the top of the coating. Components of the resulting chromatogram are detected by being sprayed with, or exposed to, the vapours of reagents, which yield coloured spots. Alternatively, certain fluorescent indicators may be used. TLC is employed chiefly to determine the number of components in a sample, to detect a given compound or compounds in a mixture, and to find conditions suitable for running a column chromatogram. Silver nitrate impregnated silica gel can also be used for TLC separations (Roelofs and Arn, 1968). With large (20 x 100 cm) and thick plates (1 to 2 mm), one can recover resolved components from the plate (preparative TLC). This method of analysis and separation is capable of high resolution and is quite rapid. Since only small amounts of material are exposed on an open surface, TLC is limited to relatively non-volatile substances. The techniques and applications of TLC are discussed in the texts of Bobbitt (1963), Labler and Schwarz (1965), Randerath (1966), Kirchner (1967), Bobbit *et al.*, (1968a) and Stahl (1969).

iii. Paper chromatography. In paper chromatography, filter paper is the solid support. This method is considerably slower than TLC and gives poorer resolution in most cases.

iv. Gas chromatography. In gas chromatography, a sample (a gas, a pure volatile liquid, a solution of a volatile liquid, or a solution of a volatile solid) is injected into a heated zone (injection port) with a hypodermic syringe. The sample (usually 1 to 5 μl for analytical purposes) is vaporized in the heated injection port and swept by a flow of inert gas (carrier gas, such as helium or nitrogen) onto a separating column. The column is generally a 0.3 to 6 m length of 0.3 or 0.6 mm tubing (aluminium, stainless steel, or glass) packed with a non-volatile liquid coated on an inert solid support or with an adsorbent solid. Normally, 3 to 15% by weight of liquid a phase is coated on an inert solid support. The various components of the sample have different affinities toward the column packing and separate into discrete moving bands as they are swept through the column by the carrier gas. As each component emerges from the column, it enters a detector which, *via* a Wheatstone bridge network, generates an electrical signal proportional to the concentration of the component in the carrier gas. This signal feeds a potentiometric recorder which produces the chromatogram. General textbooks on GC include those of Keulemans (1959), Pecsok (1959), Bayer (1961), Burchfield and Storrs (1962), Knox (1962), Purnell (1962), Littlewood (1962), Dal Nogare and Juvet (1962), Ettre and Zlatkis (1967) Schupp (1968a), Ettre and McFadden (1969), McNair and Bonelli (1969a), and Zlatkis and Pretorius (1971).

Gas chromatography (also called vapor phases chromatography (VPC)) is one of the most useful methods for separating volatile organic compounds because of:

(1) Speed. Analysis times of minutes duration are common. Preparative scale separations, or resolution of complex samples may require hours.

(2) Resolution. GC can often separate mixtures that are impossible to resolve by other techniques.

(3) Sensitivity. GC is one of the most sensitive analytical tools available to the chemist. The sensitivity depends upon the detector used. Simple thermal conductivity cells can detect down to 10^{-7} g, flame ionization detector can easily measure 10^{-9} g and electron capture and phosphorus detectors measure 10^{-12} g of sample.

(4) Precision of analysis. Qualitative and quantitative precision of ±1% is obtainable.

(5) Simplicity. Results are relatively easy to interpret. The equipment is reliable and simple to operate.

GC has been employed in virtually all of the work reported on semiochemicals. The major uses of GC in this work are: to fractionate a mixture and collect the various components, to determine the purity or homogeneity of a sample, to detect the presence of a given compound in a mixture, to determine the relative amounts of the various compounds in a mixture, and to compare synthetic with natural material.

Preparative GC, the fractionation and collection of materials, is one of the most convenient techniques for purification of semiochemicals. Formation of aerosols (suspensions of fine particles of compound in the carrier gas) is the most common difficulty encountered in collecting samples from a gas chromatograph. The usual method of sample collection is simple condensation inside a glass tube inserted into the GC outlet. Brownlee and Silverstein (1968) have described a device using thin-wall glass capillary tubes and a linear thermal gradient to minimize aerosol formation. The device (Fig. 1) consists of two parallel 250 mm long aluminium tubes (A), one 1.7 mm i.d., 3.7 mm o.d. to accept the thin-wall glass capillary (30 cm by 1.6 mm o.d.), and the other, 3.4 mm i.d., 6.34 mm o.d. to accept a 30 cm length of 3 mm o.d. glass tubing. The aluminium tubes are insulated from the ambient air by a 35 mm by 160 mm porcelain or ceramic tube (B). At one end, the aluminium tubes pass through a 38 mm cube aluminium block (C) containing a thermocouple probe and a 100 watt cartridge heater. At the other end, the tubes pass through a 38 mm by 38 mm by 50 mm aluminium block (D) that has a 178 mm by 19 mm aluminium rod (E) extending down into a cooling bath (F). Both blocks are mounted on, but insulated (G) from, a length of aluminium angle (H), which is supported rigidly in a horizontal position by a bracket (I) fastened with two screws to the side of the gas chromatograph and by another bracket (J). Provision is made for vertical and lateral alignment of the aluminium tubes with the exit port of the gas chromatograph. The exit port is fitted with a Swagelok fitting containing a Teflon (DuPont) ferrule (K) to ensure a gas-tight joint with the inserted glass collector tube. A short length of Teflon (DuPont) rod (L) with a 1.6 mm or 3 mm hole, mounted on the distal ends of the aluminium tubing, prevents frost accumulation, which would otherwise contaminate the collector tubes on introduction. This device very efficiently (> 90%) permits recovery of microgram samples in the glass capillaries, which can be sealed and stored until needed for further investigation.

Other types of samples collecting devices are discussed by Teranishi *et al.* (1967), and Schupp (1968c).

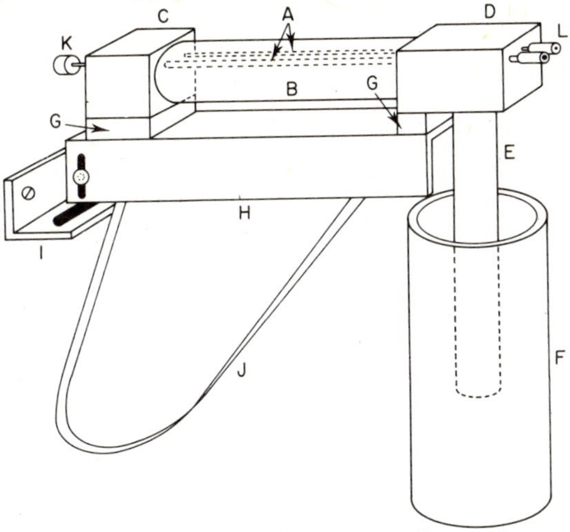

FIG. 1. Brownlee–Silverstein thermal gradient collector for gas chromatograph effluents.

The elapsed time from injection of the sample to recording of the peak maximum of a component band is called the *retention time*. This time is characteristic for a given compound and depends upon the column dimensions, packing, and temperature, and flow rate of the carrier gas through the column. These parameters are easily controlled and must always be reported along with the retention times. Although two compounds having different adjusted retention times (time from the air or solvent peak to the centre of the peak in question) can unambiguously be said to be different, the fact that two samples have identical retention times on a given column does not prove that they are identical. The terpene alcohols **23** and **24**, encountered in *Ips* spp. bark beetles, were observed to have identical retention times on the first three columns used and a difference of only 3–5% on two others (Young *et al.*, 1974).

23 **24**

To prove identity, the compounds should be compared on as many different types of columns as is convenient, and pure samples of the compounds should be collected for mass and infrared spectrometry.

If two compounds have the same retention times and also have identical mass and infrared spectra, then the probability that the samples are identical is very high. Similarly, if a given substance shows only one peak on several different types of columns, particularly of high resolving power, it is likely pure.

The area of a peak is proportional to the amount and type of material generating the peak. In every determination it is necessary to calibrate the system by relating the amount of material to the area of the peak produced. By comparing calibrated peak areas, one can determine the relative amounts of the various components in a mixture. Peak areas can be determined manually, or by mechanical or electronic integration. If fairly small samples are used and the peaks are not distorted, the area of a peak is proportional to the product of the peak height and the uncorrected retention time.

To avoid overloading columns, preliminary purification (distillation, column chromatography, solution chilling, etc.) is generally advisable. By reversing the flow of carrier gas through the column (back-flushing) after the desired components have been separated, high boiling compounds, which would otherwise be retained for a long time, can be removed from the column.

It is likely that after a GC fractionation of a complex mixture, the active sample (monitored by bioassay) will still contain impurities having similar retention times. In such a case it will be necessary to fractionate further on a second, third, or even fourth column. When fractionations must be carried out on a series of different liquid phases, one usually starts with a non-polar column. such as SE-30, and then progresses to more polar columns until the pure semiochemical is isolated.

The selection of a liquid phase is based mainly on experience and/or trial and error. In general, one chooses a liquid phase having a polarity similar to the polarity of the sample components. Silicone gum rubber SE-30 is the liquid phase of greatest utility and has been used more often in semiochemical studies than any other. It separates compounds on the basis of differences in boiling points. Carbowax 20M, Apiezon L, diethyleneglycol succinate (DEGS), fluorinated silicone (QF-1), and "free fatty acid phase" (FFAP) are useful liquid phases covering a range of polarities. For recent discussions on the choice of liquid phases, and solid supports, see Horvath (1967), Schupp (1968d), and McNair and Bonelli (1969b).

As a result of their high polarity, low volatility, or thermal instability, many samples cannot be injected directly into a GC and must first be converted to a suitable volatile derivative. This pretreatment has been discussed by Cavagnol and Betker (1967) and Schupp (1968e). Carboxylic acids have been converted to their corresponding methyl esters by treatment with diazomethane, fractionated by GC and then either reconverted to the active form by alkaline hydrolysis (Silverstein *et al.*, 1967a) or compared with known methyl esters (Meinwald, Y. C., and Eisner, T., 1964; Eisner, T. and Meinwald, Y. C., 1965; Cavill and Williams, 1967). Alcohols can be converted to their corresponding trimethyl silyl ethers (Meinwald, J. *et al.*, 1968b), or trichloroacetyl esters (Kuwahara and Casida, 1973).

The types of GC detectors most often used are based on thermal conductivity and flame ionization. The former measures the thermal conductivity of gases and uses helium as the carrier gas. When a carrier gas contains sample vapour, the thermal conductivity of the mixture is lower than that of pure helium. Thermal conductivity detectors are simple, inexpensive, rugged, respond to all compounds and are non-destructive of the sample. Flame ionization detectors (FID) measure by electrical conductivity, the amount of ionization produced when an organic compound is eluted into a flame produced by burning hydrogen in air or oxygen. A flame ionization detector is much more sensitive than a thermal conductivity detector, and can use nitrogen as the carrier gas rather than the more expensive helium. To overcome the destructive loss of sample in a FID, the carrier gas flowing from the column is passed through an effluent splitter and only 1 to 10% of the total gas stream is passed through the flame. Brownlee and Silverstein (1968) describe such a capillary variable-ratio effluent splitter. For recent description of the various types of GC detectors, see Gudzinowicz (1967), Schupp (1968b), Malone and McFadden (1969) and McNair and Bonelli (1969c).

Complete and up-to-date listings of equipment and supplies can be found in the annual "Laboratory Guide" published by the American Chemical Society through the journal *Analytical Chemistry* and the annual "Guide to Scientific Instruments" published by the American Association for the Advancement of Science through the journal *Science*.

IV. Structure determination

The remarkable developments over the past few decades in chromatographic techniques for isolating compounds have been accompanied by an even more impressive evolution of instrumentation for elucidation of chemical structure. Because of the small amount of semiochemical generally available, structure determination rests heavily on spectrometric methods.

One usually determines the mass (MS), nuclear magnetic resonance

(NMR), infrared (IR) and ultraviolet (UV) spectra of a compound and from the complementary information thus obtained postulates a partial or total structure. An experienced practitioner can often identify molecules of a fair degree of complexity *with no other information*. With refined techniques and sophisticated instrumentation, one can obtain mass spectra on 1 to 10 μg samples, NMR spectra on 100 μg samples (20 μg with Fourier transform), IR spectra on 5 μg samples, and UV spectra on 1 μg samples. Identification on the milligram scale is routine. Of course, not all molecules yield so readily and chemical manipulation becomes necessary. But the information obtained from the four spectra permits intelligent selection of chemical treatment, and the spectrometric method can be applied to the resulting products. For texts covering both chemical and spectrometric aspects of organic structure determination see: Bentley (1963a, b), Pasto and Johnson (1969), Weiss (1970), Veibel (1971), and Bentley and Kirby (1972).

Proof of structure by synthesis is absolutely necessary to prevent misassignment, a problem not unknown in semiochemical studies. To confirm the structure, one unambiguously synthesizes a compound having the postulated structure and compares the spectrometric, chromatographic, and biological properties of the pure synthetic material with those of the isolated material. The final test is confirmation of biological activity of the synthetic material *in the field*. It should be noted that if the natural semiochemical exists as only one of the possible optically active isomers (due to one or more asymmetric carbon centres in the molecule) and the semiochemical is a mixture of the possible optically active isomers, then the biological activity of the synthetic material may be less than that of the natural material. See for example Riley *et al.* (1974).

It is possible that even after taking all the necessary precautions one may still encounter problems. For example, the semiochemical may be a minor component not separated from "pure" material. The sex pheromone of the gypsy moth, *Porthetria dispar*, was originally identified as **25** (Jacobson *et al.*, 1960) and synthetic **25** was reported to be "... equally attractive [as the natural attractant] to male gypsy moths when tested in the field at 10^{-7} μg/trap." (Jacobson, 1962). Other synthetic preparations (Eiter *et al.*, 1967; and Jacobson *et al.*, 1970d)

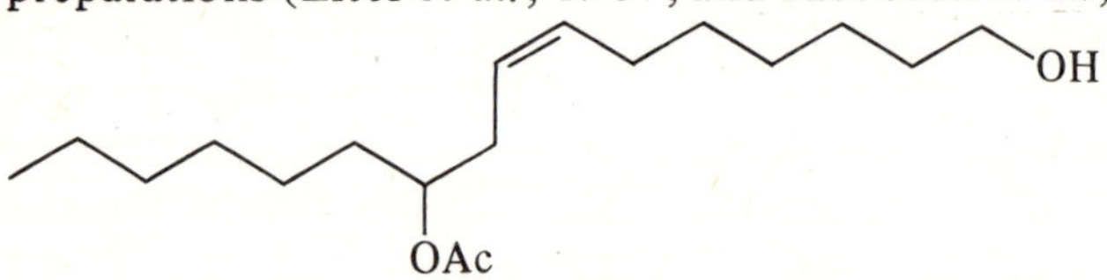

25

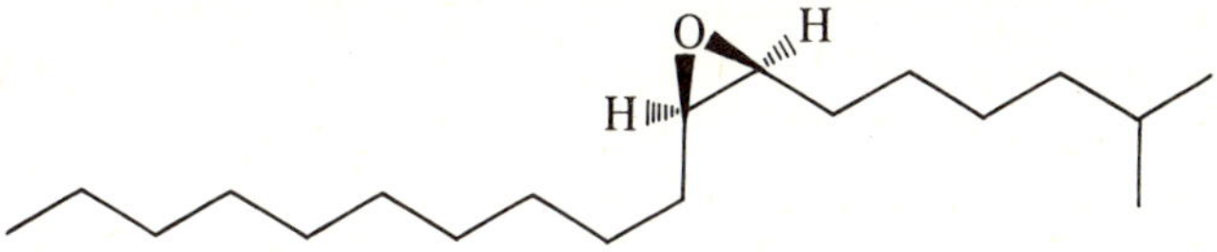

26

however were inactive. The activity of the first synthetic **25** must have been due to an unknown contaminant in the starting materials used. In their reinvestigation of the problem, Bierl *et al.* (1970) showed that although **25** was indeed present in gypsy moth extracts, the active material was **26**.

A. *Spectrometric methods*

For general texts on interpretation of spectra see: Nachod and Phillips (1962); Bentley (1963a); Baker, A. J., and Cairns (1965); Brand and Eglington (1965); Dyer (1965); Freeman (1965); Mathieson (1965); Williams, D. H., and Fleming (1966); Simon and Clerc (1967); Scheinmann (1970); Siggia (1972); and Silverstein *et al.*, (1974). Literature reviews on all aspects of spectrometric (and other) analytical methods are published by the American Chemical Society through the journal *Analytical Chemistry* in April of even numbered years.

a. Mass spectrometry

A mass spectrometer bombards a substance under investigation (in the vapour phase) with an electron beam, and quantitatively records the result as a sequence of positive ion fragments; this record is a mass spectrum. The most important single piece of information available from a mass spectrum is the molecular weight of the compound. This can usually be obtained because the intact molecule (minus one electron) will usually be recorded as one of the peaks in the mass spectrum. Most of the instruments currently in use can measure the mass of the charged molecule or its fragments to the nearest unit, and additional information is needed to develop a molecular formula. However, high resolution mass spectrometry is becoming more readily available (at a price!), and a molecular weight can thereby be determined to three or four decimal places. With this information alone, a molecular formula can be derived. For those of us who have sweated over the classical cryoscopic or ebullioscopic determinations to get within 3% of the molecular weight, spectrometry is indeed a "brave new world." In addition, of course, the vast amount of information from the fragmentation pattern is also available from a mass spectrum. For texts on interpretation see: Beynon (1960); Biemann (1962); McLafferty (1963, 1966); Budzikiewicz *et al.* (1964a, b, 1967); Burlingame (1970); Milne (1971); Hamming and Foster (1972); Hill (1972); and Johnstone (1972).

Gas chromatographs can be interfaced (coupled) with mass spectrometers so that MS analyses can be obtained directly on the various separated compounds as they elute from the GC. The GC-MS techniques are discussed in detail by Watson (1969). The GC-MS technique coupled with computer programs (e.g. Knock *et al.*, 1970; Hertz *et al.*, 1971; Groten, 1971; and Karasek, 1972) for comparing the data obtained with a file of reference spectra facilitates rapid identification of compounds. Of course, identification is limited to those compounds whose mass spectra are available for comparison. Computer programs are also being written to enable one to take unknown spectra and derive structural information (see Jurs *et al.*, 1969, 1970; Jurs, 1970; Crawford and Morrison, 1971; and references cited therein).

For listings of references mass spectra, see: Cornu and Massot (1966, 1967); McLafferty and Penzelik (1967); Stenhagen *et al.*, (1968, 1969); A.P.I. (MS) and A.S.T.M. (1963, 1969).

b. Nuclear magnetic resonance spectrometry

Nuclear magnetic resonance spectrometry became available during the 1950s. In this brief period, it has evolved into what is probably the most useful and versatile tool of all for structure determination. The energy used to perturb the molecule under investigation is very small. Approximate values for NMR spectrometry are 10^{-6} kcal compared with ~ 10 kcal for IR spectrometry, ~ 100 kacl for UV spectrometry, and ~ 1600 kcal for mass spectrometry. The only "damage" the applied energy does to the molecule is to "flip the nuclear spin" of the protons. (Other nuclei can also be examined, but proton magnetic resonance is most commonly used). The energy required to thus perturb a proton can be correlated with the environment of the proton. The change in energy (measured as a resonance frequency) with respect to an arbitrarily chosen position is termed *chemical shift*. For a given proton, the magnitude and direction of the chemical shift is related to the proximity of functional groups. An NMR spectrum appears as a series of peaks plotted against frequency. The ratio of the areas enclosed by the peaks (measured by electronic integration) is a ratio of the number of protons in each group giving rise to the peaks. The spins of the protons in one group may couple (interact) with the spins of neighbouring protons giving rise to split peaks. The magnitude and multiplicity of these peaks is related to the number and relative position of the neighbouring protons. By examining the chemical shifts and splitting patterns of the peaks one can determine much about the molecular structure.

In the NMR spectra of complex compounds, some peaks may overlap making interpretation difficult, if not impossible. This problem can sometimes be alleviated by use of the recently discovered (Sanders, J. K. M., and Williams, 1971; Sievers and Rondeau, 1971) paramagnetic shift reagents, such as tris(1,1,1,2,2,3,3-heptafluoro-7,7-dimethyl-4, 6-octanedione)europium III ($Eu(fod)_3$). These reagents complex with atoms having unshared electrons (such as oxygen and nitrogen) and greatly affect the chemical shifts of nearby protons. (For a listing of articles on applications of paramagnetic lanthanide shift reagents, see Horrocks *et al.*, 1971, and for reviews, see Begue, 1972; Von Ammon, and Fischer, 1972).

Unfortunately NMR requires a relatively large sample. With a 100 MHz spectrometer and microcell, a good spectrum with integration can be obtained on a 1 mg sample (~150 mol wt). By using a computer of average transients (CAT), the sample can be reduced to 100 μg, but with an increase in the spectrometer time required to obtain a spectrum and a loss in resolution. The CAT stores the sum of repetitive NMR scans within its memory and improves the signal noise to ratio (sensitivity) as the square root of the number of scans. There is a practical limit to NMR signal enhancement of several hundred scans.

The recent development of Fourier transform NMR spectroscopy which uses interferometry to extract the frequency *versus* absorbance information from the sample, promises to reduce sample size requirements to about 20 μg. Its advantage for small sample NMR is higher speed per scan (0.5 sec instead of 500 sec), and higher resolution than CAT. This higher speed allows a greater number of scans to be recorded in a given time and hence a greater increase in the sensitivity. Fourier transform spectroscopy has the disadvantage that the output of the spectrometer is not directly interpretable. An inverse Fourier transform must be performed by a computer before the conventional NMR spectrum is obtained. For introductions to Fourier transform NMR see Hill and Freeman (1970); Farrar and Becker (1971); Bell (1972).

With the advent of more sensitive NMR spectrometers has come the capability for routinely determining the carbon-13 NMR spectra of organic molecules. This new capability promises to add a new dimension to structure elucidation, namely a look into the carbon skeleton of a molecule. For texts on C-13 NMR see: Levy and Nelson (1972); Stothers (1972).

For texts on interpretation of NMR spectra, see: Pople *et al.* (1959);

Roberts, J. D. (1959, 1962); Wiberg and Nist (1962); Bhacca and Williams (1964); Bible (1965) Emsley *et al.* (1965); Mathieson (1967); Becker (1969); Bovey (1969); Jackman and Sternhell (1969); and Abraham (1971). Hall (1969) discusses microtechnique in conjunction with GC.

Compilations of NMR spectra can be found in Varian Associates (1962, 1963); Bhacca *et al.* (1962, 1963); Howell *et al.* (1964, 1966); Hershenson (1965); Bovey (1967); Brügel (1967); A.P.I. (NMR); Simons and Zanger (1972); Humble Oil; and Sadtler.

c. Infrared spectrometry

Infrared radiation in the range of 10 000 to 100 cm^{-1} (1 to 100 μm) is absorbed by an organic molecule and converted into energy of molecular vibration. Many of the absorption bands in the region 4000 to 667 cm^{-1} (2.5 to 15.0 μm), which is of most interest to organic chemists, are characteristic of specific types of bonds and are useful for determining the presence of absence of functional groups.

Particularly useful for semiochemical studies is an ultramicro IR cell and micropositioner for obtaining normal contrast solution spectra on 5 μg of sample, described by Price *et al.* (1967). Techniques for combining GC and IR spectrometry are reviewed by Freeman (1969). See also the recent article by Brown, R. A., *et al.* (1971). Computer programs have been developed for retrieval of information from IR data files (Anderson and Covert, 1967; Erley, 1968), and for determination of structural features from unknown spectra (Jurs *et al.*, 1969, and references cited therein).

For texts discussing IR spectrometry see: Bellamy (1958); Bauman (1962); Brugel (1962); Nakanishi (1962); Potts (1963); Rao (1963); Colthup *et al.* (1964); Phillips (1964); Syzmanski (1964a); Miller (1965); Cross, A. D., and Jones, R. A. (1969); Stine (1969); Avram and Mateescu (1971); Parker (1971); Avram and Mateescu (1972); and Conley (1972).

Compilations of IR spectra can be found in Dobriner *et al.* (1953); Brown, *et al.*, C. R. (1954); Roberts, G. *et al.* (1958); Hershenson (1959); Thomas (1960); Hershenson (1964); Szymanski (1964b); Aldrich; A.P.I. (IR); A.S.T.M. (IR); M.C.A. (IR); and Sadtler.

d. Ultraviolet spectrometry

Molecular absorption in the ultraviolet and visible region of the spectrum is dependent upon the electronic structure of the molecule. UV spectra indicate the presence or absence of specific kinds of conjugated systems and are routinely measured on as little as 1 μg of sample.

For texts discussing ultraviolet spectrometry see: Gillam and Stern (1957); Lang (1961); Rao (1961); Bauman (1962); Jaffe and Orchin (1962); Murrell (1963); Scott (1964); White (1965); Herzberg (1966) Stern and Timmons (1971).

Compilations of UV spectra can be found in Hershenson (1956); Friedel and Orchin (1958); Kamlet (1960); Ungnade (1960); Hershenson (1961); Scott (1964); Neudert and Ropke (1965); Hershenson (1966); Wheeler and Kaplan (1966); Freedman and Craig; Dacons and Rice; Phillips *et al.* (1969); A.P.I. (UV); A.S.T.M. (UV); Lang; M.C.A. (UV); Sadtler; and S.D.C.

e. Other physical techniques

Related to IR spectrometry is Raman spectroscopy, which can give useful information about functional groups, especially symmetrical ones. Relatively large sample sizes (>40 mg) are required. This technique is not widely used in organic chemistry.

None of the spectral techniques outlined thus far provides any information concerning the absolute stereochemistry of assymmetric centres in a molecule. A molecule with asymmetric centres will cause a beam of polarized light to be rotated. The measurement of optical rotation as a function of the wave length of plane polarized light is termed optical rotatory dispersion (ORD). See Djerassi (1960); Crabbé (1965); Snatzke (1967); and Crabbé (1972) for texts on ORD. The measurement of the unequal absorption of left and right circularly polarized light is termed circular dichroism (CD). See Crabbé (1965); Velluz *et al.*, (1965); Snatzke (1967); and Crabbé (1972) for texts on CD. By using the information from these complementary techniques, it is possible to predict with a high degree of accuracy the absolute stereochemistry of an asymmetric centre or molecule. Nakanishi *et al.* (1971) have recently determined the absolute configuration of the C_{18} insect juvenile hormone (**27**) by a new CD method.

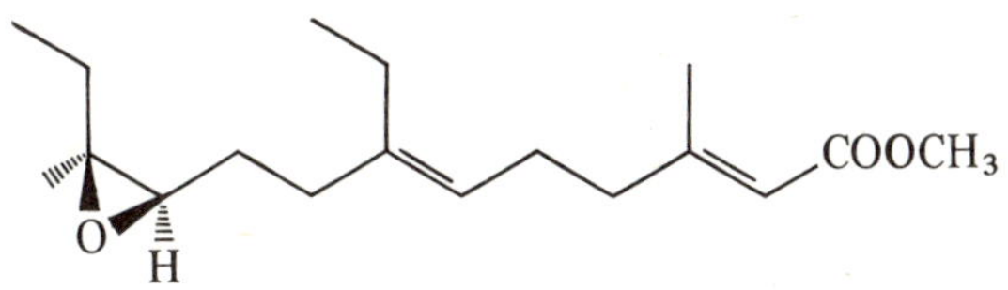

27

The structure and absolute stereochemistry of a molecule can be unambiguously determined by employing X-ray or electron diffraction techniques. Because the equipment is not generally available, analysis of the diffraction patterns is extremely complex, and relatively large samples are required, these techniques are usually applied only to extremely complex molecules. The structures of pederin (14) a defensive allomone (Furusaki *et al.*, 1968), and ecdysone (**28**), an insect molting hormone (Huber and Hoppe, 1965), were ultimately determined by X-ray analysis.

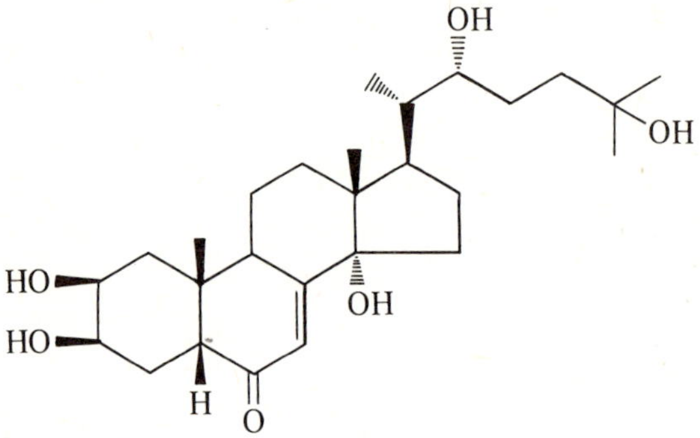

28 Ecdysone

B. Chromatographic comparison

a. Gas chromatographic retention indicies

Kovats (1958) proposed a system of GC retention indices (I) which can be used to compare compounds and even postulate structures of unknown compounds. The retention index indicates where a compound will appear on a chromatogram with respect to straight-chain alkanes, using adjusted retention times. By definition, the index for a normal alkane is 100 times the number of carbons in the alkane (e.g. $I_{ethane} = 200$, $I_{octane} = 800$), regardless of the column used or the conditions. To determine I for a particular compound under given conditions, the compound is analyzed with two *n*-alkanes, chosen so that one elutes before the compound and one after. The logs of the retention time of the alkanes are plotted against their I values (to give a straight line) and by using the log of the retention time of the compound, its I value can be read from the graph. I/100 is the number of carbon atoms in a hypothetical *n*-alkane whose retention time is identical to that of the given compound. McReynolds (1967) lists data for 400 compounds on 77 different liquid phases.

Functional groups will have characteristic index increment values (for a given liquid phase) which are additive with the carbon skeleton

I value. The difference in I values (ΔI) determined on a polar and non-polar liquid phase is also a characteristic of the structure of the compound and can be predicted by adding individual ΔI increment values for each functional group or structural feature. By comparing experimental and calculated ΔI values, structural information can be derived.

For discussions of retention indices, see Kovats (1966) and Ettre (1967). Rohrschneider (1966) has extended this concept to characterize liquid phases by the ΔI values of five selected compounds.

b. Thin-layer chromatography

Since chromatographic mobility under a wide variety of conditions can be readily determined, TLCs are often used to compare known and unknown compounds. Coloured derivatives, such as the yellow 2,4-dinitrophenylhydrazones of ketonic compounds, are conveniently used for identification. Harrison, V. J. and Weatherston (1967) have compared the R_f values for derivatives of many of the benzoquinones observed as insect defensive allomones. For references on the TLC behaviour of a wide variety of compounds, see the compendium by Stahl (1969).

C. Microchemistry

When the spectral data for an unknown compound are insufficient for deduction of the total structure, it becomes necessary to perform chemical manipulations. The compound is usually converted to a more readily identified molecule or degraded to give products that are known or more easily identified by spectroscopic and/or chromatographic techniques. Information about the nature of the functional groups or other structural features present, derived from the spectra, will determine the choice of reactions.

a. Reaction gas chromatography

Reaction gas chromatography has become one of the most convenient methods for performing chemical reactions at the microgram level. (For reviews of this technique see Beroza and Coad, 1966, 1967; Beroza and Inscoe, 1969; Beroza, 1970). In this technique, the unknown compound is injected into the gas chromatographic system and is transformed or retained, somewhere between the introduction point (usually a pre-column or injection port) and the detector. The products that elute can, of course, be collected for further analysis, such as by GC-MS or GC-IR.

Reactions involving hydrogen and a catalyst usually give hydrocarbons, which are often easy to identify by comparison of GC retention times or mass spectra. Three types are commonly employed:

(1) Hydrogenation (addition of hydrogen to multiple bonds, e.g. the unsaturated, straight-chain acetates or alcohols, often found as moth sex pheromones) afford the saturated analogs that can be more readily identified.

(2) Dehydrogenation (hydrogen removal) of cyclohexane-containing compounds to give aromatic ones is used to determine the carbon skeleton of some carbocyclic molecules.

(3) Hydrogenolysis (bond cleavage of functional groups with hydrogen addition at the point of cleavage).

Carbon-skeleton chromatography combines these processes to strip off all functional groups to give the parent hydrocarbon (with multiple bonds saturated), or its next lower homolog (when an oxygen or nitrogen function is on an end carbon atom). Elucidation of the structure of brevicomin (**29**) a sex pheromone component of the western pine beetle, *Dendroctonus brevicomis*, by Silverstein *et al.* (1968) was greatly facilitated when carbon-skeleton chromatography of **29** afforded nonane (**30**).

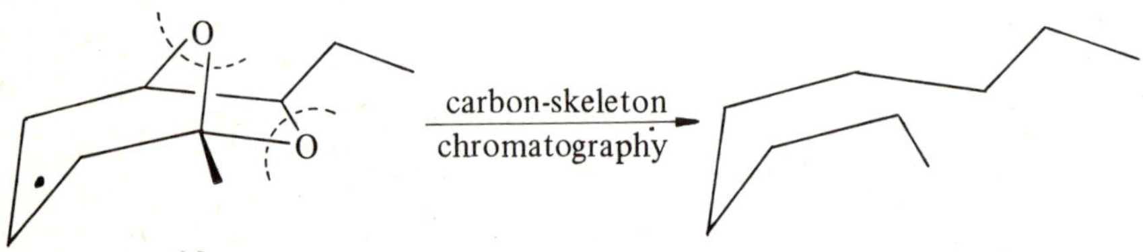

29 **30**

Compounds containing certain functional groups or structures can be "subtracted" (retained) by a chemical placed in the GC pathway. Care must be exercised in the use of this method since not enough is yet known about the nature of all groups removed by a particular chemical. The structure of the gypsy moth sex pheromone **26** was provided in part when it was subtracted by phosphoric acid, which is known to remove epoxides (Bierl *et al.*, 1970). Regnier and Huang (1970) have studied the identification of some oxygen-containing functional groups by this method.

b. "Wet" chemistry

Traditionally, crystalline derivatives are prepared and then compared with those of known compounds. The texts by Cheronis and Ma (1964) and Shriner *et al.* (1964), and the Chemical Rubber Company compilation (C.R.C., 1967; Grasselli, 1973) give physical constants for a wide variety of compounds and their derivatives. Most of the preparations described in these books are based on relatively large amounts of material (~ 100 mg), well beyond the amounts of semiochemicals encountered. Ashworth (1964, 1965) describes titrimetric methods of organic analysis, but once again on a relatively "large" scale.

Recent advances in analytical techniques have made it possible to perform functional group analysis and compound identification at the microgram level. Walsh and Merritt (1960), and Merritt (1969) describe methods for qualitative functional group analysis of gas chromatographic effluents. Basically the GC effluent is split into a number of equal streams each of which is allowed to bubble through a vial containing an appropriate classification reagent. Sensitivites at the 2-10 μg level are claimed (Merritt, 1969). A plot of the logarithm of the adjusted retention volume (related to retention time) *versus* the carbon number of an homologous series is linear. Thus it is possible to identify a compound by interpolation on the corresponding plot when the functionality and retention volume are known. When two classification reagents are positive for a given fraction it is possible to distinguish between a two component fraction and a bifunctional compound. Walsh and Merritt (1960) have examined only molecules containing less that 10 carbon atoms. The major advantage of this method over GC-MS, GC-IR, and other spectral methods is the low cost.

Feigl's (1966) valuable text lists many methods for detection of a wide variety of characteristic functional groups, structural types and individual compounds at the 0.01 to 100 μg level (mostly ~ 10 μg). Quantitative methods for elemental and functional group analysis at the 30 to 50 μg level are described by Belcher (1966).

The nature of functional groups in a molecule can also be determined by conversion or degradation to another structure. For example, if a biologically active molecule loses its activity after hydrolysis and regains it upon acetylation, it likely contains an acetate group. Similarly if the molecule loses its activity after acetylation and regains it upon subsequent hydrolysis, it likely contains a hydroxyl group. Loss of activity after hydrogenation is indicative of unsaturation.

Microozonolysis (followed by reduction, most conveniently with triphenylphosphine) has been widely used to determine the position of carbon-carbon double bonds by examination of the aldehydic and/or ketonic fragments produced. Moore and Brown, W. V. (1971), have studied the identification of various fragments commonly encountered in microozonolysis of semiochemicals. Microozonolysis apparatuses are described by Beroza and Bierl (1966, 1967, 1969). A part structure of the boll weevil, *Anthonomus grandis*, sex pheromone component **19** was determined when it yielded **31** on microozonolysis (Tumlinson *et al.*, 1969).

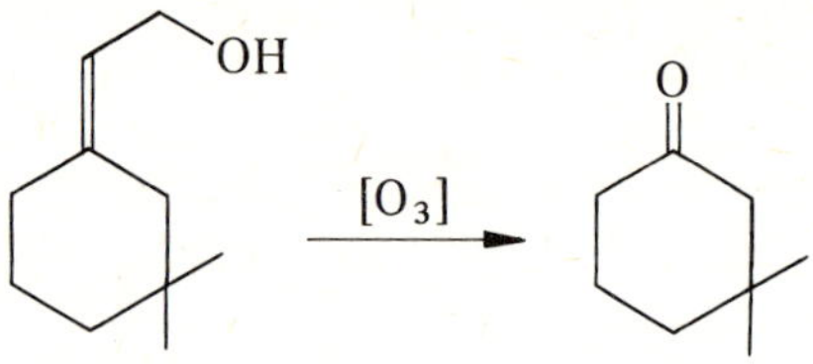

19 **31**

Rodin *et al.* (1969) determined the structure of **32** a sex pheromone component of the beetle *Trogoderma inclusum*, partly on the basis of ozonolysis, which gave **33**, and hydrogenolysis, which gave a mixture of **34** and **35**.

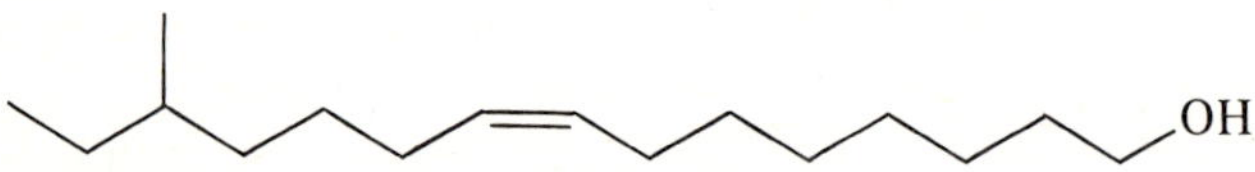

32

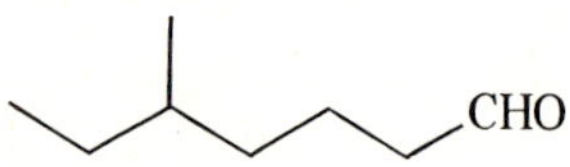

33

34

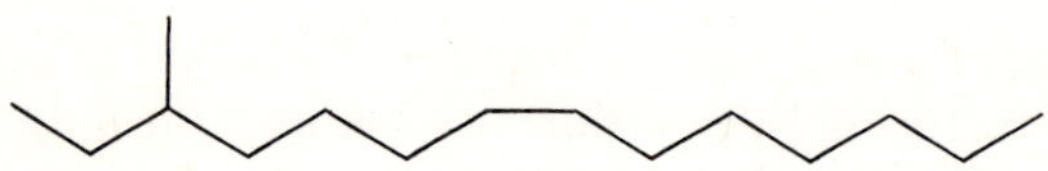

35

To illustrate range and accuracy of their sprays, insects have been prompted to discharge their defensive allemones into chemically treated paper that changes colour (Eisner, T. *et al.*, 1961, 1963b).

Although these colour changes will give information as to the types of compounds found in the spray, other more sensitive and less wasteful methods are to be recommended for identification purposes.

Once the functionality of the semiochemical is established, chemical treatment of the inactive extracts may result in the conversion of semiochemical precursors to the active compound. Identification of the precursor may be easier, and should greatly facilitate structure elucidation of the semiochemical. Bierl *et al.* (1970) applied this imaginative approach in the structure determination of the gypsy moth, *Porthetria dispar*, sex pheromone. When the pheromone was shown to be an epoxide, an olefin precursor was postulated and then confirmed after epoxidation of the monoolefin fraction of the extract yielded active material. Isolation and identification of **36** as the precursor led directly to **26** as the sex pheromone.

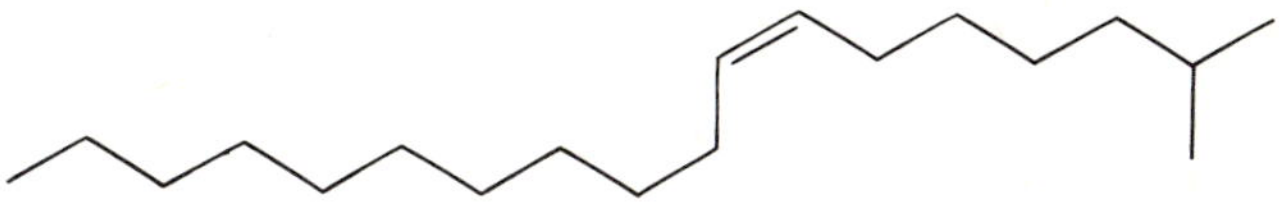

36

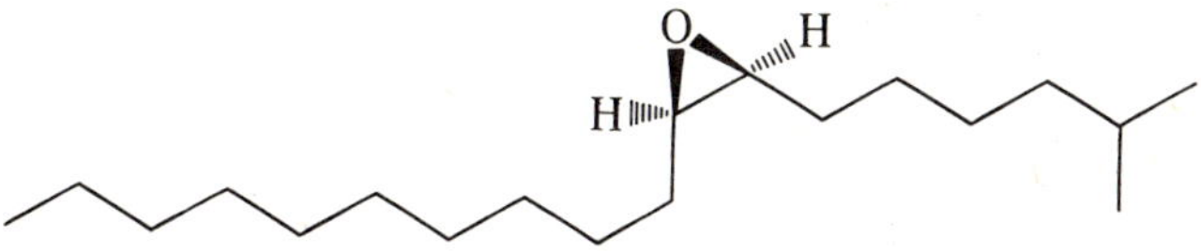

36

Table 1 lists various functional groups and their possible precursors.

TABLE I Functional groups of semiochemicals and possible precursors

Functional groups	Possible precursors
Alcohol	Aldehyde, ketone, carboxylic acid, ester
Aldehyde	Alcohol, carboxylic acid
Amine	Amide, salt
Carboxylic acid	Alcohol, ester, salt
Epoxide	Olefin
Ester	Alcohol, carboxylic acid, phenol
Ether	Alcohol, phenol
Ketone	Alcohol

c. Reaction thin-layer chromatography

Chemical reactions on semiochemicals of low volatility can be performed on TLC plates to give information about the nature of the functional groups. For example, a substance can be spotted on a plate and then over-spotted with a suitable chemical reagent. When the reaction has been completed, the plate is developed and the adsorbent at the R_f region where the semiochemical normally appears is eluted. If the biological activity of the eluate is lost, a reaction has occurred and the suspected functional group is presumed to be present. Bierl *et al.* (1971) used reaction TLC to help determine the presence and position of various epoxides, including that in the gypsy moth, *Porthetria dispar*, sex pheromone, **26**. Beroza (1970) has reviewed reaction TLC.

D. Empirical methods

From a knowledge of chemotaxonomy, one can sometimes make guesses about the possible nature of semiochemicals. For example, many moths release C_{12} to C_{16} unsaturated alcohols or acetates as sex pheromones; insect alarm pheromones are often simple C_6 to C_{12} ketones; and many beetles discharge benzoquinones as defensive allomones. Eisner (1970, pp. 205–207) discusses the chemotaxonomy of arthropod defensive allomones.

One empirical method is to screen structural variations of compounds found in nature. Although, for example, quite a few moth sex attractants have been discovered this way (Green *et al.*, 1969; Roelofs and Comeau, 1970, 1971a), a demonstration that an organism responds to a particular synthetic chemical does not by itself constitute rigorous proof that the compound is actually the pheromone of the organism. Presence of the semiochemical in the organism and its use by the organism must be demonstrated. Some investigators, moreover have abused the model compound approach by comparing the model compound with an inadequately defined, isolated "fraction."

The EAG and GC-EAG techniques of Roelofs and Comeau (1971) for identification of insect sex attractants also exploit chemotaxonomy. In the EAG method, the various C_{12} to C_{16} unsaturated alcohols or acetates are screened for the maximum EAG response. That compound showing the greatest response is then field tested, and, if attractive, is assumed to be the major sex attractant. In the GC-EAG method, crude insect extracts are fractionated by GC at one minute intervals and the collected fractions are assayed for maximum EAG response. The initial assumptions are that the chain is straight, the functional group is an alcohol or an acetate, and there is one double bond. If these assumptions are correct, the retention time of the fraction can be quickly matched with those of reference compounds. A match obtained, the reference compound is field tested. No match resulting, the investigator reviews his initial assumptions. Some risk may be involved in comparing the "retention time" of a one minute fraction with that of a pure compound as a criterion of identification. Roelofs *et al.* (1971) has extended this technique to doubly unsaturated moth pheromones on the assumption that the response from a doubly unsaturated compound will be the "summation" of the responses from two singly unsaturated molecules.

The "differential diagnosis" approach of Vité and Renwick (1971) as applied to bark beetle attractants involves GC analysis of hindgut volatiles of male and female insects at various stages of development. Compounds that appear to be of interest are isolated, field bioassayed, and identified. The major advantages of this screening method are that it avoids production of large amounts of source material and simplifies the isolation procedure. This method has disadvantages as it may miss highly active compounds present in small amounts, or multicomponent and synergisitc systems, some components of which may be contributed by the host. Furthermore, the field bioassay periods (10 to 30 minutes) necessitated by the small amounts of material, may be too short to give statistically valid results. The selection of compounds for identification may be subject to anthropocentric bias in the absence of systematic fractionation and monitoring.

V. Synthesis

After a structure has been proposed for the semiochemical, it is necessary to synthesize that molecule for confirmation of structure. Since each molecule presents a unique synthetic problem, only a few general principles are outlined here.

The first step is to determine whether or not the proposed structure is known. This is done by conducting a literature search, by molecular formula and name, usually through the abstract journals *Chemical Abstracts* and *Chemisches Zentralblatt* and Beilstein's *Handbuch der organischen Chemie* which give references to the original literature. If the structure is known, one may use the original synthetic procedure or an approach that takes advantage of more recent synthetic methods, or for recent preparations one may request a sample from the original investigator.

If the structure is unknown, then one is faced with planning and executing a synthesis. Since any given organic structure can be synthesized by a variety of different routes, the best synthesis will be the one that involves conversion of inexpensive, readily available starting material in the fewest steps and in the highest overall yield to the desired molecule. In his text, Ireland (1969) discusses the philosophy of organic synthesis and amply illustrates his arguments with examples.

In devising a synthesis, the chemist usually plans backwards step by step; he asks what simpler intermediate molecule(s) can be converted to the final product? The answer to this question will depend upon the chemist's knowledge of the synthetic reactions available. He repeats this process until he arrives at readily available starting materials. A synthesis usually involves two general types of reactions; skeleton construction from smaller units, and alteration or protection of functional groups. In assembling the molecule the chemist considers reactions that have been shown to work in analogous situations. Usually he has to modify experimental conditions to fit the particular situation. Some other important factors that should be considered in the planning include:

(1) Reactivity of functional groups. This may affect the choice of reactions. The most sensitive groups should be added last, or protected until the end.

(2) Stereochemistry. If there are asymmetric centres in the target molecule, it is important to choose reactions of known stereospecificity, so that a multiplicity of stereoisomers is avoided and the stereochemical outcome of the reaction is known. Starting materials are usually not asymmetric molecules; thus when an asymmetric centre is generated from symmetrical reagents a racemic (optically inactive) mixture is formed. If an optically active product is desired, one must start with optically active material or separate the racemic mixture into its two optically active forms by resolution at some stage of the synthesis. It has been pointed out that the *cis* or *trans* nature of olefins is important to the biological activity of, for example, moth sex pheromones. Therefore the geometric stereochemistry of the reactions must also be controlled.

(3) Side reactions. Side products often pose separation problems, which can sometimes be ameliorated by choosing symmetrical intermediates or by altering reaction conditions.

(4) Yields. If the estimated overall yields of two or more possible synthetic schemes are nearly the same, one chooses the scheme that has the lowest yield steps at an early stage.

(5) Number of steps. It is often better to choose a sequence with fewer steps, even at the expense of yields.

(6) Cost and availability of starting material. When only small amounts of material are desired, it may be worthwhile to save time by purchasing an intermediate, even if higher costs are involved.

(7) Flexibility of the proposed synthetic sequence. It is unusual for every step in a proposed synthesis to work. To avoid "dead-end" sequences, intermediates should be chosen so that detours can be used when necessary.

After the synthetic sequence to be investigated has been chosen, the chemist is then faced with working out the experimental details of the individual steps. Fortunately there are reference books that give worked-out synthetic procedures for specific or typical reactions. The most important of these is *Organic Synthesis*, published annually, with collective volumes appearing every 10 years. The series *Synthetic Methods of Organic Chemistry* (commonly referred to as Theilheimer) published annually, scans the organic chemical literature for useful general or specific synthetic reactions. Fieser and Fieser have compiled an especially valuable collection of over 1500 reagents in common use, complete with leading references. Other series of value to the synthetic chemist include: *Advances in Organic Chemistry: Methods and Results, Methoden der organischen Chemie* (commonly referred to as Houben-Weyl), *Newer Methods of Preparative Organic Chemistry*, and *Organic Reactions*. Texts giving leading references to synthetic methods include those of Wagner and Zook (1953), Sandler and Karo (1968, 1971), Buehler and Pearson (1970), Carruthers (1971), Harrison, I. T. and Harrison, S. (1971) and, House (1972). Other useful texts and series on organic chemistry, too numerous to mention here are covered in the series of articles by Hancock (1968).

Finally, it should be mentioned that computer techniques are now being applied to organic synthesis. Corey and coworkers (Corey and Wipke, 1969; Corey, 1971; Corey *et al.*, 1972a, b, c, d) have written computer programs to choose promising routes for the synthesis of organic molecules. The approach is to work backward from the desired structure using well established reaction types. See also Hendrickson (1971). In the end, however, it is the chemist at the bench who must put the molecule together.

Acknowledgements

The authors wish to thank Drs Martin Birch, John Borden, Harry Shorey, and John MacConnell for their helpful criticisms of this manuscript.

References

Abraham, R. J. (1971). Analysis of high resolution NMR. New York, American Elsevier Publishing Co.

Acree, F. Jr. (1953). The isolation of gyptol, the sex attractant of the female gypsy moth. *J. Econ. Entom.*, **46**, 313–315.

Adams, T. S. and Mulla, M. S. (1968). Ovarian development, pheromone production and mating in the eye gnat *Hippelates collusor* Town. *J. Insect Physiol.*, **14**, 627–635.

Adeesan, C., Tamhankar, A. J. and Rahalkar, G. W. (1969). Sex pheromone gland in the potato tuberworm moth, *Phthorimaea operculella*. *Ann. Entom. Soc. Amer.*, **62**, 670–671.

Adler, V. E., M. Beroza, Bierl, B. A. and Sarmiento, R. (1972). Electro-antennograms and field attraction of the gypsy moth sex attractant disparlure and related compounds. *J. Econ. Entom.*, **65**, 679–681.

Aldrich. (1971). The Aldrich library of infrared spectra. Milwaukee, Wisc., Aldrich Chemical Company.

AliNaizee, M. T. and Stafford, E. M. (1971). Evidence of a sex pheromone in the omnivorous leaf roller, *Platynota stultana* (Lepidoptera: Tortricidae): Laboratory and field testing of male attraction to virgin females. *Ann. Entom. Soc. Amer.*, **64**, 1330–1335.

Anderson, D. H. and Covert, G. L. (1967). Computer search system for retrieval of infrared data. *Anal. Chem.*, **39**, 1288–1293.

A.P.I. (IR). Catalog of infrared spectrograms, American Petroleum Institute, Project No. 44, Pittsburgh, Pa., Carnegie Institute of Technology.

A.P.I. (MS). Catalog of mass spectra data. American Petroleum Institute Research Project No. 44, Carnegie Institute of Technology, Pittsburgh, Pa., and The Manufacturing Chemists Association Research Project, Agricultural and Mechanical College of Texas, College Station, Texas.

A.P.I. (NMR). Nuclear magnetic resonance spectra data. American Petroleum Institute, Project No. 44, Chemical Thermodynamics Properties Center, Agricultural and Mechanical College of Texas, College Station, Texas.

A.P.I. (UV). Ultraviolet spectral data. American Petroleum Institute, Project No. 44, Chemical Thermodynamics Properties Center, Agricultural and Mechanical College of Texas, College Station, Texas.

Aplin, R. T. and Birch, M. C. (1968). Pheromones from the abdominal brushes of male Noctuid Lepidoptera. *Nature*, **217**, 1167–1168.

Ashworth, M. R. F. Titrimetric organic analysis, Part I: Direct methods (1964). Part II: Indirect methods (1965). New York, Wiley-Interscience.

A.S.T.M. (1963). Index of mass spectral data. A.S.T.M. Special Technical Publication No. 356, Philadelphia, American Society for Testing and Materials.

A.S.T.M. (1969). Index of mass spectral data (AMD II), Philadelphia, American Society for Testing and Materials.

A.S.T.M. (IR). ASTM-Wyandotte index, molecular formula list of compounds, names and references to published infrared spectra. ASTM Special Technical Publications 131 (1962) and 131-A (1963), Philadelphia, American Society for Testing and Materials. (43,500 spectra).

A.S.T.M. (UV). ASTM index to ultraviolet and visible spectra. ASTM Technical Publication No. 357 (1963), Philadelphia, American Society for Testing and Materials.

Atkins, M. D. (1961). A study of the flight of the Douglas-fir beetle, *Dentroctonus pseudotsugae* Hopk. (Coleoptera: Scolytidae). III. Flight capacity. *Can. Entom.*, **93**, 467–474.

August, C. J. (1971). The role of male and female pheromones in the mating behaviour of *Tenebrio molitor*. *J. Insect Physiol.*, **17**, 739–751.

Avram, M. and Mateescu, G. D. (1971). Infrared spectroscopy: Application in organic chemistry. New York, Wiley-Interscience.

Avram, M. and Mateescu, G. D. (1972). Infrared spectroscopy. New York, Wiley-Interscience.

Ayre, G. L. and Blum, M. S. (1971). Attraction and alarm of ants (*Componotus* spp. –– Hymenoptera: Formicidae) by pheromones. *Physiol. Zool.*, **44**, 77–83.

Baker, A. J. and Cairns, T. (1965). Spectroscopic techniques in organic chemistry. London, Heyden and Sons.

Baker, J. T. and Jones, P. A. (1969). Volatile constituents of the scent gland reservoir of the nymph of the Coreoid, *Pternistria bispina* Stal. *Austral. J. Chem.*, 22, 1793–1796.

Baldwin, W. F., Knight, A. G. and Lynn, K. R. (1971). A sex pheromone in the insect *Rhodnius prolixus* (Hempitera: Reduviidae). *Can. Entom.*, **103**, 18–22.

Bartell, R. J. and Shorey, H. H. (1969b). A quantitative bioassay for the sex pheromone of *Epiphyas postvittana* and factors limiting male responsiveness. *J. Insect Physiol.*, **15**, 33–40.

Bartell, R. J. and Shorey, H. H. (1969)a. Pheromone concentrations required to elicit successive steps in the mating sequence of males of the light-brown apple moth *Epithyas postvittana. Ann. Entom. Soc. Amer.*, **62**, 1206–1207.

Batiste, W. C. (1970). A timing sex-pheromone trap with special reference to codling moth collections. *J. Econ. Entom.*, **63**, 915–918.

Batiste, W. C. and Joos, J. (1972). Codling moth: A new pheromone trap. *J. Econ. Entom.*, **66**, 1741–1742.

Bauman, R. P. (1962). Absorption spectroscopy. New York, Wiley.

Bayer, E. (1961). Gas chromatography. New York, American Elsevier.

Beard, R. L. (1963). Insect toxins and venoms. *Ann. Rev. Entom.*, **8**, 1–18.

Becker, E. D. (1969). High resolution NMR: Theory and chemical applications. New York, Academic Press.

Bedard, W. D. and Browne, L. E. (1969). A delivery-trapping system for evaluating insect chemical attractants in nature. *J. Econ. Entom.*, **62**, 1201–1203.

Bedard, W. D., Tilden, P. E., Rodin, J. O. and Wood, D. L. (1971). In preparation Paper presented at Western Insect Work Conference, Glenwood Springs, Colorado, March 1971.

Bedard, W. D., Tilden, P. E., Wood, D. L., Silverstein, R. M., Brownlee, R. G. and Rodin, J. O. (1969). Western pine beetle: field response to its sex pheromone and a synergistic host terpene, mycrene. *Science*, **164**, 1284–1285.

Begue, J. P. (1972). NMR paramagnetic shifts induced by the chelates of rare earths. Use in organic chemistry. *Bull. Soc. Chim. Fr.*, 2073–2082.

Belcher, R. (1966). Submicro methods of organic analysis. New York, American Elsevier.

Bell. R. J. (1972). Introductory Fourier transform spectroscopy. New York, Academic Press.

Bellamy, L. J. (1958). The infrared spectra of complex molecules. 2nd edn. New York, Wiley.

Bennett, R. B. and Borden, J. H. (1971). Flight arrestment of tethered *Dendroctonus pseudotsugae* and *Trypodendon lineatum* (Coleoptera: Scolytidae) in response to olfactory stimuli. *Ann. Entom. Soc. Amer.* **64**, 1273–1286.

Bentley, K. W. Ed. (1963a). Elucidation of structure by physical and chemical methods, Part I. New York, Wiley-Interscience.

Bentley, K. W. Ed. (1963b). *ibid.*, Part II.

Bentley, K. W. and Kirby, G. W., eds. (1972). Elucidation of organic structures by physical and chemical methods. In three parts. New York, Wiley-Interscience.

Berger, R. S. (1966). Isolation, identification, and synthesis of the sex attractant of the cabbage looper, *Trichoplusia ni. Ann. Entom. Soc. Amer.*, **59**, 767–771.

Berger, R. S. (1968). Sex pheromone of the cotton leafworm. *J. Econ. Entom.*, **61**, 326–327.

Berger, R. S. and Canerday, T. D. (1968). Specificity of the cabbage looper sex attractant. *J. Econ. Entom.*, **61**, 452–454.

Berger, R. S., McGough, J. M. and Martin, D. F. (1965). Sex attractants in *Heliothis zea* and *H. virescens. J. Econ. Entom.*, **58**, 1023–1024.

Berger, R. S., McGough, J. M., Martin, D. F. and Ball, L. R. (1964). Some properties and field evaluation of the pink bollworm sex attractant. *J. Econ. Entom.*, **57**, 606–609.

Bergström, G. (1973). Studies on natural odoriferous compounds. VI. Use of a precolumn tube for the quantitative isolation of natural, volatile compounds for gas chromatography/mass spectrometry. *Chemica Scripta*, **4**, 135–138.

Bergström, G. and Lofquist, J. (1973). Chemical congruence of the complex odoriferous secretion from Dufour's gland in three species of ants of the genus *Formica. J. Insect Physibl.*, **19**, 877–907.

Berkson, J. (1953). A statistically precise and relatively simple method of estimating the bioassay with quantal response, based on the logistic factor. *J. Amer. Stat. Ass.*, **48**, 565–599.

Bernardi, R., Cardini, C., Ghiringhelli, D., Salva, A., Baggini, A. and Pavan, M. (1967). On the components of secretion of mandibular glands of the ant *Lasius* (*Dendrolasius*) *fuliginosus. Tetrahedron Letters*, 3893–3896.

Beroza, M. (1960). Insect attractants. *Soap Chem. Specialities*, **36**, 74, 76, 101, 103.

Beroza, M. (1967). Non-persistent inhibitor of the gypsy moth sex attractant in extracts of the insect. *J. Econ. Entom.*, **60**, 875–876.

Beroza, M. (1970a). Determination of the chemical structure of organic compounds at the microgram level by gas chromatography. *Accts. Chem. Res.*, **3**, 33–40.

Beroza, M. (1970b). *Chimia Supplement* (*Column Chromatography*), 92–103.

Beroza, M. (1971a). Identification of trace amounts of organic compounds. *J. Anim. Morphol. Physiol.*, **54**, 251–258.

Beroza, M. (1971b). Insect sex attractants. *Amer. Sci.*, **59**, 320–325.

Beroza, M. (1972a). Attractants and repellents for insect pest control. *Symp. Pest Control Strategies Future*, 226–253.

Beroza, M. (1972b). Insect sex attractant pheromones, a tool for reducing insecticide contamination in the environment. *Toxicol. Environ. Chem. Rev.*, **1**, 109–134.

Beroza, M. and Bierl, B. A. (1966). Apparatus for ozonolysis of microgram amounts of compound. *Anal. Chem.*, **38**, 1976–1977.

Beroza, M. and Bierl, B. A. (1967). Rapid determination of olefin position in organic compounds in microgram range by ozonolysis and gas chromatography. *Anal. Chem.*, **39**, 1131–1135.

Beroza, M. and Bierl, B. A. (1969). Ozone generator for microanalysis. *Mikrochim. Acta*, 720–723.

Beroza, M. and Coad, R. A. (1966). Reaction gas chromatography. *J. Gas Chromatog.*, **4**, 199–216.

Beroza, M. and Coad, R. A. (1967). Reaction gas chromatography. *In* Ettre, L. S. and Zlatkis, A. eds. The practice of gas chromatography. New York, Wiley-Interscience. pp. 461–510.

Beroza, M., Hood, C. S., Trefrey, D., Leonard, D. E., Knipling, E.F., Klassen, W. and Stevens, L. J. (1974). Large field trial with microencapsulated sex pheromone to prevent mating of the gypsy moth. *J. Econ. Entom.*, **67**, 659–664.

Beroza, M. and Inscoe, M. N. (1969). Precolumn reactions for structure determination. *In* Etter, L. S. and McFadden, W. H., eds. Ancillary techniques of gas chromatography. New York, Wiley-Interscience. pp. 89–144.

Beroza, M., Muschik, G. M. and Gentry, C. R. (1973). Small proportion of opposite geometric isomer increases potency of synthetic pheromone of oriental fruit moth. *Nature, New Biology*, **244**, 149–150.

Beynon, J. H. (1960). Mass spectroscopy and its applications to organic chemistry. New York, American Elsevier.

Bhacca, N. S., Johnson, L. F. and Shooley, J. N. NMR spectra catalog, Vol. 1, (1962), Vol. II, (1963). New York, National Press.

Bhacca, N. S. and Williams, D. H. (1964). Applications of NMR spectroscopy in organic chemistry. San Francisco, Holden-Day.

Bible, R. H. Jr. (1965). Interpretation NMR spectra. An empirical approach. New York, Plenum Press.

Biemann, K. (1962). Mass spectrometry. New York, McGraw-Hill.

Bierl, B. A., Beroza, M. and Collier, C. W. (1970). Potent sex attractant of the gypsy moth: Its isolation, identification, and synthesis. *Science*, **170**, 87–89.

Bierl, B. A., Beroza, M. and Aldrich, M. H. (1971). Determination of epoxide position and configuration at the microgram level, and recognition of epoxides by reaction thin-layer chromatography. *Anal. Chem.*, **43**, 636–641.

Birch, A. J., Chamberlain, K. B., Moore, B. P. and Powell, V. H. (1970). Termite attractants in *Santalum spicatum. Austral. J. Chem.*, **23**, 2337–2341.

Birch, M. C. (1970). Precourtship use of abdominal brushes by the nocturnal moth *Phlogophora meticulosa* (L.) (Lepidoptera: Noctuidae). *Anim. Behav.*, **18**, 310–316.

Birch, M. C. (1971). Intrinsic limitations in the uses of electroantennograms to bioassay male pheromones in Lepidoptera. *Nature*, **233**, 57–58.

Birch, M. C. (1974). Pheromones. New York, North-Holland (American Elsevier).

Block, R. C. (1960). Laboratory method for screening compounds as attractants to gypsy moth males. *J. Econ. Entom.*, **53**, 172–173.

Blum, M. S. (1961). The presence of 2-hexenal in the scent gland of the Pentatomid *Brochymena quadripustulata. Ann. Entom. Soc. Amer.*, **54**, 410–412.

Blum, M. S. (1964). Insect defensive secretions: Hex-2-enal-1 in *Pelmatosilpha coriacea* (Rehn) (Blattaria) and its repellent value under natural conditions. *Ann. Entom. Soc. Amer.*, **57**, 600–602.

Blum, M. S. (1966). Chemical releasers of social behavior. VIII. Crital in the mandibular gland secretion of *Lestrimelitta limao* (Hymenoptera: Apoidea: Melittidae). *Ann. Entom. Soc. Amer.*, **59**, 962–964.

Blum, M. S. (1969). Alarm pheromones. *Ann. Rev. Entom.*, **14**, 57–80.

Blum, M. S. (1970). The chemical basis of insect sociality. *In* Beroza, M., ed. Chemicals controlling insect behavior. New York, Academic Press, pp. 61–94.

Blum, M. S. (1971a). Dimensions of chemical sociality. *In* Tahori, A. S., ed. Chemical releasers in insects. New York, Gordon and Breach. pp. 147–162.

Blum, M. S. (1971b). Arthropod defensive secretions. *In* Tahori, A. S., ed. Chemical releasers in insects. New York, Gordon and Breach. pp. 163–175.

Blum, M. S. and Brand, J. M. (1972). Social insect pheromones: their chemistry and function. *Amer. Zool.*, **12**, 553–576.

Blum, M. S. and Crain, R. D. (1961). The occurrence of para-quinones in the abdominal secretions of *Eleodes hispilabris* Say (Coleoptera: Tenebrionidae). *Ann. Entom. Soc. Amer.*, **54**, 474–477.

Blum, M. S., Crain, R. D. and Chidester, J. B. (1961). *Trans*-2-hexenal in the scent gland of the hemipteran *Acanthocephala femorata* (F). *Nature*, **189**, 245–246.

Blum, M. S., Doolittle, R. E. and Beroza, M. (1971). Alarm pheromones: Utilization in evaluation of olfactory theories. *J. Insect Physiol.*, **17**, 2351–2361.

Blum, M. S., Moser, J. C. and Cordero, A. D. (1964). Chemical releasers of social behavior. II. Source and specificity of the odor trail substances in four Attine genera. (Hymenoptera: Formicidae). *Psyche*, **71**, 1–7.

Blum, M. S., Padovani, F. and Amante, E. (1968). Alkanones and terpenes in the mandibular glands of *Atta* species. *Comp. Biochem. Physiol.*, **26**, 291–299.

Blum, M. S., Padovani, F., Curley, A., and Hawk, R. E. (1969). Chemical releasers of social behavior. XIV. Benzaldehyde defensive secretion of a harvester ant. *Comp. Biochem. Physiol.* , **29**, 461–465.

Blum, M. S., Traynham, J. G., Chidester, J. B. and Boggus, J. D. (1960). *n*-Tridecane and *trans*-2-heptenal in the scent gland of the rice stink bug *Oebalus pugnax* (F.). *Science*, **132**, 1480–1481.

Blum, M. S. and Warter, S. L. (1966). Chemical releasers of social behavior. VII. The isolation of 2-heptanone from *Conomyrma pyramica* (Hymenoptera: Formicidae: Dolichoderinae) and its modus operandi as a releaser of alarm and digging behavior. *Ann. Entom. Soc. Amer.*, **59**, 774–779.

Blum, M. S., Warter, S. L., Monroe, R. S. and Chidester, J. B. (1963). Chemical releasers of social behavior. I. Methyl *n*-amyl ketone in *Iridomyrmex pruinosus* (Roger) (Formicidae: Dolichoderinae). *J. Insect Physiol.*, **9**, 881–885.

Blum, M. S., Warter, S. L. and Traynham, J. G. (1966). Chemical releasers of social behavior. VI. The relation of structure to activity of ketones as releasers of alarm for *Iridomyrmex pruinosus* (Roger). *J. Insect Physiol.*, **12**, 419–427.

Blum, M. S. and Woodring, J. P. (1962). Secretion of benzaldehyde and hydrogen cyanide by the millipede *Pachydesmus crassicutis* (Wood). *Science*, **138**, 512–513.

Bobbitt, J. M. (1963). Thin-layer chromatography. New York, Reinhold.

Bobbitt, J. M., Schwarting, A. E. and Gritter, R. J. (1968a). Introduction to chromatography. New York, Reinhold, pp. 21–83.

Bobbitt, J. M., Schwarting, A. E. and Gritter, R. J. (1968b). *ibid.* pp. 21–38, 84–105.

Boch, R. and Shearer, D. A. (1965). Alarm in the beehive. *Amer. Bee J.* **105**, 206–207

Boch, R. and Shearer, D. A. (1971). Chemical releasers of alarm behavior in the honey bee *Apis mellifera. J. Insect. Physiol.*, **17**, 2277–2285.

Boch, R., Shearer, D. A. and Petrasovits, A. (1970). Efficacies of two alarm substances of the honey bee. *J. Insect Physiol.*, **16**, 17–24.

Boch, R., Shearer, D. A. and Stone, B. C. (1962). Identification of isoamyl acetate as an active component in the sting pheromone of the honey bee. *Nature*, **195**, 1018–1020.

Boch, R., Shearer, D. A. and Young, J. C. (1975). Honey bee pheromones: Field tests of natural and artificial queen substance. *J. Chem. Ecol.*, **1**, in press.

Bodenstein, W. G. (1970). Distribution of female sex pheromone in the gut of *Periplaneta americana. Ann. Entom. Soc. Amer.*, **63**, 336–337.

Boeckh, J. (1962). Elektrophysiologische Untersuchungen an einzelnen geruchsrezeptoren auf der Antenne des Totengräbers (Nechrophorus). *Z. Vergl. Physiol.*, **46**, 212–248.

Boeckh, J. (1967). Reaktionsschwelle, Arbeitsbereich und Spezifität eines Geruchsrezeptors auf der Heuschreckenantenne. Z. *Vergl. Physiol.*, **55**, 378–406.

Boeckh, J., Kaissling, K. E. and Schneider, D. (1965). Insect olfactory receptors. *Cold Spring Harbor Symp. Quant. Biol.*, **30**, 263–280.

Boeckh, J., Priesner, E., Schneider, D. and Jacobson, M. (1963). Olfactory receptor response to the cockroach sexual attractant. *Science*, **141**, 716–717.

Boeckh, J., Sass, H. and Wharton, D. R. A. (1970). Antenna receptors: reactions to female attractant in *Periplaneta americana. Science*, **168**, 589.

Borden, J. H. (1967). Factors influencing the response of *Ips confusus* (Coleoptera: Scolytidae) to male attractant. *Can. Entom.*, **99**, 1164–1193.

Borden, J. H. and Bennett, R. B. (1969). A continuously recording flight mill for investigating the effect of volatile substances on the flight of tethered insects. *J. Econ. Entom.*, **62**, 782–785.

Borden, J. H., Brownlee, R. G. and Silverstein, R. M. (1968b). Sex pheromone of *Trypodendron lineatum* (Coleoptera: Scolytidea): Production, bio-assay, and partial isolation. *Can. Entom.*, **100**, 629–636.

Borden, J. H., Silverstein, R. M. and Brownlee, R. G. (1968a). Sex pheromone of *Dendroctonus pseudotsugae* (Coleoptera: Scolytidae): Production, bioassay, and partial isolation. *Can. Entom.*, **100**, 597–603.

Borden, J. H. and Stokkink, E. (1971). Secondary attraction in the Scolytidae: An annotated bibliography. Information report BC-X-57, Canadian Department of Fisheries and Forestry.

Bossert, W. H. and Wilson, E. O. (1963). An analysis of olfactory communication among animals. *J. Theor. Biol.*, **5**, 443–469.

Born, P. R. (1972). High pressure liquid chromatography: Biochemical and biomedical applications. New York, Academic Press.

Bosman, T. and Brand, J. M. (1971). Biological studies of the sex pheromone of *Kotochalia junodi* Heyl. (Lepidoptera: Psychidae) and its partial purification. *J. Entom. Soc. South Afr.*, **34**, 73–78.

Bovey, F. A. (1967). NMR Data tables for organic compounds, Vol. I. New York, Wiley-Interscience.

Bovey, F. A. (1969). Nuclear magnetic resonance spectroscopy. New York, Academic Press.

Bowers, W. S., Nault, L. R., Webb, R. E. and Dutky, S. R. (1972). Aphid alarm pheromone: Isolation, identification, synthesis. *Science*, **177**, 1121–1122.

Brady, U. E. and Smithwick, E. B. (1968). Production and release of sex attractant by the female Indian-meal moth, *Plodia interpunctella. Ann. Entom. Soc. Amer.*, **61**, 1260–1265.

Brady, U. E., Tumlinson III, J. H., Brownlee, R. G. and Silverstein, R. M. (1971). Sex stimulant and attractant in the Indian meal moth and in the almond moth. *Science*, **171**, 802–804.

Brand, J. C. D. and Eglington, G. (1965). Applications of spectroscopy to organic chemistry. London, Oldbourne.

Brower, L. P. and Brower, J. van Z. (1964). Birds, butterfiles, and plant poisons. A study in ecological chemistry. *Zoologica*, **49**, 137–159.

Brower, L. P., Brower, J. and Cranston, F. P. (1965). Courtship behavior of the queen butterfly, *Danaus giippus berenice* (Craner). *Zoologica*, **50**, 1–39.

Brown, C. R., Ayton, M. W., Goodwin, T. C. and Derby, T. J. (1954). Infrared – A bibliography. Washington, D.C., Library of congress, Technical information Division.

Brown, R. A., Kelliher, J. M., Heigl, J. J. and Warren, C. W. (1971). Rapid scan infrared spectrometer for operation with support coated open tubular or packed column gas chromatographs. *Anal. Chem.*, **43**, 353–358.

Brown, W. L. Jr., Eisner, T. and Whittaker, R. H. (1970). Allomones and kairomones, transspecific chemical messengers. *Bioscience*, **20**, 21–22.

Browne, L. E., Birch, M. C. and Wood, D. L. (1973). Novel trapping and delivery systems for airborne insect pheromones. *J. Insect Physiol.*, **20**, 183–193.

Brownlee, R. G. and Silverstein, R. M. (1968). A micro-preparative gas chromatograph and a modified carbon skeleton determinator. *Anal. Chem.*, **40**, 2077–2079.

Bruce, H. M. (1970). Pheromones. *Brit, Med. Bull.*, **26**, 10–13.

Brügel, W. (1962). An introduction to infrared spectroscopy. London, Methuen.

Brügel, W. (1967). Nuclear magnetic resonance spectra and chemical structures. New York, Academic Press.

Budzikiewicz, H. C., Djerassi, C. and Williams, D. H. (1964a). Interpretation of mass spectra of organic compounds. San Fransisco, Holden-Day.

Budzikiewicz, H. C., Djerassi, C. and Williams, D. H. (1964b). Structural elucidation of natural products by mass spectrometry. Vol. I. Alkaloids. Vol. II. Steroids, terpenoids, sugars, and miscellaneous classes. San Fransisco, Holden-Day.

Budzikiewicz, H. C., Djerassi, C. and Williams, D. H. (1967). Mass spectrometry of organic compounds. San Fransisco, Holden-Day.

Buehler, C. A. and Pearson, D. E. (1970). Survey of organic synthesis. New York, Wiley-Interscience.

Burchfield, H. P. and E. E. Storrs. (1962). Biochemical application of gas chromatography. New York, Academic Press.

Burkholder, W. E. (1970). Pheromone research with stored-product Coleoptera. *In* Wood, D. L., Silverstein, R. M. and Nakijima, M., eds. Control of insects behavior by natural products. New York, Adacemic Press, pp. 1–20.

Burkholder, W. E. and Dicke, R. J. (1966). Evidence of sex pheromones in females of several species of Dermestidae. *J. Econ. Entom.*, **59**, 540–543.

Burlingame, A. ed. (1970).Topics in organic mass spectrometry. New York, Wiley-Interscience.

Budenandt, A. (1959). Wirkstoffe des Insektenreiches. *Naturwissenschaften*, **46**, 461–471.

Butenandt, A. (1963). Bombycol, the sex attractive substance of the silkworm, *Bombyx mori. J. Endocrinal.*, **27**, ix–xvi.

Butenandt, A., Beckmann, R. and Hecker, E. (1961a). Über den Sexuallockstoffe des Seidenspinners, I: Der biologische Test und die Isolierung des rienen Sexualstoffes Bombykol. *Hoppe Seyler's Z. Physiol. Chem.*, **324**, 71–83.

Butenandt, A., Beckmann, R. and Stamm, D. (1961b). Über den Sexuallockstoffe des Seidenspinners, II: Konstitution und Konfiguration des Bombykols. *Hoppe Seyler's Z. Physiol. Chem.*, **324**, 84–87.

Butenandt, A., Beckmann, R., Stamm, D. and Hecker, E. (1959). Über den Sexuallockstoffe des Seidenspinners *Bombyx mori.* Reindarstellung und Konstitution. *Z. Naturforsch.*, **14b**, 283–284.

Butenandt, A., Hecker, E., Hopp, M. and Koch, W. (1962). Über den Sexuallockstoffe des Seidenspinners, IV: Die Synthese des Bombykols und der *cis-trans*-Isomeren Hexadecadien-(10,12)-ole-(1). *Justus Liebig's Ann. Chim.*, **658**, 39–64.

Butenandt, A., Linzen, B. and Lindauer, M. (1959). Über einen Duftstoff aus der Mandibeldrüse der Blattschneiderameise *Atta sexdens rubropilosa* Forel. *Arch. Anat. Microscop. Morphol.*, **48**, 13–20.

Butler, C. G. (1967). Insect Pheromones. *Biol. Rev. Cambridge Phil. Soc.*, **42**, 42–87.

Butler, C. G. (1969). Some pheromones controlling honeybee behaviour. *International Union for the Study of Social Insects, 6th Congress*, pp. 19–32.

Butler, C. G. (1970). Chemical communication in insects: Behavioral and ecological aspects. *In* Johnson, J. W. Jr., Moulton, D. G. and Turk, A. eds. Advances in chemoreception, Vol. I, Communication by chemical signals. New York, Appleton-Century-Crofts, pp. 35–78.

Butler, C. G. (1973). The queen and the "spirit of the hive." *Proc. Roy. Entom. Soc.*, **37**, 59–65.

Butler, C. G. and Fairey, E. M. (1964). Pheromones of the honeybee: Biological studies of the mandibular gland secretion of the queen. *J. Apicult. Res.*, **3**, 65–67.

Calam, D. H. and Scott, G. C. (1969). The scent gland complex of the adult cotton strainer bug. *J. Insect Physiol.*, **15**, 1695–1702.

Callow, R. K. (1970a). Chemistry of communication in the honey bee. *Central Assoc. of Bee-Keepers*, Apr., 1–7.

Callow, R. K. (1970b). Pheromones. *May Baker Lab. Bull.*, **9**, 43–45.

Callow, R. K., Chapman, R. J. and Paton, P. N. (1964). Pheromones of the honeybee: Chemical studies of the mandibular gland secretion of the queen. *J. Apicult. Res.*, **3**, 77–89.

Calmon, C. and Kressman, T. R. E., eds. (1957). Ion exchangers in organic and biochemistry. New York, Wiley-Intersciences.

Carr, R. V. and Levin, M. D. (1967). Pheromones in the honeybee. *Amer. Bee J.*, **107**, 410–411.

Carruthers, W. (1971). Some modern methods of Organic Synthesis. London. Cambridge University Press.

Carthy, J. D. (1951). The orientation of two allied species of British ant. I. Visual direction finding in *Acanthomyops* (*Lasius*) *niger*. *Behav.*, **3**, 275–303.

Casida, J. E., Coppel, H. C. and Watanabe, T. (1963). Purification and potency of the sex attractant from the introduced pine sawfly, *Diprion similis*. *J. Econ. Entom.*, **56**, 18–24.

Cassidy, H. G. (1951). Adsorption and chromatography. New York, Wiley-Interscience.

Cassidy, H. G. (1957). Fundamentals of chromatography. New York, Wiley-Interscience.

Cavagnol, J. C. and Betker, W. R. (1967). Sampling: Preparation of volatile derivatives and sample introduction. *In* Ettre, L. S. and Zlatkis, A. eds. The practice of gas chromatography. New York, Wiley-Interscience, pp. 71–127.

Cavill, G. W. K. (1960). The cyclopentanoid monoterpenes. *Rev. Pure Appl. Chem.*, **10**, 169–183.

Cavill, G. W. K. (1970). Chemistry of insect secretions. *J. Proc. Roy. Soc. N.S.W.*, **103**, 109–118.

Cavill, G. W. K., Clark, D. V. and Whitfield, F. B. (1968). Insect venoms, attractants and repellents. XI. Massiolactone from two species of Formicine ants and some observations on constitutents of the bark oil of *Cryptocarya massoia*. *Austral. J. Chem.*, **21**, 2819–2823.

Cavill, G. W. K. and Hinterberger, H. (1960a). The chemistry of ants. IV. Terpenoid constituents of some *Dolichoderus* and *Iridomyrmex* species. *Austral. J. Chem.*, **13**, 514–519.

Cavill, G. W. K. and Hinterberger, H. (1960b). Dolichoderine ant extractives. *Proc. 11th Internat. Cong. Entom.*, **3**, 53–59.

Cavill, G. W. K. and Robertson, P. L. (1965). Ant venoms, attractants and repellents. *Science*, **149**, 1337–1345.

Cavill, G. W. K. and Williams, P. J. (1967). Constituents of Dufour's gland in *Myrmecia gulosa. J. Insect Physiol.*, **13**, 1097–1103.

Chadha, M. S., Eisner, T. and Meinwald, J. (1961). Defense mechanisms of arthopods. III. Secretion of 2-hexenal by adults of the cockroach *Cutilia soror* (Brunner). *Ann. Entom. Soc. Amer.*, **54**, 642–643.

Chadha, M. S., Eisner, T., Monro, A. and Meinwald, J. (1962). Defense mechanisms of arthropods. VIII Citronellal and citral in the mandibular gland secretion of the ant *Acanthomyops claviger* (Roger). *J. Insect Physiol.*, **8**, 175–179.

Chang, V. C. S. and Curtis, G. A. (1972). Pheromone production by the New Guinea sugarcane weevil. *Envir. Entom.*, **1**, 476–481.

Chapman, J. A. (1966). The effect of attack by the ambrosia beetle, *Trypodendron lineatum* (Oliver) on log attractiveness. *Can. Entom.*, **98**, 50–59.

Chapman, J. A. (1967). Response behaviour of Scolytid beetles and odour meteorology. *Can. Entom.*, **99**, 1132–1137.

Chaudhury, M. F. B., Ball, H. J. and Jones, C. M. (1972). A sex pheromone of the female face fly, *Musca autumnalis*, and its role in sexual behavior. *Ann. Entom. Soc. Amer.*, **65**, 607–612.

Cheronis, N. D. and Ma, T. S. (1964). Organic functional group analysis by micro and semimicro methods. New York, Wiley-Interscience.

Chopard, L. (1962). Les zygènes papillous à acide cyanhydrique. *Nature* (*Paris*), 206–207.

Colthup, N., Daly, L. and Wiberley, S. (1964). Introduction to infrared and raman spectroscopy. New York, Academic Press.

Comfort, A. (1971). Likelihood of human pheromones. *Nature*, **230**, 432–433, 479.

Cone, W. W., McDonough, L. M., Maitlen, J. C. and Burdajewicz, S. (1971). Pheromone studies of the two spotted spider mite. 1. Evidence of a sex pheromone. *J. Econ. Entom.*, **64**, 355–358.

Conley, R. T. (1972). Infrared spectroscopy, 2nd ed. Boston, Allyn and Bacon.

Coppel, H. C., Casida, J. E. and Dauterman, W. C. (1960). Evidence for a potent sex attractant in the introduced pine sawfly *Diprion similis. Ann. Entom. Soc. Amer.*, **53**, 510–512.

Corey, E. J. (1971). General methods for the construction of complex molecules. *XXIII International Congress of Pure and Applied Chemistry*, **2**, 45–68.

Corey, E. J., Cramer III, R. D. and Howe, W. J. (1972a). Computer-assisted synthetic analysis for complex molecules. Methods and procedures for machine generation of synthetic intermediates. *J. Amer. Chem. Soc.*, **94**, 440–459.

Corey, E. J. and Petersson, G. A. (1972b). An algorithm for machine perception of synthetically significant rings in complex cyclic organic structures. *J. Amer. Chem. Soc.*, **94**, 460–465.

Corey, E. J. and Wipke, W. T. (1969). Computer-assisted design of complex organic syntheses. *Science*, **166**, 178–192.

Corey, E. J., Wipke, W. T., Cramer III, R. D. and Howe, W. J. (1972c). Computer-assisted synthetic analysis. Facile man-machine communication of chemical structure by interactive computer graphics. *J. Amer. Chem. Soc.*, **94**, 421–430.

Corey, E. J., Wipke, W. T., Cramer III, R. D. and Howe, W. J. (1972d). Techniques for perception by a computer of synthetically significant structural features in complex molecules. *J. Amer. Chem. Soc.*, **94**, 431–439.

Cornu, A. and Massot, R. (1966). Compilation of mass spectral data. London, Heyden.

Cornu, A. and Massot, R. (1967). Compilation of mass spectral data – First supplement. London, Heyden.

Crabbé, P. (1965). Optical rotatory dispersion and circular dichroism in organic chemistry. San Fransisco, Holden-Day.

Crabbé, P. (1972). ORD and CD in chemistry and biochemistry. An introduction. New York and London, Academic Press.

Crawford, L. R. and Morrison, J. D. (1971). Computer methods in analytical mass spectrometry. Development of programs for analysis of low resolution mass spectra. *Anal. Chem.*, **43**, 1790–1795.

C. R. C. (1967). Handbook of tables for organic compound identification. 3rd edn. Cleveland, Ohio, The Chemical Rubber Co.

Crewe, R. M., Brand, J. M., Fletcher, D. J. C. and Eggers, S. H. (1970). The mandibular gland chemistry of some South African species of *Crematogaster* (Hymenoptera: Formicidae). *J. Georgia Entom. Soc.*, **5**, 42–47.

Cross, A. D. and Jones, R. A. (1969). Introduction to practical infrared spectroscopy. 3rd edn. New York, Plenum.

Cross, W. H., Hardee, D. D., Nichols, F., Mitchell, H. C., Huddleston, P. M. and Tumlinson, J. H. (1969). Attraction of female boll weevils to traps baited with males or extracts of males. *J. Econ. Entom.*, **62**, 154–161.

Curtis, R. F., Ballantine, J. A., Keverne, J. A., Bonsall, R. W. and Mitchell, R. P. (1971). Identification of primate sexual pheromones and the properties of synthetic attractants. *Nature*, **232**, 396–398.

Cuthbert, F. P. Jr. and Reid, W. J. Jr. (1964). Studies of sex attractant of banded cucumber beetle. *J. Econ. Entom.*, **57**, 247–250.

Dacons, J. C. and Rice, R. G. Organic electronic spectra data. New York, Wiley-Interscience. Covers 1964–1965.

Dahm, K. H., Richter, I., Meyer, D. and Röller, H. (1971). The sex attractant of the Indian-meal moth *Plodia interpunctella* (Hübner). *Life Sci.*, **10**, 531–539.

Dal Nogare, S. and Juvet, R. S. Jr. (1962). Gas-liquid chromatography. Theory and practice. New York, Wiley-Interscience.

Dateo, G. P. and Roth, L. M. (1967). D-Gluconic acid: Isolation from the defensive secretion of the cockroach *Eurycotis decipiens*. *Science*, **155**, 88–89.

Daterman, G. E. (1970). An improved technique for mating European pine shoot moth, *Rhyacionia bouliana* (Lepidoptera: Olethreutidae) in the laboratory. *Can. Entom.*, **102**, 541–545.

Daterman, G. E. (1972). Laboratory bioassay for sex pheromone of the European pine shoot moth, *Rhyacionia bouliana*. *Ann. Entom. Soc. Amer.*, **65**, 119–123.

Daterman, G. E. and McComb, D. (1970). Female sex attractant for survey trapping European pine shoot moth. *J. Econ. Entom.*, **63**, 1406–1409.

Davis, F. M. and Henderson, C. A. (1967). Attractiveness of virgin female moths of the southwestern corn borer. *J. Econ. Entom.*, **60**, 279–280.

Debolt, J. W. (1970). Dispensers for baiting traps equipped with blacklight lamps with synthetic sex pheromone of the female cabbage looper. *J. Econ. Entom.*, **63**, 141–144.

Dethier, V. G. (1957). Chemoreception and the behavior of insects, *In* Glass, B., ed. *Survey of biological progress*, **3**, 149–183.

Djerassi, C. (1960). Optical rotatory dispersion. New York, McGraw-Hill.

Dobriner, K., Katzenellenbogen, E. R. and Jones, R. N. (1953). Infrared absorption spectra of steroids – An atlas. Vol. I. New York, Wiley-Interscience.

Dyer, J. R. (1965). Applications of absorption spectroscopy of organic compounds. Englewood Cliffs, New Jersey, Prentice-Hall.

Eisner, T. (1958). Spray mechanism of the cockroach *Diploptera punctata. Science*, **128**, 148–149.

Eisner, T. (1970). Chemical defense against predatation in arthropods. *In* Sondheimer, E. and Simenone, J. B. eds. Chemical ecology. New York, Academic Press. pp. 157–217.

Eisner, T. and Blumberg, D. (1959). Quinone secretion, a widespread defensive mechanism of arthropods. *Anat. Rec.*, **134**, 558–559.

Eisner, T. and Eisner, H. E. (1965). Mystery of a millipede. *Natural History*, **74**, 30–37.

Eisner, T., Eisner, H. E., Hurst, J. J., Kafatos, F. C. and Meinwald, J. (1963a). Cyanogenic glandular apparatus of a millipede. *Science*, **139**, 1218–1220.

Eisner, T., Hendry, L. B., Peakall, D. B., and Meinwald, J. (1971). 2,5-Dichlorophenol (from injested herbicide?) in defensive secretion of grasshopper. *Science*, **172**, 277–278.

Eisner, T., Kluge, A. F., Ikeda, M. I., Meinwald, Y. C. and Meinwald, J. (1971). Sesquiterpenes in the osmeterial secretion of a Papilionid butterfly, *Battus polydamas. J. Insect Physiol.*, **17**, 245–250.

Eisner, T. and Meinwald, J. (1966). Defensive secretions of arthropods. *Science*, **153**, 1341–1350.

Eisner, T. and Meinwald, Y. C. (1965). Defensive secretion of a caterpillar (Papillo). *Science*, **150**, 1733–1735.

Eisner, T., Meinwald, J., Monro, A. and Ghent, R. (1961). Defense mechanisms of arthropods. I. The composition and function of the spray of the whip scorpion *Mastigoproctus giganteus* (Lucas) (Arachnida: Pedipalpida). *J. Insect Physiol.*, **6**, 273–298.

Eisner, T., Meinwald, Y. C., Alsop, D. W. and Carrel, J. E. (1968). Defense mechanisms of arthropods. XXI. Formic acid and *n*-nonyl acetate in the defensive spray of two species of *Helluomorphoides. Ann. Entom. Soc. Amer.*, **61**, 610–613.

Eisner, T., Swithenbank, C. and Meinwald, J. (1963b). Defense mechanisms of arthropods. VII. Secretion of salicylaldehyde by a Carabid beetle. *Ann. Entom. Soc. Amer.*, **56**, 37–41.

Eiter, K. (1970). Insektensexuallockstoffe. *Fortschr. Chem. Org. Natur.*, **28**, 204–255.

Eiter, K., Truscheit, E. and Boness, M. (1967). Synthesen von D,L-10-Acetoxy-hexadecen-(7-*cis*)-ol-(1), 12-Acetoxy-octadecen-(9-*cis*)-ol-(1) ("Gyplure") und 1-Acetoxy-10-propyl-tridecadien-(5-*trans*,9). *Justus Liebig's Ann. Chim.*, **709**, 29-45.

Emsley, J. W., Feeney, J. and Sutcliffe, L. H. (1966). High resolution nuclear magnetic resonance spectroscopy. Oxford, Pergammon Press Ltd.

Engelhardt, M., Rapoport, H. and Sokoloff, A. (1965). Odorous secretion of normal and mutant *Tribolium confusam. Science*, **150**, 632–633.

Erley, D. S. (1968). Fast searching systems for the ASTM infrared data file. *Anal. Chem.*, **40**, 894–898.

Ettre, L. S. (1967). The interpretation of analytical results: Qualitative and quantitative analysis. *In* Ettre, L. S. and Zlatkis, A. eds. The practice of gas chromatography. New York, Wiley-Interscience. pp. 381–387.

Ettre, L. S. and McFadden, W. H. eds. (1969). Ancillary techniques of gas chromatography. New York, Wiley-Interscience.

Ettre, L. S. and Zlatkis, A. eds. (1967). The practice of gas chromatography. New York, Wiley-Interscience.

Farkas, S. R. and Shorey, H. H. (1972). Chemical trail-following by flying insects: A mechanism for orientation to a distant odor source. *Science*, **178**, 67–68.

Farkas, S. R. and Shorey, H. H. (1973). Odor-following and anemotoxis. *Science*, **180**, 1302.

Farrar, T. C. and Becker, E. D. (1971). Pulse and Fourier transform NMR: Introduction to theory and methods. New York, Academic Press.

Fatzinger, C. W. (1972). Bioassay, morphology, and histology of the female sex pheromone gland of *Dioryctria abietella* (Lepidoptera: Pyralidae (Phycitinae)). *Ann. Entom. Soc. Amer.*, **65**, 1208–1214.

Fatzinger, C. W. and Asher, W. C. (1971). Mating behavior and evidence for a sex pheromone of *Dioryctria abietella* (Lepidoptera: Pyralidae (Phycitinae)). *Ann. Entom. Soc. Amer.*, **64**, 612–620.

Feigl, F. (1966). Spot tests in organic analysis. 7th English ed. New York, American Elsevier.

Fieser, M. and Fieser, L. F. Reagents for organic synthesis. Vol. 1 (1967), Vol. 2 (1969), Vol. 3 (1971). New York, Wiley.

Finney, D. J. (1964). Probit analysis 2nd edn. London, Cambridge University Press.

Fletcher, B. S. (1969). The structure and function of the sex pheromone glands of the male Queensland fruit fly, *Dacus tryoni. J. Insect Physiol.*, **15**, 1309–1322.

Fletcher, L. W., O'Grady, J. J., Claborn, H. V. and Graham, O. H. (1966). A pheromone from male screw worm flies. *J. Econ. Entom.*, **59**, 142–143.

Francia, F. C. and Graham, K. (1967). Aspects of orientation behavior in the ambrosia beetle *Trypodendron lineatum* (Oliver). *Can. J. Zool.*, **45**, 985–1002.

Franke-Grossman, H. (1963). Some new aspects in forest entomology. *Ann. Rev. Entom.*, **8**, 415–438.

Freedman, L. D. and Craig, J. C. eds. Organic electronic spectral data. Vol. VI. New York, Wiley-Interscience. Covers 1962–1963.

Freeman, S. K. ed. (1965). Interpretive spectroscopy. New York, Reinhold.

Freeman, S. K. (1969). Gas chromatography and infrared and raman spectrometry. *In* Ettre, L. S. and McFadden, W. H. eds. Ancillary techniques of gas chromatography. New York, Wiley-Interscience. pp. 227–267.

Friedel, R. A. and Orchin, M. (1959). Ultraviolet spectra of aromatic compounds. Revised edn. New York, Wiley.

Fukami, H. (1972). Insect sex pheromones. *J. Synth. Org. Chem. Japan*, **30**, 801–812.

Furuski, A., Watanabe, T., Matsumoto, T. and Yanagiya, M. (1968). The crystal and molecular structure of pederin di-*p*-bromobenzoate. *Tetrahedron Letters*, 6301–6304.

Gabba, A. and Pavan, M. (1970). Researches on trail and alarm substances in ants. *In* Johnson, J. W. Jr., Moulton, D. G. and Turk, A. Eds. Advances in chemoreception, Vol. I, Communication by chemical signals. New York, Appleton-Century-Crofts, pp. 161–195.

Gara, R. I. (1967a). Studies on the attack behavior of the southern pine beetle. I. The spreading and collapse of outbreaks. *Contrib. Boyce Thompson Inst.*, 23, 349–354.

Gara, R. I. (1967b). A field olfactometer for studying the response of the southern pine beetle to attractants. *J. Econ. Entom.*, **60**, 1180–1181.

Gary, N. E. (1962). Chemical mating attractants in the queen honey bee. *Science*, **136**, 773–774.

Gary, N. E. (1970). Pheromones of the honey bee, *Apis mellifera* L. *In* Wood, D. L., Silverstein, R. M. and Nakajima, M. eds. Control of insect behavior by natural products. New York, Academic Press, pp. 29–53.

Gaston, L. K., Fukoto, T. R. and Shorey, H. H. (1966). Sex pheromones of Noctuid moths. IX. Isolation techniques and quantitative analysis for the pheromones with special reference to that of *Trichoplusia ni* (Lepidoptera: Noctuidae). *Ann. Entom. Soc. Amer.*, **59**, 1062–1066.

Gaston. L. K. and Shorey, H. H. (1964). Sex pheromones of Noctuid moths. IV. An apparatus for bioassaying the pheromones of six species. *Ann. Entom. Soc. Amer.*, **57**, 779–780.

Gentry, C. R., Lawson, F. R. and Hoffmann, J. D. (1964). A sex attractant in the tobacco budworm *Heliothis virescens* (F). *J. Econ. Entom.*, **57**, 819–821.

George, D. A. and McDonough, L. M. (1972). Multiple sex pheromones of the codling moth, *Laspeyresia pomonella* (L). *Nature*, **239**, 109.

George, J. A. (1965). Sex pheromone of the oriental fruit moth *Grapholitha molesta* (Busck) (Lepidoptera: Tortricidae). *Can. Entom.*, **97**, 1002–1007.

Gillam, A. E. and Stern, E. S. (1957). An introduction to electronic absorption spectroscopy in organic chemistry. 2nd edn. London, Edward Arnold.

Glass, E. H., Roelofs, W. L., Arn, H. and Comeau, A. (1970). Sex pheromone trapping red-banded leaf roller moths and development of a long-lasting polyethylene wick. *J. Econ. Entom.*, **63**, 370–373.

Graham, H. M., McGough, J. M. and Jacobson, M. (1966). Influence of solvents on effectiveness of sex lure for pink bollworm. *J. Econ. Entom.*, **59**, 761–762.

Granett, J. (1973). A disparlure-baited box trap for capturing large numbers of gypsy moths. *J. Econ. Entom.*, **66**, 359–362.

Grant, G. G. (1970). Evidence for a male sex pheromone in the Noctuid, *Trichoplusia ni. Nature*, 227, 1345–1346.

Grant, G. G. (1971). Electroantennogram responses to the scent brush secretions of several male moths. *Ann. Entom. Soc. Amer.*, **64**, 1428–1431.

Grant, G. G., Brady, U. E. and Brand, J. M. (1972). Male armyworm scent brush secretion: Identification and electroantennogram study of major components. *Ann. Entom. Soc. Amer.*, **65**, 1224–1227.

Grasselli, J. G. ed. (1973). Atlas of spectral data and physical constants for organic compounds. Cleveland, Ohio, CRC Press.

Gravitz, N. and Willson, C. (1968). A sex pheromone from the citrus mealybug. *J. Econ. Entom.*, **61**, 1458–1459.

Green, N., Beroza, M. and Hall, S. A. (1960). Recent developments in chemical attractants for insects. *Adv. Pest Control Res.*, **3**, 129–179.

Green, N., Jacobson, M. and Keller, J. C. (1969). Hexalure, an insect sex attractant discovered by empirical screening. *Experientia*, **25**, 682–683.

Groten, S. L., (1971). Computer techniques for identifying low resolution mass spectra. *Anal. Chem.*, **43**, 1362–1370.

Grubb, T. C. Jr. (1973). Odor-following and anemotaxis. *Science*, **180**, 1302.

Gudzinowicz, B. J. (1967). Detectors. *In* Ettre, L. S. and Zlatkis, A. eds. The practice of gas chromatography. New York, Wiley-Interscience, pp. 239–332.

Guerra, A. (1968). New techniques to bioassay the sex attractant of pink bollworms with olfactometers. *J. Econ. Entom.*, **61**, 1252–1254.

Guerra, A. (1969). Catches of male pink bollworms in several modifications of the Frick trap baited with sex attractant. *J. Econ. Entom.*, **62**, 1514–1515.

Gupta, A. P. (1961). A critical review of the studies on the so-called stink or repugnatorial glands of Heteroptera with further comments. *Can. Entom.*, **93**, 482–486.

Hadden, N., Baumann, F., MacDonald, F., Munk, M., Stevenson, R., Gere, D., Zamoroni, F. and Majors, R. (1971). Basic liquid chromatography. Walnut Creek, California, Varian Associates.

Hall, G. E. (1969). Gas chromatography and nuclear magnetic resonance spectroscopy. *In* Ettre, L. S. and McFadden, W. H. eds. Ancillary techniques of gas chromatography. New York, Wiley-Interscience, pp. 269–298.

Hamming, M. C. and Foster, N. G. (1972). Interpretation of mass spectra of organic compounds. New York, Academic Press.

Hammond, A. M. and Hensley, S. D. (1970). Bioassay for the sex attractant in the sugarcane borer. *Ann. Entom. Soc. Amer.*, 63, 64–66.

Hancock, J. E. H. (1968). An introduction to the literature of organic chemistry. (In three parts). *J. Chem. Educ.*, **45**, 193–199, 260–266, 336–339.

Happ, G. M. (1970). Maturation of the response of male *Tenebrio molitor* to the female sex pheromone. *Ann. Entom. Soc. Amer.*, **63**, 1782.

Happ, G. M., Schroeder, M. E. and Wang, J. C. H. (1970). Effects of male and female scent on reproductive maturation in young female *Tenebrio molitor. J. Insect Physiol.*, **16**, 1543–1548.

Happ, G. M. and Wheeler, J. (1969). Bioassay, preliminary purification, and effect of age, crowding, and mating on the release of sex pheromone by female *Tenebrio molitor. Ann. Entom. Soc. Amer.*, **62**, 846–851.

Hardee. D. E. (1972). A review of literature on the pheromone of the boll weevil, *Anthonomus grandis* Boheman (Coleoptera: Curculionidae). *U.S.D.A. Coop. Econ. Ins. Rpt.*, **22**, 200–207.

Hardee, D. D., McKibben, G. H., Gueldner, R. C., Mitchell, E. B., Tumlinson, J. H. and Cross, W. H. (1972). Boll weevils in nature respond to grandlure, a synthetic pheromone. *J. Econ. Entom.*, **65**, 97–100.

Hardee, D. D., Mitchell, E. B., and Huddleston, P. M. (1967). Procedure for bioassaying the sex attractant of the boll weevil. *J. Econ. Entom.*, **60**, 169–171.

Hardee, D. D., Wilson, N. M., Mitchell, E. B. and Huddleston, P. M. (1971). Factors affecting activity of grandlure, the pheromone of the boll weevil in laboratory bioassays. *J. Econ. Entom.*, **64**, 1454–1456.

Hardy, E. (1947). Scents from insects. Butterfly and moth perfumes and their extraction. *Perfum. Essent. Oil Rec.*, **38**, 403–405.

Harrison, I. T. and Harrison, S. (1971). Compendium of organic synthetic methods. New York, Wiley-Interscience.

Harrison, V. J. and Weatherston, J. (1967). Thin layer chromatography of simple naturally occurring benzoquinones. *J. Chromatog.*, **31**, 258–259.

Hatanaka, A. (1963). Chemie über den Sexuallockstoff der Insekten. *Botyu Kagaku*, **28**, 110–112.

Hendricks, D. E., Hollingsworth, J. P. and Hartstack, A. W. Jr. (1972). Catch of tobacco budworm moths influenced by color of sex-lure traps. *Envir. Entom.*, **1**, 48–51.

Hendrickson, J. B. (1971). A systematic characterization of structures and reactions for use in organic synthesis. *J. Amer. Chem. Soc.*, **93**, 6847–6854.

Henneberry, T. J. and Howland, A. F. (1966). Response of male cabbage loopers to blacklight with or without the presence of the female sex pheromone. *J. Econ. Entom.*, **59**, 623–626.

Henneberry, T. J., Howland, A. F. and Wolf, W. W. (1967). Combinations of blacklight and virgin females as attractants to cabbage looper moths. *J. Econ. Entom.*, **60**, 152–156.

Henneberry, T. J., Shorey, H. H. and Kishaba, A. N. (1966). Mating frequency and copulatory aberrations, response to female sex pheromone, and longevity of male cabbage loopers treated with TEPA. *J. Econ. Entom.*, **59**, 573–576.

Henzell, R. F. and Lowe, M. D. (1970). Sex attractant of the grass grub beetle. *Science*, **168**, 1005–1006.

Henzell, R. F., Lowe, M. D., Taylor, H. J. and Boston, E. (1969). Laboratory demonstration of a chemical sex attractant in *Costelytra zealandica. N.Z. J. Sci.*, **12**, 252–257.

Herbert, R. B. (1968). Arthropod hormones, pheromones, and defensive secretions. *J. Leeds Univ. Union Chem. Soc.*, **10**, 35–38.

Hershenson, H. M. (1956). Ultraviolet and visible spectra index for 1930–1954. New York, Academic Press.

Hershenson, H. M. (1959). Infrared absorption spectra index for 1945–1957. New York, Academic Press.

Hershenson, H. M. (1961). Ultraviolet and visible spectra index for 1955–1959. New York, Academic Press.

Hershenson, H. M. (1964). Infrared absorption spectra index for 1958–1962. New York, Academic Press.

Hershenson, H. M. (1965). Nuclear magnetic resonance and electron spin resonance spectra index for 1958–1963. New York, Academic Press.

Hershenson, H. M. (1966). Ultraviolet and visible spectra index for 1960–1963. New York, Academic Press.

Hertz, H. S., Hites, R. A. and Biemann, K. (1971). Identification of mass spectra by computer-searching a file of known spectra. *Anal. Chem.*, **43**, 681–691.

Herzberg, G. (1966). Molecular spectra and molecular structure, III. Electronic spectra and electronic structure of polyatomic molecules. New York, D. van Nostrand.

Hill, H. C. (1972). Introduction to mass spectrometry. 2nd edn. New York, Heyden and Son.

Hill, H. and Freeman, R. (1970). Introduction to Fourier transform NMR. Palo Alto, Calif., Varian Associates.

Holbrook, R. F., Beroza, M. and Burgess, E. D. (1960). Gypsy moth (*Porthetria dispar*) detection with the natural female sex lure. *J. Econ. Entom.*, **53**, 751–756.

Hölldobler, B. and Wilson, E. O. (1970). Recruitment trails in the harvester ant *Pogonomyrmex badius. Psyche*, **77**, 385–399.

Horrocks, W.dew. Jr., Sipe III, J. P., and Luber, J. R. (1971). Lanthanide shift reagents. The X-ray structure of the eight-coordinate bis (4-picoline) adduct of 2,2,6,6-tetramethylheptane-3,5-dionato-holmium, $Ho(dpm)_3(4\text{-pic})_2$. *J. Amer. Chem. Soc.*, **93**, 5258–5260.

Horvath, C. (1967). Columns in gas chromatography. *In* Ettre, L. S. and Zlatkis, A. eds. The practice of gas chromatography. New York, Wiley-Interscience. pp. 169–198.

House, H. O. (1972). Modern synthetic reactions. 2nd edn. New York, Benjamin.

Howell, J. F. (1972). An improved sex attractant for codling moths. *J. Econ. Entom.*, **65**, 609–611.

Howell, J. F. and Davis, H. G. (1972). Protecting codling moths captured in sex-attractant traps from predaceous yellow jackets. *Envir. Entom.*, **1**, 122–123.

Howell, M. G., Kende, A. S. and Webb, J. S. eds. Formula index to NMR literature data. Vol. 1, (1964); Vol. 2, (1966). New York, Plenum Press.

Howland, A. F., Debolt, J. W., Wolf, W. W., Toba, H. H., Gibson, T. and Kishaba, A. N. (1969). Field and laboratory studies of attraction of the synthetic sex pheromone to male cabbage looper moths. *J. Econ. Entom.*, **62**, 117–122.

Hsiao, H. S. (1972). The attraction of moths (*Trichoplusia ni*) to infrared radiation. *J, Insect Physiol.*, **18**, 1705–1714.

Huber, R. and Hoppe, W. (1965). Zur chemie des Ecdysones, VII. Die Kristall- und Molekuelstruktur- analyse des Insektenverpuppungs Hormons Ecdyson mit der automatisierten Faltmolekülmethode. *Chem. Ber.*, **98**, 2403–2424.

Hughes, P. R. and Pitman, G. B. (1970). Method for observing and recording the flight behavior of tethered bark beetles in response to chemical messengers. *Contrib. Boyce Thompson Inst.*, **24**, 329–336.

Humble Oil. A catalogue of the nuclear magnetic resonance spectra of hydrogen in hydrocarbons and their derivatives. Bay Town, Texas, Humble Oil and Refining Co.

Hurst, J. J., Meinwald, J. and Eisner, T. (1964). Defense mechanisms of arthropods XII. Glucose and hydrocarbons in the quinone-containing secretion of *Eleodes longicollis*. *Ann. Entom. Soc. Amer.*, **57**, 44–46.

Ignoffo, C. M., Berger, R. S., Graham, H. M. and Martin, D. F. (1963). Sex attractant of *Trichoplusia ni* (Hübner). *Science*, **141**, 902–903.

Ikan, R., Bergmann, E. D., Yinon, U. and Shulov, A. (1969). Identification, synthesis and biological activity of an 'assembling scent' from the beetle *Trogoderma granarium*. *Nature*, **223**, 317.

Ireland, R. E. (1969). Organic synthesis. New York, Prentice-Hall.

Ishii, S. (1970). Aggregation of the German cockroach *Blattella germanica* (L). *In* Wood, D. L., Silverstein, R. M., and Nakajima, M. eds. Control of insects behavior by natural products. New York, Academic Press, pp. 93–109.

Jackman, L. M. and Sternhell, S. (1969). Applications of nuclear magnetic resonance spectroscopy. 2nd edn. New York, Pergamon Press.

Jacobson, M. (1962). Insect sex attractants, III: The optical resolution of D,L-10-acetoxy-*cis*-7-hexadecen-1-ol. *J. Org. Chem.*, **27**, 2670–2671.

Jacobson, M. (1963). Recent progress in the chemistry of insect attractants. *Adv. Chem. Series*, **41**, 1–10.

Jacobson, M. (1965). Insect sex attractants. New York, Wiley-Interscience.

Jacobson, M. (1966a). Natural insect attractants and repellents, new tools in pest control. *Adv. Chem. Series*, **53**, 17–26.

Jacobson, M. (1966b). Chemical insect attractants and repellents. *Ann. Rev. Entom.*, **11**, 403–422.

Jacobson, M. (1968). Sex pheromone of the pink bollworm moth. Biological masking by its geometrical isomer. *Science*, **163**, 190–191.

Jacobson, M. (1970). Methodology for isolation, identification, and synthesis of sex pheromones in the Lepidoptera. *In* Wood, D. L., Silverstein, R. M. and Nakajima, M. eds. Control of insect behavior by natural products, New York, Academic Press, pp. 111–132.

Jacobson, M. (1972). Insect sex pheromones. New York, Academic Press.

Jacobson, M. and Beroza, M. (1963). Chemical insect attractants. *Science*, **140**, 1367–1373.

Jacobson, M. and Beroza, M. (1964). Insect attractants. *Sci. Amer.*, **211**, 20–27.

Jacobson, M., Beroza, M. and Jones, W. A. (1960). Isolation, identification and synthesis of the sex attractant of gypsy moth. *Science*, **132**, 1011–1012.

Jacobson, M., Beroza, M. and Yamamoto, R. T. (1963). Isolation and identification of the sex attractant of the American cockroach. *Science*, **139**, 48–49.

Jacobson, M., Green, N., Warthen, D., Harding, C. and Toba, H. H. (1970a). Sex pheromones of the Lepidoptera. Recent progress and structure relationships, *In* Beroza, M. ed. Chemicals controlling insect behavior. New York and London, Academic Press, pp. 3–20.

Jacobson, M., Lilly, C. E. and Harding, C. (1968a). Sex attractant of sugar beet wireworm. Identification and biological activity. *Science*, **159**, 208–210.

Jacobson, M., Redfern, R. E., Jones, W. A. and Aldridge, M. H. (1970b). Sex pheromones of the southern armyworm moth: isolation, identification, and synthesis. *Science*, **170**, 542–544.

Jacobson, M. and Smalls, L. A. (1966). Masking of the american cockroach sex attractant. *J. Econ. Entom.*, **59**, 414–416.

Jacobson, M. and Smalls, L. A. (1967). Sex attraction masking in the Cynthia moth. *J. Econ. Entom.*, **60**, 296.

Jacobson, M., Sonnet, P. E., Adler, V. E. and Cook, D. (1970c). Inactivity of a preparation reported to be highly active as a gypsy moth sex attractant. *Ann. Entom. Soc. Amer.*, **63**, 614–615.

Jacobson, M., Toba, H. H., Debolt, J. and Kishaba, A. N. (1968b). Insect sex attractants. VIII. Structure activity relationships in sex attractants for male cabbage loopers. *J. Econ. Entom.*, **61**, 84–85.

Jacobson, M., Waters, R. M. and Schwarz, M. (1970d). Gypsy moth sex attractants: A reinvestigation. *J. Econ. Entom.*, **63**, 943–945.

Jaffe, H. H. and Orchin, M. (1962). Theory and applications of ultraviolet spectroscopy. New York, Wiley.

Jantz, O. K. and Rudinsky, J. A. (1965). Laboratory and field methods for assaying olfactory responses of the Douglas fir beetle *Dendroctonus pseudotsugae* Hopkins. *Can. Entom.*, **97**, 935–941.

Johnstone, R. A. W. (1972). Mass spectrometry for organic chemists. London, Cambridge Univ. Press.

Jones, P. A., Coppel, H. C. and Casida, J. E. (1965). Collection of additional sex attractant from the virgin female introduced pine sawfly. *J. Econ. Entom.*, **58**, 465–466.

Jones, W. A., Jacobson, M. and Martin, D. F. (1966). Sex attractant of the pink bollworm moth: isolation, identification, and synthesis. *Science*, **152**, 1516–1517.

Jurs, P. C. (1970). Mass spectral feature selection and structural correlations using computerized learning machines. *Anal. Chem.*, **42**, 1633–1638.

Jurs, P. C., Kowalski, B. R., Isenhour, T. L. and Reilley, C. N. (1969). An investigation of combined patterns from diverse analytical data using computerized learning machines. *Anal. Chem.*, **41**, 1949–1953.

Jurs, P. C., Kowalski, B. R., Isenhour, T. L. and Reilley, C. N. (1970). Computerized learning machines applied to chemical problems. Molecular structure parameters from low resolution mass spectrometry. *Anal. Chem.*, **42**, 1387–1394.

Kaae, R. S. and Shorey, H. H. (1972). Sex pheromones of Lepidoptera. XXIX. An improved double-cone trap for males of *Trichoplusia ni. Envir. Entom.*, **1**, 675–677.

Kaae, R. S., Shorey, H. H. and Gaston, L. K. (1973). Pheromone concentrations as a mechanism for reproductive isolation between Lepidopterous species. *Science*, **179**, 487–488.

Kamlet, M. J. ed. (1960). Organic electronic spectral data. Vol. I. New York, Wiley-Interscience. Covers 1946–1952.

Kaplan, L. A. and Wheeler, O. H. eds. (1966). Organic electronic spectral data. Vol. III. New York, Wiley-Interscience. Covers 1956–1957.

Karasek, F. W. (1972). GC/MS/Computers. *Anal. Chem.*, **44**(4), 32A–42A.

Karlson, P. (1967). The chemistry of insect hormones and pheromones. *Pure Appl. Chem.*, **14**, 75–87.

Karlson, P. and Butenandt, A. (1959). Pheromones (ectohormones) in insects. *Ann. Rev. Entom.*, **4**, 39–58.

Karlson, P., Lüscher, M. and Hummel, H. (1968). Extraction and bioassay of the trail pheromone of the termite, *Zootermopsis nevadensis. J. Insect Physiol.*, **14**, 1763–1771.

Karlson, P. and Schneider, D. (1973). Sexualhormone der Schmetterlinge als Modelle Chemischer Kommunikation. *Naturwiss.*, **60**, 113–121.

Keulemans, A. I. M. (1959). Gas chromatography. 2nd edn. New York, Reinhold.

Keys, R. E. and Mills, R. B. (1968). Demonstration and extraction of a sex attractant from female Angoumois grain moths. *J. Econ. Entom.*, **61**, 46–49.

Killenen, R. G. and Ost, R. W. (1971). Pheromone-maze trap for cabbage looper moths. *J. Econ. Entom.*, **64**, 310–311.

Kirchner, J. G. (1967). Thin-layer chromatography. New York, Wiley-Interscience.

Kirkland, J. J. (1971). Modern practice of liquid chromatography. New York, Wiley-Interscience.

Kishaba, A. N., Toba, H. H., Wolf, W. W. and Vail, P. V. (1970). Response of laboratory reared male cabbage looper to synthetic sex pheromone in the field. *J. Econ. Entom.*, **63**, 178–181.

Kliefoth, R. A., Vité, J. P. and Pitman, G. B. (1964). A laboratory technique for testing bark beetle attractants. *Contrib. Boyce Thompson Inst.*, 22, 283–290.

Klein, M. G., Ladd, T. L. Jr. and Lawrence, K. O. (1972). The influence of height of exposure of virgin female Japanese beetles on captures of males. *Envir. Entom.*, **1**, 600–601.

Klun, J. A. (1968). Isolation of a sex pheromone of the European corn borer. *J. Econ. Entom.*, **61**, 484–487.

Klun, J. A., Chapman, O. L., Mattes, K. C., Wojkowski, P. W., Beroza, M. and Sonnet, P. E. (1973). Insect sex pheromones: Minor amount of opposite geometrical isomer critical to attractions. *Science*, **161**, 661–662.

Klun, J. A. and Robinson, J. F. (1970). Inhibition of European corn borer mating by *cis*-11-tetradecenyl acetate, a borer sex stimulant. *J. Econ. Entom.*, **63**, 1281–1283.

Knock, B. A., Smith, I. C., Wright, D. E., Ridley, R. G. and Kelly, W. (1970). Compound identification by computer matching of low resolution mass spectra. *Anal. Chem.*, **42**, 1516–1520.

Knox, J. H. (1962). Gas chromatography. New York, Wiley.

Kovats, E. (1958). Gas-chromatographische Charakterisierung organischer Verbindungen. Teil I: Retentionsindices aliphatischer Halogenide, alkohole, und Ketone. *Helv. Chim. Acta*, **41**, 1915–1932.

Kovats, E. (1966). Gas chromatographic characterization of organic substances in the retention index system. *In* Giddings, J. C. and Keller, R. A. eds. Advances in chromatography. New York, Dekker, pp. 229–247.

Kuwahara, Y. and Casida, J. E. (1973). Quantitative analysis of the sex pheromone of several Phyctid moths by electron-capture gas chromatography. *Agr. Biol. Chem.*, **37**, 681–684.

Kuwahara, T., Kitamura, C., Takahashi, S., Hara, H., Ishi, S. and Fukami, H. (1971). Sex pheromone of the almond moth and the Indian meal moth: *cis*-9, *trans*-12-tetradecadienyl acetate. *Science* **171**, 801–802.

Labler, L. and Schwarz, V. L. (1965). Thin-layer chromatography. New York, American Elsevier.

Lacher, V. (1964). Elektrophysiologische Untersuchungen an einzeln Rezeptoen für Geruch, Kohlendioxid, Luftfeuchtigkeit und Temperatur auf den Antennen der Arbeitsbiene und der Drohne (*Apis mellifica* L.). *Z. Vergl. Physiol.*, **48**, 587–623.

Ladd, T. L. Jr. (1970). Sex attraction in the Japanese beetle. *J. Econ. Entom.*, **63**, 905–908.

Ladisch, R. F. and McQue, B. (1953). Methods of obtaining quinones from flour beetles. *Science*, **118**, 324–325.

Lang, L. ed. Absorption spectra in the ultraviolet and visible region. Volumes I to VII, New York, Academic Press. Volumes VIII to XI, Budapest, Hungary, Akademiai Kiado.

Lange, R. and Weissinger, M. (1972). Untersuchungen am Kiefernknospentriebwickler über die Wirkung des Sexualpheromons unter verschiedenen Freilandbedingungen. *Z. Ang. Entom.*, **70**, 365–370.

Law, J. H. and Regnier, F. E. (1971). Pheromones. *Ann. Rev. Biochem.*, **40**, 533–548.

Lawrence, L. A. and Bartell, R. J. (1972). The effect of age on calling behavior of virgin females of *Epiphyas postvittana* (Lepidoptera) and on their pheromone content and ovarian development. *Ent. Exp. & Appl.*, **15**, 455–464.

Lebedeva, K. V. (1968). Attractants. *Zh. Vses. Khim. Obshchest.*, 272–281.

Lederer, E. (1950). Odeurs et parfumes der animeux. *Fortschr. Chem. Org. Natur.* **6**, 87–153.

Levinson, H. Z. (1972). Zur Evolution und Biosynthese der terpenoiden Pheromone und hormone. *Naturwiss.*, **59**, 477–484.

Levy, G. C. and Nelson, G. L. (1972). Carbon-13 nuclear magnetic resonance for organic chemists. New York, John Wiley and Son.

Lilly, C. E. and McGinnis, A. J. (1965). Reaction of male click beetles in the laboratory to olfactory pheromones. *Can. Entom.*, **97**, 317–321.

Lilly, C. E. and Shorthouse, J. D. (1971). Responses of males of the 10-lined June beetle, *Polyphylla decemlineata* (Coleoptera: Scarabaeidae), to female sex pheromone. *Can. Entom.*, **103**, 1757–1761.

Littlewood, A. B. (1970). Gas chromatography, principles, techniques, and application. 2nd ed. New York, Academic Press.

Loconti, J. D. and Roth, L. M. (1953). Composition of the odorous secretion of *Tribolium castaneum. Ann. Entom. Soc. Amer.*, **46**, 281–289.

MacConnell, J. G. and Silverstein, R. M. (1973). Recent results in insect pheromone chemistry. *Angew. Chem. Int. Ed.*, **12**, 644–654.

MacGregor, E. G. (1948). Odour as a basis for orientated movement in ants. *Behav.*, **1**, 267–296.

Madsen, H. F. and Vakenti, J. M. (1972). Codling moths: Female-baited and synthetic pheromone traps as population indicators. *Envir. Entom.*, **1**, 554–557.

Magnetti, P. and Tibaldi, E. (1969). Chemical language in animals. *Cron. Chim.*, **24**, 3–7.

Maloeuf, N. S. R. (1938). Secretions from ectodermal glands of arthropods. *Quart. Rev. Biol.*, **13**, 169–195.

Malone, C. T. and McFadden, W. H. (1969). Special identification detectors. *In* Ettre, L. S. and McFadden, W. H. eds. Ancillary techniques of gas chromatography. New York, Wiley-Interscience, pp. 341–373.

Manley, C. and Farrier, M. H. (1969). Attraction of the Nantucket pine tip moth, *Rhyacionia frustrana* to incandescent light, and blacklight, and to virgin females. *Ann. Entom. Soc. Amer.*, **62**, 443.

Marinsky, J. A. ed. Ion exchange. A series of advances. Vol. 1 (1966), Vol. 2 (1969). New York, Dekker.

Maschwitz, U. W. (1966). Alarm substances and alarm behavior in social insects. *Vitamins and Hormones*, **24**, 267–290.

Mathieson, D. W. ed. (1965). Interpretation of organic spectra. New York, Academic Press.

Mathieson, D. W. ed. (1967). Nuclear magnetic resonance for organic chemists. New York, Academic Press.

Matsumura, F. M., Coppel, H. C. and Tai, A. (1968). Isolation and identification of termite trail following pheromones. *Nature*, **219**, 963–964.

Mayer, M. S. (1973). Electrophysiological correlates of attraction with *Trichoplusia ni. J. Insect Physiol.*, **19**, 1191–1198.

Mayer, M. S. and Thaggard, C. W. (1966). Investigation of an olfactory attractant specific for males of the housefly *Musca domestica. J. Insect Physiol.*, **12**, 891–897.

M.C.A. (IR). Catalog of infrared spectral data. Manufacturing Chemists Association Research Project, Chemical and Petroleum Research Laboratories, Pittsburgh, Pa., Carnegie Institute of Technology, to June 30, 1960; Chemical Thermodynamics Properties Center, Agricultural and Mechanical College of Texas, College Station, Texas, from July 1, 1960.

M.C.A. (UV) (1960). Ultraviolet spectral data. Manufacturing Chemists Association Research Project, Pittsburgh, Pa., Carnegie Institute of Technology.

McCullough, T. (1966). Compounds found in the defensive scent glands of *Harpalus caliginosus* (Coleoptera: Carabidae). *Ann. Entom. Soc. Amer.*, **59**, 1020–1021.

McDonough, L. M., George, D. A., Butt, B. A., Jacobson, M. and Johnson, C. R. (1969). Isolation of the sex pheromone of the codling moth. *J. Econ. Entom.*, **62**, 62–65.

McDonough, L. M., George, D. A. and Landis, B. J. (1970). Partial structure of two sex pheromones of the corn earworm, *Heliothis zea. J. Econ. Entom.*, **63**, 408–412.

McGovern, W. L., Cross, W. H., Leggett, J. E. and Mitlen, N. (1973). Boll weevils: A technique for measuring the production of pheromone and the competitiveness of $_{32}$P tagged adults in the field. *J. Econ. Entom.*, **66**, 809–810.

McGurk, D. J., Frost, J., Eisenbraun, E. J., Vick, K., Drew, W. A. and Young, J. (1966). Volatile components in ants. Identification of 4-methyl-3-heptanone from *Pogonomyrmex* ants. *J. Insect Physiol.*, **12**, 1435–1441.

McGurk, D. J., Frost, J., Waller, G. R., Eisenbraun, E. J., Vick, K., Drew, W. A. and Young, J. (1968). Iridodial isomer variation in Dolichoderines. *J. Insect Physiol.*, **14**, 841–845.

McKibben, G. H. (1972). A device for injecting grandlure into cigarette filters. *J. Econ. Entom.*, **65**, 1509–1510.

McLafferty, F. W. ed. (1963). Mass spectrometry of organic ions. New York, Academic Press.

McLafferty, F. W. (1966). Interpretation of mass spectra. New York, Benjamin.

McLafferty, F. W. and Penzelik, J. (1967). Index and bibliography of mass spectrometry, 1963–1965. New York, Wiley-Interscience.

McNair, H. M. and Bonelli, E. J. (1969a). Basic gas chromatography. Walnut Creek, California, Varian Aerograph.

McNair, H. M. and Bonelli, E. J. (1969b). *ibid.* pp. 37–79.

McNair, H. M. and Bonelli, E. J. (1969c). *ibid.* pp. 81–121.

McReynolds, W. O. (1967). Gas chromatographic data. Evanston, Ill., Preston Technical Abstracts Co.

Meinwald, J., Chadha, M. S., Hurst, J. J. and Eisner, T. (1962). Defense mechanisms of arthropods. IX. Anisomorphal, the secretion of a phasmid insect. *Tetrahedron Letters*, **1**, 29–33.

Meinwald, J., Chambers, A. M., Pliske, T. E. and Eisner, T. (1968a). Pheromones. III. Identification of *trans*, *trans*-10-hydroxy-3,7-dimethyl-2,6-decadienoic acid as a major component in hairpencil secretion of the male monarch butterfly. *Tetrahedron Letters*, 4893–4896.

Meinwald, J., Erikson, K., Hartshorn, M., Meinwald, Y. C. and Eisner, T. (1968b). Defensive mechanisms of arthropods. XXIII. An allenic sesquiterpenoid from the grasshopper *Romalea microptera. Tetrahedron Letters*, 2959–2962.

Meinwald, J., Meinwald, Y. C., Wheeler, J. W., Eisner, T. and Brower, L. P. (1966). Major components in the exocrine secretion of a male butterfly (Lycorea). *Science*, **151**, 583–585.

Meinwald, J., Meinwald, Y. C. and Mazzocchi, P. H. (1969). Sex pheromone of the queen butterfly: chemistry. *Science*, **164**, 1174–1175.

Meinwald, J. and Muetterties, E. (1971). Pheromones. *In* Houben-Weil. Methoden der organischen Chemie. 1971. Stuttgart, Germany, Thiem Verlag.

Meinwald, Y. C. and Eisner, T. (1964). Defense mechanisms of arthropods XIV. Caprylic acid, an accessory component of the secretion of *Eleodes longicollis. Ann. Entom. Soc. Amer.*, **57**, 513–514.

Meinwald, Y. C., Meinwald, J. and Eisner, T. (1966). 1,2-Dialkyl-4(3H)-quinazolinones in the defensive secretion of a millipede *Glomeris marginata. Science*, **154**, 390–391.

Merritt, C. Jr. (1969). Chemical identification of gas chromatographic fractions. *In* Ettre, L. S. and McFadden, W. H. eds. Ancillary techniques of gas chromatography. New York, Wiley-Interscience, pp. 325–340.

Michael, R. P., Keverne, E. B. and Bonsell, R. W. (1971). Pheromones: Isolation of male sex attractants from a female primate. *Science*, **172**, 964–966.

Miller, R. G. J. ed. (1965). Laboratory methods in infrared spectroscopy. London, Heyden.

Milne, G. W. A. (1971). Mass spectrometry: Techniques and applications. New York, Wiley-Interscience.

Minks, A. K. (1971). Decreased sex pheromone production in an in-bred stock of the summerfruit tortrix moth, *Adoxophyes orana*. *Ent. Exp. & Appl.*, **14**, 361–364.

Minks, A. K., Roelofs, W. L., Ritter, F. J. and Persoons, C. J. (1973). Reproductive isolation of two Tortricid moth species by different ratios of a two-component sex attractant. *Science*, **180**, 1073–1074.

Mitchell, E. B., Hardee, D. D., Cross, W. H., Huddleston, P. M. and Mitchell, H. C. (1972). Influence of rainfall, sex ratio, and physiological condition of boll weevils on their response to pheromone traps. *Envir. Entom.*, **1**, 438–440.

Mitchell, E. R., Webb, J. C., Baumhover, A. H., Hines, R. W., Stanley, J. W., Endris, R. G., Lindquist, D. A. and Masuda, S. (1972). Evaluation of cylindrical electric grids as pheromone traps for loopers and tobacco hornworms. *Envir. Entom.*, **1**, 365–367.

Mitchell, W. C. and Mau, R. F. L. (1971). Response of the female southern green stink bug and its parasite, *Trichopoda pennipes*, to male stink bug pheromones. *J. Econ. Entom.*, **64**, 856–859.

Moeck, H. A. (1970). An olfactometer for the bioassay of attractants for Scolytids. *Can. Entom.*, **102**, 792–796.

Monro, A., Chadha, M., Meinwald, J. and Eisner, T. (1962). Defense mechanisms of arthropods. VI. Para-benzoquinones in the secretion of five species of millipedes. *Ann. Entom. Soc. Amer.*, **55**, 261–262.

Moody, D. S., White, J. R. and Bottrell, D. G. (1972). A machine for dispensing grandlure on cigarette filters used to bait boll weevil traps. *J. Econ. Entom.*, **65**, 1215–1217.

Moore, B. P. (1964). Volatile terpenes from Nasutitermes soldiers (Isoptera: Termitidae). *J. Insect Physiol.*, **10**, 371–375.

Moore, B. P. (1966). Isolation of the scent trail pheromone of an Australian termite. *Nature*, **211**, 746–747.

Moore, B. P. (1969). Biochemical studies in termites. *In* Krishna, K. and Weesner, F. M. eds. Biology of termites. Vol. I. New York, Academic Press. pp. 407–432.

Moore, B. P. and Brown, W. V. (1971). Gas-liquid chromatographic identification of ozonolysis fragments as a basis for micro-scale structure determination. *J. Chromatog.*, **60**, 157–166.

Moorehouse, J. E., Yeadon, R., Beevor, P. S. and Nesbitt, B. F. (1969). Studies of insect chemical communication. *Nature*, **223**, 1174–1175.

Moreno, D. S., Carman, G. E. and Bain, N. S. (1972). Extraction and bioassay of and effect of solvents on the sex pheromone of the yellow scale. *Ann. ent. Soc. Amer.*, **65**, 1061–1064.

Morse, R. A., Shearer, D. A., Boch, R. and Benton, A. W. (1967). Observations on alarm substances in the genus *Apis*. *J. Apicult. Res.*, **6**, 113–118.

Moser, J. C. (1970). Pheromones of social insects. *In* Wood, D. L., Silverstein, R. M. and Nakajima, M. eds. Control of insect behavior by natural products. New York, Academic Press, pp. 161–178.

Moser, J. C. and Blum, M. S. (1963). Trail marking substance of the Texas leaf cutting ant. Source and potency. *Science*, **140**, 1228.

Moser, J. C., Brownlee, R. G. and Silverstein, R. M. (1968). Alarm pheromones of the ant *Atta texanna*. *J. Insect Physiol.*, **14**, 529–535.

Muto, F. (1968). Insect attractants. *Noyaku Seisan Gijutsu*, **19**, 1–19.

Murrell, J. N. (1963). The theory of electronic spectra of organic molecules. London, Methuen.

Myers, J. (1972). Pheromones and courtship behavior in butterflies. *Amer. Zool.*, **12**, 545–551.

Mykytowycz, M. (1970). The role of skin glands in mammalian communication. *In* Johnson, J. W. Jr., Moulton, D. G. and Turk, A. eds. Advances in chemoreception, Vol. I, Communication by chemical signals. New York, Appleton-Century-Crofts, pp. 327–360.

Nachod, F. C. and Phillips, W. D. eds. (1962). Determination of organic structures by physical methods. New York, Academic Press.

Nakajima, M. (1970). Studies on sex pheromones of the stored grain moths. *In* Wood, D. L., Silverstein, R. M. and Nakajima, M. eds. Control of insect behavior by natural products. New York, Academic Press, pp. 209–224.

Nakanishi, K. (1962). Infrared absorption spectroscopy. San Fransisco, Holden-Day.

Nation, J. L. (1972). Courtship behavior and evidence for a sex attractant in the male Caribbean fruit fly, *Anastrepha suspensa. Ann. Entom. Soc. Amer.*, **65**, 1364–1367.

Negata, K., Tamaki, Y., Noguchi, H. and Yushima, T. (1972). Changes in sex pheromone activity in adult females of the smaller tea tortrix moth *Adoxophyes fasciata. J. Insect Physiol.*, **18**, 339–346.

Neudert, W. and Ropke, H. (1965). Atlas of steroid spectra. New York, Springer-Verlag.

Nielson, D. G. and Balderston, C. P. (1973). Evidence for intergeneric sex attraction among Aegeriids. *Ann. Entom. Soc. Amer.*, **66**, 227–228.

Nijholt, W. W. and Chapman, J. A. (1968). A flight trap for collecting living insects. *Can. Entom.*, **100**, 1151–1153.

Novak, V. J. A. (1966). Insect hormones. London, Methuen.

Onsager, J. A., McDonough, L. M. and George, D. A. (1968). A sex pheromone in the Pacific coast wire worm. *J. Econ. Entom.*, **61**, 691–693.

Ouye, M. T. and Butt, B. A. (1962). A natural sex lure extracted from female pink bollworms *Pectinophora gossypiella* (Saunders). *J. Econ. Entom.*, **55**, 419–421.

Packard, A. S. (1896). Literature on defensive or repugnatorial glands of insects. *J.N.Y. Entom. Soc.*, **4**, 26–32.

Pain, J. (1970). Pheromones and behaviour in some Hymenoptera. *Probl. Actuels Endocrinol. Nutr.*, **14**, 237–258.

Pain, J. (1971). Phéromones et comportement chez quelques Hyménoptères. *Apidologie*, **2**, 319–355.

Pain, J. (1973). Pheromones and Hymenoptera. *Bee World*, **54**, 11–24.

Parker, F. S. (1971). Applications of infrared spectroscopy in biochemistry, biology and medicine. New York, Plenum Publishing Corp.

Pasteels, J. M. (1972). Sex-specific pheromones in a termite. *Experentia*, **28**, 105–106.

Pasto, D. J. and Johnson, C. R. (1969). Organic structure determination. Englewood Cliffs, New Jersey, Prentice-Hall.

Patrick, J. C. and Hensley, S. D. (1970). Recapture of males released at different distances from a trap baited with virgin female sugarcane borers. *J. Econ. Entom.*, **63**, 1341–1342.

Pavan, M. (1958). Significato chemico e biologico di alcuni velenti di insetti. *Artigianelli, Pavia*, **15**, 1–75.

Pavan, M. (1959). Biochemical aspects of insect poisons. 1958 Int. Congr. Biochem., XII: Biochem. of insects. London, Pergamon Press, pp. 15–36.

Pavan, M. and Dazzini, M. V. (1971). Toxicology and pharmacology – arthropoda. *Chemical Zoology*, **6**, 365–409.

Payne, T. L. (1970). Electrophysiological investigations on response to pheromones in bark beetles. *Contrib. Boyce Thompson Inst.*, **24**, 275–282.

Payne, T. L. (1971). Bark beetle olfaction. I. Electroantennogram responses of the southern pine beetle. (Coleoptera: Scolytidae) to its aggregation pheromone frontalin. *Ann. Entom. Soc. Amer.*, **64**, 266–268.

Payne, T. L., Shorey, H. H. and Gaston, L. K. (1970). Sex pheromones of noctuid moths. Factors influencing antennal responsiveness in males of *Trichoplusia ni*. *J. Insect Physiol.*, **16**, 1043–1055.

Pecsok, R. L. ed. (1959). Principles and practices of gas chromatography. New York, Wiley.

Percy, J. E., Gardiner, E. J. and Weatherston, J. (1971). Studies of physiologically active arthropod secretions. VI. Evidence for sex pheromones in female *Orgyia leucostigma* (Lepidoptera: Lymantridae). *Can. Entom.*, **103**, 706–712.

Perez, R. and Long, W. H. (1964). Sex attractant and mating behavior in the female sugar cane borer, *Diatraea saccharalis* (F). *J. Econ. Entom.*, **57**, 688–690.

Perry, E. S. and Weissberger, A. eds. (1965). Distillation. 2nd edn. New York, Wiley-Interscience.

Perry, S. G., Amos, R. and Brewer, P. I. (1972). Practical liquid chromatography. New York, Plenum Publishing Company.

Phillips, J. P. (1964). Spectra structure correlation. New York, Academic Press.

Phillips, J. P., Lyle, R. E. and Jones, P. R. (1969). Organic electronic spectral data. Vol. V. New York, Wiley-Interscience. Covers 1960–1961.

Phillips, J. P. and Nachod, F. C. ed. (1963). Organic electronic data. Vol. IV. New York, Wiley-Interscience. Covers 1958–1959.

Pinder, A. R. and Staddon, B. W. (1965). *trans*-4-Oxohex-2-enal in the odoriferous secretion of *Sigara fallini* (Fieb.) (Hemiptera-Heteroptera). *Nature*, **205**, 106.

Pitman, G. B. (1971). *trans*-Verbenol and alpha-pinene: Their utility in manipulation of the mountain pine beetle. *J. Econ. Entom.*, **64**, 426–430.

Pitman, G. B., Kliefoth, R. A. and Vité, J. P. (1965). Studies on the pheromone of of *Ips confusus* (LeConte). II. Further observations on the site of production. *Contrib. Boyce Thompson Inst.*, **23**, 13–18.

Pitman, G. B. and Vité, J. P. (1969). Aggregation behaviour of *Dendroctonus ponderosae* (Coleoptera:Scolytidae) in response to chemical messengers. *Can. Entom.*, **101**, 143–149.

Pitman, G. B. and Vité, J. P. (1970). Field response of *Dendroctonus pseudotsugae* (Coleoptera: Scolytidae) to synthetic frontalin. *Ann. Entom. Soc. Amer.*, **63**, 661–664.

Pitman, G. B., Vité, J. P., Kinzer, G. W. and Fentiman, A. F. Jr. (1968). Bark beetle attractants. *trans*-Verbenol isolated from *Dendroctonus*. *Nature*, **218**, 168–169.

Pliske, T. E. and Eisner, T. (1969). Sex pheromones of the queen butterfly: Biology. *Science*, **164**, 1170–1172.

Pople, J. A., Schneider, W. G. and Bernstein, H. J. (1959). High resolution nuclear magnetic resonance. New York, McGraw-Hill.

Potts, W. J. Jr. (1963). Chemical infrared spectroscopy, Vol. 1, Techniques. New York, Wiley.

Price, G. D., Sunas, E. C. and Williams, J. F. (1967). Microcell for obtaining normal contrast infrared solution spectra at the five-microgram level. *Anal. Chem.*, **39**, 138–140.

Priesner, E (1968). Die interspezifischen Wirkung der Sexualstoffe der Saturniidae (Lepidoptera). *Z. Vergl. Physiol.*, **61**, 263–297.

Purnell, H. (1962). Gas chromatography. New York, Wiley.

Pyatnova, Yu. B., Ivanov, L. L. and Kyskina, A. S. (1969). Insect sex attractants. *Russ. Chem. Rev.*, **38**, 126–139.

Ralls, K. (1971). Mammalian scent marking. *Science*, **171**, 443–449.

Randerath, K. (1966). Thin-layer chromatography. New York, Academic Press.

Rao, C. N. R. (1961). Ultra-violet and visible spectroscopy chemical applications. London, Butterworths.

Rao, C. N. R. (1963). Chemical applications of infrared spectroscopy. New York, Academic Press.

Read, J. S., Warren, F. L. and Hewitt, P. H. (1968). Identification of the sex pheromone of the false codling moth (*Argyroploce leucotreta*). *Chem. Comm.*, 792–793.

Redfern, R. E., Butt, B. A. and Cantu, E. (1970). Bioassay of the sex pheromone of the southern armyworm. *J. Econ. Entom.*, **63**, 658.

Regnier, F. E. (1971). Semiochemicals. Structure and function. *Biol. Reprod.*, **4**, 309–326.

Regnier, F. E. (1972). Semiochemicals. *In* Waller, G. R. ed. Biochemical applications of mass spectrometry. New York, Wiley-Interscience. pp. 723–734.

Regnier, F. E. and Law, J. H. (1968). Insect pheromones. *J. Lipid Res.*, **9**, 541–551.

Regnier, F. E. and Huang, J. C. (1970). Identification of some oxygen-containing functional groups by reaction gas chromatography. *J. Chromatog. Sci.*, **8**, 267–271.

Regnier, F. E. and Wilson, E. O. (1968). The alarm-defense system of the ant *Acanthomyops claviger. J. Insect Physiol.*, **14**, 955–970.

Regnier, F. E. and Wilson, E. O. (1969). The alarm-defense system of the ant *Lasius alienus. J. Insect Physiol.*, **15**, 893–898.

Reichstein, T., Euw, J. Von, Parsons, A. and Rothschild, M. (1968). Heart poisons in the monarch butterfly. *Science*, **161**, 861–866.

Rejesus, R. S. and Reynolds, H. T. (1970). Demonstration of the presence of a female sex pheromone in the cotton leaf perforator. *J. Econ. Entom.*, **63**, 961–964.

Renwick, J. A. A. (1970). Chemical aspects of bark beetle aggregation. *Contrib. Boyce Thompson Inst.*, **24**, 337–341.

Renwick, J. A. A. and Vité, J. P. (1970). Systems of chemical communication in *Dendroctonus. Contrib. Boyce Thompson Inst.*, **24**, 283–292.

Richards, O. W. (1927). Sexual selection and allied problems in the insects. *Biol. Rev. Cambridge Phil. Soc.*, **2**, 298–360.

Riley, R. G., Silverstein, R. M. and Moser, J. C. (1974). Biological responses of *Atta texana* to its alarm pheromone and the enantiomer of the pheromone. *Science*, **183**, 760–762.

Ritter, F. J., Rotgans, I. E. M., Talman, E., Verweil, P. E. J. and Stein, F. (1973). 5-Methyl-3-butyl-octahydroindolozine, a novel type of pheromone attractive to pharaoh's ants. (*Monomorium pharaonis* (L)). *Experientia*, **29**, 530–531.

Roach, S. H., Agee, H. R. and Ray, L. (1972). Influence of position and color of male-baited traps on captures of boll weevils. *Envir. Entom.*, **1**, 530–532.

Roberts, G., Gallagher, B. S. and Jones, R. N. (1958). Infrared absorption spectra of steroids – An atls. Vol. I. New York, Wiley-Interscience.

Roberts, J. D. (1959). Nuclear magnetic resonance. Applications to organic chemistry. New York, McGraw-Hill.

Roberts, J. D. (1962). An introduction to the analysis of spin-spin splitting in high-resolution nuclear magnetic resonance. New York, Benjamin.

Rodin, J. O., Silverstein, R. M., Burkholder, W. E. and Gorman, J. E. (1969). Sex attractant of female dermestid beetle *Trogoderma inclusam* LeConte. *Science,* **165**, 904–906.

Roeder, K. D. (1966). A differential anemometer for measuring the turning tendency of insects in stationary flight. *Science*, **153**, 1634–1636.

Roelofs, W. L. (1971). Insect electroantennograms. *XXIII International Congress of Pure and Applied Chemistry*, Boston.

Roelofs, W. L. and Arn, H. (1968). Sex attractant of the red banded leaf roller moth. *Nature*, **219**, 513.

Roelofs, W. L. and Cardé, R. T. (1971a). Sex attractants in Lepidoptera. *In* Tahori, A. S. ed. Chemical releasers in insects. New York, Gordon and Breach Science Publishers. pp. 91–112.

Roelofs, W. L. and Cardé, R. T. (1971b). Sex pheromone perception: Electroantennogram responses of the red-handed leaf roller moth. *J, Insect Physiol.*, **17**, 1969–1982.

Roelofs, W. L. and Cardé, R. T. (1971c). Hydrocarbon sex pheromone in tiger moths (Arctiidae). *Science*, **171**, 684–686.

Roelofs, W. L., Cardé, R. T., Bartell, R. J. and Tierney, P. G. (1972). Sex attractant trapping of the European corn borer in New York. *Envir. Entom.*, **1**, 606–608.

Roelofs, W. L. and Comeau, A. (1970). Lepidopterous sex attractants discovered by field screening tests. *J. Econ. Entom.*, **63**, 969–974.

Roelofs, W. L. and Comeau, A. (1971). Sex pheromone perception: Electroantennogram responses of the red-banded leaf roller moth. *J. Insect Physiol,*, **17**, 1969–1982.

Roelofs, W. L., Comeau, A., Hill, A. and Milicevic, G. (1971). Sex attractant of the codling moth: Characterization with electroantennogram technique. *Science*, **174**, 297–299.

Roelofs, W. L. and Feng, K. C. (1967). Isolation and bioassay of the sex pheromone of the red banded leaf roller *Argyrotaenia velutinanc* (Lepidoptera: Tortricidae). *Ann. Entom. Soc. Amer.*, **60**, 1199–1203.

Rogoff, W. M., Beltz, A. D., Johnsen, J. O. and Plapp, F. W. (1964). A sex pheromone in the housefly *Musca domestica* L. *J. Insect Physiol.*, **10**, 239–246.

Rohrschneider, L. (1966). Eine Methode zur Charakterisierung von Gaschromatographische Trenn flüssigkeiten. *J. Chromatog.*, **22**, 6–22.

Roth, L. M. and Dateo, G. P. (1966). A sex pheromone produced by males of the cockroach *Nauphoeta cinera. J. Insect Physiol.*, **12**, 255–265.

Roth, L. M. and Eisner, T. (1962). Chemical defenses of arthropods. *Ann. Rev. Entom.*, **7**, 107–136.

Roth, L. M. and Howland, R. B. (1941). Studies on the gaseous secretion of *Tribolium confusum* Duval. I. Abnormalities produced in *Tribolium confusum* Duval by exposure to a secretion given off by the adults. *Ann. Entom. Soc. Amer.*, **34**, 151–175.

Roth, L. M., Niegisch, W. D. and Stahl, W. H. (1956). Occurrence of 2-hexenal in the cockroach, *Eurycotis floridana. Science*, **123**, 670–671.

Roth, L. M. and Stay, B. (1958). The occurrence of para-quinones in some arthropods, with emphasis on the quinone-secreting tracheal glands of *Diploptera punctata. J. Insect Physiol.*; **1**, 305–318.

Roth. L. M. and Willis, E. R. (1952). A study of cockroach behavior. *Amer. Midland Naturalist*, **47**, 66–129.

Rowley, W. A., Graham, C. L. and Williams, R. E. (1968), A flight mill system for the laboratory study of mosquito flight. *Ann. Entom. Soc. Amer.*, **61**, 1507–1514.

Rudinsky, J. A., Furniss, M. M., Kline, L. N. and Schmitz, R. F. (1972). Attraction and repression of *Dendroctonus pseudotsugae* (Coleoptera: Scolytidae) by three synthetic pheromones in traps in Oregon and Idaho. *Can. Entom.*, **104**, 815–822.

Rudinsky, J. A. and Michael, R. R. (1972). Sound production in Scolytidae: Chemostimulus of sonic signal by the Douglas-fir beetle. *Science*, **175**, 1386–1390.

Ruttner, F. and Kaissling, K. E. (1968). Über die interspezifische Wirkung des sexuallockstoffes von *Apis mellifica* und *Apis cerana. Z. Vergl. Physiol.*, **59**, 362–370.

Saario, C. A., Shorey, H. H. and Gaston, L. K. (1970). Sex pheromones of Noctuid moths. XIX. Effect of environmental and seasonal factors on capture of males of *Trichoplusia ni* in pheromone baited traps. *Ann. Entom. Soc. Amer.*, **63**, 667–672.

Sadtler Standard Spectra. An annually updated collection of infrared, nuclear magnetic resonance, and ultraviolet spectra. Philadelphia, Pa., Sadtler Research Laboratories.

Sanders, C. J. (1971b). Sex pheromone specificity and taxonomy of budworm moths (*Choristoneura*). *Science*, **171**, 911–913.

Sanders, C. J. (1971a). Laboratory bioassay of the sex pheromone of the female eastern spruce budworm, *Choristoneura fumiferana* (Lepidoptera: Torticidae). *Can. Entom.*, **103**, 631–637.

Sanders, J. K. M. and Williams, D. H. (1971). Tris-(dipivalomenthanato)-europium, a paramagnetic shift reagent for use in NMR spectroscopy. *J. Amer. Chem. Soc.*, **93**, 641–645.

Sandler, S. R. and Karo, W. Organic functional group preparations. Vol. I (1968), Vol. II (1971). New York, Academic Press.

Scheinmann, F. ed. (1970). An introduction to spectroscopic methods for the identification of organic compounds. Volume I. Nuclear magnetic resonance and infrared spectroscopy. New York, Pergamon Press.

Schildknecht, H. (1961). Über Insekten-und Pflansensbwehrstoffezihre, Isolierung und Aufklärung. *Angew. Chem.*, **73**, 629.

Schildknecht, H. (1970). The defensive chemistry of land and water beetles. *Angew. Chem. Internat. Ed.*, **9**, 1–9.

Schildknecht, H. (1971). Evolutionary peaks in the defensive chemistry of insects. *Endeavour*, **30**, 136–141.

Schildknecht, H., Birringer, H. and Maschwitz, U. (1967). Testosterone as protective agent of the water beetle *Ilybius. Agnew. Chem. Internat. Ed.*, **6**, 558–559.

Schildknecht, H. and Holoubek, K. (1961). Die Bombardierkäfer und ihre Explosionchemie. V. Mitteilung über Insekten-abswehrstoffe. *Agnew. Chem.*, **73**, 1–7.

Schlunegger, U. P. and Leuthdold, R. H. (1972). Rapid identification of volatile compounds in insects with a direct gas chromatography/mass spectrometry procedure: 3-octanon and 3-octanol in *Crematogaster scutellaris* and *C. ashmedi. Insect Biochem.*, **2**, 150–152.

Schneider, D. (1957b). Elektrophysiologische Untersuchungen von Chem- und Mechanorezeptoren der Antenne des Seidenspinners *Bombyx mori* L. *Z. Vergl. Physiol.*, **40**, 8–41.

Schneider, D. (1962). Electrophysiological investigation of the olfactory specificity of sexually attracting substances in different species of moths. *J. Insect Physiol.*, **8**, 15–30.

Schneider, D. (1964). Insect antennae. *Ann. Rev. Entom.*, **9**, 103–161.

Schneider, D. (1965). Chemical sense communication in insects. *Symp. Soc. Exp. Biol.*, **20**, 273–297.

Schneider, D. (1969). Insect olfaction: Deciphering system for chemical messages. *Science*, **163**, 1031–1037.

Schneider, D., Block, B. C., Boeckh, J. and Priesner, E. (1967). Die Reaktion der männlichen Seidenspinner auf Bombykol und seine Isomeren: Extrakten des nogram und Verhalten. *Z. Vergl. Physiol.*, **54**, 192–209.

Schneider, D. and Hecker, E. (1956). Zur Elektrophysiologie der Antenne des Seidenspinners *Bombyx mori* bei Reizung mit angereicherten Extrakten des Sexualstoffes. *Z. Naturforsch.*, **11b**, 121–124.

Schnieder, D., Lacher, V. and Kaissling, K. E. (1964). Die Reaktionwiese und das Reaktionsspektrum von Reichzellen bei *Antheraea pernyi* (Lepidoptera: Saturniidae). *Z. Vergl. Physiol.*, **48**, 632–662.

Scnieider, D. and Seibt, U. (1969). Sex pheromone of the queen butterfly: Electroantennogram responses. *Science*, **164**, 1173–1174.

Schupp, O. E. (1968a). Gas chromatography. New York, Wiley-Interscience.

Schupp, O. E. (1968b). *ibid.* pp. 119–158.

Schupp, O. E. (1968c). *ibid.* pp. 162–166.

Schupp, O. E. (1968d). *ibid.* pp. 173–247.

Schupp. O. E. (1968e). *ibid.* pp. 258–285.

Schwinck, I. (1954). Experimentelle Untersuchungen über geruchssinn und Strömungswahrnehmung in der Orientierung bei Nachschmetterlingen. *Z. Vergl. Physiol.*, **37**, 19–56.

Schwinck, I. (1955). Weitere Untersuchungen zur Frage der Geruchsorientierung der Nachtschmetterlinge: Partielle Fühleramputation bei Spinnermännchen inbesondere am Seidenspinner *Bombyx mori* L. *Z. Vergl. Physiol.*, **37**, 439–458.

Schwinck, I. (1958). A study of olfactory stimuli in the orientation of moths. *Proc. Xth Internat. Congr. Entom.*, **2**, 577–582.

Scott, A. I. (1964). Interpretation of the ultraviolet spectra of natural products. New York, MacMillan.

S.D.C. Scientific Documentation Centre Ltd. Halbeath House, Dunfermline, Fife, United Kingdom.

Sekul, A. A. and Cox, H. C. (1965). Sex pheromone in the fall armyworm, *Spodoptera frugiperda* (J. E. Smith). *Bioscience*, **15**, 670–671.

Sekul, A. A. and Cox, H. C. (1967). Response of males to the female sex pheromone of the fall armyworm, *Spodoptera frugiperda* (Lepidoptera: Noctuidae): A laboratory evaluation. *Ann. Entom. Soc. Amer.*, **60**, 691–693.

Sekul, A. A. and Sparks, A. N. (1967). Sex pheromone of the fall armyworm moth. Isolation, identification and synthesis. *J. Econ. Entom.*, **60**, 1270–1272.

Sharma, R. K., Shorey, H. H. and Gaston, L. K. (1971). Sex pheromones of Noctuid moths. XXIV. Evaluation of pheromone traps for males of *Trichoplusia ni. J. Econ. Entom.*, **64**, 361–364.

Sharma, R. K., Rice, R. E., Reynolds, H. T. and Hannibal, R. M. (1973). Effect of trap design and size of hexalure dispensers on catches of pink bollworm males. *J. Econ. Entom.*, **66**, 377–379.

Shearer, D. A. and Boch, R. (1965). 2-Heptanone in the mandibular gland secretion of the honey bee. *Nature*, **206**, 530.

Shearer, D. A., Boch, R., Morse, R. A. and Laigo, F. M. (1970). Occurrence of 9-oxodec-*trans*-2-enoic acid in queens of *Apis dorsata, Apis cerana,* and *Apis mellifica. J. Insect Physiol.*, **16**, 1437–1441.

Shorey, H. H. (1964). Sex pheromones of Noctuid moths. II. Mating behavior of *Trichoplusia ni* (Lepidoptera: Noctuidae) with special references to the role of the sex pheromone. *Ann. Entom. Soc. Amer.*, **57**, 371–377.

Shorey, H. H. (1970). Sex pheromones of Lepidoptera. *In* Wood, D. L., Silverstein, R. M. and Nakajima, M. eds. Control of insect behavior by natural products. New York, Academic Press, pp. 249–284.

Shorey, H. H. (1973). Behavioral responses to insect pheromones. *Ann. Rev. Entom.*, **18**, 349–380.

Shorey, H. H. and Bartell, R. J. (1970). Role of a volatile female sex pheromone in stimulating male courtship behaviour in *Drosophila melanogaster. Anim. Behav.*, **18**, 159–164.

Shorey, H. H., Bartell, R. J. and Browne, L. B. B. (1969). Sexual stimulation of males of *Lucilia cuprina* (Calliphoridae) and *Drosophila melanogaster* (Drosophilidae) by the doors of aggregation sites. *Ann. Entom. Soc. Amer.*, **62**, 1419–1421.

Shorey, H. H. and Gaston. L. K. (1964). Sex pheromones of Noctuid moths. III. Inhibition of male responses to the sex pheromone in *Trichoplusia ni* (Lepidoptera: Noctuidae). *Ann. Entom. Soc. Amer.*, **57**, 775–779.

Shorey, H. H. and Gaston. L. K. (1965a). Sex pheromones of Noctuid moths. VII. Quantitative aspects of the production and release of pheromone by females of *Trichoplusia ni* (Lepidoptera: Noctuidae). *Ann. Entom. Soc. Amer.*, **58**, 604–608.

Shorey, H. H. and Gaston. L. K. (1965b). Sex pheromones of Noctuid moths. VIII. Orientation to light by pheromone-stimulated males of *Trichoplusia ni* (Lepidoptera: Noctuidae). *Ann. Entom. Soc. Amer.*, **58**, 833–836.

Shorey, H. H. and Gaston. L. K. (1967a). Sex pheromones of Noctuid moths. XII. Lack of evidence for masking of sex pheromone activity in extracts from females of *Heliothis zea* and *H. virescens* (Lepidoptera: Noctuidae): *Ann. Entom. Soc. Amer.*, **60**, 847–849.

Shorey, H. H. and Gaston, L. K. (1967b). Pheromones. *In* Kilgore, W. W. and Doutt, R. L. eds. Pest Control. New York, Academic Press, pp. 241–265.

Shorey, H. H., Gaston, L. K. and Fukoto, T. R. (1964). Sex pheromones of Noctuid moths. I. A quantitative bioassay for the sex pheromone of *Trichoplusia ni* (Lepidoptera: Noctuidae). *J. Econ. Entom.*, **57**, 252–254.

Shorey, H. H., Gaston, L. K. and Jefferson, R. N. (1968a). Insect sex pheromones *In* Mercalf, R. L. ed. Advances in pest control. Vol. 8. New York, Wiley-Interscience, pp. 57–126.

Shorey, H. H., Gaston. L. K. and Roberts, J. S. (1965). Sex pheromones of Noctuid moths. VI. Absence of behavioral specificity for the female sex pheromones of *Trichoplusia ni* versus *Autographa californica* and *Heliothis zea* versus *H. virescens.* (Lepidoptera: Noctuidae). *Ann. Entom. Soc. Amer.*, **58**, 600–603.

Shorey, H. H., Gaston, L. K. and Saario, C. A. (1967). Sex pheromones of Noctuid moths. XIV. Feasibility of behavioral control by disrupting pheromone communication in cabbage loopers. *J. Econ. Entom.*, **60**, 1541–1545.

Shorey, H. H., McFarland, S. U. and Gaston, L. K. (1968c). Sex pheromones of Noctuid moths. XIII. Changes in pheromonal quantity, as related to reproductive age and mating history, in females of seven species of Noctuidae (Lepidoptera). *Ann. Entom. Soc. Amer.*, **61**, 372–376.

Shorey, H. H., Morin, K. L. and Gaston. L. K. (1968b). Sex pheromones of Noctuid moths. XV. Timing of development of pheromone responsiveness and other indicators of reproductive age in males of eight species. *Ann. Entom. Soc. Amer.*, **61**, 857–861.

Shriner, R. L., Fuson, R. C. and Curtin, D. Y. (1964). The systematic identification of organic compounds. 5th edn. New York, Wiley.

Sievers, R. E. and Rondeau, R. E. (1971). New superior paramagnetic shift reagents for NMR spectral clarification. *J. Amer. Chem. Soc.*, **93**, 1522–1524.

Siggia, S. ed. (1972). Instrumental methods of organic functional group analysis. New York, Wiley-Interscience.

Silverstein, R. M. (1968). Spectrometric identification of insect sex attractants. *J. Chem. Ed.*, **45**, 794–797.

Silverstein, R. M. (1970a). Attractant pheromones of Coleoptera. *In* Beroza, M. ed. Chemicals controlling insect behavior. New York, Academic Press, pp. 21–40.

Silverstein, R. M. (1970b). Methodology for isolation and identification of insect pheromones – examples from Coleoptera. *In* Wood, D. L., Silverstein, R. M., and Nakajima, M. eds. Control of insect behavior by natural products. New York, Academic Press, pp. 285–299.

Silverstein, R. M. (1971a). Recent and collaborative studies of insect pheromones. *In* Tahori, A. S. ed. Chemical releasers in insects. New York, Gordon and Breach Science Publishers, pp. 69–89.

Silverstein, R. M. (1971b). Some collaborative pheromone investigations. *XXIII International Congress of Pure and Applied Chemistry*, **3**, 3–15.

Silverstein, R. M., Bassler, G. C. and Morrill, T. C. (1974). Spectrometric identification of organic compounds. 3rd edn. New York, Wiley.

Silverstein, R. M., Brownlee, R. G., Bellas, T. E., Wood, D. L. and Browne, L. E. (1968). Brevicomin: Principle sex attractant in the frass of the female western pine beetle. *Science*, **159**, 889–891.

Silverstein, R. M. and Rodin, J. O. (1966). Insect pheromone collection with absorption columns. I. Studies on model organic compounds. *J. Econ. Entom.*, **59**, 1152–1154.

Silverstein, R. M., Rodin, J. O., Burkholder, W. E. and Gorman, J. E. (1967a). Sex attractant of the black carpet beetle. *Science*, **157**, 85–87.

Silverstein, R. M., Rodin, J. O. and Wood, D. L. (1967b). Methodology for isolation and identification of insect pheromones with reference to studies on California five-spined *Ips*. *J. Econ. Entom.*, **60**, 944–949.

Simon, W. and Clerc, T. (1967). Strukturaufklärung organischer Verbindungen mit spektroskopischen Methoden. Frankfurt am Main, Germany, Akademische Verlagsgesellschaft.

Simons, W. W. and Zanger, M. (1972). The Sadtler guide to NMR spectra. Philadelphia, Pa., Sadtler Research Labs. Inc.

Skrikevicius, A. and Vaitkeviciene, G. (1970). Insects possessing sexual pheromones. *Acta Entom. Lithuanica*, **1**, 163–174.

Smith, H. W. and Furniss, M. M. (1966). An automatically recording insect flight mill. *Can. Entom.*, **98**, 249–252.

Smythe, R. V., Coppel, H. C., Lipton, S. H. and Strong, F. M. (1967). Chemical studies of attractants associated with *Reticulitermes flavipes* and *R. virginicus*. *J. Econ. Entom.*, **60**, 228–233.

Snatzke, G. ed. (1967). Optical rotatory dispersion and circular dichroism in organic chemistry. London, Heyden.

Snow, J. W. and Copeland, W. W. (1969). Fall armyworm: Use of virgin female traps to detect males and to determine seasonal distribution. U.S.D.A. Production Research Report No. 110.

Solomon, J. D., Doolittle, R. E. and Beroza, M. (1972). Isolation and analysis of the carpenterworm sex pheromone. *Ann. Entom. Soc. Amer.*, **65**, 1058–1061.

Solomon, J. D. and Morris, R. C. (1966). Sex attraction of the carpenterworm moth. *J, Econ. Entom.*, **59**, 1534–1535.

Soo Hoo, C. F. and Roberts, R. J. (1965). Sex attraction in *Rhopaea* (Coleoptera: Scarabaeidae). *Nature*, **205**, 724–725.

Sower, L. L., Coffelt, J. A. and Vick, J. W. (1973). Sex pheromone: a simple method of obtaining relatively pure material from females of five species of moths. *J. Econ. Entom.*, **66**, 1220–1222.

Sower, L. L., Gaston, L. K. and Shorey, H. H. (1971). Sex pheromones of Noctuid moths. XXVI. Female release rate, male response threshold, and communication distance for *Trichoplusia ni*. *Ann. Entom. Soc. Amer.*, **64**, 1448–1456.

Sower, L. L., Shorey, H. H. and Gaston, L. K. (1970). Sex pheromones of Noctuid moths. XXI. Light dark cycle regulation and light inhibition of sex pheromone release by females of *Trichoplusia ni*. *Ann. Entom. Soc. Amer.*, **63**, 1090–1092.

Staddon, B. W. and Weatherston, J. (1967). Constituents of the stink gland of the water bug *Illycoris cimicoides*. *Tetrahedron Letters*, 4567–4571.

Stahl, E. ed. (1969). Thin-layer chromatography. 2nd edn. New York, Springer-Verlag.

Stállberg-Stenhagen, S. (1972). Studies on natural odoriferous compounds. V. Splitter-free all glass intake system for glass capillary gas chromatography of volatile compounds from biological material. *Chemica Scripta*, 2, 97–100.

Steinbrecht, R. A. (1964). Die Abhängigkeit der Lockwirkung des Sexualduftorgans weiblicher Seidenspinner (*Bombyx mori*) von Alter und Kopulation. *Z. Vergl. Physiol.*, **48**, 341–356.

Stenhagen, A., Abrahamsson, S. and McLafferty, F. W. eds. (1968). Atlas of mass spectral data. (Three volumes) New York, Wiley-Interscience.

Stenhagen, A., Abrahamson, S. and McLafferty, F. W. eds. (1969). Archives of mass spectral data. New York, Wiley-Interscience.

Stern, E. S. and Timmons, C. J. (1971). Electronic absorption spectroscopy in organic chemistry. New York, St. Martins Press.

Stine, K. E. (1969). Modern practices in infrared spectroscopy. Beckman Instruments.

Stothers, J. B. (1972). Carbon-13 NMR spectroscopy. New York, Academic Press.

Strang, G. F. (1970). A study of honey bee drone attraction in the mating response *J. Econ. Entom.*, **63**, 641–645.

Struble, D. L. (1970). A sex pheromone in the forest tent caterpillar. *J. Econ. Entom.*, **63**, 295–296.

Struble, D. L. and Jacobson, L. A. (1970). A sex pheromone in the red-backed cutworm. *J. Econ. Entom.*, **63**, 841–844.

Stuart, A. M. (1970). The role of chemicals in termite communication. *In* Johnson, J. W. Jr., Moulton, D. G. and Turk, A. eds. Advances in chemoreception Vol. I. Communication by chemical signals. New York, Appleton-Century-Crofts, pp. 79–106.

Stürckow, B. (1965). The electroantennogram (EAG) as an assay for the reception of odours by the gypsy moth. *J. Insect Physiol.*, **11**, 1573–1584.

Stürckow, B. (1970). Responses of olfactory and gustatory receptor cells in insects. *In* Johnson, J. W. Jr., Moulton, D. G., and Turk, A. eds. Advances in chemoreception, Vol. 1. New York, Appleton-Century-Crofts, p. 107.

Sudha, G. and Nayar, K. K. (1970). Pheromones in insects. *J. Anim. Morphol. Physiol.*, **17**, 111–120.

Szymanski, H. A. (1963). Infrared band handbook. Supplements 1, 2 (1964), 3, 4 (1966). New York, Plenum Press.

Szymanski, H. A. (1964). Interpreted infrared spectra. Vol. I and II. New York, Plenum Press.

Takahashi, F., Masui, A. and Kuwahara, Y. (1972). Studies on sex pheromone of Pyralidae. IV. The male response to the female sex pheromone of the almond moth, *Cadra cautella* Walker (Phycitinae). *Botyu Kagaku*, **37**, 56–60.

Takahashi, S. and Kitamura, C. (1972). Bioassay procedure of the sex stimulant of the American cockroach, *Periplaneta americana* (L.) (Orthoptera: Blattidae). *Appl. Entom. Zool.*, **7**, 133–141.

Tardif, J. G. R. and Stevens, L. J. (1972). A simple method of pinioning insects. *J. Econ. Entom.*, **65**, 1199–1200.

Tashiro, H., Beavers, J. B. and Moreno, D. (1969a). Comparative response of two strains of California red scale, *Aonidiella aurantii*, males to pheromone extract and to females of the reciprocal strain. *Ann. Entom. Soc. Amer.*, **62**, 279–280.

Tashiro, H. and Chambers, D. L. (1967). Reproduction in the California red scale, *Aonidiella aurantii* (Homoptera: Diaspididae). I. Discovery and extraction of a female sex pheromone. *Ann. Entom. Soc. Amer.*, **60**, 1166–1170.

Tashiro, H., Chambers, D. L., Moreno, D. and Beavers, J. (1969b). Reproduction in the California red scale *Aonidiella aurantii*. III. Development of an olfactometer for bioassay of the female sex pheromone. *Ann. Entom. Soc. Amer.*, **62**, 935–940.

Taylor, W. I. (1969). Insect terpenoids and nepetalactone. *In* Taylor, W. I. ed. Cyclopentanoid terpene derivatives. New York, Dekker, pp. 203–238.

Teranishi, R., Lundin, R. E., McFadden, W. H. and Scherer, J. R. (1967). Ancillary systems. *In* Ettre, L. S. and Zlatkis, A. eds. The practice of gas chromatography. New York, Wiley-Interscience, pp. 408–411.

Theilheimer, W. Synthetic methods of organic chemistry. (A continuing series). New York, Karger.

Thomas, M. B. B. ed. (1960). An index of published infra-red spectra. Vols. I and II. New York, British Information Service.

Toba, H. H., Kishaba, A. N. and Wolf, W. W. (1968). Bioassay of the synthetic female sex pheromone of the cabbage looper. *J. Econ. Entom.*, **61**, 812–816.

Toba, H. H., Kishaba, A. N. and Wolf, W. W. (1969). A polythylene bag for dispensing the synthetic sex pheromone of the cabbage looper. *J. Econ. Entom.*, **62**, 517–518.

Traynier, R. M. M. (1970). Habituation of the response to sex pheromone in two species of Lepidoptera, with reference to a method of control. *Ent. Exp. & Appl.*, **13**, 179–187.

Trier, K. H. and Metzger, R. (1972). Möglichkeiten des Einsatzes von Lockstoffen in der Schädlings-bekämpfung. *Biol. Rundsch.*, **10**, 188–199.

Tschinkel, W. R. and Close, P. G. (1973). The trail pheromone of the termite *Trinervitermes trinervoides. J. Insect Physiol.*, **19**, 707–721.

Tumlinson, J. H., Gueldner, R. C., Hardee, D. D., Thompson, A. C., Hedin, P. A. and Minyard, J. P. (1970). The boll weevil sex attractant. *In* Beroza, M. ed. Chemicals controlling insect behavior. New York, Academic Press, pp. 41–59.

Tumlinson, J. H., Hardee, D. D., Gueldner, R. C., Thompson, A. C., Hedin, P. A. and Minyard, J. P. (1969). Sex pheromones produced by male boll weevil: Isolation, identification, and synthesis. *Science*, **166**, 1010–1012.

Tumlinson, J. H., Hardee, D. D., Minyard, J. P., Thompson, A. C., Gast, R. T. and Hedin, P. A. (1968). Boll weevil sex attractant: Isolation studies. *J. Econ. Entom.*, **61**, 470–474.

Tumlinson, J. H., Moser, J. C., Silverstein, R. M., Brownlee, R. G. and Ruth, J. M. (1972). A volatile trail pheromone of the leaf-cutting ant, *Atta texana. J. Insect Physiol.*, **18**, 809–814.

Tumlinson, J. H., Moser, J. C., Silverstein, R. M., Brownlee, R. G. and Ruth, J. M. (1971). Identification of the trail pheromone of a leaf cutting ant, *Atta texana. Nature*, **234**, 348–349.

Ungnade, H. G. ed. (1960). Organic electronic spectral data. Vol. II. New York, Wiley-Interscience. Covers 1953–1955.

Vadasz, C. (1971). Pheromones. Biologically active substances of interspecific regulation. *Biol. Kozlem*, **19**, 85–113.

Varian Associates. High resolution NMR spectra catalogue. Vol. I. (1962), Vol. 2. (1963). Palo Alto, Calif., Varian Associates.

Veibel, S. (1971). The identification of organic compounds. 7th edition. Kobenhaun, Denmark, G.E.C. Gads Forlag.

Velluz, L., Legrand, M. and Grosjean, M. (1965). Optical circular dichroism. New York, Academic Press.

Vick, K. W., Burkholder, W. E. and Gorman, J. E. (1970). Interspecific response to sex pheromones of *Trogoderma* species (Coleoptera: Dermestidae). *Ann. Entom. Soc. Amer.*, **63**, 379–381.

Vick, K. W., Drumond, P. C., Sower, L. L. and Coffelt, J. A. (1973). Sex pheromone habituation: The effects of habituation on the pheromone response level of *Trogoderma inclusum* (Coleoptera: Dermestidae). *Ann. Entom. Soc. Amer.*, **66**, 667–670.

Vinson, S. B. (1972). Courtship behavior and evidence for a sex pheromone in the parasitoid *Campoletis sorensis* (Hymenoptera: Ichneumonidae). *Envir. Entom.*, **1**, 409–414.

Vité, J. P. (1970). Sex attractants in frass from bark beetles. *Science*, **156**, 105.

Vité, J. P. and Gara, R. I. (1961). A field method for observation on olfactory responses of bark beetles (Scolytidae) to volatile materials. *Contrib. Boyce Thompson Inst.*, **21**, 175–182.

Vité, J. P. and Pitman, G. B. (1969). Insect and host odours in the aggregation behavior of the western pine beetle. *Can. Entom.*, **101**, 113–117.

Vité, J. P. and Renwick, J. A. A. (1970). Differential diagnosis and isolation of population attractants. *Contrib. Boyce Thompson Inst.*, **24**, 175–182.

Vogel, A. I. (1956). Practical organic chemistry. 3rd edn. London, Longmans.

Von Ammon, R. and Fischer, R. D. (1972). Shift reagents in NMR spectroscopy. *Angew. Chem. Int. Ed.*, **11**, 675–692.

Wagner, R. B. and Zook, H. D. (1953). Synthetic organic chemistry. New York, Wiley.

Walsh, J. T. and Merritt, C. Jr. (1960). Qualitative functional group analysis of gas chromatographic effluents. *Anal. Chem.*, **32**, 1378–1381.

Warren, F. L. (1972). Insect hormones and pheromones – correlation of chemical structure and insect-plant relation. *Chem. New Zealand*, **36**, 37–51.

Waters, R. M. and Jacobson, M. (1968). A rack for holding male gypsy moths during laboratory bioassays of airborne attractants. *J. Econ. Entom.*, **61**, 873.

Watson, J. T. (1969). Gas chromatography and mass spectroscopy. *In* Ettre, L. S. and McFadden, W. H. eds. Ancillary techniques of gas chromatography. New York, Wiley-Interscience, pp. 145–225.

Weatherston, J. (1967). The chemistry of arthropod defensive substances. *Quart. Rev.*, **21**, 287–313.

Weatherston, J. and Percy, J. E. (1968). Studies of physiologically active arthropod secretions. I. Evidence for a sex pheromone in female *Vitula edmandsae* (Lepidoptera: Phycitidae). *Can. Entom.*, **100**, 1065–1070.

Weatherston, J. and Percy, J. E. (1969). Studies of physiologically active arthropod secretions. III. Chemical, morphological, and histological studies of the defense mechanism of *Uroblaniulus canadensis* (Say) (Diplopoda: Julida). *Can. J. Zool.*, **47**, 1389–1394.

Weatherston, J. and Percy, J. E. (1970). Arthropod defensive secretions. *In* Beroza, M. ed. Chemicals controlling insect behavior. New York, Academic Press, pp. 95–144.

Weatherston, J., Roelofs, W. L., Comeau, A. and Sanders, C. J. (1971). Studies of physiologically active arthropod secretions. X. Sex pheromone of the eastern spruce budworm, *Choristoneura fumiferana* (Lepidoptera: Tortricidae). *Can. Entom.*, **103**, 1741–1747.

Weiss, F. T. (1970). Determination of organic compounds: Methods and procedures. New York, Wiley-Interscience.

Wharton, D. R. A., Black, E. D., Merritt, C. Jr., Wharton, M. L., Bazinet, M. and Walsh, J. T. (1962). Isolation of sex attractant of the American cockroach, *Periplaneta americana* (L.). *Science*, **137**, 1062–1063.

Wharton, D. R. A., Miller, G. L. and Wharton, M. L. (1954a). The odorous attractant of the American cockroach, *Periplaneta americana* (L.). I. Quantitative aspects of the response to the attractant. *J. Gen. Physiol.*, **37**, 461–469.

Wharton, D. R. A., Miller, G. L. and Wharton, M. L. (1954b). The odorous attractant of the American cockroach, *Periplaneta americana* (L.). II. A bioassay method for the attractant. *J. Gen. Physiol.*, **37**, 471–481.

Wheeler, O. H. and Kaplan, L. A. (1966). Organic electronic spectral data. Vol. III. New York, Wiley-Interscience. Covers 1956–1957.

White, R. G. (1965). Handbook of ultraviolet methods. New York, Plenum Press.

Whittaker, R. H. and Feeny, P. P. (1971). Allelochemics: Chemical interactions between species. *Science*, **171**, 757–770.

Whitten, W. K. and Bronson, F. H. (1970). The role of pheromones in mammalian reproduction. *In* Johnson, J. W. Jr., Moulton, D. G. and Turk, A. eds. Advances in chemoreception, Vol. I, Communication by chemical signals. New York, Appleton-Century-Crofts, pp. 309–325.

Wiberg, K. B. (1960a). Laboratory technique in organic chemistry. New York, McGraw-Hill, pp. 20–74.

Wiberg, K. B. (1960b). *ibid.* pp. 149–165.

Wiberg, K. B. and Nist, B. J. (1962). The interpretation of NMR spectra. New York, Benjamin.

Williams, C. M. and Bowers, W. S. (1968). Conference on insect-plant interactions. *Bioscience*, **18**, 791–792.

Williams, D. H. and Fleming, I. (1966). Spectroscopic methods in organic chemistry. New York, McGraw-Hill.

Wilson, E. O. (1959). Source and possible nature of the odor trail of fire ants. *Science*, **129**, 643–644.

Wilson, E. O. (1963a). The social biology of ants. *Ann. Rev. Entom.*, **8**, 345–368.

Wilson, E. O. (1963b). Pheromones. *Sci. Amer.*, **208**, 100–114.

Wilson, E. O. (1965). Chemical communication in the social insects. *Science*, **149**, 1064–1071.

Wilson, E. O. (1968). Chemical systems. *In* Sebeok, T. A. ed. Animal communication, techniques of study and results of research. Univ. of Indiana Press. pp. 75–102.

Wilson, E. O. (1970). Chemical communication within animal species. *In* Sondheimer, E. and Simeone, J. B. eds. Chemical ecology. New York, Academic Press, pp. 133–155.

Wilson, E. O. and Bossert, W. H. (1963). Chemical communication among animals. *Rec. Prog. Hormone Res.*, **19**, 673–716.

Wolf, W. W., Hartsock, J. G., Ford, J. H., Henneberry, T. J., Hills, O. A. and Debolt, J. W. (1969). Combined use of sex pheromone and electric traps for cabbage looper control. *Trans. Amer. Soc. Agr. Eng.*, **12**, 329–335.

Wolf, W. W., Kishaba, A. N., Howland, A. F. and Henneberry, T. J. (1967). Sand as a carrier for synthetic sex pheromone of cabbage loopers used to bait blacklight and carton traps. *J. Econ. Entom.*, **60**, 1182–1184.

Wong, T. T. Y., Cleveland, M. L., Ralston, D. F. and Davis, D. G. (1971). Virgin female traps to determine activity and populations of red-banded leaf roller. *J. Econ. Entom.*, **64**, 132–134.

Wood, D. L. (1970). Pheromones of bark beetles. *In* Wood, D. L., Silverstein, R. M. and Nakajima, M. eds. Control of insect behavior by natural products. New York, Academic Press, pp. 301–316.

Wood, D. L. (1971). Selection and colonization of ponderosa pine bark beetles. *Symp. Roy. Ent. Soc. London.*

Wood, D. L., Browne, L. E., Bedard, W. D., Tilden, P. E., Silverstein, R. M. and Rodin, J. O. (1968). Response of *Ips confusus* to synthetic sex pheromones in nature. *Science*, **159**, 1373–1374.

Wood, D. L., Browne, L. E., Silverstein, R. M. and Rodin, J. O. (1966). Sex pheromones of bark beetles. I. Mass production, bioassay, source, and isolation of the sex pheromone of *Ips confusus*. *J. Insect Physiol.*, **12**, 523–536.

Wood, D. L. and Bushing, R. W. (1963). The olfactory response of *Ips confusus* (Le Conte) (Coleoptera: Scolytidae). to the secondary attraction in the laboratory. *Can. Entom.*, **95**, 1066–1078.

Wood, D. L., Silverstein, R. M. and Rodin, J. O. (1967a). Sex attractants in frass from bark beetles. *Science*, **156**, 105.

Wood, D. L., Stark, R. W., Silverstein, R. M. and Rodin, J. O. (1967b). Unique synergistic effects produced by the principle sex attractant compounds of *Ips confusus* (Le Conte). *Nature*, **215**, 206.

Wright, R. H. (1958). The olfactory guidance of flying insects. *Can. Entom.*, **90**, 81–89.

Wright, R. H. (1964). After pesticides – what? *Nature*, **204**, 121–125.

Wright, R. H. (1966). An insect olfactometer. *Can. Entom.*, **98**, 282–285.

Yamada, M. (1970). Electrophysiological investigation of insect olfaction. *In* Wood, D. L., Silverstein, R. M. and Nakajima, M. eds. Control of insect behavior by natural products. New York, Academic Press, pp. 317–330.

Yamamoto, R. (1963). Collection of the sex attractants from female American cockroaches. *J, Econ. Entom.*, **56**, 119–120.

Yinon, U. and Shulov, A. (1967). A bioassay of the pheromone of *Trogoderma granarium* males as an attractant for both sexes of the species. *Entom. Exp. Appl.*, **10**, 453–462.

Young, J. C., Brownlee, R. G., Rodin, J. O., Hildebrand, D. N., Silverstein, R. M., Wood, D. L., Birch, M. C. and Browne, L. E. (1974). Identification of Linalool produced by two species of bark beetles of the genus *Ips. J. Insect Physiol.*, **19**, 1615–1622.

Zenther-Møller, O. and Rudinsky, J. A. (1967). Studies on the site of sex pheromone production in *Dendroctonus psuedot: sugae* (Coleoptera: Scolytidae). *Ann. Entom. Soc. Amer.*, **60**, 575–582.

Zlatkis, A. and Pretorius, V. (1971). Preparative gas chromatography. New York, Wiley-Interscience.

Zweig, G. and Sherma, J. eds. (1972). Handbook of chromatography. 2 vols. Cleveland, Ohio, CRC Press.

4 The Application of Transmission Electron Microscopy on the Study of the Olfactory Epithelium of Vertebrates

HELMUT ALTNER

IRIS KOLNBERGER
Facbereich Biologie
Universität Regensburg
West Germany

I. Introduction

The development of techniques apt for the investigation of biological substrates in the transmission electron microscope elicited a great number of descriptive studies on the olfactory organs in vertebrates. More essentially, the new techniques can be used as a tool for answering functional questions. For about one hundred years it has been asked whether morphologically different receptor types exist within the olfactory epithelium of one species. The question was whether structures could be detected of which the properties could give some information about the transducer mechanism. Further, since Adrian (1955) had observed signs of interaction at the receptor level, one might now ask, by which structural specializations such influences could be mediated. Moreover, electron microscopy could be expected to be a favourable means for checking the hypothesis of whether receptors regenerate after destruction of the olfactory mucosa. Related to this question is that of whether a constant regeneration takes place in the olfactory epithelium and what kind of changes happen on regeneration. This idea was mainly introduced by the finding of a regeneration in the taste buds of rats by Beidler and Smallman (1965). While the merely descriptive aspects are reviewed in the first section of this chapter the functional aspects are discussed particularly in the following two parts. Attempts to check former quantitative investigations, e.g. of the receptor packing and to measure the size of the receptive surface are subjects of the last section.

II. Structural analysis of the cell types in the nasal and vomero-nasal olfactory mucosa

Techniques

The success of an electron microscopical structural analysis depends mainly on the fixation techniques applied. Therefore the first ultra-structural investigations impart only limited information because of the insufficiently developed techniques. Methodological aspects are intensively discussed by Andres (1969) and Seifert (1970). From their observations as well as from the methodological work by Maunsbach (1966a, b) and Sjöstrand (1967) it is evident that the following conditions of the procedure are of particular importance: (1) Application of perfusion fixation instead of immersion, (2) Osmolarity of the buffer used for fixation and washing, and (3) Purity of the glutaraldehyde used for fixation.

Especially from the investigations of Andres (1966, 1969) Drenck-hahn (1970) and Seifert (1970) it can be concluded that in birds and mammals good results are obtained by using perfusion fixation. According to Seifert (1970) the following procedure can be recommended:

Anaesthetic: Nembutal 50 mg/l, 10–100 mg/kg i.v. or i.p. (with Heparin added to prevent coagulation)

Perfusion: I. 40.0 g Macrodex (Knoll), 9.0 g NaCl, 0.3 g KCl, 2 g $CaCl_2$, 0.05 g $MgCl_2$ in 1000 ml Veronal-acetate-buffer pH 7.4–7.8, 37°C, 10–20 s.

II. 2.5% glutaraldehyde in 0.1 M phosphate buffer (Sörensen) pH 7.4–7.8, 37°C, 30–60 s., 3–6°C, 20 min. 120–140 mmHg until followed by 30–40 mmHg.

III. 6.2% sucrose in 0.1 M phosphate buffer pH 7.4–7.8, 10 minutes.

Postfixation: 2% OsO_4 and 5% sucrose in 0.1 M phosphate buffer.

In smaller animals, however, perfusion is hardly practicable. In these cases and in others when special problems are to be dealt with such as an optimal preservation of cilia (Okano *et al.*, 1967), immersion fixation has proved to produce acceptable results. Altner *et al.* (1970), Kolnberger and Altner (1971) and Kolnberger (1971a) use 6% glutaraldehyde in 0.1 M cacodylate buffer pH 7.3 (4°C or temperatures up to 18°C in the beginning and then cooled to 4°C, 2–3 h) with good success in the vomero-nasal organ of amphibia, reptiles and mammals.

Other authors use a variety of fixatives. Since many publications most reasonably restrict figures to high-power micrographs of selected structures it is difficult to evaluate the fixation procedures if they are not discussed by the authors.

It may be stated that several authors immerse the tissue in OsO_4-solutions without preceding aldehyde fixation: Reese (1965: frog) uses 1.0% OsO_4 in sodium phosphate or veronal acetate buffer pH 7.6–7.8 with NaCl and $CaCl_2$ added to give an osmolarity equivalent to 0.65% NaCl. Yamamoto *et al.* (1965: frog) fix in 1% OsO_4 in 0.02 M phosphate buffer (Millonig, 1962) 4°C. pH 7.4. Thornhill (1967: lamprey) uses 2% OsO_4 in 0.02 M phosphate buffer (Millonig, 1962) adapted to the osmolarity of lamprey blood by the addition of sucrose. Graziadei (1966a: mole), Graziadei and Bannister (1967: duck) and Graziadei and Tucker (1968: turtle) also use 2% OsO_4 in phosphate buffer.

According to the improvement of fixation procedures it seems as if some structural observations discussed in earlier investigations would be interpreted more reservedly today. The following major artifacts can be observed in olfactory epithelia most frequently: (1) Inflation of cilia and microvilli (Figs. 1, 2), (2) expansion of the extracellular spaces by shrinkage of sensory cell pericarya, (3) swelling of the apical parts of dendrites and supporting cells.

In summing up it can be stated that several fixation techniques may be used with good results. For special purposes such as the investigation of the intercellular space (see p. 194) diverse methods are to be checked with respect to the possibility of artifacts of minor extension. The discussion about the terminal film elucidates the difficulty of judging to which extent structures might be changed by fixation procedures. Andres (1966, 1969) claims that there exists a rather dense terminal film covering two mucus layers of lesser viscosity (Fig. 3). The terminal film represents the bordering area against the air which fills the nasal cavity. It is characterized by

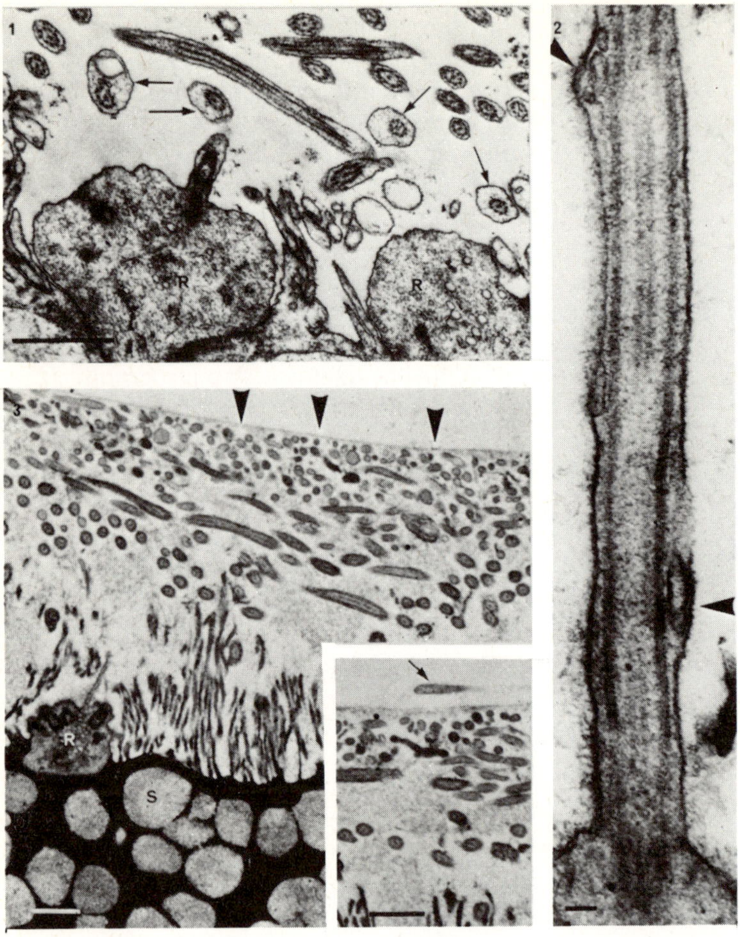

FIG. 1. Section through the terminal swellings of two nasal olfactory receptors (R) in the frog (Rana esculenta). Several dilated cilia and microvilli are visible (arrows). These dilations can be regarded, at least partly, to be caused by inadequate fixation. Osmium impregnation (Friend). Scale 1 μm.

FIG. 2. Longitudinal section of an olfactory cilium (slowworm, *Anguis fragilis*). In the proximal portion of the ciliary shaft small vesicles are found between the outer tubules and the ciliary membrane (arrow heads). Glutaraldehyde fixation. Scale 1000 Å.

FIG. 3. Nasal olfactory epithelium of a reptile (*Anguis fragilis*). The surface of the mucus appears as a dense line (terminal film, arrow heads). Inset: the distal segment of a cilium (arrow) is visible outside the terminal film. R terminal swelling of a receptor, S supporting cell containing secretory granules. Glutaraldehyde fixation. Scale 1 μm.

filamentous material, which is supposed to be secreted by the Bowman's glands and which is observed to aggregate around the distal segments of the olfactory cilia (Andres, 1969). Reese's (1965) observation of an existing thin surface layer of watery mucus would, therefore, be due to a destruction of the terminal film. Seifert (1970), however, notices the terminal film ("Deckhäutchen") very clearly but is of the opinion that this structure is an artifact. He argues that distal segments of sensory cell cilia which are recorded to lie over this layer (Fig. 3) could not be in contact to air but mostly be covered by mucus.

Results

Cell forms in the nasal and vomero-nasal sensory epithelium. Most investigations of the olfactory mucosa with the electron microscope were mainly aimed at defining the cell forms within the epithelium and drawing up an inventory of the different cell forms (Bloom and Engström, 1952; Bloom, 1954; de Lorenzo, 1957, 1963, 1970; Brettschneider, 1958; Yasutake, 1959; Trujillo-Cenoz, 1961; Andres, 1965, 1966, 1969; Bannister, 1965, 1968; Bronstein and Ivanov, 1965; Graziadei, 1965, 1966a, 1966b; Okano, 1965; Okano and Sugawa, 1965; Reese, 1965; Yamamoto *et al.,* 1965; Bronstein and Pjatkina, 1966, 1969; Frisch, 1967, 1969; Graziadei and Bannister, 1967; Okano *et al.,* 1967; Seifert and Ule, 1967; Thornhill, 1967; Wilson and Westerman, 1967; Altner and Müller, 1968; Graziadei and Tucker, 1968, 1970; Seifert, 1968, 1970; Cauna *et al.,* 1969; Luckhaus, 1969; Altner *et al.,* 1970; Drenckhahn, 1970; Kauer and Moulton, 1970; Kolb, 1971; Kolnberger, 1971a; Schulte and Holl, 1971; Schulte, 1972). The occurrence of sensory cells, supporting and basal cells was well established by light microscopical studies. In using the electron microscope Andres (1965, 1966) was able to describe another cell type, the blastema cells, in the nasal mucosa of the rat, the cat and the dog. These cells, scattered among the basal cells are considered as precursors of regenerating receptors (see page 194). It was observed that they divide occasionally. Resembling mature receptors in occurrence and arrangement of the substantial contents they can be distinguished by the irregular arrangement and less tight packing of the membranes of the endoplasmic reticulum (Andres, 1966). In reptiles (Kolnberger, unpublished) and birds (Drenckhahn, 1970), however, such blastema cells have not been observed.

The shape and the contents of the receptor cells are rather uniform in the different orders of vertebrates. The receptors can be divided

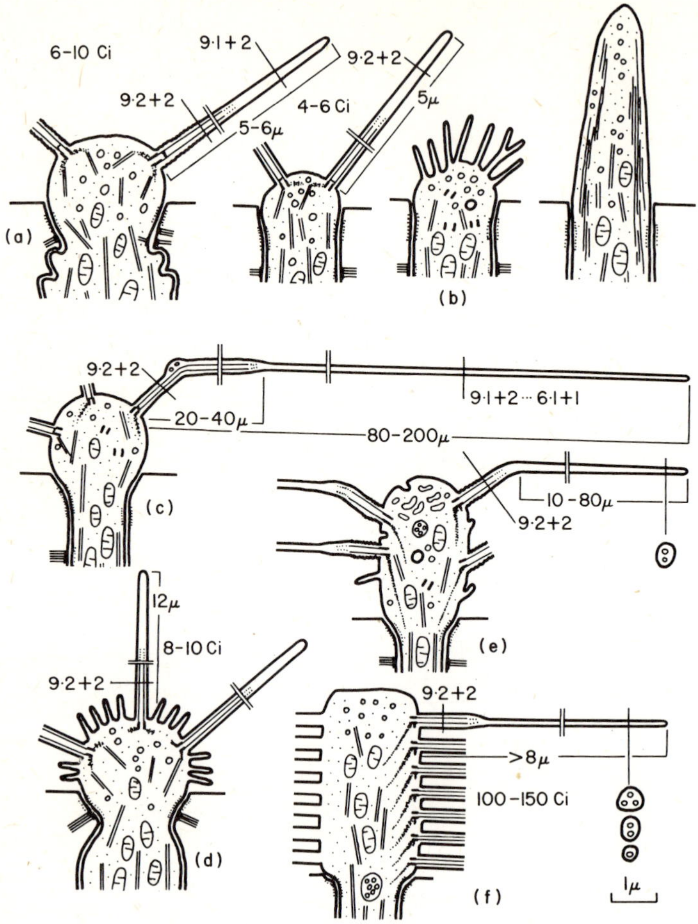

FIG. 4. Schematic diagram of nasal (a-f) and vomero-nasal (g-k) receptor endings in vertebrates Number and length of cilia (Ci) are indicated as well as number and pattern of ciliary tubules. a, lamprey (*Lampetra*) after Thornhill (1967), b, minnow (*Phoxinus*) after Bannister (1965) and Altner (unpublished), c, frog (*Rana*) adapted from Reese (1965), and Yamamoto *et al.* (1965), d, duck (*Anas*) after Graziadei and Bannister (1967), e, cat (*Felis*) after Andres (1969), f, dog (*Canis*) after Okano *et al.* (1967), g, frog (*Rana*), h, lizard (*Lacerta*), i, viper (*Vipera*), k, tree-shrew (*Tupaia*).

into a cylinderic dendrite, the pericaryon and a rather thin axon, measuring only about 0.2 μm in diameter. Specially the terminal swelling of the dendrite, the olfactory knob, and the structure of the cilia originating from this part of the sensory cell have been studied

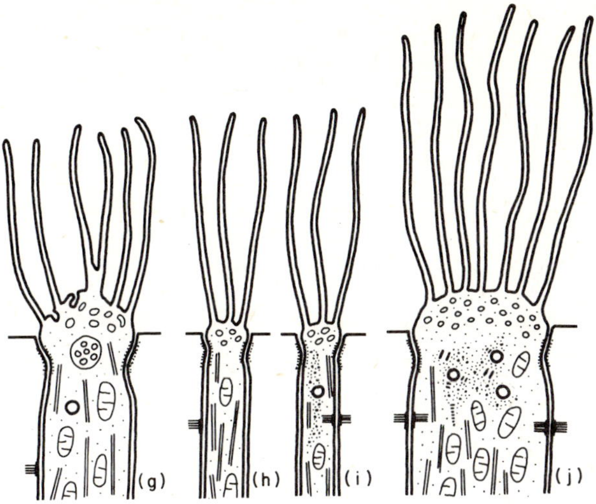

in detail in several species (Bannister, 1965; Okano, 1965; Reese, 1965; Andres, 1966, 1969; Bronstein and Pjatkina, 1966, 1969; Frisch, 1967, 1969; Okano *et al.,* 1967; Graziadei and Bannister, 1967; Thornhill, 1967; Drenckhahn, 1970). There are some differences between nasal and vomero-nasal receptors worth mentioning (Fig. 4). While nasal receptors, as a rule, extend over the surface of the flattened supporting cell apices forming the olfactory knob, vomero-nasal sensory cells do not have such protrusions. Moreover, in the vomero-nasal organ cilia are lacking (Fig. 5) (Altner and Müller, 1968; Bannister, 1968; Altner *et al.,* 1970; Graziadei and Tucker, 1970; Andres, 1970; Kauer and Moulton, 1970; Kolnberger, 1971a; Kolnberger and Altner, 1971; Moulton, 1971). In the nasal mucosa receptors without cilia have been observed only in fish (Figs 6, 7) (Bannister, 1965; Schulte, 1972).

The centrioles and cilium structure precursor bodies, which are accumulated in the dendrites of the vomero-nasal receptors of reptiles and mammals (Fig. 5) can be regarded as representing stages in ciliary structure reduction. In several species of *Lacerta* a complete absence of these structures has been recorded (Altner *et al.,* 1970; Kolnberger, 1971a; Kolnberger and Altner, 1971).

The receptors in both organs are alike in that they contain light vesicles (Figs 4, 5, and 11) or tubules (de Lorenzo, 1963, 1970; Andres, 1965: "Labyrinth") in the apical region of the dendrite. Though highly characteristic for all vertebrate receptors the function

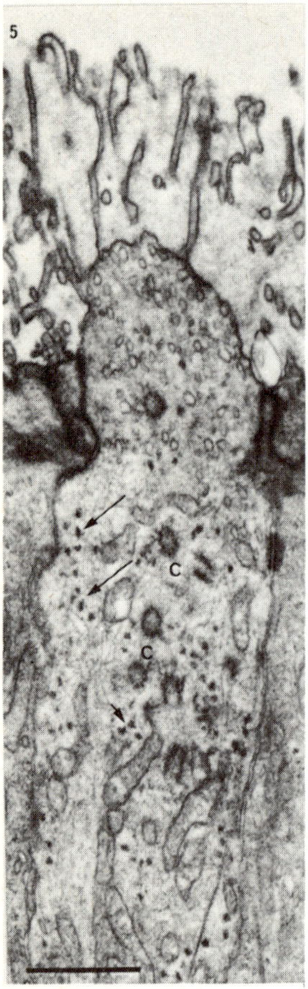

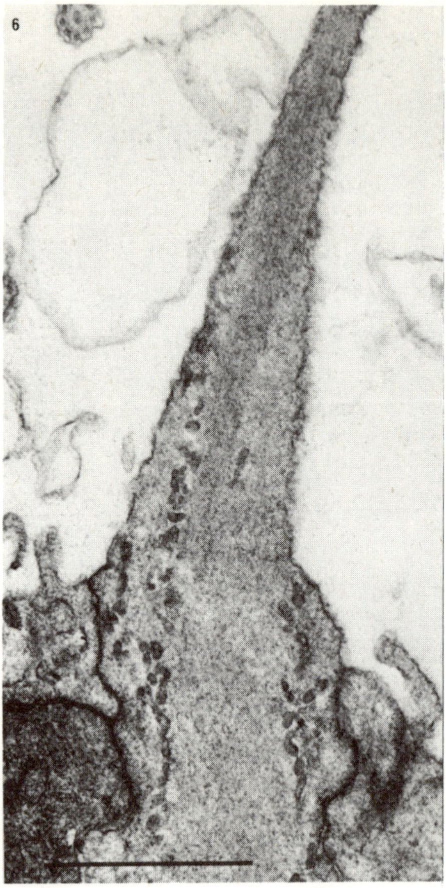

FIG. 5. Distal dendritic segment of a vomero-nasal receptor from a mammal (*Tupaia belangeri*). Note that cilia are absent. Light vesicles, centrioles (C) and ciliary-structure precursor bodies (arrows) are visible. Scale 1 μm.

FIG. 6. Vertical section through the dendrite of a rod-shaped receptor ending in the nasal mucosa of a teleost (*Phoxinus laevis*). Scale 1 μm.

of these vesicles is unknown. They do not engulf material by pinocytosis as can be demonstrated by tracer experiments (Kolnberger, 1971b). The pericarya of both nasal and vomero-nasal receptors contain, beside mitochondria, Golgi fields and cisternae of the granular endoplasmic reticulum. Additionally in the vomero-nasal receptors the smooth endoplasmic reticulum plays an important role (Figs 8–10). In these cells it forms extensive stacks of membranes,

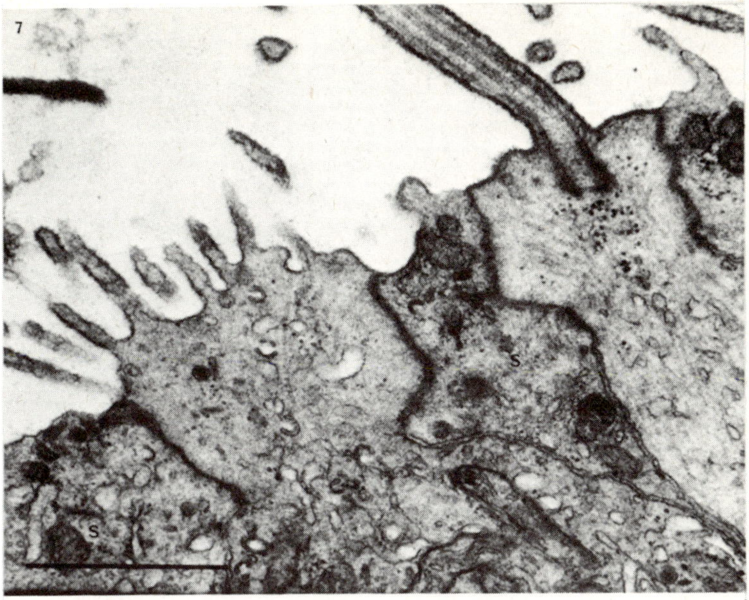

FIG. 7. Olfactory epithelium of the teleost *Phoxinus laevis*. A receptor with microvilli (left) and a ciliated receptor (right) separated by supporting cell processes (S) are visible. Scale 1 μm.

often with complicated membrane arrangements (Kolnberger, 1971a; Bannister and Cuschieri, 1972). In the nasal mucosa the smooth endoplasmic reticulum is on the contrary highly developed in the supporting cells.

Some remarks have to be made on modifications of the ciliary shaft in nasal receptors. There are two modifications of this region (Figs 2, 4): (1) Alterations of number and arrangement of the tubules within the shaft, and (2) Dilations of the shaft.

The changes of the inner structural components of the cilia yield a reduction in number of tubules accompanied by a narrowing of the cilium (Fig. 4). The proximal section, which is rather short, contains the usual pattern of tubules, (9.2)+2. This pattern deviates in the longer distal segment: in mammals we find only 4, 3 or as a rule 2 separate tubules (Andres, 1966, 1969; Okano *et al.*, 1967). In the dog a single tubule was observed in some cases (Okano *et al.*, 1967).

Dilations of the ciliary shaft have been described by some authors (Okano, 1965; Reese, 1965; Frisch, 1967, 1969; Andres, 1969). Frisch (1969, p. 30) for instance stresses the argument, that these dilations "reflect a real difference between the membranes of the

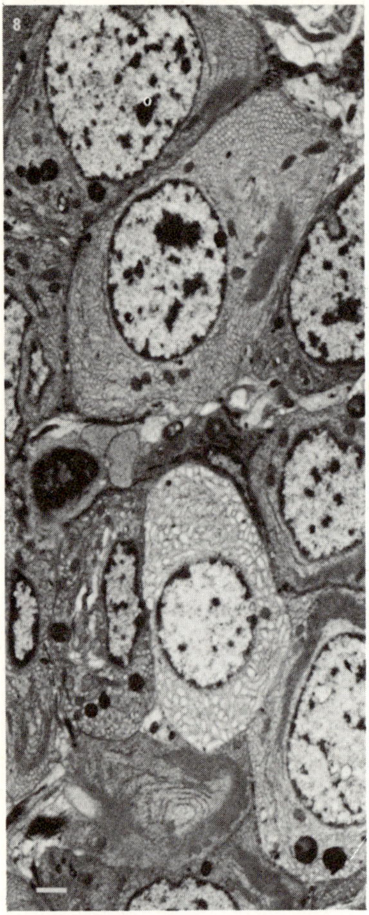

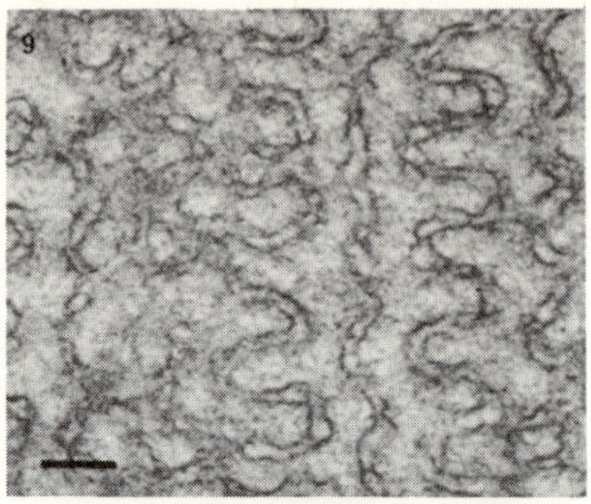

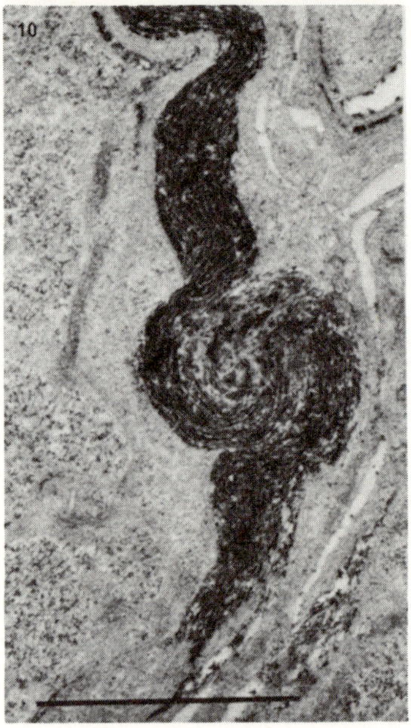

FIG. 8. Low power electron micrograph of the vomero-nasal epithelium of a reptile (*Anguis fragilis*). Numerous receptor cell pericarya are visible. Smooth endoplasmic reticulum is predominant in the cytoplasm of these cells. Scale 1 μm.

FIG. 9. Regularly arranged profiles of the agranular endoplasmic reticulum within a vomero-nasal receptor cell of the field-mouse (*Apodemus sylvaticus*). Scale 1000 Å.

FIG. 10. Cisternae of the smooth surfaced endoplasmic reticulum in a vomeronasal receptor of the slow-worm (*Anguis fragilis*) are heavily stained after osmium impregnation. Scale 1 μm.

adjacent respiratory cilia". From this observation he concludes to a specialized site, which could be implicated in the first steps of the transducer process.

Since cilia are very sensitive to the fixation procedure (see p. 166, Fig. 1), it cannot be excluded that some of the structural features which have been described will have to be regarded as artifacts. Moreover, as the cilia including the basal apparatus seem to be dispensable (see p. 169; Tucker, 1967) one cannot regard without reservation olfactory cilia as morphologically specialized for stimulus perception. Several authors raise the question whether the olfactory cilia are motile (Reese, 1965; Frisch, 1967; Vinnikov, 1969). Earlier data are reviewed by Moulton and Beidler (1967). From the presence of a central pair of tubules in the proximal segment of the cilia one would tend to accept this view. The structural deviations in the distal segment, on the other hand, seem to be in contradiction. It must be borne in mind that a conclusive answer cannot be derived from the structure alone.

The question of distinct receptor types. Considerations of physiologists concerning the coding of olfactory quality have led to the postulation of different types of chemoreceptors. Independently of such argument morphologists have discussed different receptor types within the olfactory mucosa of individual species for about one hundred years. Since the discussion is still going on, though now based in part on fine structural data, these attempts have to be reviewed.

The most sophisticated system for differentiating among types as through light microscopical criteria is that, based on trypan blue supravital staining. Neuhaus (1955) observed different numbers of knots and thickenings in the apical part of sensory cells in mammals. Thus four main receptor types were claimed to exist in the olfactory epithelium of the dog. From ultrastructural investigations we may assume that the knots described by Neuhaus (1955) are caused by accumulations of mitochondria (Andres, 1966). In all likelihood these differences are thus rendered irrelevant. Teleosts can be regarded as an exception, insomuch as they possess at least two fundamentally distinct receptor types (Figs 4, 6, and 7). The physiological properties of these types are unknown.

Attempts have been made to define different receptor classes using the following morphological criteria: (1) The size of the sensory cells and their peripheral processes (Le Gros Clark, 1956; Andres, 1966). (2) The number of olfactory cilia (Le Gros Clark, 1957). (3) The number of microtubules within the dendritic shaft or within the axons of the sensory cells (Andres, 1966). (4) The number of mitochondria within the dendrite (de Lorenzo, 1963).

Andres (1966) mentions that there is no precise correlation between the number of microtubules and the diameter of the shaft, but depending on the size of the receptor cell he observes between 45 and 130 tubules. He formulates the hypothesis that these numbers correspond to the number of microtubules within the axonic processes of the receptor cell. These vary between one and twenty. About 15 receptor types could be distinguished on this basis. Andres (1966) reasons that these receptor types could be correlated to the types postulated by Amoore *et al.* (1964), even if some of them would turn out to be development stages.

If we take into consideration the possibility of structural changes in the dendrites during the replacement of the receptors the differences which might serve as criteria to distinguish receptor types lose much of their significance.

In conclusion it must be emphasized that it is improbable that differences in the fine structure of the cell should really reflect differences in the reaction spectrum. Any such differences are to be expected to be properties of the excitable membranes and would thus appear only at the molecular level. We should not expect such differences to become accessible to the methods of structural research available now.

III. Cell contacts and extracellular space in olfactory epithelia

State of the discussion

A detailed study of the cell contacts in the apical border of the olfactory epithelium is important, (1) for the discussion on the location of the receptor sites on the membrane of the sensory cells, and (2) to clarify the problem whether there are interactions between the sensory cells.

Receptor sites are assumed to lie on the membranes of the cilia and microvilli of the sensory cells (Hopkins, 1926; Altner *et al.*, 1970; Graziadei and Tucker, 1970; Reese and Brightman, 1970). However, the existence of additional sites of interaction between the molecules of odorous substances and the membrane of the sensory cells within the epithelium cannot be excluded as a matter of fact. The taste buds of vertebrates which have been investigated more extensively may be regarded in analogy: Physiological and morphological data lead to the conclusion that in the taste buds there exist receptor sites not only on the membrane of the apical structures of the sensory cells but also on the surface of the pericarya (Beidler, 1970; Mistretta, 1971). However there is discussion as to whether the two regions are separated from each other by a diffusion barrier built up by specialized junctional complexes such as tight junctions (Farbman, 1965; Murray and Murray, 1970).

Tight junctions have also been described in the olfactory epithelium (Reese, 1965; Seifert and Ule, 1967; Brightman and Reese, 1969; Drenckhahn, 1970 ("Nexus"); Reese and Brightman, 1970; Themann *et al.*, 1971). The presence of these junctions raises the question whether interaction occurs between the receptor cells and if so, of what kind. An electrical coupling between the cells in the olfactory epithelium is discussed by Reese and Brightman (1970). Graziadei (1971) stresses the argument that the sensory cells are so arranged that interactions of neurons may be assumed; it is even within anatomical possibility that receptors are connected to superunits and/or integrating units.

Techniques

Progress towards a solution of these problems requires a detailed study of the junctional complexes between the cells constituting the olfactory epithelium, and this not only by mere study of the appearance of this region under different fixation conditions. It is important to investigate the junction complexes experimentally by the application of tracers (Reese and Brightman, 1970; Kolnberger, 1971b; Kolnberger and Altner, 1972) or by using methods which have been developed for the characterization of junctions (Kolnberger, 1971b; Kolnberger and Altner, 1972). Freeze fracturing has proved to be another valuable tool (Friend and Gilula, 1972). Unfortunately, this technique has not yet been applied on the study of olfactory epithelia.

Two general categories of junctions have to be distinguished (Friend and Gilula, 1972), those in which apposing membranes seem to share a structural component (gap, tight, septate junctions) and those with no such mutual bond. Only the group first mentioned is significant for the question discussed here. Recent reports on the incidence and location of tight and gap junctions (Friend and Gilula, 1972) render the information from earlier publications on junctions in the olfactory epithelium incomplete.

In this connection one has to be aware that the appearance of junctions may change with fixation conditions (Figs 10–14). It was demonstrated in reptiles that five-layered membrane junctions occur

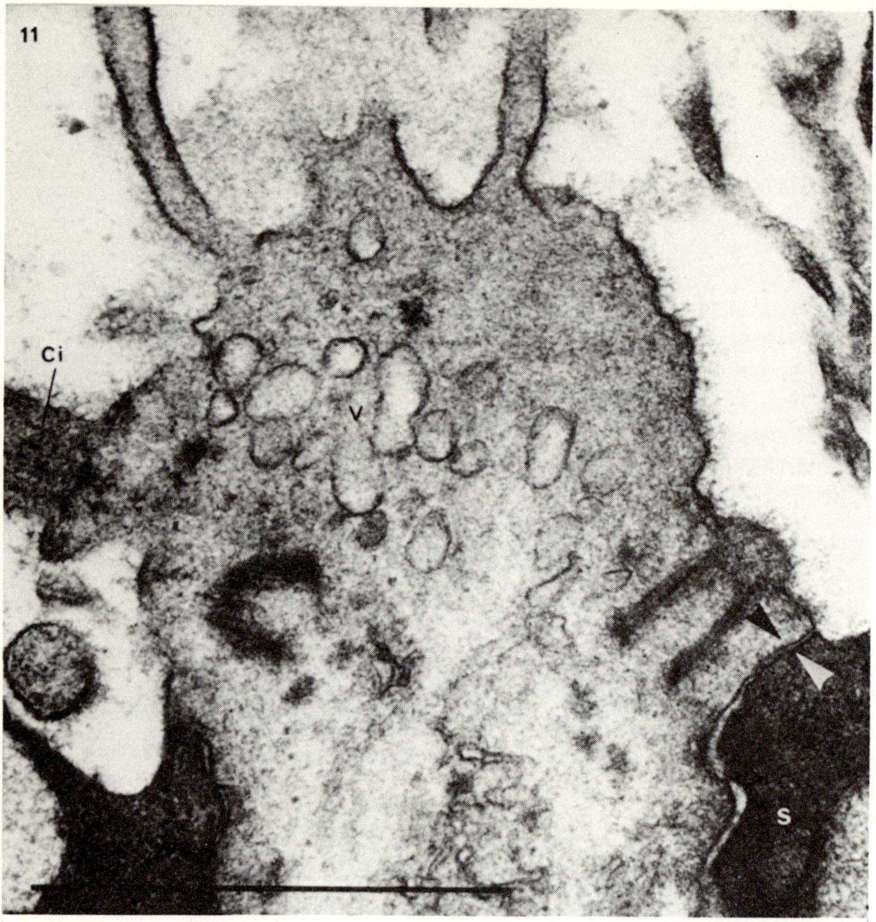

FIG. 11. Section through tip of nasal olfactory receptor in the slow-worm (*Anguis fragilis*). Light vesicles (V) and two basal bodies of olfactory cilia can be seen. The junctional complex between receptor and supporting cell (S) is indicated by arrow heads. Cicilium. Glutaraldehyde fixation. Scale 1000 Å.

rather frequently after osmium fixation as compared with the seven layers in material fixed with aldehyde (Kolnberger, 1971b; Kolnberger and Altner, 1972).

Tracer substances may be applied in two different ways: (1) From the blood vessels (Reese and Brightman, 1970; Themann *et al.*, 1971), (2) from the surface of the epithelium (Kolnberger, 1971b; Kolnberger and Altner, 1972). Though no other significant differences in procedure could be determined, the results are at least partly contradictory.

Reese and Brightman (1970) injected elasmobranchs intravenously with 0.25–1 g horseradish peroxidase (Sigma type II) per kg body weight one hour before fixation. Peroxidase activity was demonstrated in the olfactory epithelium histochemically according to Karnovsky (1967). Kolnberger (1971b) and Kolnberger and Altner (1972) use 3% horseradish peroxidase in 0.7% NaCl. Drops of the solution are put on the surface of the living vomero-nasal epithelium of lizards (*Anguis, Lacerta*) for 5–70 min. The histochemical procedure was the same as applied by Reese and Brightman (1970). Further experiments using lanthanum nitrate as a tracer are reported by Kolnberger (1971b) and Kolnberger and Altner (1972).

Another approach to clarify the structural properties of the junctional complexes is possible by using the E-PTA and Bi-J-techniques which are described thoroughly by Bloom and Aghajanian (1968) and Pfenninger *et al.* (1969) respectively.

Results

In summing up the results one must realize that experimental studies have been few. Their results should not be generalized at this stage, especially since important techniques such as freeze fracturing have yet to be applied.

From descriptive papers it is not clear whether one has to assume a single definite type of junctional complex for all olfactory epithelia. Published data do not strongly suggest this interpretation. The lack of clarity may result from differences in preparation or in fixation techniques (Figs 11–14). While several authors report that sensory and supporting cells in the olfactory epithelium are connected by tight junctions (see p. 175) other authors claim that junctions with five-layered membrane appositions do not exist (Thornhill, 1967; Altner *et al.*, 1970).

The results of experimental work with tracers are also divergent. Reese and Brightman (1970) observed that horseradish peroxidase is prevented from reaching the surface of the nasal epithelium of the guitar fish, when injected intravenously. Kolnberger and Altner (1972) on the other hand stated that the same tracer penetrated to the intercellular spaces within the epithelium of the vomero-nasal organ of reptiles when applied to the surface of the living epithelium (Fig. 15). Since under similar conditions electro-olfactograms can be recorded from the vomero-nasal epithelium (Müller, W., 1971) it

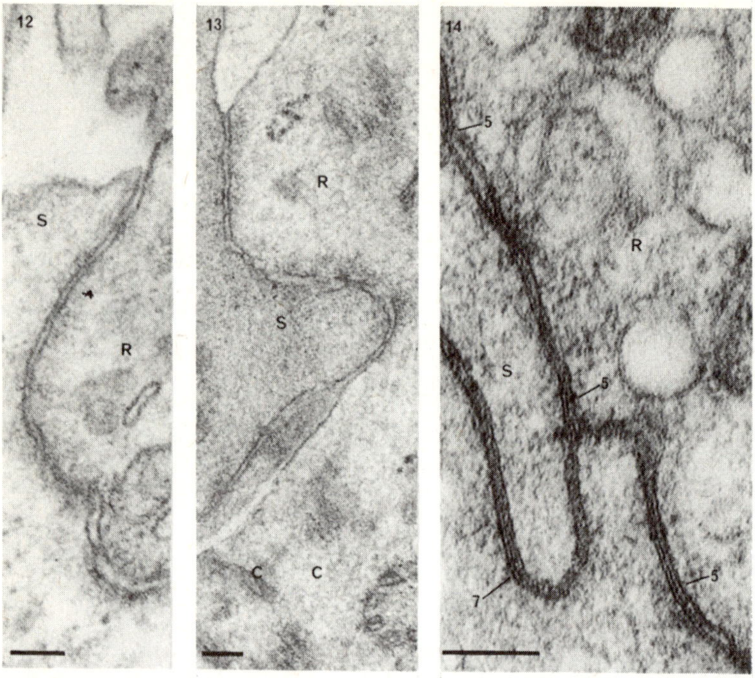

FIG. 12. Apical part of the junctional complex (longitudinal section) between receptor (R) and supporting cell (S) in the vomero-nasal mucosa of a lizard (*Lacerta sicula*). The membranes are separated. Glutaraldehyde fixation. Scale 1000 Å.

FIG. 13. Apical part of the junctional complex (longitudinal section) between receptor (R) and supporting cell (S) in the vomero-nasal mucosa of a field-mouse. (*Apodemus sylvaticus*). The membranes are separated. C centriole. Glutaraldehyde fixation. Scale 1000 Å.

FIG. 14. Cross section through vomero-nasal receptor (R) and supporting cell processes (S) in a reptile (*Gymnodactylus kotschyi*). Numbers indicate regions where the membranes are fused (tight junctions: 5) or closely apposed (7). In other regions the inner leaflets are obviously separated by the normal intercellular cleft. It may be asked whether membrane fusion is caused by fixation. OsO_4-fixation. Scale 1000 Å.

can be assumed that the cellular function was not impaired prior to fixation. The junctional complex, however, can be regarded to be altered under pathological conditions. In the hyperergic reacting olfactory mucosa of the rabbit Themann *et al.* (1971) observed that under such conditions, and when applied from the blood vessels, horseradish peroxidase is found on the surface of the mucosa.

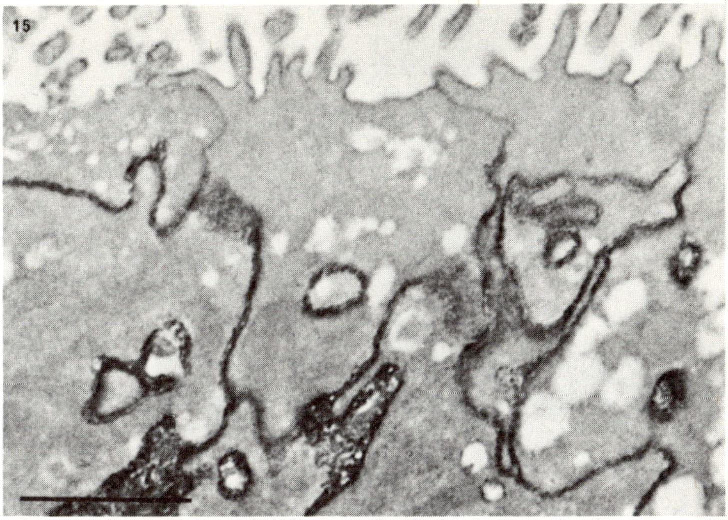

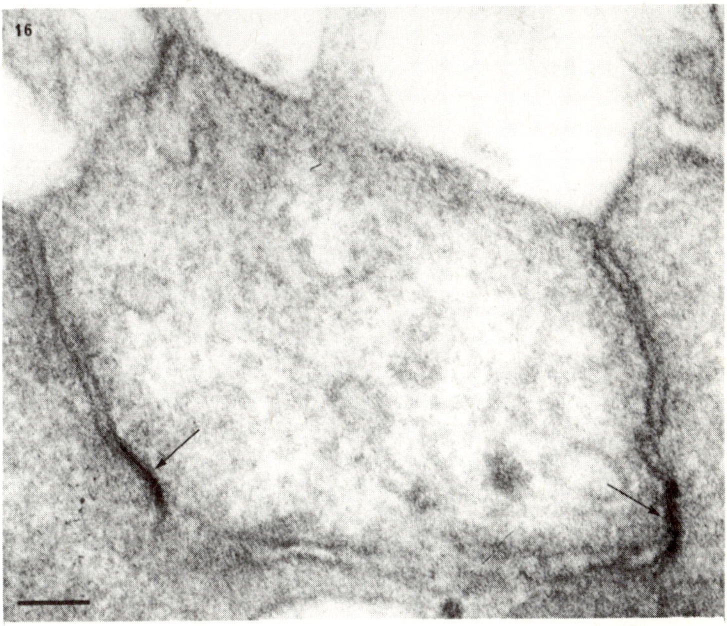

FIG. 15. Section through vomero-nasal mucosa of a reptile (*Anguis fragilis*). Horseradish peroxidase has been applied to the surface of the living epithelium. The tracer has moved into the clefts between sensory and supporting cells. Scale 1 μm.

FIG. 16. Section through tip of vomero-nasal receptor of *Anguis fragilis*. Lanthanum administered with fixing fluid is deposited in the apical part of the junctional complex (arrows). Scale 1 μm.

Experiments with lanthanum nitrate, bismuth iodide (Bi-J) and ethanolic phosphotungstic acid (E-PTA) (Kolnberger, 1971b; Kolnberger and Altner, 1972) can be regarded, as supporting the view, that the intercellular space is open to diffusion. Lanthanum is deposited between the membranes in the apical part of the junctional complex (Fig. 16). With the Bi-J and E-PTA techniques intercleft lines are demonstrated throughout the apical region of the epithelium. The type of junction present, however, cannot be determined with certainty by these results.

The supporting cell processes do not cover the pericarya of the receptor cells completely. The membranes of adjacent sensory cells can be seen to run parallel over considerable distances. In our own investigations (Altner *et al.*, 1970; Kolnberger 1971b) we did not find any signs of specialized contacts between receptor cells which can be regarded to function as synapses. Neither dense projections nor vesicles reacting with Zn-J-OsO_4 could be found. On the other hand Graziadei (1971) detected by an extensive spatial reconstruction that neurons are interconnected by processes in some portions of the olfactory mucosa of several vertebrates. Thus interconnected rings are formed which join in a complex pattern. The conclusion drawn is that the supporting cells in no way have the function of isolating the neurons but can be regarded as having a secretory and most probably a supportive function. The arrangement of receptors and the fact of their interconnection is regarded as "the anatomical possibility that the brain may collect information from single receptors or from groups of receptors acting as super-units and/or integrating units" (Graziadei, 1971, p. 467). Both structural and physiological work can be expected to contribute to a further elucidation of this problem.

IV. Regeneration of receptors and sensory cell activity phases in the olfactory mucosa

Since earlier light microscopical investigations it has been claimed by several authors that a regeneration of sensory cells occurs within the olfactory epithelium (Smith, 1938, 1951; Schultz, 1941, 1960), though this opinion has not been unquestioned (cf. Moulton and Beidler, 1967). Strictly speaking, one has to distinguish two different problems: (1) Does a regeneration of olfactory receptors occur after lesions in the olfactory epithelium and/or bulb? (2) Does a regular renewal of sensory cells occur within the undamaged olfactory epithelium?

Indeed, it would not be impossible for the mechanisms responsible for both phenomena to be the same. On the other hand one has to realize that the capacity for regeneration in the various orders of vertebrates may well be different and that the kind of lesion may influence the type of reaction of the tissues. It may be mentioned that degeneration phenomena in the olfactory sensory epithelium of mammals caused by rhinitis is described by Seifert (1970). Regeneration after a rhinitic degeneration has not been studied, however.

In earlier light microscopical studies degeneration mostly was produced by application of zinc sulphate (Smith, 1938, 1951; Schultz, 1941, 1960). Mulvaney and Heist (1971a), in a reinvestigation using rabbits, report that well differentiated olfactory cells reappear eight days after treatment with 1% $ZnSO_4$ (for technical details see Schultz, 1941, 1960). Thirty days after destruction the regeneration is essentially completed. The exact source of cells for the regenerative process, however, remains still in doubt. Submucosal elements particularly Bowman's glands are suspected to furnish new cells, since the entire epithelium including the basal cells is supposed to be stripped away by $ZnSO_4$ treatment (Mulvaney and Heist, 1971a). On the other hand islands of undamaged receptors may persist (cf. Moulton and Beidler, 1967).

Jackson and Lee (1965), in a light microscopical study, report that cholchicine injections into the olfactory bulb evoke drastic degenerative effects on the olfactory receptors. Regeneration was not the point of their studies. An analysis of the effect of colchicine treatment on a cytological level by electron microscopy is still missing. It may be speculated that colchicine prevents the division of the blastema cells thus causing a remarkably high level (up to 20%) of degeneration. Partial degeneration of sensory cells in the olfactory epithelium of mammals has been observed after severing the olfactory nerve and/or the olfactory bulb (Nagahara, 1940; Le Gros Clark, 1956; de Lorenzo, 1957; Martinez and Jackson, 1960). In this case too a reinvestigation would be desirable.

The idea of a regular renewal of sensory cells is supported by results from mainly three different kinds of structural analysis: (1) From radioautographic observations (Thornhill, 1970; Graziadei and Metcalf, 1971), (2) from ultrastructural characterization of cell types within the epithelium (Andres, 1965, 1969; Thornhill, 1967; Kolnberger, 1971a) and (3) from the study of centriole number and centrio location within the sensory cells (Heist and Mulvaney, 1968; Mulvaney and Heist, 1971b).

As can be expected the results of the radioautographic studies give unequivocal information about the turnover of cells. In both studies tritiated thymidine has been used in the following ways: injected intraperitoneally (Lamprey; Thornhill, 1970) or in the dorsal lymph sac (Rana; Graziadei and Metcalf, 1971) respectively. Cells with labelled nuclei were studied by light and electron microscopy. The results essentially support the conception which has been developed from the other two kinds of investigation as mentioned above.

Thus it can be taken as proven that the sensory epithelium contains a population of immature cells, which undergo division and maturation processes thus supplementing the pool of sensory cells. The fully differentiated sensory cells which have to be regarded as neurons according to their embryological origin do not divide. The interpretation of the basal cells (Thornhill, 1967) or blastema cells (Andres, 1965, 1969), being at least partly immature sensory cells, can be therefore regarded as correct. Moreover, it has been observed that fully differentiated sensory cells undergo remarkable changes in their apical parts. Andres (1969) in the nasal epithelium of the cat and Kolnberger (1971a) in the vomero-nasal organ of *Tupaia* describe apices of sensory cells which differ from the "normal" appearance in higher electron density, changes in the labyrinth (or olfactory vesicles) within this region and a loss of cilia and microvilli, accompanied by a deformation of the apical protrusion (Fig. 17). These structural transformations can be interpretated as stages of degeneration. It is not clear, however, whether this phenomenon is restricted to the distal dendritic segment and whether it is followed by a restoration.

Kolnberger (1971a) observed that in reptiles the agranular endoplasmic reticulum in many—but not all—sensory cell pericarya shows a definite affinity for binding osmium (Osmium impregnation according to Friend and Murray, 1965; cf. also Friend, 1969, and Friend and Brassil, 1970). It is still in question, whether this reaction corresponds to degenerative processes.

Heist and Mulvaney (1968) observed that centrioles reduplicate in the cell bodies of nasal receptors of mammals and then migrate to the terminal swelling. In another paper (Mulvaney and Heist, 1971b) these authors show that this migration is a developmental phenomenon which persists in fully developed tissue. Further it can be shown by measuring the receptor density at different levels in the olfactory epithelium that a significant number of developing sensory cells is present. The density of receptors at the level of the terminal swelling is $1.87 (\pm 0.46) \times 10^5$ in thirty days old and $1.26 (\pm 0.49) \times 10^5$ in adult rabbits, and reaches $2.15 (\pm 0.13) \times 10^5$ and $1.54 (\pm 0.33) \times 10^5$ respectively in the level of the upper dendritic region.

The average density at the dendritic level is significantly higher indicating that there are receptors which have not yet reached the surface. These observations are in agreement with those from centriole distribution.

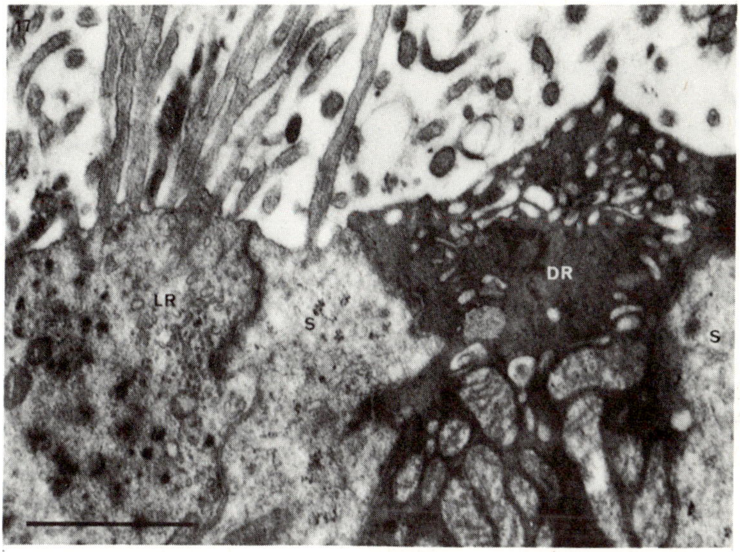

FIG. 17. Electron dense "dark" receptor (DR) and normal "light" receptor (LR) separated by a supporting cell (S) in the vomero-nasal mucosa of a tree-shrew (*Tupaia belangeri*). Note the lack of microvilli and the abundance of dilated vesicles in the electron dense dendrite. Scale 1 μm.

The fact of cell renewal in the olfactory epithelium seems to be well established; although as already mentioned, several aspects need further study. One has to realize, however, that new problems are raised with the acceptance of this concept. The process has to be explained as to how the axons of the differentiating receptors make contact with the olfactory bulb. Correspondingly old synaptic contacts must be dissolved with the degeneration of sensory cells. The consequences for sensory coding remain uncertain for the present since information concerning receptor specificity in vertebrates is still scanty (Gesteland *et al.*, 1963; Shibuya and Shibuya, 1963; Altner and Boeckh, 1967; Shibuya and Tucker, 1967).

V. Cell countings and morphometric approach

Data concerning receptor packing within the olfactory mucosa have been estimated in several cases on the basis of light microscopical investigations (Allison and Warwick, 1949; Teichmann, 1954, 1959; Müller, A., 1955; Schilling, 1970). Electron microscopy has proved an invaluable tool here. Countings have been made of the number of dendrites per unit area in cross sections (Kolnberger, 1971a; Mulvaney and Heist, 1971b) and of the total number of axons in relation to the area of sensory epithelium (Gasser, 1956; Drenckhahn, 1970). Moreover it is possible to estimate the receptor packing from longitudinal sections (Kolnberger, 1971a). It has to be taken into consideration, of course, that the receptor density varies from region to region within

TABLE I. Average diameter of dendrites and average receptor density in the nasal and vomero-nasal mucosa of vertebrates. Asterisks indicate that the values have been calculated based on data given by these authors. LM light microscopical investigation. From Kolnberger, 1971a.

Species	Diameter of dendrites (μm)	Receptor density (number/mm^2)	Author
A. nasal mucosa	(1)	(2)	
teleosts	1.56	46 000–95 000	(1) Bannister (1965) (2) Teichmann (1954, 1959) LM
frog (*Rana*)	1.5–2	110 000–160 000	(1) Yamamoto *et al.* (1965) (2) Yamamoto *et al.* (1965)*
reptile (*Anguis*)	1.5	25 000	–
gull (*Larus*)	3–4	40 000–50 000	Drenckhahn (1970)
mammals (dog, cat, rabbit)	1–2	50 000–215 000	(1) Okano *et al.* (1967) (2) Drenckhahn (1970), Mulvaney and Heist (1971) and Okano *et al.* (1967)*
B. vomero-nasal mucosa	(1)	(2)	
frog (*Rana*)	1.2	325 000	–
reptiles (*Lacerta, Vipera*)	0.5–0.6	1 000 000–1 200 000	(1) Altner and Müller (1968)
tree-shrew (*Tupaia*)	2–3.5	25 000	–
mouse lemur (*Microcebus*)	–	90 000	(2) Schilling (1970) LM

the olfactory mucosa of a given species. Further, in examining cross sections differences in density are to be expected depending on the level of the section within the dendritic region. This is due to the fact that there are growing dendrites which have not yet reached the surface of the epithelium. It has already been mentioned (see p. 249) that the determination of the density of dendrites has furnished evidence for the concept of sensory cell replacement.

Table I gives a survey of data recently published. There it can be seen that the receptor density in the vomero-nasal organ of reptiles is extremely high.

A morphometric evaluation of electron micrographs can be useful, e.g. for determination of the total surface of the olfactory cilia or of the thinner distal segments of these cilia. This procedure is explained at length by Seifert (1970). The applicability of data of this kind, however, seems to be limited at the moment, since the location and extension of the receptor sites, as already pointed out, cannot be established by electron microscopical techniques.

References

Adrian, E. D. (1955). Synchronized activity in the vomero-nasal nerves with a note on the function of the organ of Jacobson. *Arch. Ges. Physiol. Menschen. Tiere.* **260**, 118–192.

Allison, A. C. and Warwick, R. T. T. (1949). Quantitative observations on the olfactory system of the rabbit. *Brain* 72, 186–197.

Altner, H. and Boeckh, J. (1967). Über das Reaktions-spektrum von Rezeptoren aus der Riechschleimhaut von Wasserfroschen (Rana esculenta). *Z. Vergl. Physiol.* **55**, 299–306.

Altner, H. and Müller, W. (1968). Elektrophysiologische und elektronenmikroskopische Untersuchungen an der Riechschleimhaut des Jacobsonschen Organs von Eidechsen (Lacerta). *Z. Vergl. Physiol.* **60**, 151–155.

Altner, H., Müller, W. and Brachner, I. (1970). The ultrastructure of the vomeronasal organ in reptilia. *Z. Zellforsch. Mikrosk. Anat.* **105**, 107–122.

Amoore, J. E., Johnston, J. W., Jr. and Rubin, M. (1964). Die stereochemische Theorie des Geruchs. *Umschau Techn.* **64**, 600–604.

Andres, K. H. (1965). Differenzierung und Regeneration von Sinneszellen in der Regio olfactoria. *Naturwissenschaften* **52**, 500.

Andres, K. H. (1966). Der Feinbau der Regio olfactoria von Makrosmatikern. *Z. Zellforsch. Mikrosk. Anat.* **69**, 140–154.

Andres, K. H. (1969). Der olfaktorische Saum der Katze. *Z. Zellforsch. Mikrosk. Anat.* **96**, 250–274.

Andres, K. H. (1970). Anatomy and ultrastructure of the olfactory bulb in fisch, amphibia, reptiles, birds and mammals. *In* "Taste and Smell in Vertebrates" (Wolstenholme, G. and Knight, J., eds.), pp. 177–196. J. and A. Churchill, London.

Bannister, L. H. (1965). The fine structure of the olfactory surface of teleostean fishes. *Quart. J. micr. Sci.* **106**, 333–342.

Bannister, L. H. (1968). Fine structure of the sensory endings in the vomeronasal organ of the slow-worm Anguis fragilis. *Nature* (*Lond.*) **217**, 275–276.

Bannister, L. H. and Cuschieri, A. (1972). The fine structure and enzyme histochemistry of vomero-nasal receptor cells. *In* "Olfaction and Taste" (Schneider, D., ed.), Vol. 4.

Beidler, L. M. (1970). Physiological properties of mammalian taste receptors. *In* "Taste and Smell in Vertebrates". (Wolstenholme, G. E. W. and Knight, J., eds.), pp. 51–70. J. and A. Churchill, London.

Beidler, L. M. and Smallman, R. (1965). Renewal of cells within taste buds. *J. Cell Biol.* **27**, 263–272.

Bloom, G. (1954). Studies on the olfactory epithelium of the frog and the toad with the aid of light and electron microscopy. *Z. Zellforsch. Mikrosk. Anat. Abt. Histochem.* **41**, 89–100.

Bloom, F. E. and Aghajanian, G. K. (1968). Fine structural and cytochemical analysis of the staining of synaptic junctions with phosphotungstic acid. *J. Ultrastruct. Res.* **22**, 361–375.

Bloom, G. and Engström, H. (1952). The structure of the epithelial surface in the olfactory region. *Exptl. Cell Res.* **3**, 699–700.

Brettschneider, H. (1958). Elektronenmikroskopische Untersuchungen an der Nasenschleimhaut. *Anat. Anz.* **105**, 194–204.

Brightman, M. W. and Reese, T. S. (1969). Junctions between intimately apposed cell membranes in the vertebrate brain. *J. Cell Biol.* **40**, 648–677.

Bronstein, A. A. and Ivanov, V. P. (1965). Electron optical study of the olfactory organ in the lamprey. *Zh. Evolyuts. Fiziol. Biokhim.* **1**, 251–261.

Bronstein, A. A. and Pjatkina, G. A. (1966). Ultrastructural organization of olfactory cell hairs of the fish. *Citologija* **8**, 642–645.

Bronstein, A. A. and Pjatkina, G. A. (1969). Electron-microscopic study of sodium-ion localization in the olfactory organ of rana temporaria. *Zh. Evolyuts. Fiziol. Biokhim.* **5**, 274–278.

Cauna, N., Hinderer, K. H. and Wentges, R. T. (1969). Sensory receptor organs of the human nasal respiratory mucosa. *Amer. J. Anat.* **124**, 187–210.

Drenckhahn, D. (1970). Untersuchungen an Regio olfactoria und Nervus olfactorius der Silbermöve (Larus argentatus). *Z. Zellforsch. Mikrosk. Anat.* **106**, 119–142.

Farbman, A. J. (1965). Fine structure of the taste bud. *J. Ultrastruct. Res.* **12**, 328–350.

Friend, P. S. (1969). Cytochemical staining of multivesicular body and golgi vesicles. *J. Cell Biol.* **41**, 269–279.

Friend, P. S. and Brassil, G. E. (1970). Osmium staining of endoplasmic reticulum and mitochondria in the rat adrenal cortex. *J. Cell Biol.* **46**, 252–266.

Friend, P. S. and Gilula, N. B. (1972). Variations in tight and gap junctions in mammalian tissues. *J. Cell Biol.* **53**, 758–776.

Friend, P. S. and Murray, M. J. (1965). Osmium impregnation of the golgi apparatus. *Amer. J. Anat.* **117**, 135–150.

Frisch, D. (1967). Ultrastructure of mouse olfactory mucosa. *Amer. J. Anat.* **121**, 87–120.

Frisch, D. (1969). Notes on the fine structure of olfactory epithelium. *In* "Olfaction and Taste" (Pfaffman, C., ed.) (Vol. 3), pp. 22–33. Rockefeller

University Press, New York.

Gasser, H. S. (1956). Olfactory nerve fibres. *J. gen. Physiol.* **39**, 473–496.

Gesteland, R. C., Lettvin, J. Y., Pitts, W. H. and Rojas, A. (1963). Odour specificities of the frog's olfactory receptors. *In* "Olfaction and Taste" (Zotterman, Y., ed.) Vol. 1, pp. 19–44. Pergamon Press, Oxford.

Graziadei, P. (1965). Electron microscopic observations on the olfactory mucosa of the cat. *Experientia* (*Basel*) **21**, 1–4.

Graziadei, P. (1966a). Electron microscope observations of the olfactory mucosa of the mole. *J. Zool.* (*Lond.*) **149**, 89–94.

Graziadei, P. (1966b). Electron microscope observations of the olfactory mucosa of vertebrates. *In* 6th Internat. Congr. for electron microscopy, pp. 487–488.

Graziadei, P. (1971). Topological relations between olfactory neurons. *Z. Zellforsch.* **118**, 449–466.

Graziadei, P. and Bannister, L. H. (1967). Some observations on the fine structure of the olfactory epithelium in the domestic duck. *Z. Zellforsch.* **80**, 220–228.

Graziadei, P. and Metcalf, J. F. (1971). Autoradiographic and ultrastructural observations on the frog's olfactory mucosa. *Z. Zellforsch.* **116**, 305–318.

Graziadei, P. and Tucker, D. (1968). Vomero-nasal receptor's ultra-structure. *Fed. Proc.* **27**, 2065.

Graziadei, P. (1971). Topological relations between olfactory neurons. *Z. Zellforsch.* **105**, 498–514.

Heist, H. E. and Mulvaney, B. D. (1968). Centriole migration. *J. Ultrastruct. Res.* **24**, 86–101.

Hopkins, A. E. (1926). The olfactory receptors in vertebrates. *J. comp. Neurol.* **41**, 253–289.

Jackson, R. T. and Lee, Ch. (1965). Degeneration of olfactory receptors induced by colchicine. *Experimental Neurology* **11**, 483–492.

Karnovsky, M. J. (1967). The ultrastructural basis of capillary permeability studied with peroxidase as a tracer. *J. Cell Biol.* **35**, 213–236.

Kauer, J. S. and Moulton, D. G. (1970). Ultrastructure of vomero-nasal and olfactory neurosensory epithelia in rat and rabbit. *Federation Proc.* **29**, 251.

Kolb, A. (1971). Licht- und elektronenmikroskopische Untersuchungen der Nasenhöhle und des Riechepithels einiger Fledermausarten. *Z. f. Säugetierkunde* **36**, 202–213.

Kolnberger, I. (1971a). Vergleichende Untersuchungen am Riechepithel, insbesondere des Jacobsonschen Organs von Amphibien, Reptilien und Säugetieren. *Z. Zellforsch.* **122**, 53–67.

Kolnberger, I. (1971b). Über die Zugänglichkeit des Interzellularraums und Zellkontakte im Riechepithel des Jacobsonschen Organs. *Z. Zellforsch.* **122**, 564–573.

Kolnberger, I. and Altner, H. (1971). Ciliary-structure precursor bodies as stable constituents in the sensory cells of the vomero-nasal organ of reptiles and mammals. *Z. Zellforsch.* **118**, 254–262.

Kolnberger, I. and Altner, H. (1972). The occurrence of diffusion barriers and interreceptor contacts in the olfactory epithelium of the vomero-nasal organ. *In* "Olfaction and Taste" (Schneider, D., ed.), Vol. 4.

Le Gros Clark, W. E. (1956). Observation on the structure and organization of the olfactory receptors of the rabbit. *Yale J. Biol. Med.* **29**, 83–95.

Le Gros Clark, W. E. (1957). Inquiries into the anatomical basis of olfactory discrimination. *Proc. Roy. Soc.* (*London*), Ser. B, **146**, 299–319.

Lorenzo, A. J. De (1957). Electron microscopic observations of the olfactory mucosa and olfactory nerve. *J. biophys. biochem. Cytol.* **3**, 839–850.

Lorenzo, A. J. De (1963). Studies on the ultrastructure and histophysiology of cell membranes, nerve fibers and synaptic junctions in chemoreceptors. *In* "Olfaction and Taste" (Zotterman, Y., ed.), Vol. 1, pp. 5–18. Pergamon Press, Oxford-London-New York-Paris.

Lorenzo, A. J. De (1970). The olfactory neuron and the blood-brain barrier. *In* "Taste and Smell in Vertebrates" (Wolstenholme, G. and Knight, J., eds.), pp. 151–176. J. and A. Churchill, London.

Luckhaus, G. von (1969). Licht- und elektronenmikroskopische Befunde an der Lamina epithelialis des Vomeronasalorgans vom Kaninchen. *Anat. Anz.* **124**, 477–489.

Martinez, J. R. and Jackson, R. T. (1960). Olfactory nerve degeneration. *Assoc. Southeast. Biol. Bull.* **7**, 34.

Maunsbach, A. B. (1966a). The influence of different fixation methods on the ultrastructure of rat kidney proximal tubule cells. I. *J. Ultrastruct. Res.* **15**, 242–282.

Maunsbach, A. B. (1966b). The influence of different fixation methods on the ultrastructure of rat kidney proximal tubule cells. II. *J. Ultrastruct. Res.* **15**, 283–309.

Millonig, G. (1962). Further observations on a phosphate buffer for osmium solutions in fixation. *In* "Proceedings of the 5th International Congress for Electron microscopy, Philadelphia" (Breese, S. S., Jr., ed.), Vol. 2, P-8. Academic Press, New York.

Mistretta, Ch. M. (1971). Permeability of tongue epithelium and its relation to taste. *Am. J. Physiol.* **220**, 1162–1167.

Moulton, D. G. (1971). Detection and recognition of odour molecules. *In* "Gustation and Olfaction". (Ohloff, G., and Thomas, A. F. eds.) pp. 1–25. Academic Press, London-New York.

Moulton, D. G. and Beidler, L. M. (1967). Structure and function in the peripheral olfactory system. *Physiol. Rev.* **47**, 1–52.

Müller, A. (1955). Quantitative Untersuchungen am Riechepithel des Hundes. *Z. Zellforsch.* **41**, 335–350.

Müller, W. (1971). Vergleichende elektrophysiologische Untersuchungen an den Sinneszellen des Jacobsonschen Organs und der Nase von Amphibien (Rana), Reptilien (Lacerta) und Säugetieren (Mus). *Z. vergl. Physiol.* **72**, 370–385.

Mulvaney, B. D. and Heist, H. E. (1971a). Regeneration of rabbit olfactory epithelium. *Am. J. Anat.* **131**, 241–252.

Mulvaney, B. D. and Heist, H. E. (1971b). Centriole migration during regeneration and normal development of olfactory epithelium. *J. Ultrastruct. Res.* **35**, 274–281.

Murray, R. G. and Murray, A. (1970). The analogy and ultrastructure of taste endings. *In* "Taste and Smell in Vertebrates" (Wolstenholme, G. E. W. and Knight, J., eds.), pp. 3–30. J. and A. Churchill, London.

Nagahara, Y. (1940). Experimentelle Studien über die histologischen Veränderungen des Geruchsorgans nach der Olfactoriusdurchschneidung. *Japan. J. Med. Sci. Pt. V, Pathol.* **6**, 165–199.

Neuhaus, W. (1955). Die Form der Riechzellen des Hundes. *Naturwissenschaften* **42**, 374–375.

Okano, M. (1965). Fine structure of the canine olfactory hairlets. *Arch. histol.*

jap. **26**, 169–185.

Okana, M. and Suguwa, Y. (1965). Ultrastructure of the respiratory mucous epithelium of the canine nasal cavity. *Arch. histol. jap.* **26**, 1–21.

Okano, M., Weber, A. F. and Frommes, S. P. (1967). Electron microscopic studies of the distal border of the canine olfactory epithelium. *J. Ultrastruct. Res.* **17**, 487–502.

Pfenninger, K. H., Sandri, C., Akert, K. and Eugster, C. H. (1969). Contribution to the problem of structural organization of the presynaptic area. *Brain Res.* **12**, 10–18.

Reese, T. S. (1965). Olfactory cilia in the frog. *J. Cell Biol.* **25**, 209–230.

Reese, T. S. and Brightman, M. W. (1970). Olfactory surface and central olfactory connections in some vertebrates. *In* "Taste and Smell in Vertebrates" (Wolstenholme, G., and Knight, J., eds.), pp. 115–149. J. and A. Churchill, London.

Schilling, A. (1970). L'organe de Jacobson du lémurien malagache Microcebus murinus (Miller, 1777). *Mem. Mus. Nation. Hist. Nat.*, Ser. A, **61**, 203–280.

Schulte, E. (1972). Untersuchungen an der Regio olfactoria des Aals, Anguilla anguilla L. I. Feinstruktur des Riechepithels. *Z. Zellforsch.* **125**, 210–228.

Schulte, E. and Holl, A. (1971). Feinstruktur des Riechepithels von Calamoichthys calabaricus J. A. Smith (Pisces, Brachiopterygii). *Z. Zellforsch.* **120**, 261–279.

Schultz, E. W. (1941). Regeneration of olfactory cells. *Proc. Soc. exp. Biol. (N.Y.)* **46**, 41–43.

Schultz, E. W. (1960). Repair of the olfactory mucosa with special reference to regeneration of olfactory cells (sensory neurones). *Amer J. Path.* **37**, 1–19.

Seifert, K. (1968). Die Feinstruktur des Riechsaumes. *Arch. Klin. exp. Ohren-, Nasen- und Kehlkopfheilk.* **192**, 182–213.

Seifert, K. (1970). Die Ultrastruktur des Riechepithels beim Makrosmatiker. *In* "Normale und pathologische Anatomie". H. 21. Georg Thieme, Stuttgart.

Seifert, K. and Ule, G. (1967). Die Ultrastruktur der Riechschleimhaut der neugeborenen und jugendlichen weiben Maus. *Z. Zellforsch.* **76**, 147–169.

Shibuya, T. and Shibuya, S. (1963). Olfactory epithelium: unitary responses in the tortoise. *Science* **140**, 495–496.

Shibuya, T. and Tucker, D. (1967). Single unit responses of olfactory receptors in vultures. *In* "Olfaction and Taste" (Hayashi, T., ed.), Vol. 2, pp. 219–233. Pergamon Press, Oxford.

Sjöstrand, F. S. (1967). "Electron Microscopy of Cells and Tissues". Academic Press, New York and London.

Smith, C. G. (1938). Changes in the olfactory mucosa and the olfactory nerves following intranasal treatment with one per cent zinc sulphate. *Can. med. J.* **39**, 138–140.

Smith, C. G. (1951). Regeneration of sensory olfactory epithelium and nerves in adult frogs. *Anat. Rec.* **109**, 661–671.

Teichman, H. (1954). Vergleichende Untersuchungen an der Nase der Fische. *Z. Morph. Ökol. Tiere* **43**, 171–212.

Teichman, H. (1959). Über die Leistung des Geruchssinnes beim Aal (Anguilla anguilla L.). *Z. vergl. Physiol.* **42**, 206–254.

Theman, H., Backwinkel, K. P. and Terrahe, K. (1971). Die Permeabilität der normal und hyperergisch reagierenden Nasenschleimhaut. Eine elektronenmikroskopische Studie mit Peroxidase als Markierungssubstanz. *Cytobiol.* **3**, 93–105.

Thornhill, R. A. (1967). The ultrastructure of the olfactory epithelium of the lamprey lampetra fluviatilis. *J. Cell Sci.* **2**, 591–602.

Thornhill, R. A. (1970). Cell division in the olfactory epithelium of the lamprey, lampetra fluviatilis. *Z. Zellforsch.* **109**, 147–157.

Trujillo-Cenóz, O. (1961). Electron microscope observations on chemo- and mechanoreceptor cells of fishes. *Z. Zellforsch.* **54**, 654–676.

Tucker, D. (1967). Olfactory cilia are not required for receptor function. *Fed. Proc.* **26**, 544.

Vinnikov, Ya. A. (1969). The ultrastructural and cytochemical basis of the mechanism of function of the sence organ receptors. *In* "The Structure and Function of Nervous Tissue" (Bourne, G. H., ed.), Vol. 2, pp. 265–392. Academic Press, New York and London.

Wilson, J. A. F. and Westerman, R. A. (1967). The fine structure of the olfactory mucosa and nerve in the teleost Carassius carassius L. *Z. Zellforsch.* **83**, 196–206.

Yamamoto, T., Tonosaki, A. and Kurosawa, T. (1965). Electron microscope studies on the olfactory epithelium in frogs. *Acta Anat. Nippon.* **40**, 342–353.

Yasutake, S. (1959). The fine structure of the olfactory epithelium studied with the electron microscope. *J. Kurume Med. Assoc.* **22**, 1279–1304.

5 Application of Scanning Electron Microscopy and Autoradiography in the Study of Olfactory Mucosa

P. P. C. GRAZIADEI

Department of Biological Sciences
Florida State University
Tallahassee, Florida

I. Scanning electron microscopy history and basic principles of operation

The idea that eventually led to the development of the scanning electron microscope (SEM) was introduced by Knoll (1935) and rested on the principle that when focusing an electron beam onto a specimen surface, the emitted current could be recorded as a function of the position. Three years later von Ardenne (1938) actually built the first SEM in which two magneti lenses were used to focus a small electron beam spot at the specimen. The instrument, as devised by von Ardenne, was primitive and precise focusing of the beam was not possible. However, the base was laid for further studies and improvements. Zworykin *et al.* (1942), Lecaute (1946), Davoine (1957) all contributed to the development of the instrument, and with the further work of McMullan (1953) and Smith (1956) the

SEM was made capable of obtaining resolutions approximating 250 10^{-1} nm. The Cambridge group lead by Oatley has worked on this instrument from 1948 and one of the results of these studies has been a commercially available SEM microscope with a resolution of some 60–100, 10^{-1} nm. (Smith and Oatley, 1955, Oatley, 1958; Everhart *et al.*, 1960; Pease, 1963; Nixon, 1964; Oatley *et al.*, 1965; Pease and Nixon, 1965). Further bibliographical information on the SEM can be found in Thornton (1968).

While the images produced by the SEM are similar to the ones obtained with a dissecting light microscope (LM) or to the image that our eye perceives from light-reflecting objects, the mode of image information is substantially different. The SEM has, moreover, uncommon advantages as compared to any optical system using a conventional light source, namely a far greater point to point resolution and an increased depth of field.

The following discussion of the principles of operation of the SEM will proceed from the electron source, down through the column, through the detector system and to the display system. It will end with a brief treatment of specimen-beam interactions. The approach throughout will be general, without reference to the more theoretical aspects, and will dwell rather heavily on results rather than derivation.

The SEM is essentially a device for producing a beam of electrons, focusing this beam to a small cross-sectional area on the specimen surface, collecting and displaying the current emitted from the specimen surface as a function of position on the specimen, and recording the resulting "map" of the specimen surface. The image "map" is accomplished by scanning the specimen synchronously with the scan on the display cathode ray tube (CRT), magnification of the image being the ratio of the size of the raster on the specimen to the size of the display raster. For example, if the display raster is 10 x 10 cm and the specimen raster is 1 x 1 mm the resulting linear magnification will be 10^2.

The SEM apparatus consists of an electron gun to produce the beam of electrons, a column composed of magnetic lenses to focus the image of the source on the specimen being examined, scanning coils to produce the raster on the specimen, a vacuum system to evacuate the specimen chamber and the beam space inside the column, an electron collection system to detect the radiation emitted from the specimen, and various current supplies, display units, and amplifiers (Fig. 1).

Electrons are thermally boiled off a tungsten filament which is heated by passing a high current through it. The electrons thus produced escape through a hole in the centre of the Wehnelt cylinder, a cap covering the filament and biased negatively with respect to the filament. The assembly consisting of the Wehnelt cylinder and filament is known as the cathode and is held at a high negative potential, usually from 1 to 3 kV. Electrons escaping from the hole in the Wehnelt cylinder are accelerated in the field produced by the anode, a grounded annular cylinder with a central bore which acts as an aperture through which the accelerated electrons pass. After leaving

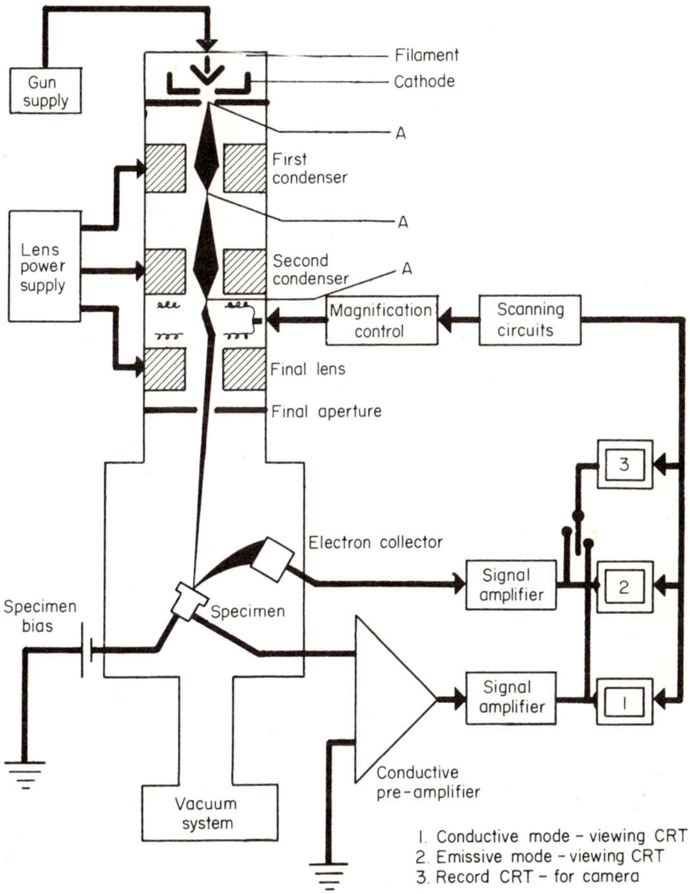

FIG. 1. A schematic diagram showing the principles of operation of a typical scanning electron microscope (SEM). The beam is brought to focus at the points labelled A (cross over points). 1, 2, 3, are cathode ray tubes which provide visual readout depending on the mode of operation (see text).

the anode the beam enters the first of the lenses used to focus the beam to a fine spot or probe.

The lenses usually employed are magnetic and depend for their action on the fact that electrons possess a small magnetic moment which can be used as a sort of "handle" to manipulate their trajectories. However, a curious property of this magnetic moment causes the actual path of the electron to be a spiral when it encounters a magnetic field. This fact is of little import to the average user of the SEM but it must be borne in mind that the typical diagrams showing the path of the electron beam do not actually show what is happening to the individual electrons inside the column of the microscope, but show rather a sort of average of the behaviour of many electrons taken together. In a properly designed lens the electron is deflected slightly towards the axis of the lens on a spiral path as it enters the lens and is deflected slightly away from the lens axis on leaving. The second deflection is always less than the first since the electron is closer to the axis as it exits and the deflecting component of the lens field is smaller there. The rotation produced on entering the lens is exactly counterbalanced on leaving it so that the net effect of the lens is to cause a deflection towards the axis of the lens and a rotation around that axis, i.e. the electron does *not* continue to spiral after leaving the lens.

Lens systems are subject to certain faults such as astigmatism, spherical and chromatic aberration, coma, and curvature of field. The lenses of the SEM are no exception to this rule. It is pertinent to discuss briefly the various defects of magnetic lenses in order to appreciate the limitations these defects place on instrument performance.

Rotational asymmetry in the lens field gives rise to a condition known as astigmatism in which the lens forms two line foci of a point object. The foci are separated by some small distance and are always at right angles to each other. Between these two foci is a so-called circle of least confusion, whose diameter is in part determined by the distance between the positions of the foci. Hence, astigmatism in the lenses results in the inability to focus a point image of a point object. Proper design and manufacture of the lenses can eliminate this lens fault to a large degree. Also, compensating mechanisms are employed in the column to correct residual astigmatism which cannot be eliminated in the manufacture of the lenses.

Spherical aberration arises from the fact that lenses are stronger in their effect on rays which enter them off the lens axis. This means that the image of a grid having a square mesh will be distorted due to the variation in magnification across the image. Each part of image will be in focus but the size of the squares will differ as distance from the lens axis increases. A special form of this defect arises in the electron microscope because of the spiral path of the electrons through the lens system. An object formed of two straight lines at right angles will be imaged as a curved swastika since the degree of rotation of the ray depends on the strength of the lens field. This effect is known as anisotropic distortion. It is impossible to eradicate spherical aberration since it is in part due to the nature of the system itself. The degree of aberration can be reduced by proper design and operation of the lens, and proper choices of accelerating voltage.

Chromatic aberration is a condition in which the image of a point source of radiation having a distribution of energy, is spread along the line on the axis, each point of the line representing the focus of a particular energy component of the radiation from the source. In practice, chromatic aberration can arise in the SEM from three different sources. The first source is instability of the high voltage which accelerates the electrons forming the irradiating beam. This can be controlled to a high degree by proper engineering of the power supplies used to furnish the accelerating potential. The second source of chromatic aberration is instability in the lens current system. This can also be taken care of to a large extent by the proper design of the supply system. The third source of aberration is the spread of thermal velocities of the electrons emitted by the cathode. This effect may be partially overcome by operating the filament under conditions which permit the largest fraction of electrons emitted to be approximately mono-energetic. Clearly there will always be a spread in energy of the emitted electrons. It is possible to block out the lower energy fraction by selecting the proper bias voltage on the Wehnelt cylinder of the electron gun.

The effects of all three aberrations (astigmatism, chromatic aberration, and spherical aberration) is to blur the image of a point object into a so-called circle of least confusion. This means that the electron beam can only be focused to a certain spot size whose diameter is determined by the additive effects of all these lens errors. This, in turn, limits the localization of the scanning beam on the specimen in which the SEM depends for its resolution. In general, the resolution of the SEM may be taken approximately as the diameter of the electron probe although surface penetration effects degrade this figure. A typical resolution for a SEM operating in the emissive mode is approximately 200, 10^{-1} nm, but it may be 100 10^{-1} nm or better when adequate conditions of microscope and specimen are met.

Once the beam is focused, the spot must be made to scan over the surface of the specimen in order to form the "map" of surface features which is the essence of the micrograph. The beam must be swept synchronously with the beam of the CRT used for visual display or photographic recording of the image. For this purpose the beam is displaced by a set of scanning coils which move the beam at a known rate across the specimen and provide for the scanning of a raster. The coils of the CRT usually connected to the same scan generator that supplies current to the beam deflecting coils. The mode of scan can be any of several different types (see below). Visual raster provides a continuous scan and is used for visual observation of the specimen, and focusing. Record raster scans a single raster at any of several pre-set speeds for recording the image, usually by photographing a CRT with a special short-persistance phosphor screen. Other modes include slow scan (a single line scan at low speed), spot (a single spot which can be positioned where desired), and external (scan generated by external inputs). Each scan mode has its particular application but visual and record raster modes are the most used in general biological application.

The SEM may also be operated in several modes depending upon the type of radiation detected to provide contrast information in the image. The *emissive mode* uses the secondary electrons emitted from the specimen as a result of bombardment by the primary electron beam. The other modes which can be used are the *absorptive mode* (specimen current due to the primary beam is used to modulate directly the CRT display beam), *conductive mode* (primary beam generates secondary currents in semiconductor materials, whereby the specimen acts as a detector to map regions below its surface), and *luminescent mode* (specimen luminesces under the influence of the primary beam and the visible light produced is detected to provide contrast information).

Of these modes the emissive and luminescent modes are more useful in biological investigations. The reflective, absorptive, and conductive modes are suited particularly to semiconductor and surface-physics investigations. The best resolution is obtained in the emissive and reflective modes, with the emissive mode being favoured for complex or re-entrant surfaces. Although the reflective mode offers greater contrast, it fails to reveal details in shadows.

Contrast in the final image is derived solely from interactions between the specimen and the electron beam. These interactions can take any or all of several forms. Usually one or another process will be the main process contributing to the final image although other types of interaction may be taking place simultaneously. There are four main types of electron-solid interactions which result from electron bombardment.

Elastic reflection of the incident electrons occurs when the electrons interact with the fields of the specimen atoms in such a way that no interchange of energy takes place, the direction of the incident electron's path being changed drastically enough in most cases to send it out of the specimen entirely. Obviously, low energy electrons or electrons incident on a very dense surface should tend to be reflected rather than absorbed in the target. The incident electron may also be reflected after an inelastic interaction with the specimen, in which case the electron loses some fraction of its initial energy and the interacting atom gains this fraction. In this case, the electron will be ejected from the specimen surface with less than its initial energy. Secondary electrons may be formed by the process of impact ionization, occurring as a result of one or more inelastic interactions in which the fraction of the incident energy gained by the specimen atoms is enough to eject one or more electrons from the target atoms. These ejected electrons may have energies of the same order as the incident electron and may proceed on into the specimen, initiating a chain of impact ionization phenomena until the energy of the ejected electrons is insufficient to cause further ionization. The result of all these interactions is a cascade of electrons mingled together with the stationary ions left behind from the ionization collisions, the mixture being approximately centred on the path of the initial incident electron. The secondary electrons formed near the surface of the specimen can diffuse outward, losing energy by ionizing atoms of the specimen, until they reach the surface. If they still have enough energy they may escape the surface and are available to be collected by a suitably constructed detector. Electrons liberated further inside the specimen may recombine with the stationary ions left as a result of the original ionization and give rise to radiant quanta of energy whose positions in the electromagnetic spectrum depend on the level to which the electrons fall. An electron falling into an outer orbit will release a photon with energy corresponding to the visible or near infrared part of the spectrum. If the electron subsequently drops into an inner state it will release x-rays whose energy is a characteristic of the target material. The x-ray thus produced is the basis of the x-ray microanalyser and in general will not be of concern so far as the SEM is concerned. The visible light produced as a result of the electron bombardment (cathodol-luminescence) is one of the major phenomena used in the SEM to produce a final image.

Two other interactions of the beam and specimen are beam-induced conductivity and radiation damage. Neither of these two processes is of immediate interest except for the effects of radiation damage on certain specimens, chiefly semiconductors. In general, biological specimens exposed to the beam are not harmed appreciably, the extent of penetration of the beam being approximately 10^{-3} cm deep.

A more important effect of irradiation by the beam is the deposition of a layer of contamination composed chiefly of hydrocarbons and polymerized diffusion-pump oil which linger in the specimen chamber. This layer is usually somewhat insulative and tends to decrease contrast by charging up under the beam and altering the trajectories of emitted electrons.

Specimens to be observed in the SEM should be either conductive or made conductive by the deposition of a thin layer of metal such as gold or gold-palladium. Insulating specimens will charge up under the beam and wash out contrast. Their value is lost except for examination under very low beam energies and very low beam currents. It is often possible to coat up to about 1/4 mm of the area of interest and reduce the resistance to ground sufficiently to prevent undesirable charge build-up in the region of interest.

II. Application of the SEM to biological material

The possibility offered by the scanning electron microscope of visualizing surfaces routinely at continuously-varying magnifications from 20X to 50 000X, and with a depth of field far greater than the one obtained by any conventional optical system, could hardly have escaped the interest of the morphologist. With the machine made available commercially in the last six years, a burst of interest has lead many to study a variety of biological specimens using the SEM both independently and as a complement to observations carried out with other well-known morphological techniques such as LM, TEM, tissue culture, etc. (surveys by Smith and Oatley, 1955; Hayes and Pease, 1968; and Wells, 1969; and specific reports by Hay and Sandberg, 1967; Neurath *et al.*, 1967; Boyde and Barber, 1968; Boyde *et al.*, 1968; Boyde, Grainger and James, 1969; Graziadei, 1969a, b, 1971a, b; Graziadei and Tucker, 1970).

The papers which appeared in the first few years have reported, together with new perspectives and unsuspected achievements, the limitations of this technique when applied to routine studies of organs and tissues of animals. An analysis of both positive and negative aspects of SEM when applied to animal tissues will be related in the following pages. Many pictures will illustrate the following paragraphs. They are intended to show the possibilities and limitations of the instrument in a variety of circumstances and requirements. The legends will be intentionally elaborate, as it is more appropriate to refer to specifics of the technique directly together with the visual result of the same. Comments, that are usually part of the text, will instead be made in the captions where they are directly related to the figures.

Because of inherent properties of the instrument, which have been briefly set forth in the previous paragraph, any biological specimen to be observed advantageously should, before any observation is performed, fulfil the following requirements: (a) The surfacc to be observed should be free from unwanted and distracting extraneous material (such as mucus, blood, tissue fluids, extraneous products of disrupted cells, etc.). (b) Water should be removed from the specimen without substantially altering the basic structures to be studied (the vacuum in the specimen chamber is in the order of 10^{-5} Torr). (c) Significant surfaces should be made electrically conductive to avoid charging up under the electron beam.

To obtain the above-mentioned requirements to a compatible degree of satisfaction, biological specimens should consequently be (a) washed in a suitable medium to insure their cleanliness, (b) properly fixed, hardened and dehydrated, (c) glued to an appropriate aluminium stub on proper orientation, and coated with a thin layer of conductive metal. It must be kept in mind that the layer of conductive material (usually gold or palladium or an alloy of carbon with gold) when used at an estimated thickness of 200, 10^{-1} nm, acts like a blanket of snow on any surface structure, obscuring details in the size range proportional to its thickness.

There are some exceptions to the rules reported above. For instance, some insects have been observed alive, and without metallic shadowing in the microscope, surviving both the high vacuum and the electron bombardment without apparent damage. Due to their physical properties, bone and enamel may be observed without fixation as may many other tissues and organs like nails, hair and the chitinous covering of many insects.

Biological surfaces can for practical reasons be divided into two main categories, namely normally-free or natural surfaces, and artifactual or artificially-induced surfaces. Both of these types of surfaces, when properly prepared, can yield relevant information.

Natural surfaces are the ones readily exposed such as the skin, the linings of the digestive system, peritoneal cavities, nasal cavities, and

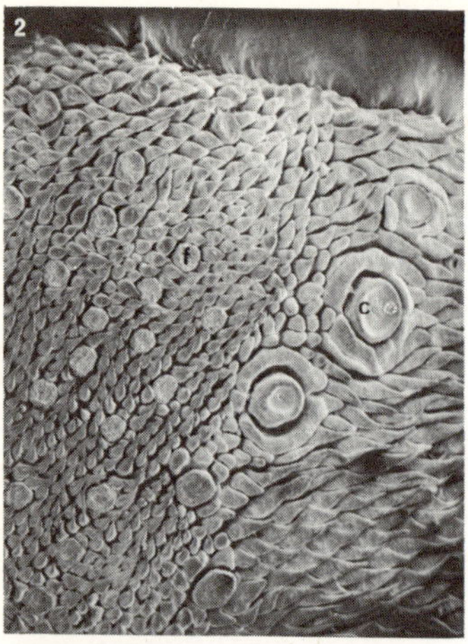

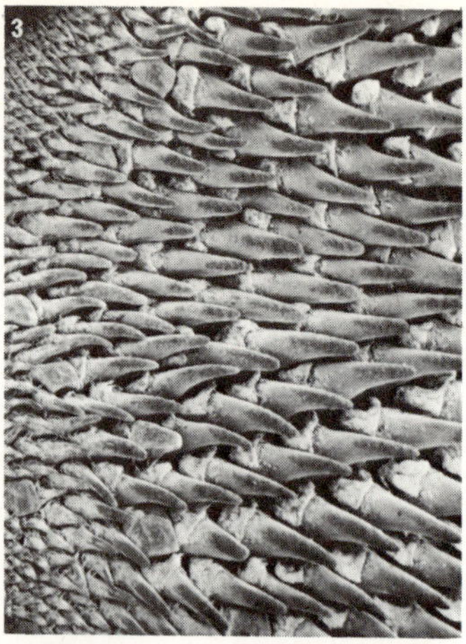

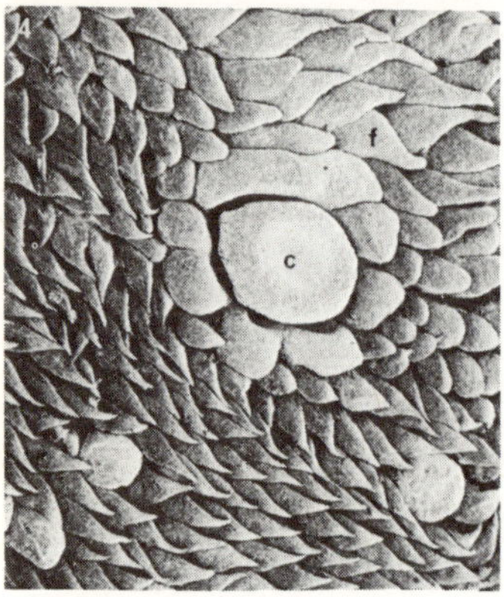

FIG. 2. SEmicrograph of the hind portion of a puppy tongue. Same preparation as in Fig. 3. Circumvallatae (c) and fungiform (f) papillae are clearly visible and can be easily numbered in such preparations. Mucus and other alimentary debris have been easily removed. At this magnification there is no sign of technical deficiencies and no obvious sign of shrinkage. However a more careful observation shows that the surface of the circumvallatae papillae is concave and the surface of the fungiform papillae is also concave and uneven due to loss of turgor during fixation and dehydration procedures. The tongue as a whole, underwent a reduction of approximately 1/3 during the alcohols and vacuum dehydration procedures. (22X)

FIG. 3. SEmicrograph of the tip of a tongue of kitten, fixed in 10% formalin in distilled water and prepared according to method n. 1. The low power picture allows a large field to be covered and many filiform papillae are illustrated. The cornified epithelium of this portion of the tongue provides a rough surface which has been easily cleaned with a soft brush and distilled water. Notice that no shrinkage is observed in so on. Of these, some are particularly hard, such as the cornified epithelium of the skin, the epithelium of the tongue of many vertebrates, the scales of the fish (Figs 2, 3, 4, 5), and the cartilagineous articular surfaces (see Woodward *et al.*, 1969). None of these surfaces require treatment other than a rough mechanical cleaning and dehydration. Micropaleontology has provided a particularly fertile field of application for the SEM (Hay and Sandeberg, 1967 and Bens and Drew, 1967) and studies of the planktonic diatomes have shown some of the most beautiful images of the SEM. Shrinkage is not usually significant in these tissues which provide by far the best subjects for the SEM. Likewise surfaces covered with a cuticular or chitinous lining provide satisfactory preparations (Figs 6, 7). Such observations mislead the beginner into believing that the SEM can be used on any tissue with comparable results. Other natural surfaces, however, are mechanically more vulnerable. The tissue has a high content of water and the surface itself is covered with a dense layer of sticky secretion which is difficult to remove by either mechanical or chemical means. There is a variety of these tissues each one provided with individual properties. A good example is the internal lining of the digestive tract of most vertebrates, usually covered with a tenacious layer of mucus (Figs 8–15). The olfactory mucosa of terrestrial vertebrates is another classic example of a delicate tissue that requires extreme care. Whole isolated cells can be observed from the blood after simply drying them on a coverslip. Comparative LM, SEM observation have been

---

these filiform papillae due to the hardness of their covering epithelium. These preparations are easy to obtain and do not require any specific care. (21X).

FIG. 4. SEmicrograph of the same tongue of Fig. 2, however at higher magnification to show, on purpose, a technical defect not obvious at low power. The surface of both circumvallata (c) and filiform (f) papillae appear rather granular and fine cytological details of the squamous epithelium are obscured. This has been obtained by coating the specimen with an excessive stratum of conductive material, in this case, gold. These preparations are excellent for low power shots, but they obviously do not yield fine results at higher magnifications. (55X).

FIG. 5. SEmicrograph of the lip of a fish (Lebistes reticulatus). Same preparation as in previous figures. The pattern of the scales is perfectly preserved in all details, and the hole represents the pore of a taste bud from which average ball like structures, the microvillar appendages of the taste tells (m). The hardy, scaly surface of the lip of fish has been cleaned from mucus and debris by means of a soft brush. The cleaning is satisfactory and can be achieved without fear of damaging the surface. (4750X)

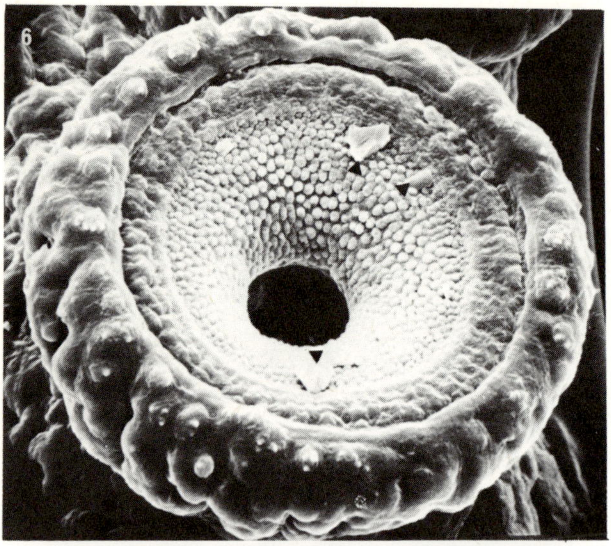

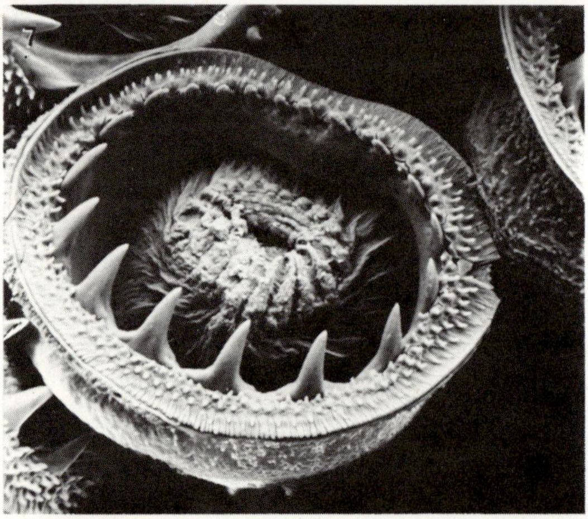

FIGS. 6 and 7. SEmicrograph of a sucker of Octopus Jubini. These prehensile organs are covered with a delicate cuticle (see also Figs 25 and 27). In spite of the abundant mucous secretion they can easily be mechanically cleaned. Some debris of dirt, possibly mucus fragments, are indicated by arrows. Notice the considerable depth of field, even more obvious in the following figure. (660X)

successfully performed with simple methodology by McDonald and Hayes (1969).

Artificial surfaces can be prepared by several procedures. (a) Material conventionally fixed and embedded for histological observations can be studied with the SEM to gain information from the surface of the sections. The embedding medium must be removed at least from the section surface and the fixation procedures adopted should be sophisticated enough not to show, at the magnification employed, obvious artifactual damage. The same procedure can be used with material prepared for TEM and embedded in resin, if the latter is properly removed after sectioning (Fig. 16). (b) Fresh or already-fixed material can be cut with a freezing microtone or with a sharp razor blade so as to expose a fresh, undamaged surface and, after proper dehydration and preparation, the section surface can be observed. The use of power enzymes to free specific structures from unwanted intercellular material has been successfully adapted in the study of connective tissue by Finlay *et al.* (1971). (c) Fresh or partly-fixed material can be dissected out or dilacerated to expose internal surfaces of the tissues (Fig. 19). Mechanical damage, however, and the presence of debris and tissue fluids often make this method rather frustrating. One example of this technique is provided by Lewis *et al.* (1969a) in their study on the neural organization of invertebrate ganglia. Visual receptors have also successfully been demonstrated by Lewis *et al.* (1969b). (d) Fracture of hardened blocks of tissues, both fixed and/or frozen will provide cleavage planes within the tissue which can be usefully studied (Figs 20–21). (e) Single cells can be dissociated from any tissue by enzymatic activity or maceration techniques. Of the latter, 30% ethanol, saturated aqueous picric acid solution or 0.5% NaOH will dissociate most cellular elements in 24 to 36 hours. The dissociated elements can then be post-fixed, dehydrated and mounted on appropriate stubs for observation. A ciliated cell from the respiratory epithelium of the nose of the cat is illustrated in Fig 19. Cultured cells can be observed as a whole when properly prepared (Boyde *et al.*, 1968) and human chromosomes were seen by Christenhess *et al.* (1968) in cultured lymphocytes fixed in sodium citrate and acetic ethyl alcohol and by Neurath *et al.* (1967) and Veurath (1968). (f) Freeze-etched surfaces are also obtainable for the study with the SEM but it is our opinion that the replica technique available with the TEM, although technically more complex, has so far provided superior results.

At the moment of this writing it appears that none of the techniques mentioned above has been worked out in sufficient detail to provide

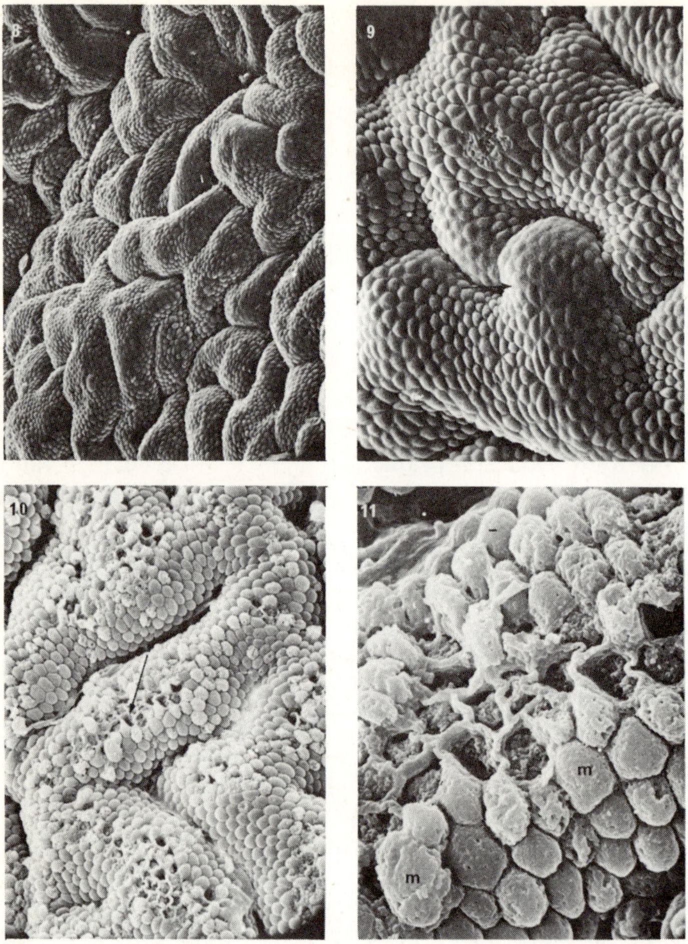

S. 8–15. This and the following SEmicrographs through Fig. 15 **are** from the stomach and intestine of a frog (*Rana pipiens*). They all represent the result of a considerable effort in having these surfaces prepared with a minimum amount of damage and with a sufficient cleanliness. This figure (Fig. 8) represents a low power view of the stomach wall where folds of the mucosa and the pattern created by single cells can be individualized. At this low power (200X) the preparation seems to be sufficiently clean and representative of the real surface of the stomach. For preparing these tissues the animal was perfused through the aorta with buffered paraformaldehyde while physiology saline was forced throughout its stomach and intestine. With this method the tissue was slowly fixed while the surfaces were continuously washed. At higher magnification, however, as can be seen in Fig. 9 (550X) a zone of abrasion is evident (large arrow) and in some areas the cells show obvious coating of foreign material (small arrow). In

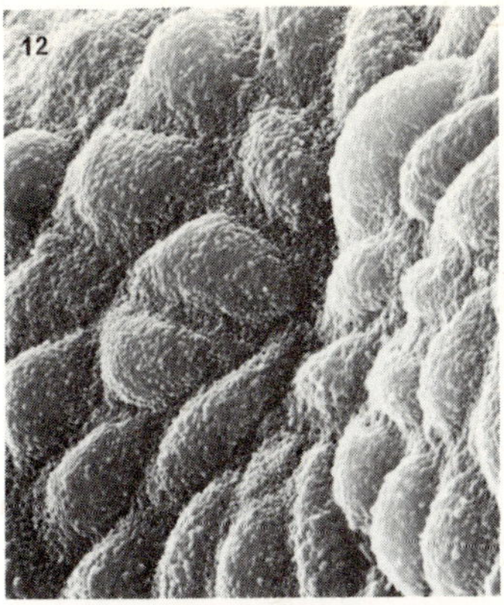

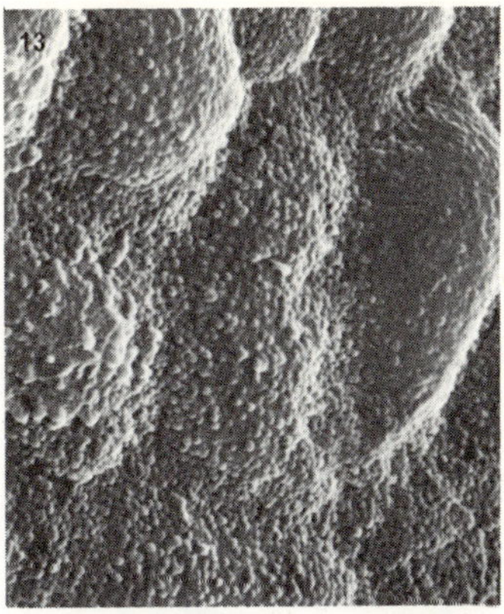

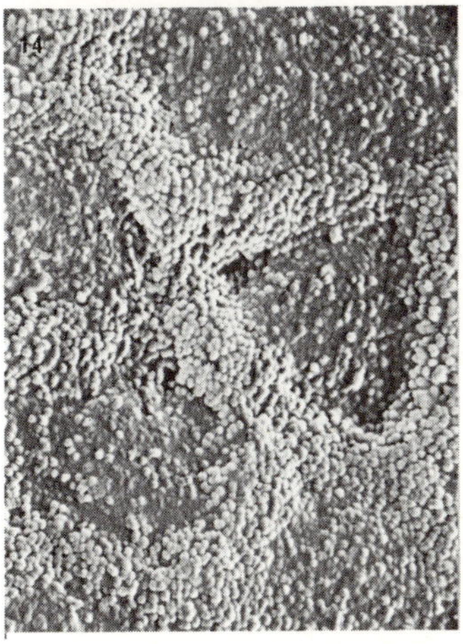

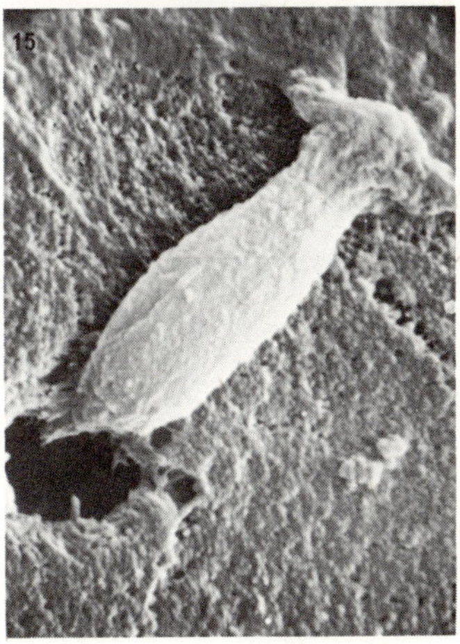

Fig. 10 (500X) the damage on the cell surfaces appears even more obvious (arrows). However, the surface boundaries of the cells are sufficiently distinct. In Fig. 11 (2000X) the damage is more obvious and again the surface of the cells shows an undesirable layer of coating material, likely to be mucus (m). Figures 12 and 13 (3000X and 5000X, respectively) show the cell surfaces where microvilli are partially obscured by the mucous secretion. This series of pictures emphasized the relative concept of "good preparation" as obtained with the SEM. A very satisfactory picture at a low power (Figs 8 and 9) becomes a second rate picture of an unsuccessful preparation when the need for finer details is desired, such as in Figs 11 and 13. Figure 14 is a detail (5500X) of the surface of four cells where the microvilli are preserved and free only at the boundaries of the cells. In the centre of the surface of each cell the mucus has covered the microvilli and the tips only are visible. Figure 15 (1100X) represents a goblet cell mucus lump which has been just extruded. The cell was situated in the intestinal wall of the frog.

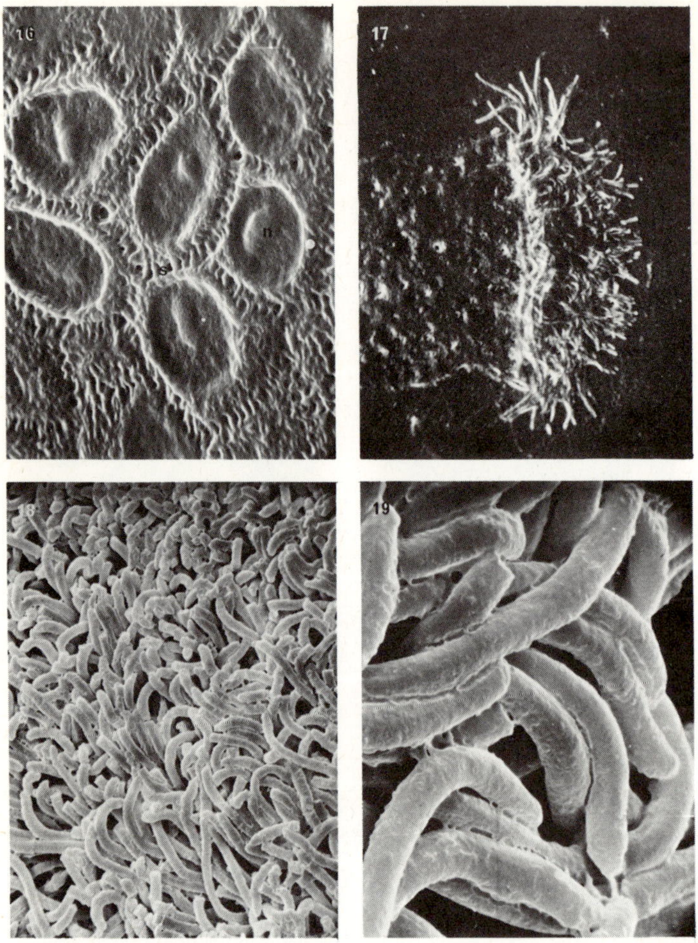

FIG. 16. SEmicrograph of the spinosum layer of the squamous epithelium covering the tongue of a rat (4000X). The tissue was prepared following glutaraldehyde fixation, for conventional TEM preparations. One section 1 μm thick has been cleared of the embedding medium (Araldite) and observed with the SEM. In these preparations the main difficulty is to free the sections from the embedding medium without damaging the tissue. In the preparation shown here the Araldite has been only partially removed. This allows the visualization of only some details of the cells—but covers most of the finer cytological characteristics. Some preparations, but embedded in paraffin are more easily cleared from the embedding medium but obvious damage to fine structures results (see text for further comments).

FIG. 17. SEmicrograph of a columnar ciliated cell of the respiratory mucosa of the nose of cat. The cell was isolated by maceration of the reliable results routinely especially outside the laboratory where they have been worked out. However, most of these techniques are susceptible of substantial refinement and, if pursued with proper care and adapted to specific requirements of the tissues and the scope of the investigation, they become powerful tools of investigation.

III. Techniques of preparation of biological material for SEM

There are several techniques that are claimed by many students to work "best" for the preservation of different sorts of biological specimens. In this section we will indicate a number of techniques and procedures, taking care to describe their essential details. No assurance, however, is given that all, or any, will always be completely satisfactory, as the needs of each of us are different and the tissues and animals we work with are many. The method that in our personal experience has been found most satisfactory is the one of repeated trials and errors, of self-criticism and endless endurance. This method is very expensive, and time consuming; so far, however, it has been found to be the best.

In dealing with biological tissues with a high water-content the following steps are generally necessary and unavoidable: (1) fixation, (2) dehydration, (3) drying, (4) mounting on a metallic stub, and (5) coating of the surface with a conductive layer of metal. Before describing specific methods in detail, a brief discussion of each of these steps is appropriate.

For study with the SEM it is of paramount importance that the surface of the specimen tissue be fixed properly with minimum shrinkage, while preservation of internal cellular details is secondary

---

epithelium in 30% ethanol after conventional fixation in buffered glutaraldehyde. Fine cytological details of the surface have not been preserved but the free surface is obviously covered with a thick carpet of short cilia. (4000X)

FIG. 18 and FIG. 19. These SEmicrographs illustrate the frog's retina receptors seen after removing the pigment epithelium. The preparation was obtained with conventional glutaraldehyde fixation in buffered medium and partial Araldite embedding of the specimen (see method at page 00). Partial embedding of the specimen with Araldite and removal of the same from the surface with propilene oxide helps considerably in removing unwanted debris from the surface and in minimizing shrinkage phenomena. (570X and 2300X respectively) (unless further use of the specimen is planned for TEM study). The fixative is therefore considered more in relation to its capacity for surface fixation and hardening of the tissue than for fixing it properly in depth. For this reason fine surface details can be preserved in rather large blocks (some 5–10 mm surface) when properly buffered OsO_4 fixatives are used. Properly buffered aldehydes are also recommended, for in these fixatives specimens can be stored for days without significant damage. (Specimens fixed for long periods in OsO_4 tend

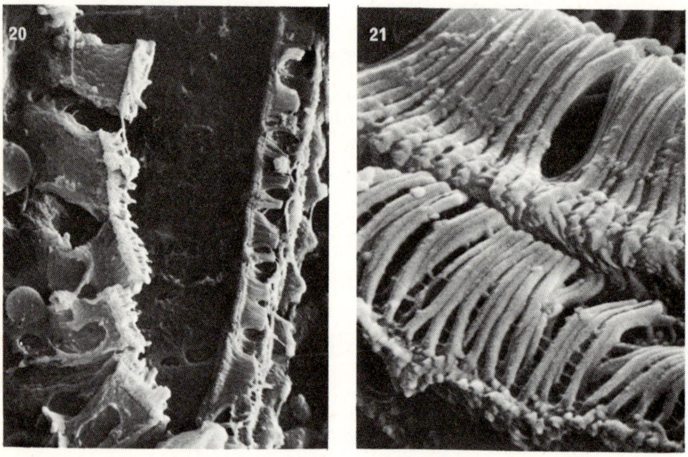

FIG. 20 and FIG. 21. SEmicrographs of the statocyst of Octopus Jubini. The tissues were fixed in buffered 5% glutaraldehyde, the ciliated epithelium fractured to show the intercellular surfaces and subsequently partially embedded in Araldite. In Fig. 20 rows of cells are separated one from the other to show single cells (1400X). In Fig. 21 (8500X) details of the ciliary processes are shown. The preparation is rather clean, cilia are not disrupted and while fine cytological details of the cells surface are not visible, the preparation allows a three dimensional appreciation of the histological structures. These pictures provide mainly complementary information when coupled with LM and TEM observations.

to crack or become too brittle to handle.) Furthermore aldehyde fixatives can be easily perfused through the blood vessels coupling the fixation with simultaneous washing of the surface under study. We recommend that both these types of fixative be tried before choosing one and adopting it for routine observations. Some fine details of olfactory cilia have been obtained after fixing the material in a 10% formalin solution either in distilled water or in frog's ringer, and such simple fixatives should not be discarded *a priori.* They should be tried when the study does not require the collection of extremely fine details. Formol acetic alcohol was, for instance, successfully used by Boyde *et al.* (1968) for fixation of cultured nerve cells.

When ultra-fine observation of the surface is instead a primary goal and high resolution micrographs are to be taken and/or when further study of the material is planned with the TEM, fixation procedures should be carried on with all the cautions used for sectioned TEM material. The following points are of major concern in finding a suitable fixative solution: (1) pH, (2) buffer, (3) tonicity, (4) concentration of the primary fixative, and (5) temperature and duration of the fixation. These points, which are well known to the traditional electron microscopist and are amply discussed in TEM technical textbooks (Pease, 1964; Sjostrand, 1967; Hayat, 1970), may play decisive roles in the preservation of fine surface-details such as microvilli (Figs 38, 39, 50). It is a safe rule to establish the degree of resolution and detail one wishes to achieve, and accordingly adapt the technique. With the SEM, large specimens very often have to be scanned for rather gross surface details, while at other times only high magnifications are of interest. A proper planning of the scope of the work saves time and money.

The mode of fixation deserves some further comments. The two common ways of fixing for SEM with chemical fixatives are as follows: (1) fixation of the specimen by immersion on a single fixative and (2) fixation by perfusion with simultaneous flushing of the surface with buffer solution and/or fixative solution. Both methods are usually quite satisfactory. However, the perfusion-flushing procedure, when practical is to be favoured. The perfusion gives a rapid fixation of the entire block, allowing homogeneous hardening of the tissue while the flushing of the surface (like the olfactory mucosa, for instance) provides a method for both fixing and cleaning from unwanted secretory products. With this method, also, two types of fixatives can be used, a perfusate with excellent hardening capacity and a surface flush with the capability of fine preservation of structure.

In routine preparations dehydration can be obtained with an ascending series of ethanol or acetone. A longer stay in the 95% and 100% ethanol seems to harden the tissues so there is less shrinkage when the pieces are passed from the alcohol to the air. Shrinkage, however, is always considerable and should be reckoned with when relative size of structures is of any concern. Comparison with fresh specimens under a light microscope and with TEM specimens provides a reliable means to measure the degree of the phenomenon.

Many authors have emphasized the importance of avoiding damage of the surface such as cracks and folds that may happen during drying of the tissue and the need to minimize such defects. We have found that specimens allowed to dry from 100% ethanol in a petri dish usually retain surface details without major damage (see Figs 22–23); more specialized techniques should be considered only when detailed and very fine observations must be carried out.

Mounting the specimens on the metal stub is usually done with Duco cement or any brand of epoxy resin. The process is easy but

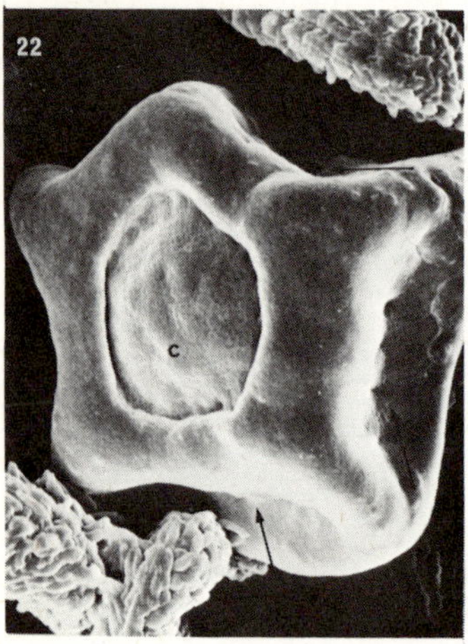

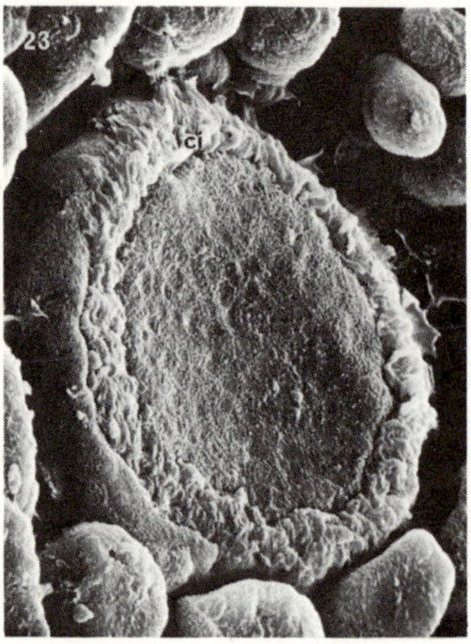

FIG. 22. The eye of a common scallop (*Pecten irradians*) is shown among the villous structures of the border of the mantle. The cornea (c) proper is well preserved and the pattern of its epithelial cells can be perceived. The lack of mucus on these surfaces makes the preparation of such organs extremely easy. However, fixation in 10% formol in sea water and simple dehydration in alcohols and air was not completely adequate to avoid obvious, however not disturbing, shrinkage phenomena observed at arrows. (210X)

FIG. 23. The taste organ of a frog represents a more difficult preparation because the tongue of these amphibia is notoriously covered by a dense stratum of mucus. Mechanical cleaning, as it was performed in the tongues of cats (see Fig. 3) was not possible due to the delicate structure of these papillae. Notice for instance the ciliated border of the papilla (Ci) which would have been damaged by mechanical cleaning. The preparation was obtained by perfusion of the tongue and contemporary washing of the surface previously with saline and then again with the fixative, a buffered aldehyde dehydration was performed through ethanols and the piece was left to dry in a petri dish. (600X) care should be exercised to provide the correct orientation of the specimen. This may require particular attention when the surface to be observed is corrugated or presents grooves or deep concave portions. Prior to observation the specimen must be coated with a layer of conductive metal in an evaporator. This process is to prevent charging of the surface in most biological specimens. The shadowing

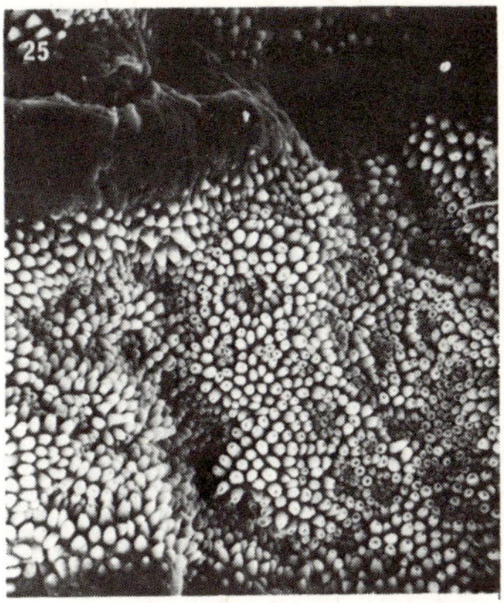

FIG. 24. The SEmicrograph represents an entire Octopus Jubini that was collected and fixed in 10% formalin. Dehydration was performed in alcohols. This picture shows how an entire animal can be observed at extremely low power (27X) and subsequently analysed in amazing detail. The following pictures are the details of parts of its suckers.

FIG. 25. This figure represents a detail of the internal surface of one sucker from the same material of the previous figure. It shows the cuticular apparatus of each cell lining the internal cavity of the sucker.

metal is deposited on the surface of the specimen following straight lines from the point of origin. Consequently, when the surface is complicated by folds some areas are less coated and others not at all. Proper orientation of the specimen on the stub should provide for an optimal coating of all the specimen surfaces to be observed. Also, the orientation of the specimen should be carefully selected to permit an optimal angle of view in the microscope. The observer should be familiar with the capabilities and limitations of the goniometer stage provided in the SEM and, when possible, personally supervise the position of the block on the stub.

In many laboratories where preparation of the material is routinely carried on by a trained technical staff, the coating of the specimen is also satisfactorily performed by technicians. This step, however, should be carefully supervised by the student himself. The amount of material evaporated on the specimen surface should be precisely controlled, and its thickness known. While a 200, 10^{-1} nm, layer is reckoned generally satisfactory, the amount of metal evaporated on the surface should be kept to a minimum to allow correct observations under the operational conditions of the microscope. Layers in excess of 200, 10^{-1} nm, tend to obscure details which may be of no concern in low power observations, but of great concern in studies based on high resolution pictures. It is obvious from these notes that the student should have a clear understanding both of the techniques of preparation and of the operational requirements of the machine in order to achieve satisfactory results.

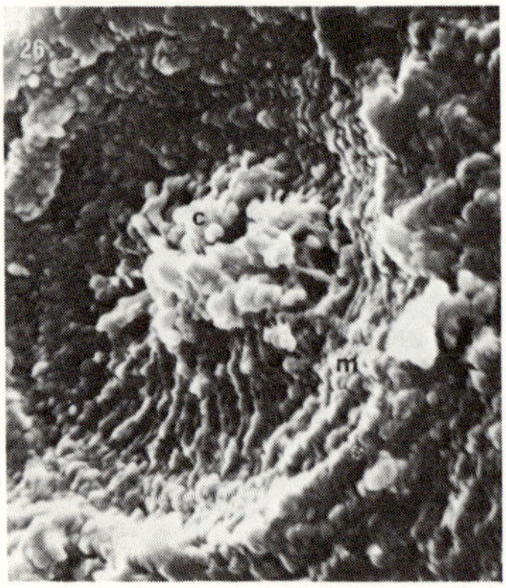

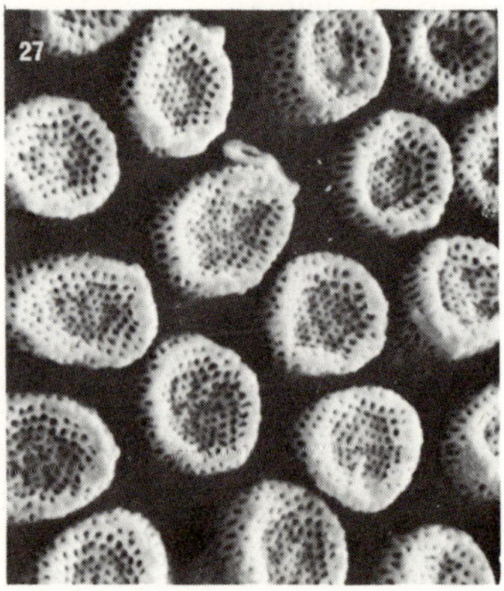

FIG. 26. Cilia and microvilli (c, m) of the sensory cells of a sensory organ of the octopus sucher emerge from a crater like concavity. These details are not easy to obtain because often mucous strands tend to obscure them. (13 000X)

FIG. 27. A further detail of the cuticular apparatus of each cell in the surface of the octopus sucker (6250X). The detail is excellent and the cleanliness of the preparation quite satisfactory. This series of form pictures illustrating an octopus at varying magnifications from 27X to 8 000X is an example of food results obtained with relatively simple technique. The animal was fixed in common formaline and dehydration performed in alcohols. The detail of the sensory organ (Fig. 26) and the cuticles at Fig. 27 suffice to demonstrate the fine details wanted without waste of time in more sophisticated techniques.

The first specific procedure to be described is also the simplest. It is one that has been found to be highly satisfactory in the preparation of olfactory mucosa as well as many other similarly delicate tissues (Figs 28–31).

(a) The surface to be studied, when not readily accessible, is surgically exposed in the anaesthetized animal and gently washed with a buffered saline solution. This must be adapted to the kind of animal under observation. One may use the same buffer solutions that will be employed subsequently for fixing the tissue with either OsO_4 or aldehydes. With tissues of some toughness, however, such as the tongues of mammals, simple 0.9% saline solutions are satisfactory. The time of the washing should be protracted till observation with a dissecting microscope indicates that the secretory products are washed away. Blocks of tissue are then excised with a sharp scalpel and again washed in the same solution kept at 4°C in a petri dish. The low temperature slows down the rate of secretion by the secreting cells present in the tissue (such as supporting cells, for instance, in the olfactory epithelium) which continuously frustrate the efforts to free the surface from that unwanted material. While the block of tissue is in the petri dish, a jet of the buffer solution can be directed on its surface with a common pipette. This helps to remove the secretion. In olfactory mucosas with long cilia the jet should be very gentle, however, to avoid damage to the fine appendages. On more consistent surfaces covered with a dense layer of mucus, such as the stomach or intestine a gentle mechanical washing with a fine, soft brush is indicated. All these procedures should be performed with the aid of a dissecting microscope.

(b) After washing, the blocks are immersed in a fixative solution which uses the same buffer as that employed for washing. The fixing agent we use for routine preparations is either 10% paraformaldehyde or 4% glutaraldehyde. OsO_4 can also be used and provides good results in 1 to 2% dilutions. When aldehydes are used the fixation time may be restricted to 2–4 h, but overnight fixation is usually preferable. Specimens left in the fixative for many days also provide very good results at the more commonly employed magnifications, 16X to 20 000X. During the entire fixation period the pieces should be left adherent to any bony underlying or cartilaginous tissue to prevent curling of the softer tissues. If this is not possible, the layer of tissue may be pinned on wax at the bottom of a petri dish; fixation in this condition allows the specimens to become fixed with a minimum of curling of the surface. After fixation the blocks are washed in the same buffer and then in several short baths of distilled water to remove salt deposits on the surface. During the washing, both in buffer and distilled water, the surface should be carefully observed with a dissecting microscope. A beam of light grazing the surface helps considerably in detecting remaining layers of mucus or any

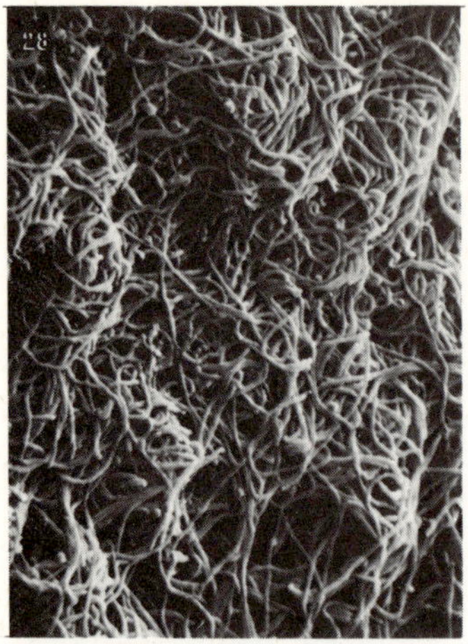

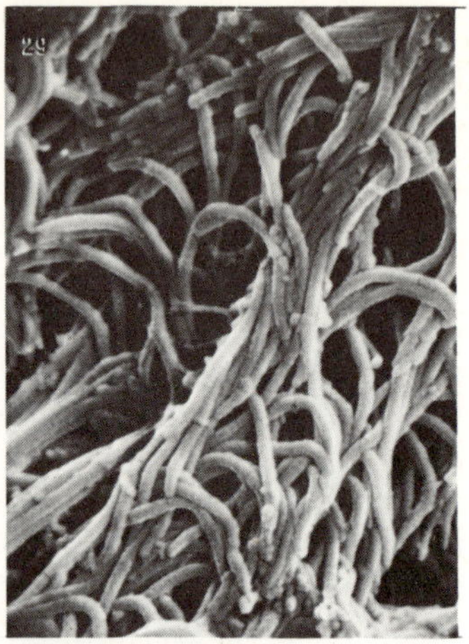

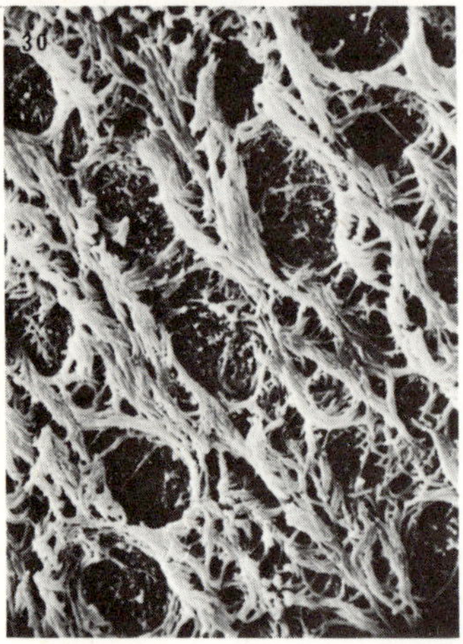

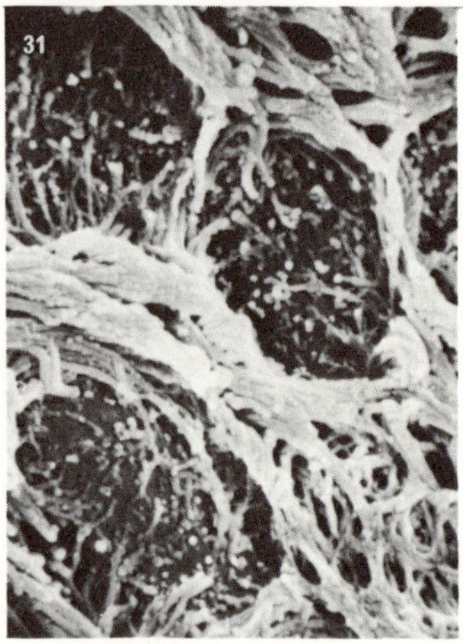

FIG. 28. The surface of olfactory mucosa is usually covered with the long olfactory cilia that appear like an irregular meshwork of interwaving filaments. In this SEmicrograph from a Box turtle the cilia completely mask the epithelial surface. The preparation was obtained after washing the nasal cavity with saline, by simple fixation in 10% formalin and dehydration in alcohols. (5200X)

FIG. 29. There is little significant variation between the surfaces of olfactory mucosas of different vertebrates when the cilia are long and cover completely all the epithelium. In this SEmicrograph taken from a frog the pattern is similar to the one of the previous figure. The technique of preparation was the same. It is noticeable in both preparations the lack of mucus which allows the visualization of the single cilia. These clean preparations are often obtained with the simple method indicated above, however, they are not to be expected as routine.

other unwanted particles which should be removed with fine forceps or a brush.

(c) From distilled water the blocks are passed in a series of ascending ethanols from 30% to 100%. Each of these steps should be long enough to result finally in complete dehydration of the block and this depends on the size of the specimen itself. It has been found, moreover, that leaving the specimen in 95% or 100% alcohol for one day each increases their hardness and reduces to some extent their shrinkage. From 100% alcohol the specimens can be directly dried in the air or passed through a bath of propylene oxide. There is no substantial difference in end result as long as the 100% alcohol is perfectly anhydrous. From the alcohol or propylene oxide the blocks are transferred to a petri dish with a filter paper on the bottom and left to dry for some hours. It is at this stage that shrinkage and curling of the specimens becomes obvious. When blocks appear dry they should be placed in a desiccator to insure their complete dehydration.

(d) After the desiccation the specimens are properly oriented and mounted on an aluminium stub. The mounting is performed with Duco cement or any type of fast drying resin. When the glue is dry it is usually appropriate to paint the edges of the tissue and the glue with a conductive layer of silver paint. This ensures proper conductivity between the surface, when coated, and the aluminium stub. Specimens properly fixed on the stub are then put in an evaporator and a layer of metal evaporated on their surface. Gold, palladium or gold with carbon are used, and the average thickness of the layer is in the range of 200, 10^{-1} nm. Details of the coating technique and of the use of the evaporator are provided in many specialized publications, e.g. Hall (1966) and Sjostrand (1967).

The method outlined above does not consider in detail the fine

---

FIG. 30 and FIG. 31. The two SEmicrographs were prepared by perfusing the animal (dog) with aldehyde fixative and contemporary washing of the olfactory surface with saline. Dehydration was performed with alcohols. During the washing of the sensory surface a particularly strong jet of liquid was directed in selected areas so as to damage and break the cilia. The result has been to shorten the cilia exposing the surface of the supporting cells. In Fig. 30 the cilia are clearly seen to surround the supporting cells according to a pattern recently described in the olfactory mucosa of many vertebrates (see Graziadei, 1971). In Fig. 31 a detail of the surface of the supporting cells shows fine microvilli and the stumps of the cilia surrounding the supporting cells surface. (2400X and 5500X respectively) aspects of physical and chemical events occurring during the different steps of preparation. It is, rather, an empirical method which provides consistent and rather fine results on routine preparations, and for this reason is recommended. It is obvious that many theoretical and practical criticisms can be raised. Most of the micrographs presented in this chapter, however, were obtained with this very method, which yielded sufficient details to provide the information needed (see for instance, Figs 24–27).

The major practical drawback of this method is the often considerable shrinkage that occurs in the specimens.

Another method which is very similar to the one just proposed in all steps but (A) and (B) also gives good results. Only the first two steps will be described.

(A) The animal is deeply anaesthetized and if necessary, the surface to be studied is surgically exposed. A major artery supplying the sector under study is then chosen and perfusion of the sector performed. Previous-treatment of the animal with heparin avoids the danger of blood clotting. The first perfusing liquid is a buffered solution isotonic with the blood; it is used to clean the blood vessels of any blood. The injecting pressure must be evaluated for each animal, as must the pH, osmolarity and composition of the perfusate. After the perfusant returns clean from the blood vessels, an identical solution with 3%–5% glutaraldehyde or paraformaldehyde added, is injected and perfusion is continued for 15 to 20 minutes. Simultaneous with the perfusion of the part with both buffer and buffer plus fixative the surface to be studied is irrigated with the buffer solution by gently sending a jet of solution through a pipette. When the surface looks clean the buffer is substituted with fixative and the procedure is continued until the perfusion is stopped. This procedure is particularly suitable for the olfactory mucosa where a continuous layer of mucus overflows the surface. Fixation from within fixes the tissue well, leaving the surface secretion, fluid enough to be removed. Removal of blocks of tissue from the animal and subsequent steps are similar to those in the first method proposed.

Another variation of either of the two preceding procedures, which is suitable for the study of both natural and artificial surfaces includes a bath in glycerol. These substances, in concentrations variable from 5% to 30%, can be used for periods of 12 to 36 hours. Preparation of the material and fixation are similar to those outlined in previous methods. The glycerol step is interposed between the brief washing in water and dehydration in ascending ethanols. This procedure facilitates, sometimes, cleaning of the surface from unwanted secretory products, as observed, for instance, in the mammary gland by Nemanic and Pitelka (1971). The same authors also used dehydration in ethanol followed by replacement of ethanol by Freon 133; and the drying procedure was accomplished with the Freon critical point method of Cohen *et al.* (1968).

When apparatuses for freeze-drying are available this method can be used for preserving specimens for SEM. This technique, which is well described and illustrated in many textbooks (Siegel, 1964; Hall, 1966; Sjostrand, 1967; Hayat, 1970), may be advantageous for avoiding chemical fixation and for preserving the *in vivo* position of structures which may be displaced by the ordinary steps of chemical fixation and dehydration with alcohols. Furthermore, it provides some advantage in minimizing shrinkage of the specimen and cracking of its surface during the dehydration. Boyde and Barber (1968), have proposed this technique to achieve better preservation of cilia and to avoid shrinkage and tearing of epithelial surfaces. Details of the method are provided by Boyde and Wood (1969) with comments on the steps required. In our personal experience this method gives results comparable to those outlined above, at least in a majority of tissues. Figures 28–31 show fairly satisfactory preservation of ciliary structures, for instance, while using less sophisticated techniques. However, when other techniques are unsuccessful the freeze drying method should be tried. Fixation in 10% formalin, followed by freezing in liquid air and freeze dry was used by Hammond (1968) in worms. Horridge and Tamm (1969) have suggested OsO_4 fixation followed by critical point drying to avoid distortion of the cilia due to surface tension forces.

In most procedures thus far used for the preparation of SEM materials, the step dealing with dehydration is one of the most critical. When the tissue is dehydrated, inter- and intra-cellular water occupying spaces in the tissue must be replaced by a low-viscosity liquid such as alcohol or propylene oxide. This in turn, is replaced before the specimen is placed in a vacuum chamber (either the evaporator bell or the SEM column). In most cases, and particularly when specimens comprise several tissues of different consistency and with different water content, shrinkage phenomena may substantially alter the overall appearance of the surface observed. These phenomena interfere with the observation and description of normal relations between structures. The method that we propose below circumvents this problem by always keeping the tissue filled with a medium that avoids both collapse and shrinkage (secondarily, it helps remove unwanted secretory products from the surface of the specimen). This procedure has been used for years in our laboratory and has been also used with slight modifications by other authors (Cleveland and Schneider, 1969).

The specimens are usually treated by the method described in the first procedure and they pass through the steps A, B, and C. After the carefully trimmed blocks are bathed in distilled water, they are then immersed in a mixture of 60 parts of distilled water and 40 parts of Durcopan A. In this 60/40 mixture the pieces stay one day. They are subsequently passed through solutions where the Durcopan A is increased from the original 40% to 100% through 10% steps. Each of these steps should last at least one day but pieces left for several days in Durcopan solution do not show obvious damage. It should be remembered that the rate of penetration of the Durcopan is dependent on the characteristics of the tissue and the duration of each step should be commensurate to these characteristics. The bath in 100% Durcopan, however, should last for at least 48 h with two changes of the medium. At this point Durcopan is substituted with Araldite 506 and gradual steps are again advisable. Araldite is used without hardener. (Ascending concentrations of Araldite in Durcopan from 20% to 100% are changed every 24 hours.) In the final 100% Araldite it is not necessary to add accelerator except in the case that further sectioning of the block is foreseen. When the tissue is completely penetrated by Araldite it is gently removed and positioned on a petri dish with the surface to be observed on top. A jet of propylene oxide is then directed on the surface to remove the excess Araldite and other debris. This procedure is rather delicate and the amount of propylene oxide as well as the length of time of the washing must be determined by experience. Usually 5–10 min. are sufficient for cleaning of the surface of the block. The use of a low power dissecting microscope is suggested because it is during the washing that there is a good chance to drive away much of the mucus covering many epithelia. After the washing the piece is left to dry in a petri dish (the drying process is rather quick due to the high volatility of the propylene oxide) and then mounted properly on the stub and coated as previously indicated for other procedures.

This method is rather lengthy, requiring several days or weeks for the preparation of suitable specimens, but it has been found successful in reducing shrinkage and distortion of the surface (see Figs 18, 19, 20, 21).

IV. Capabilities and limitations of the SEM techniques in the study of the olfactory mucosa and other tissues and organs

In this section an analysis will be made of the methods of preparation of biological specimens so far proposed. Their usefulness and

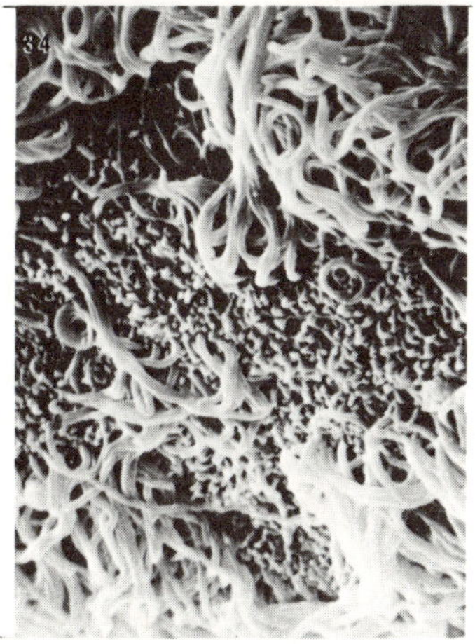

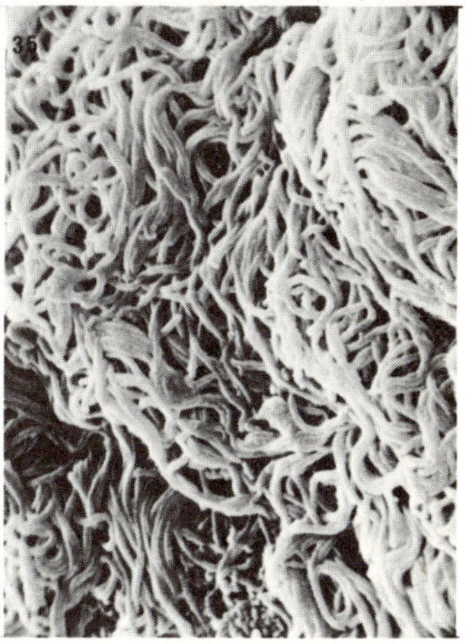

FIG. 32. Low power (20X) over-all view of the olfactory organ of a catfish. In fish the preparation of the olfactory mucosa is extremely simple and most often satisfactory due to the fact that no layer of mucus covers the sensory surface. Simple fixation in formalin 10% and dehydration in alcohols obtains extremely fine results.

FIG. 33. Olfactory organ of garfish at low power (23X). Same procedures as in the previous preparation. These low power views allow the scanning of large surface areas for further more detailed analysis of selected points.

FIG. 34 and FIG. 35. Two high power SEmicrographs (5800X) of the olfactory mucosa of garfish where olfactory cilia and microvilli can be observed. One can notice that there is no appreciable difference in the pattern of cilia between these fish and turtles or frogs (see previous Figs 28, 29).

drawbacks in the study of the olfactory mucosa will be considered in the light of our personal experiences and an evaluation of their worthiness both in research and teaching discussed.

The first technique that has been described in the preceding section is the one that should be tried first because it often provides excellent results, does not require sophisticated equipment and allows considerable flexibility in the timing of every step. The difficulty in cleaning of mucus and other secretory products from the surface represents the major limitation of this technique (and many others, too) and care should be exercised in washing the surface. With this technique the olfactory mucosa of fish, which is normally free of mucus shows quite consistent and fine details (Figs 32 to 39). There is no appreciable fracture of the surface; in spite of the rather simple drying procedure, there is no indication that the specimen is damaged by surface tension events occurring during this step. Cilia and microvilli are correctly preserved and their arrangement, so far as we can determine, seems to maintain the *in vivo* condition. When a prolonged dehydration in alcohols has been used, shrinkage of the specimen is within acceptable limits and rather even across the surface thus minimizing artefactual displacement of structures at least as observed at the currently used magnifications.

When cleaning of the surface becomes particularly difficult, the procedure based on perfusion should be attempted. The perfusion not only insures optimum fixation but allows simultaneous protracted washing of the surface, thus facilitating the cleaning process. It has often been observed that the perfusion technique is the only method that allows one to obtain a clean surface. Some epithelia, and the olfactory mucosa is one of them, continue to secrete mucus even when they are excised from the animal. Perfusion is useful because it fixes the tissue from within and prevents further secret extrusion of material before the surface is hardened by fixation. We have, then, a tissue that is fixed in its deep layers while the surface can be gently and efficiently washed and the secretion carried away. In animals such as turtles, where the secretion is rather heavy and sticky, the perfusion technique produces good results. Rather clean pictures of intestinal surfaces have been obtained with perfused material by Ceccarelli *et al.* (1968). The use of substances which solidify in the blood vessels has not been routinely applied in our laboratory. This technique, however, helps preventing the shrinkage of large pieces of tissues and should be furtherly explored.

The use of glycerol in the preparation of SEM specimens has been suggested by many and recently applied with very good results by

Nemanic and Pitelka (1971) in the mammary gland. Although not a fool proof procedure, glycerol helps considerably in cleaning the surfaces. As noted by these authors, sophisticated fixation procedures are not required; they may actually represent an inconvenience because they may fix too much and too well the structures under study. We have found the glycerol quite adequate in the preparation of some mucosas of the intestinal tract as well as in the preservation of fine surface details such as cilia and microvilli.

The freeze-drying technique has been recommended by many authors and used on a variety of tissues with remarkable results. The technique in itself does not provide a means for cleaning the surfaces from unwanted material, but it is certainly helpful in avoiding the damage induced by both chemical fixation and alcohol dehydration. In routine work and when a freeze-drying apparatus is not readily available other preparative techniques are certainly preferable for they are easy to carry on and provide excellent results. However, the freeze-dry technique should be considered as an option when the tissue may be damaged by chemicals or some component of the tissue may be lost in the steps of the chemical fixation and dehydration.

The method that uses Araldite is possibly the most lengthy one, but has the advantage that, when processing of the blocks cannot be carried out immediately, these blocks may be stored indefinitely in any one of the resin baths. This is not, however, the only advantage of the method, because we have had considerable success with this procedure not only in preserving delicate pieces without apparent shrinkage, but we have felt that the surface of some pieces comes out notably cleaner than with other procedures and without mechanical damages that either the washing or brushing usually produce. Some illustration of the results provides the demonstration that extremely delicate structures are still well preserved and with the arrangement that they presumably have *in vivo* (Figs 20–21).

When a successful preparation of a surface has been obtained and a three-dimensional view made possible, there are important questions that we should ask ourselves in the study of the olfactory mucosa or of any other tissue. Is there any bit of information which is new and worthwhile in this preparation that makes the lengthy effort scientifically justifiable, or do we gain only a complementary but redundant electron micrograph? Is there any other less expensive technique that could provide us with the same kind of understanding of the structures under study? Undoubtedly the SEM provides us with a "different" view of the specimen, and, as such, something new is always gained. However, this "something" may seem at first extremely small.

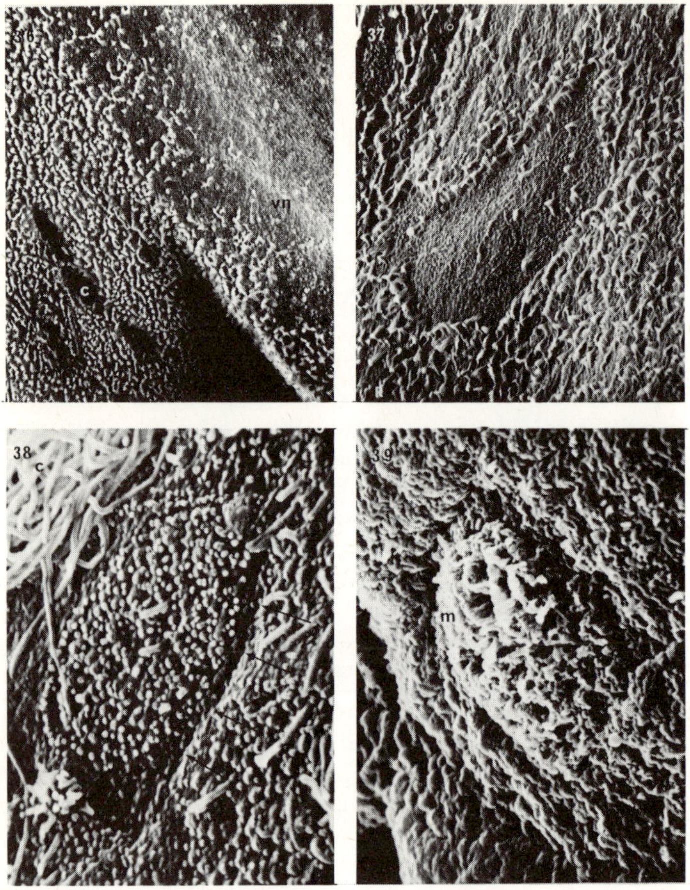

FIG. 36. Low power view (240X) of the internal surface of the nose of Box turtle to show the change in surface pattern from the respiratory, ciliated region (c) to the vomeronasal region (Vu). At this low power the cilia appear as ridges smoothly disappearing towards the vomeronasal region which is covered by cells provided only by microvilli (see Graziadei and Tucker, 1970). The preparation was obtained by washing the nasal cavity in the anaesthetized animal with saline and then fixing in situ with 10% formalin followed by dehydration in alcohols.

FIG. 37. From the same preparation of Fig. 36 this region in the respiratory area shows an oval patch of scarcely ciliated epithelium. The presence of non-ciliated cells in the respiratory regions has been frequently observed and the visualization of these variations in epithelial characteristics is made particularly easy with the SEM. (550X)

FIG. 38. The surface of a supporting cell in the garfish olfactory mucosa shows its boundaries in the form of a shallow groove (arrows). The free surface

In the example of the olfactory mucosa, which is our main concern here, we know already that the surface is covered by a carpet of cilia, as this has been established both by LM and TEM. We know the number of these cilia, their structure and the location of each of them on each cellular component of the mucosa. The new information seems then to be not especially important in scientific terms. What the SEM says is simply that there is a carpet of cilia, whose length is by all means difficult to determine, and of whose structure and dimensions are better determined by the conventional TEM. When straightforward morphological observations are performed, the SEM seems to add little if anything to what we already know, though it provides breath-taking views of the surface which undoubtedly please our eyes and help us teach students the general morphology of the organ (compare the surfaces of several ciliated epithelia in Figs 36–43).

However, when these pictures lie on our desk, specific problems concerning the functional anatomy of this and other organs, can be more effectively dealt with when the research does not limit itself to the observation of static morphology but tries to penetrate into the details of the function itself, then we see that this new technique can really be exploited to its best advantage and the information we retrieve is really complementary but not redundant at all (see Figs 40–43, 44–47, 48–51). The usefulness of the SEM is well illustrated, for instance, by some work in our laboratory dealing with the study of the olfactory mucosa. In association with Dr. Tucker we were trying to determine the presence or absence of cilia in the vomeronasal receptors of turtles. Evidence, as discussed in detail by Graziadei and Tucker (1970), was contradictory; some authors reported cilia, others, no cilia, in the same animals. Observations made with the TEM showed that many receptors were, in fact, lacking cilia. However, the VN surface is quite extensive and complete serial sectioning

---

is covered with microvilli which are readily distinguished from the olfactory cilia in the upper left corner (c). Same preparation as in Fig. 34. (7500X)

FIG. 39. One single cell, presumed to be a sensory cell, of the vomero nasal organ of turtle (same preparation as in Fig. 36) can be visualized with the surface covered with microvilli (m). These sensory cells, contrary to the olfactory receptors proper, totally lack cilia. This picture provides one example of a very limited spot of surface which is satisfactorily clean while the surrounding regions are still unsatisfactorily covered with mucous deposits. Surrounding this sensory cell are other supporting cells. Their microvilli are shorter and do not appear satisfactorily in the picture due to the presence of mucus deposits. (6300X)

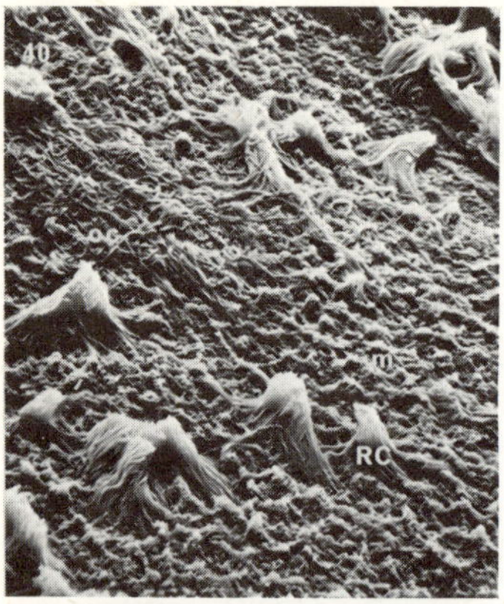

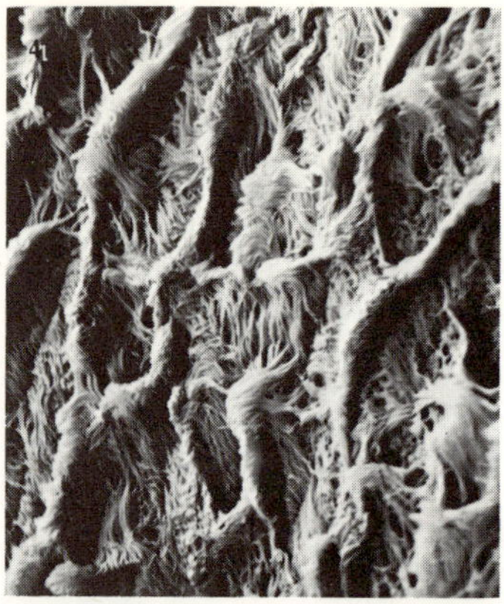

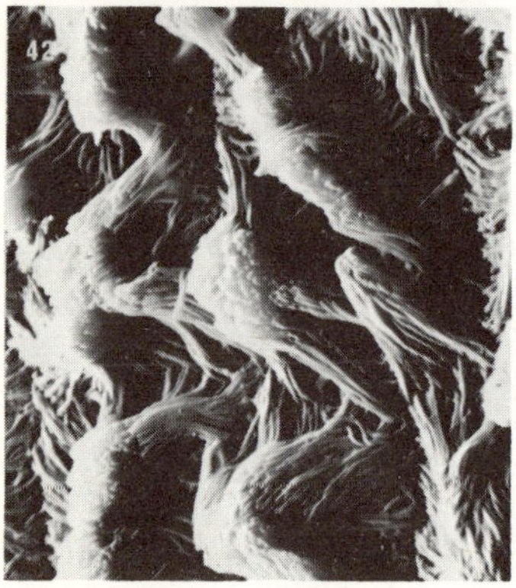

FIG. 40. Olfactory cilia (oc) and respiratory cilia (rc) are observed in this SEmicrograph from the nose of a turtle (same preparation as in Fig. 36). Their arrangement is quite different and is due to several factors. Respiratory cilia are rather short and beat synchronously while olfactory cilia are in this animal long (100 to 150 μm). and their beat is asynchronous. Microvilli (m) from supporting cells cover the surface not occupied by cilia. (2000X)

FIG. 41, FIG. 42, and FIG. 43. These three pictures taken from the same preparation of Fig. 40 show at different magnifications (2200X, 4500X, 11 500X respectively) the respiratory cilia of turtle. In Fig. 41 the low power allows the visualization of the wave-like pattern of the beating cilia. Cell boundaries are not discernible due to the thick and continuous carpet of cilia. In Fig. 42, 43 the individual cilia can be observed and while no mucus is present the arrangement of these appendages is reckoned very similar to the one of the alive specimen.

could have became a time consuming job. With the scanning EM the entire nose cavity of turtles was prepared and pictures taken at both low and high magnifications. It was clear (see Fig. 36) that the ciliated epithelium of the respiratory region ended abruptly in the vomeronasal area where the surface was covered with microvilli only. The possibility of scanning all the VN region then, provided the opportunity to establish that, although a few cilia could occasionally be present in some receptors, the vomeronasal region and the vomeronasal receptors were in essence completely unprovided with cilia.

Experimental observations are also worth mentioning. It is well known that many detergent substances as well as a strong jet of saline break the cilia from olfactory receptors close to their origin in the olfactory vesicle. Figures 30, 31 show examples of this procedure which allows the observer to see the receptors with artefactually shortened cilia surrounding the supporting cells. This peculiar arrangement has been recently described by Graziadei (1971b, c) in detail in the olfactory mucosa of several vertebrates.

The study of cilia as a means of evaluating the functional state of the receptors during experimentally-induced degeneration of them, is another example of how the SEM can provide information not easily attainable by other more conventional means.

It seems clear that the SEM can be extremely useful and even indispensable in biological studies of the olfactory mucosa and, no doubt, other organs when specific problems are set. (It is also true, however, that sometimes a different way to see things may not necessarily mean that we gain more meaningful information.)

V. Application of autoradiographic techniques to the study of the olfactory mucosa

Over the past forty years there has been a great deal of work and much speculation concerning the turnover or renewal of cells in the olfactory epithelium of adult vertebrates. Some authors have reported that no regeneration of olfactory receptors occurs following destruction of the epithelium or transection of the olfactory nerves (Kolmer, 1927; Frunwald, 1935; Le Gros Clarke, 1951). Others, however, have found evidence of a gradual reappearance of the olfactory receptors following the same procedures (Nagahara, 1940; Schultz, 1941; Smith, 1951). Morphological studies, at both LM and TEM levels, have indicated the possibility of regeneration of neurons in the olfactory epithelium of young mammals (Andres, 1965, 1966, 1969). More recent studies with autoradiographic techniques at LM level (Thornhill,

1967; Moulton, 1970) have indicated that some cells do turnover in the olfactory mucosa of different vertebrate. However at LM level their cytological characteristics and their nature could not be clearly understood. Definite evidence of receptor renewal that is continuous during the entire life of vertebrates has been provided by (Graziadei, 1971a, b, c; Graziadei and Metcalf, 1971a, b) although there is a different rate in different species and in the different ages of each species.

It is, then, obvious that the study of the olfactory mucosa, to be complete, must take into consideration this rather peculiar capacity of its neurons to die and be replaced by newly-formed cells. For this reason, it is worthwhile to include autoradiographic techniques, in this discussion of the gears for study of the olfactory mucosa.

Furthermore, a definite answer to the question of olfactory neuron regeneration in higher vertebrates is of major interest to the neurologist. The neurons of the central and peripheral nervous system of adult vertebrates are generally regarded as forming a static cell population, usually incapable of continuous renewal or regeneration. The olfactory receptor neuron may represent an exception to this general rule. The study of this phenomenon, therefore, may lead to a better understanding of basic mechanisms regulating histogenesis and regeneration in the nervous system.

Many books, articles and papers have recently been published dealing with autoradiographic techniques (Leblond *et al.,* 1959; Cameron, 1968–1970; Hughes, 1958; Fliedner *et al.,* 1968; Messier and Leblond, 1960). Most recently, Sidman has treated the subject with reference specifically to the nervous system (Sidman, 1970). The reader who is interested in an extensive survey of the subject is referred to the above cited. Our experience with respect to the study of the olfactory mucosa, however, represents a rather new and unexplored field, and we will limit our exposition to this subject.

Unless specific species are required for reasons inherent to the experiment, when autoradiographs of olfactory mucosa are planned it is advisable to carefully choose an appropriate animal species. The basic configuration of the organ is similar in all vertebrates, but for autoradiography some species have advantages over others in the way each cell type disposes itself in the epithelium. We have found that in *Rana pipiens* the supporting cell nuclei are discretely aligned above the neuron nuclei, and these are in turn, distinctly separate from the so-called basal cells. Bullfrogs do not have the same orderly disposition of cellular elements for instance. In mammals, Gerbils have been preferred to rats for the same reason. Some illustrations of olfactory mucosa from Terrapane Carolin, cat and Gopher tortoise are provided to clarify this point (Figs 44, 45–46).

Recognition of the different cell groups is made easier by these organizational differences, and statistical counts can be performed with comparable ease. We have found it possible, in many instances, to avoid tedious TEM autoradiographs for routine observations because the LM autoradiographs were sufficiently precise and the diagnosis of each element by and large correct. It must be stated, however, that with the exception of very clear cases (Figs 52, 53 and 56), EM recognition of each element is advisable to avoid the danger of erroneous conclusions. We have noticed, for instance, that in many instances the nuclei of supporting cells are very close to the basal lamina of the epithelium and very often the neuronal nuclei are among supporting cell nuclei. A further reason why it is advisable to have the control of EM autoradiographs is the limited dimension of the cellular components of the olfactory mucosa. It is well known that olfactory neurons have a diameter at their perikary on of only 4–5 μm, their dendrite is only 2 μm and even the supporting cells, while extending all through the thickness of the epithelium, have a transverse diameter of only 5–7 μm. In ordinary LM sections, 8 to 10 μm in thickness, one element may be superimposed on the other creating difficulty in establishing an accurate localization of the silver granules. One-micron sections of Araldite embedded material allow better resolution than ordinary histological preparations, and we should remember that the optimum resolution obtainable with H^3 is in the order of 1 μm at LM level. The major obstacle of cell recognition should consequently be kept in mind before collecting data concerning the specific duplicative capacity of the cellular elements in the olfactory mucosa.

The usefulness of tritiated thymidine in combination with autoradiographic techniques for the study of chromosome replication, cell renewal and cell migration is generally recognized (Hughes, 1958). Thymidine is a specific precursor of nuclear DNA and is incorporated into the nucleus of the cell only during the DNA-synthetic period of the cell cycle. The radioactive isotope, tritium, may be substituted for a hydrogen atom of the thymidine molecule by a catalytic exchange reaction, thus labelling the thymidine with the radioactive isotope and forming tritiated thymidine (^3H – TdR). The beta particle (electron of nuclear origin) produced by decay of the tritium nucleus has an average energy of only 5.7 KeV and is one of the least energetic known. The low energy and short range of the tritium beta particle are advantageous because radiation toxicity is minimal in the

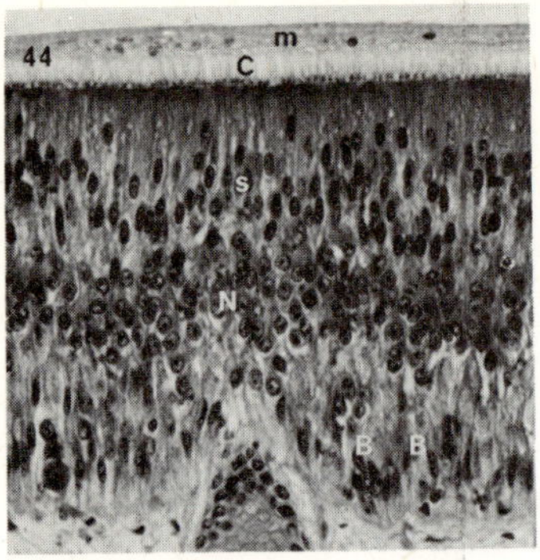

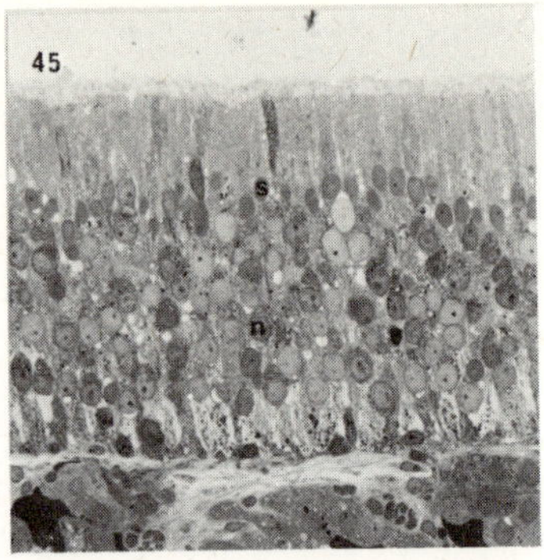

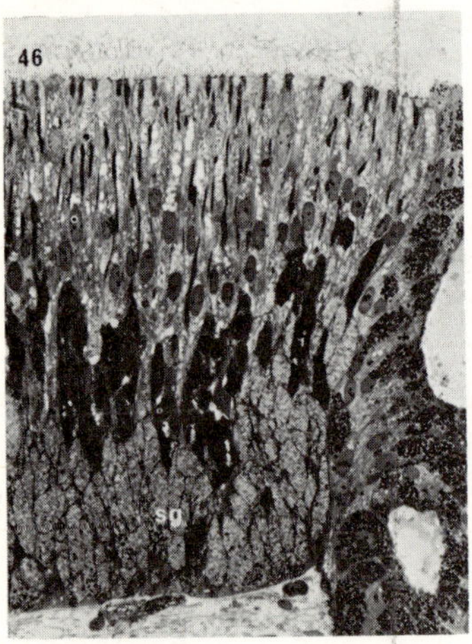

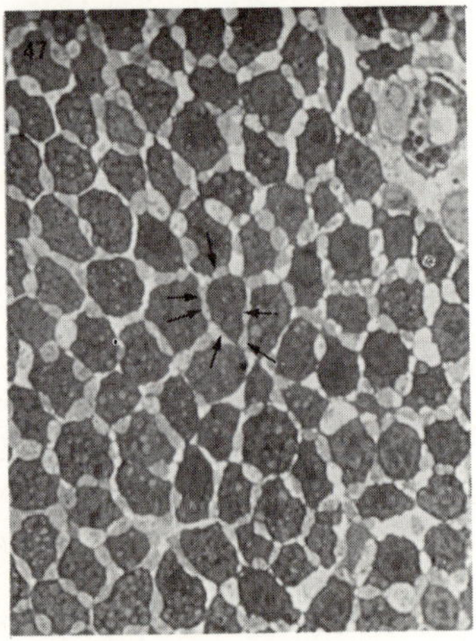

FIG. 44. This picture and the following 45, 46, 47 are LM pictures to point out the obvious different kind of information that can be collected with a different, however complementary, technique. In Fig. 44 (a 10 μm section of turtle olfactory mucosa stained with iron haematoxylin) the cilia (c) are clearly seen covered with a thick layer of mucus (m). However, no indication is provided in this picture about their arrangement, namely wavelike of random. On the other hand this picture provides a clear understanding of the composition of the olfactory mucosa where supporting (S) sensory (V) and basal (B) can be individualized in discrete strata.

FIG. 45. In this LM picture of cat olfactory mucosa fixed in glutaraldehyde and Araldite embedded and cut at 1 μm with a glass knife the cilia at the surface are hardly discernible while the structure of the cellular components is better preserved than in the previous turtle illustration. The nuclei of the supporting cells (s) are clearly distinguishable from the neurons nuclei (n). It is proper to mention that the latter show a very noticeable variation in their structure. As it is properly demonstrated in another part of this chapter the olfactory neurons continuously turnover in adult vertebrates. This means that neurons of tissue under study and maximum resolution can be obtained for autoradiography.

When ^3H – TdR is administered to animals it is rapidly distributed through all the tissues of the body and is taken up by those cells which are duplicating their DNA content in preparation for cell division. Resting cells do not incorporate thymidine. The nuclei of dividing cells become tagged with the radioactive label, and since DNA is metabolically stable, the label is retained through subsequent divisions, migrations and differentiation of the cell.

The technical details that will be outlined below refer to an ideal experiment carried out on the frog, *Rana pipiens.* The reference to a specific animal may help explain some technical details, but the same rules can be applied to animals of any species from mammals to fish. Actually, all of our experiments that have been carried out on mice, gerbils and rats had satisfactory results. Normally for our study small animals were usually preferred because of the cost of radioactive material, which is injected in amounts proportional to the body weight of the subject. Mature frogs, approximately 40 gms each, were injected in the dorsal lymph sac with 1 μC ^3H – TdR/g of body

---

different age are represented in the olfactory mucosa and this ageing process is in all likelihood expressed in the morphology of their nuclei.

FIG. 46. In this LM picture of turtle olfactory mucosa prepared as in Fig. 45 one can see far more details than in the corresponding LM picture of a similar animal prepared with classic histological methods (Fig. 44). One can also see the conspicuous differences in thickness of the epithelium and in its structure between two species (compare cat in Fig. 45 and this picture from turtle). Close to the basal lamina of this olfactory mucosa there is not the usual layer of so-called basal cells but an assembly of cell processes belonging to the supporting cells which are full of secretory granules (sg). These variations among species which can be easily detected at LM level do not correspond to any surface difference where the SEM only provides a view of a carpet of cilia randomly arranged and not properly specific of any species. (400X)

FIG. 47. This LM preparation from frogs olfactory mucosa was fixed in glutaraldehyde and Araldite embedded. Cutting at 1 μm was oriented parallel to the mucosa surface. It shows the patterned arrangement of the dendrites (some of them indicated by small arrows), seen as clear profiles regularly surrounding the darker supporting cells. This typical arrangement of the receptors was also observed with the SEM in dog's olfactory mucosa (see Figs. 30, 31) where the cilia properly shortened were arranged around the cellular areas of supporting cells covered with short microvilli. The two preparations with LM and SEM are one example of complementary data which can be collected in the study of the olfactory mucosa. (1000X)

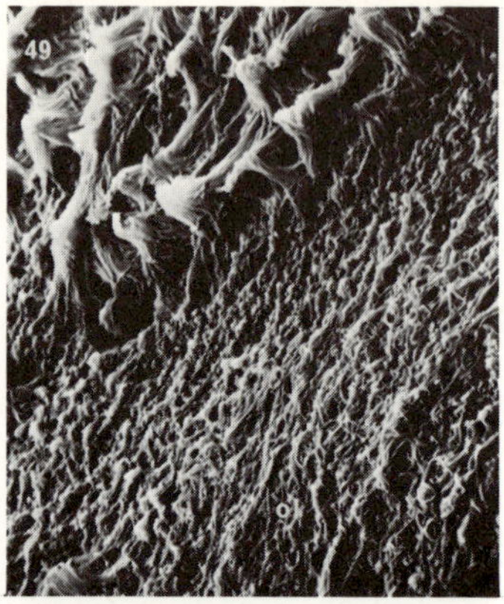

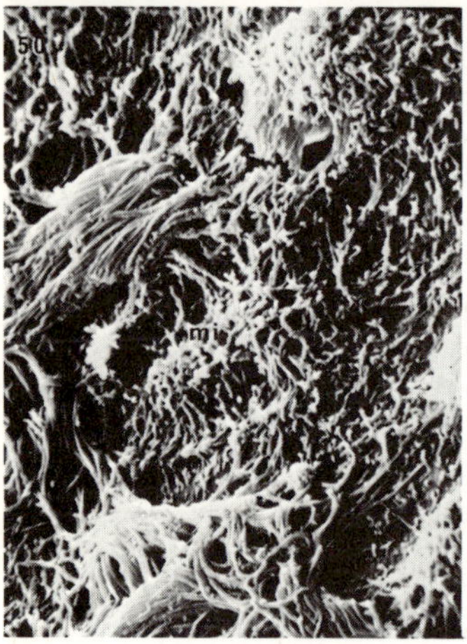

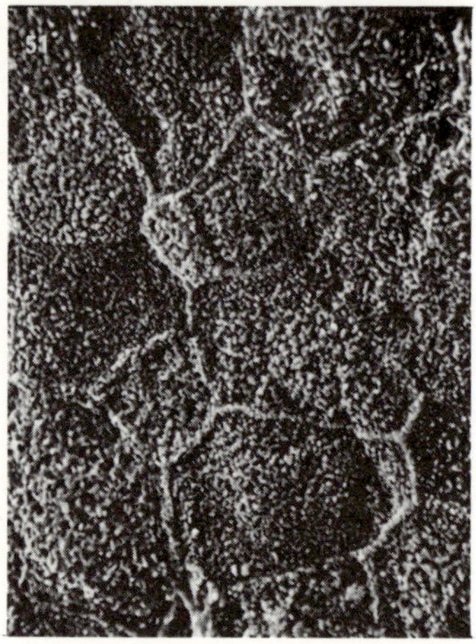

FIG. 48. SEmicrograph of the olfactory mucosa region from the same animal as in Fig. 40 to show the often frustrating results of areas where the mucus has not been completely removed. Cilia can be seen but the results cannot be compared with the successful ones of Figs 28 or 40–43. These negative results are often the most frequent, while the good ones are the most frequently published and mistakenly taken for granted. (1000X).

FIG. 49. SEmicrograph of the transition zone between olfactory (ol) and respiratory (r) regions in the nose of a turtle. Same preparation as in Fig. 36. Notice the obvious contrast between the olfactory cilia apparently randomly arranged and the wave like pattern of the respiratory cilia. (2000X)

FIG. 50. In the respiratory area of the same preparation of the previous figure there are often patches of cells not provided with cilia. They are usually covered with microvilli (mi) that may clump as indicated in this figure. (4700X) weight. Injections were given once or twice a day for one to ten days. The dose and the number of injections depend on the specific experiment. Intravenous injection can also be performed. It is assumed that thymidine is incorporated into the nuclear fraction in the first 30–60 min. after injection. Labelled cells are then the ones which have undergone DNA synthesis during that period of time only. We have found no side effects with the doses used, even when administered for 30 to 50 days.

The animal may be sacrificed at any time after the last injection. We used either perfusion under anaesthesia, or decapitation, rapid dissection and immersion of the desired part in fixative. Both give satisfactory results. However, perfusion, particularly in mammals, is recommended, as it provides more time between the death of the animal and the time the blocks are carved out and properly stored in the fixative. In frogs, the olfactory mucosa was removed from the nose in large pieces and carved in (0.5 mm/1 mm) pieces directly in the fixative. Glutaraldehyde and/or paraformaldehyde in cacodylate or phosphate buffer were used for fixation, followed by postfixation in OsO_4 in the same buffer (Graziadei and Metcalf, 1971a, for fixation details).

For both LM and TEM observations the pieces were dehydrated and embedded in Araldite, and from the blocks alternate series of sections were cut at a thickness of 1 μm for LM and 600–800 A for TEM. This procedure of alternating sections for both LM and TEM has the advantage of allowing observation of the same element with both microscopes, providing a way to confirm, when needed, the nature of every cell that has been labelled. Before starting the autoradiographic procedures some of the sections from each block were used for morphological observations to insure that proper fixation and orientation of the blocks has been obtained.

For autoradiography at the LM level the basic procedure described by Rogers (1967) was employed. One micron sections were transferred with a nylon brush to a drop of distilled water on a subbed slide, and flattened by warming on a hot plate at 60° to 80°C. In the

---

FIG. 51. SEmicrograph of a detail in the same preparation from Figure 49. In some region of the respiratory part of the nose the cells often show a regular array of short microvilli such as represented in this picture. The cell boundaries are noticeable for the microville seem to form thin edges along them. The significance of these regions is unclear. They may represent zones where the primitive typical respiratory ciliated epithelium is changed due to traumatic of pathological events. (2800X) darkroom, the slides were dipped in Kodak NTB-2 Nuclear Track emulsion (Eastman Kodak Co.), melted at 42–43°C. They were dried at room temperature for two hours, and were stored in light-proof Bakelite boxes containing Drierite, in a refrigerator at 5° to 10°C. After exposure for two weeks, the slides were developed in Kodak Dektol developer (1 part stock solution diluted with 2 parts distilled water) for 2 minutes at 17°–19°C. The developed slides were rinsed in distilled water and fixed for 8 minutes in Kodak Fixer, washed in running tap water for 15 minutes, rinsed in distilled water and dried at room temperature.

Mallory Azur 11 – methylene blue staining solution for Araldite sections (Richardson *et al.*, 1960) was modified to obtain optimal staining of the emulsion coated slides. It was found that the presence of borate in the staining solution often caused loss of the emulsion from the slide. Therefore borate was omitted from the solution. The stain was prepared by dissolving 0.1 g of Azur 11 and 0.1 g of methylene blue in 100 ml of distilled water. Prior to use, 0.6 g of Tris hydroxymethylamine is added; the pH of the final solution is 9.5. Slides may be stained by flooding the slide with staining solution and warming on a hot plate (60° to 80°C) for a few seconds, then rinsing with distilled water. Large numbers of slides may be stained simultaneously by immersing them in the staining solution at 45°C for 10 to 30 minutes. After rinsing in distilled water the slides are dried thoroughly in an oven at 68°C. Cover slips were mounted with Protex (Scientific Products Co.). The sections were photographed with a Zeiss Ultraphot LM using Polaroid type 55 P/N 50 ASA film.

For autoradiography at TEM level, the procedures described by Stevens (1966) were employed. Thin sections (600 to 800 A) were picked up on formvar coated grids and stored in a dust proof container until ready for coating with emulsion. The loop and glass rod technique was used to apply Ilford L-4 emulsion to each grid individually. The grids were attached to the top of the glass rod with a small piece of double-stick Scotch tape. A wire loop, slightly larger than the grids was dipped into the diluted emulsion and the excess was removed by touching the edge of the loop to a piece of tissue paper. The emulsion was allowed to dry until interference colour could be detected, then the loop was passed over the grid to cover the tissue sections. After coating, the grids then were attached to the edge of a piece of double-stick Scotch tape lying lengthwise down the middle of a glass microscope slides. The slides with attached grids were sealed in a light-proof slide box containing Drierite and placed in a refrigerator to expose for 8 weeks.

All solutions used for developing and staining the grids were filtered through a millipore filter before use. Prior to development the grids were placed in 100% alcohol to harden the emulsion. The grids were then developed in Microdol X for 6 min. at 24°C temperature, rinsed briefly in distilled water, fixed for 10 min, in 15% sodium thiosulphate, then rinsed again in two changes of distilled water, for 5 minutes each. After development, the grids were stained in alcoholic uranyl acetate and lead citrate for two minutes in each. They were examined and photographed in a Philips EM 300 at 60 KV.

Small numbers of grids were processed in groups of three or four through the developing and fixing solutions, using watch glasses or a porcelain spot plate to hold the solutions. For processing large numbers, the grids were placed in a Ladd grid holder in which holes had been bored to permit solutions to flow freely around the grids. The solutions were placed in petri dishes and the grid holder containing the grids were passed through the solutions.

The illustrations that accompany this section (Figs 52 to 57), shows autoradiographs that can be routinely obtained from the olfactory mucosa of frog or any other vertebrate. We will reiterate here that while fixation of the olfactory mucosa by immersion is often preferred when removal of the tissue can be accomplished quickly, as is the case in some fish or amphibia, perfusion should be kept in mind when the isolation of the olfactory organ requires complex, time-consuming procedures. Perfusion, moreover, seems sometimes to provide some better fixation of the epithelial layer and allows time to dissect carefully different areas of the sensory surface.

VI. Some general conclusions on the use of SEM and autoradiography in the study of the olfactory mucosa

While both LM and TEM techniques provide the observer with only a two dimensional, artificial view of the structure under study (Figs 6 to 7), the SEM provides to our eye a three-dimensional image, that it is thus far easier to comprehend because it is similar to the ones our brain is accustomed to perceiving from the environment. I would like to emphasize that, when the SEM is used for morphological observations, the information it provides, however, is mainly of complementary nature and should be at all times integrated with LM and TEM observations. However, we should also recognize that this instrument provides us with another and new dimension in our studies, and as such, renders more easy our understanding of the structures observed. For this reason the instrument has a considerable

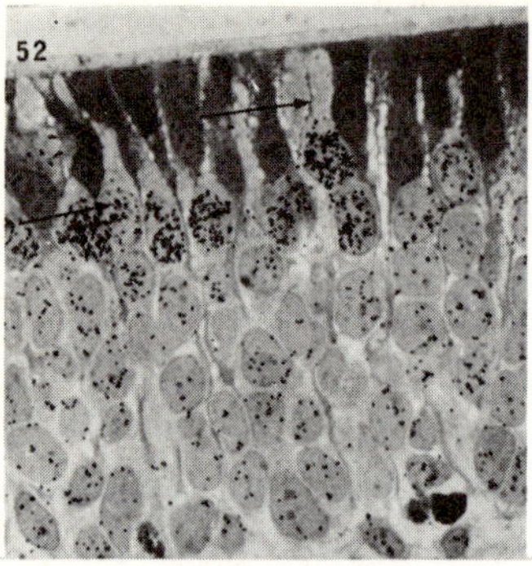

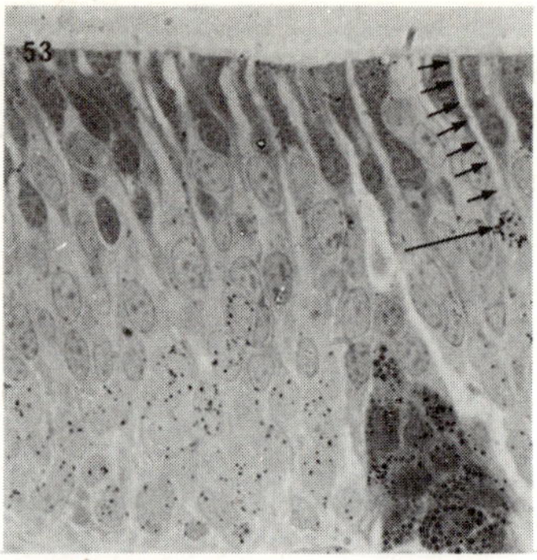

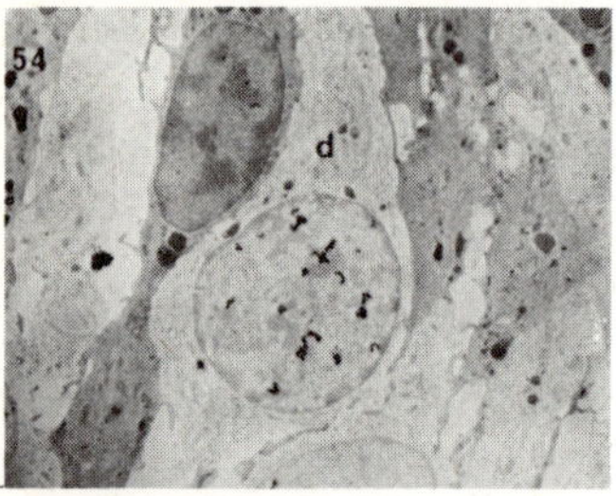

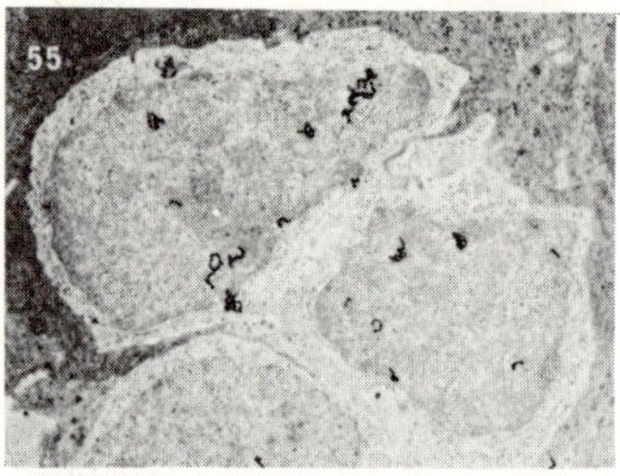

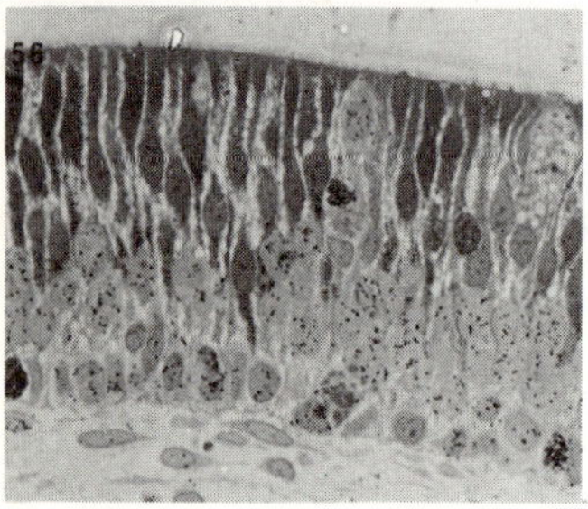

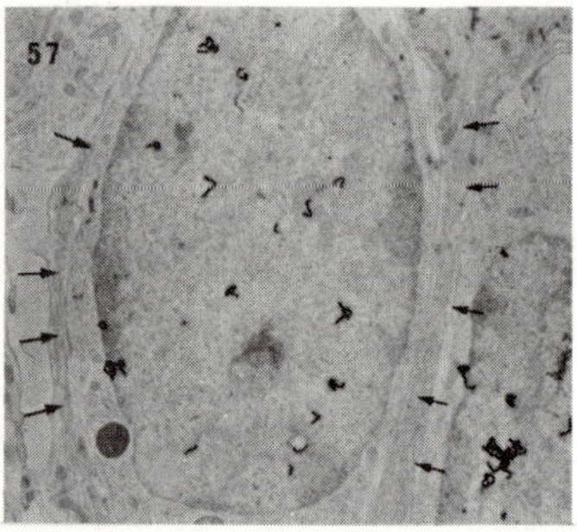

FIG. 52 and FIG. 53. The olfactorys mucosa of the right and left side of the same frog on which the right olfactory nerve was cut 42 days prior to sacrifice and H^3 tymidine injected for ten days from the 6th to the sixteenth day after surgery. In the operated side (Fig. 52) all the existing neurons at the moment of the operation degenerated in a period of approximately 5 to 10 days and new neurons developed from basal cells. We can see here that practically all the basal cells, differentiating neurons and some mature neurons with their dendrites (indicated by arrows) have a labelled nucleus. On the left side (Fig. 53) only one mature neuron recognized by its dendrite reaching the surface is labelled (arrow) and only the basal cells are otherwise labelled. Notice that none of the supporting value in teaching and in providing explanatory views of organs that would have required, previously, a three-dimensional reconstruction from a tiresome process of serial sectioning.

In the case of the olfactory mucosa the SEM has certainly provided a unique tool for the study of large surface areas, allowing the observation of ciliated versus microvillar surfaces as well as revealing the different patterns that cilia assume in the olfactory versus respiratory areas (see Figs 36 to 51). In the study of pathological as well as experimental olfactory mucosa surfaces the SEM seems indeed to open a new and more approachable way to document changes, eliminating the complex techniques of serial reconstruction at the TEM level.

The instrument seems to be less helpful, at this moment, when asked to detect gross differences among species in the organization of normal olfactory mucosa surfaces. The presence of cilia, which are known to be rather long and in the range of several tens of microns, scarcely allows the detection of specific species differences. In our studies on a variety of vertebrate species the intricate meshwork of cilia does not reflect the differences in organization of the olfactory epithelium itself that the LM or TEM obviously shows (see Figs 44,

---

cells nuclei is labelled. Autoradiography on thin (1 μm) sections allows precise recognition of most elements. (The magnification of the two pictures is respectively 800X and 680X)

FIG. 54 and FIG. 55. TEmicrographs of the same preparation used in Fig. 52 to show at TEM level the labelling of a mature neuron with its dendrite (d) as observed in Fig. 54 and of two basal cells as observed in Fig. 55. (Magnification is respectively 7980X and 13 950X)

FIG. 56. Shows another frog treated as the one from previous Fig. 52 but killed 35 days after surgery. In the operated side no neurons have yet reached the surface with their dendritic process. However, all basal cells and differentiating receptors are labelled. Again not one supporting cell is labelled. (520X)

FIG. 57. Detail from the previous preparation of Fig. 56 where a thin section has been cut to illustrate in detail a labelled differentiating receptor with a tagged nucleus (another is partially enclosed in the picture) and surrounded by fine profiles of supporting cells indicated by small arrows. The use of Glutaraldehyde and/or osmium fixed and Araldite embedded material allows the simultaneous observation of the same specimen both in LM and TEM on successive thick (1μm) and thin (500 A) sections. (17 000X)

45, 46). This is why we believe that SEM cannot be considered a self sufficient technique in the way TEM has been sometimes considered.

We have found the SEM technique extremely useful, however, when it is specifically coupled with LM and/or TEM for the solution of specific problems related to the function of the olfactory mucosa. The study of degenerative changes of the olfactory epithelium has been pursued with LM, TEM and autoradiographic techniques, and has been extremely useful to observe with SEM the surface changes of the epithelium following degeneration of the sensory neurons and during regeneration of the same.

The observation that olfactory neurons are normally turning over in the adult vertebrates, originating from repeated division and differentiation of the so-called basal cells, has provided further impetus for the application of autoradiographic techniques to the study of the olfactory organ. So far only observations of a preliminary nature have been accomplished, but it seems clear that these specialized neurons are part of a fast-renewing population of cells. From this derives the concept that their contacts with the nerve cells in the olfactory bulb must also be limited in time and continuously changing. In spite of this, a reliable system of detection and discrimination of odours enables the animal to perceive coherent olfactory information all through its life. It seems important to point out that the dynamics of the olfactory organs should take a great deal of our attention in the future. The opportunity should not be overlooked also in these studies, to understand better the phenomena related to neuron differentiation as they present themselves in adult animals and in a structure anatomically relatively simple such as the olfactory epithelium.

Acknowledgements

This research was supported (in part) by Florida State University and the following grants: USPHS, 5RO1 NS 08943–03 and the Psychobiology Research Centre grants USPHS, NB–7438 and USPHS, MH–11218. The author wishes to thank J. F. Metcalf, J. Kirkland and C. Lewis for manuscript preparation, R. D. Parker for scanning electron micrographs and A. S. Black for photographic assistance.

References

Andres, K. H. (1965). Differenzierung und regeneration von sineszellen in der regio olfactoria. *Naturwissenschaften* 17, 500.

Andres, K. H. (1966). Der feinbau der regio olfactoria von makrosmatikern. *Z. Zellforsch.* **69**, 140–154.

Andres, K. H. (1969). Der olfatorische saum der katze. *Z. Zellforsch.* **96**, 250–274.

Ardenne, M. von (1938a). Das elektronen rastermikroskop praktische ausfuhrung. Zeitschrift fur Technische Physik **19**, 407–416.

Ardenne, M. von (1938b). The scanning electron microscope theoretical fundamental. *Z. Physik.* **109**, 553–572.

Bens, E. M. and Drew, C. M. (1967). Diatomaceous earth: scanning electron microscope of 'chromosorb P'. *Nature, Lond.* **216**, 1046–1048.

Boyde, A., James, D. W.,Tresme, R. L. and Willis, R. A. (1968). Outgrowth from chick embryo spinal cord *in vitro*, studied with the scanning electron microscope. *Z. Zellforsch. mikrosk. Anat.* **90**, 1–18.

Boyde, A. and Barber, V. (1968). Scanning electron microscopical studies of cilia. *Z. Zellforsch. mikrosk. Anat.* **84**, 269–284.

Boyde, A. and Jones, S. J. (1968). Scanning electron microscopy of cenemtum and sharpey fibre bone. *Z. Zellforsch.* **92**, 536–548.

Boyde, A. and Wood, C. (1969). Preparation of animal tissues for surface-scanning electron microscopy. *J. Microscopy* **90**, 221–249.

Boyde, A., Grainger, F. and James, D. W. (1969). Scanning electron microscopic observations of chick embryo fibroblasts *in vitro*, with particular reference to the movement of cells under others. *Z. Zellforsch* **94**, 46–55.

Cameron, I. L. (1968). "Methods in Cell Physiology", Vol. III. Academic Press, New York and London.

Cameron, I. L. (1970). Cell renewal in the organs and tissues of the nongrowing adult mouse. *Texas Rep. Biol. Med.* **28**, 3.

Ceccarelli, B., Clementi, F., DeGiule, C. and Marini, D. (1968a). Aspetto di alcune mucose dell'apparato digerente al microscopio eletrronico a scansione. *Bull. Soc. Ital. Biol. Sper.* **45**, 646–648.

Cohen, A. L., Marlow, D. P. and Garner, G. E. (1968). A rapid critical point method using fluorocarbons as intermediate and transitional fluids. *J. Microscopie* **7**, 331–342.

Christenhess, R., Buchner, T. and Pfeiffer, R. A. (1968). Visualization of human somatic chromosomes by scanning electron microscopy. *Nature, Lond.* **216**, 379–380.

Cleveland, P. H. and Schneider, C. W. (1969). A simple method of preserving ocular tissue for scanning electron microscopy. *Vision Res.* **9**, 1401–1402.

Davoine, F. (1957). Secondary electron emission of metals under mechanical strain. Ph. D. Dissertation L'University de Lyon.

Everhart, T. E., Smith, K. C. A., Wells, O. C. and Oatley, C. W. (1960). Recent developments in scanning electron microscopy. *In* "Proceedings 4th International Conference on Electron Microscopy" (Bargmann, W. *et al.*, eds.), pp. 269–273. Springer-Verlag, Berlin.

Finlay, J. B., Hunter, J. A. A. and Steven, F. S. (1971). Preparation of human skin for high-resolution scanning electron microscopy using phosphate buffered crude bacterial a-amylase. *J. Microscopy* **93**.

Fliedner, T. M., Haas, R. J., Stehle, H. and Adams, A. (1968). Complete labelling of all cell nuclei in new born rats with H^3-thymidine. *Laboratory Investigation* **18**, 249–259.

Frunwald, V. (1935). Die folgen des einseitigen nasenverschlusses auf die reichschleimhaut und den bulbus und tractus olfactorius. *Arch. Ohren-, Nasen-u Khlkopfheik.* **139**, 153–173.

Graziadei, P. P. C. (1969a). The ultrastructure of vertebrate taste buds. *In* "Olfaction and Taste III" (Pfaffmann, C., ed.), pp. 315–330. Rockefeller University Press, New York.

Graziadei, P. P. C. (1969b). The ultrastructure of taste buds in mammals. *In* "Second Symposium on Oral Sensation and Perception" (Bosma, J., ed.), pp. 5–35. C. C. Thomas, Springfield, Ill.

Graziadei, P. P. C. and Tucker, D. (1970). Vomeronasal receptors in turtles. *Z. Zellforsch.* **105**, 498–514.

Graziadei, P. P. C. (1971a). Topological relations between olfactory neurons. *Z. Zellforsch.* **118**, 449–466.

Graziadei, P. P. C. (1971b). The ultrastructure of vertebrates olfactory mucosa. *In* "Ultrastructure of Animal Tissues and Organs" (Friedmann, I., ed.). The North-Holland Publishing Co., Amsterdam.

Graziadei, P. P. C. (1971c). The olfactory mucosa of vertebrates. *In* "Handbook of Sensory Physiology, Vol. IV" (Beidler, L. M., ed.), pp. 29–58. Springer-Verlag, Berlin-Heidelberg-New York.

Graziadei, P. P. C. and Metcalf, J. F. (1971a). Autoradiographic and ultrastructural observations on the frog's olfactory mucosa. *Z. Zellforsch.* **116**, 305–318.

Graziadei, P. P. C. and Metcalf, J. F. (1971b). Autoradiographic study of frog's olfactory mucosa. *Amer. Zool.* **10**, 716.

Hall, C. E. (1966). "Introduction to Electron Microscopy", 2nd Edn. McGraw-Hill, New York.

Hammond, R. A. (1968). Observations on the body surface of some acanthocephalans. *Nature, Lond.* **218**, 872–873.

Hay, W. W. and Sandberg, P. A. (1967). The scanning electron microscope, a major breakthrough for micropaleontology. *Micropalentology* **13**, 407–418.

Hayat, M. A. (1970). "Principles and Techniques of Electron Microscopy, Vol. I". Van Nostrand Reinhold Company, New York.

Hayes, T. L. and Pease, R. F. W. (1968). The scanning electron microscope: principles and applications in biology and medicine. *Adv. biol. Med. Phys.* **12**, 85.

Horridge, G. A. and Tamm, S. L. (1969). Critical point drying for scanning electron microscopic study of ciliary motion. *Science, N.Y.* **163**, 817–818.

Hughes, W. L. (1958). Chromosomal replication and the dynamics of cellular proliferation autoradiographic observations with tritiated thymidine. *In* "Symp. Chemical Basis of Development" (McElroy, W. D. and Glass, B., eds.), pp. 136–152. The Johns Hopkins Press, Baltimore, Md.

Knoll, M. (1935). Aufladepotenitel und sekundaremission elektronenbestrahlter korpe. *Z. Technische Physik* **16**, 767.

Kolmer, W. (1927). Geruchsorgan. *In* Von Mollendorff, "W. Handbuch der mikroskopischen Anatomie des Menschen," Bd. III, S. 192–249. Springer-Verlag, Berlin.

Leblond, C. P., Messier, B. and Kopriwa, B. (1959). Thymidine H^3 as a tool for the investigation of the renewal of cell populations. *Laboratory Investigation* **8**, 296–308.

Lecaute (1946). Quoted by Oatley *et al.*, 1965.

LeGros, Clarke, W.E. (1951). The projection of the olfactory epithelium on the olfactory bulb in the rabbit. *J. Neurol. Neurosurg. Psychiat.* **14**, 1–10.

Lewis, E. R., Everhart, T. E. and Zeevi, Y. Y. (1969a). Studying neural organization in aplysia with the scanning electron microscope. *Science, N.Y.* **165**, 1140–1143.

Lewis, E. R., Zeevi, Y. Y. and Werblin, F. S. (1969b). Scanning electron microscopy of vertebrate visual receptors. *Brain Research* **15**, 559–562.

Messier, B. and Leblond, C. P. (1960). Cell proliferation and migration as revealed by radioautography after injection of thymidine-H^3 into male rats and mice. *Amer. J. Anat.* **106**, 247–265.

Moulton, D. G. (1970). Olfaction in Mammals–two aspects: proliferation of cells in the olfactory epithelium and sensitivity to odours. *In* "Taste and Smell in Vertebrates" (Wolstenholme, G. E. W. and Knight, J., eds.), pp. . J. and A. Churchill, London.

McDonald, L. W. and Hayes, T. L. (1969). Correlation of scanning electron microscope and light microscope images of individual cells in human blood and blood clots. *Experimental and Molecular Pathology* **10**, 186–198.

McMullan, D. (1953). The scanning electron microscope and the electron-optical examination of surfaces. *Electronic Engineering* **25**, 46–50.

Nagahara, Y. (1940). Experimentelle Studien uber die histologischen Veranderungen des Geruchsorgans nach der Olfactoriusdurchschneidung. Beitrage zur Kenntnis des feineren Baus des Geruchsorgans. *Japan J. Med. Sci. V. Pathol.* **5**, 165–199.

Nemanic, M. K. and Pitelka, D. R. (1971). A scanning electron microscope study of the lactating mammary gland. *J. Cell Biol.* **48**, 410–415.

Neurath, P. W., Ampola, M. G. and Vetter, H. G. (1967). Scanning electron microscopy of chromosomes. *The Lancet*, pp. 1366–1367.

Neurath, P. W. (1968). Scanning electron microscopy of chromosomes. *The Lancet*, p. 114.

Nixon, W. C. (1964). Scanning electron microscopy. *J. Roy. Microscopical Soc.* **83**, 213–216.

Oatley, C. W. (1958). The scanning electron microscope. *New Scientist* **5**, 153–155.

Oatley, C. W., Nixon, W. C. and Pease, R. F. W. (1965). "Advances in Electronics and Electron Physics" (Morton, L., ed.), pp. 181–247. Academic Press, New York and London.

Pease, D. C. (1964). "Histological Techniques for Electron Microscopy". Academic Press, New York and London.

Pease, R. F. W. (1963). High resolution scanning microscopy. Ph.D. Dissertation, Cambridge, Massachusetts.

Pease, R. F. W. and Nixon, W. C. (1965). High resolution scanning electron microscopy. *J. Sci. Inst.* **42**, 81–85.

Richardson, K. C., Jarrett, L. and Finke, E. H. (1960). Embedding in epoxy resins for ultrathin sectioning in electron microscopy. *Stain Technol.* **35**, 313–323.

Rogers, A. W. (1967). "Techniques of Autoradiography". Elsevier Publishing Co., Amsterdam.

Schultz, E. W. (1941). Regeneration of the olfactory cells. *Proc. Soc. exp. Biol.* **46**, 41–43.

Sidman, R. L. (1970). Autoradiographic methods and principles for study of the nervous system with thymidine–H^3. *In* "Contemporary Research Methods in Neuroanatomy" (Nauta, J. H. and Ebbesson, S. O. E., eds.), pp. 252–274.

Springer-Verlag, New York.

Siegel, B. M. (1964). The Physics of the Electron Microscope. *In* "Modern Developments in Electron Microscopy" (Siegel, B. M., ed.), pp. 1–97. Academic Press, New York and London.

Sijostrand, F. S. (1967). Electron Microscopy of Cells and Tissues. Academic Press, New York and London.

Smith, C. G. (1951). Regeneration of sensory olfactory epithelium and nerves in adult frogs. *Anat. Rec.* **109**, 661–671.

Smith, K. C. A. and Oatley, C. W. (1955). The scanning electron microscope and its fields of application. *Brit. J. Appl. Phys.* **6**, 391–399.

Stevens, A. R. (1966). High resolution autoradiography. *In* "Methods in Cell Physiology", Vol. II (Prescott, D. M., ed.), pp. 255–310. Academic Press, New York and London.

Thornhill, R. A. (1967). The ultrastructure of the olfactory epithelium of the lamprey fluvialilis. *J. Cell Sci.* **2**, 591–602.

Thornton, P. R. (1968). "Scanning Electron Microscopy". Chapman and Hall Ltd., London.

Wells, O. C. (1969). Bibliography on the scanning electron microscope. *In* Proc. 10th IEEE Ann. Symp. on Electron Ion and Laser Beam Technology (Malton, L., ed.). San Francisco Press, San Francisco, Calif.

Woodward, D. H., Gryfe, A. and Gardner, D. L. (1969). Comparative study of scanning electron microscopy of synovial surfaces of four mammalian species. *Experientia* **25**, 1301–1303.

Yasuzumi, F., Yamaguchi, S. and Johnson, G. H. (1971). Electron Microscopic studies on the Palpi of Cybister fimbriolatus fimbriolatus (Say) I. Examination of the Cuticular surfaces by means of the scanning and transmission electron microscopes. *Z. Zellforsch.* **117**, 476–484.

Zworykin, V. K., Hillier, J. and Snyder, R. L. (1942). A scanning electron microscope. *Bulletin of the American Society for the Testing of Materials* **117**, 15.

6 Correlative Anatomical Methods for Studying the Intrinsic Organization of the Olfactory Bulb

R. N. L. BROOKE

A. J. PINCHING

Department of Anatomy
The University of Oxford,
South Parks Road
Oxford

I. Introduction

The considerable advances in electron microscopic technique for the study of the central nervous system in recent years have enabled detailed ultrastructural studies to be carried out on many parts of the brain; sites most extensively studied have tended to be those for which there is a large background of light microscopic information—particularly from Nissl staining. Golgi impregnation and silver degeneration techniques. The olfactory bulb is among the more privileged sites in the brain from this point of view, as much as anything on account of its clear and characteristic lamination. In the latter half of the nineteenth century, attention was paid to the structure of the bulb by almost every distinguished neuroanatomist of the time, and the advent of the Golgi impregnation methods allowed an extensive analysis of the neurons of the bulb by, among others, the groups of Golgi, Cajal, van Gehuchten and Kölliker; the findings of these observers are most adequately summarized by Cajal (1911), where further references may be found. Recent work extending and clarifying the light microscopy of the olfactory bulb may be found in the section by Lohman and Heimer. Several electron microscopic studies have capitalized on this background of knowledge (for instance, Hirata, 1964; Andres, 1965; Rall *et al.*, 1966; Price and Powell, 1970 a–d; Pinching and Powell, 1971a–d); the emphasis in the latter two groups of studies has been on correlating the evidence obtained from past light microscopy with electron microscopy, and extending light microscopic observations to clarify questions arising from electron microscopy. The ultimate aim has been an understanding of the organizational principles of the neuronal interconnections of the bulb. Since the light microscopic techniques being applied to the bulb are being dealt with elsewhere, only brief comments will be made on the technical and conceptual aspects of these techniques, and these will concentrate on special considerations arising from the study of the intrinsic organization of the bulb. Moreover, since books have recently been published in which the general techniques for studying the central nervous system have been discussed at length (e.g. Nauta and Ebbesson, 1970–technical; Peters, Palay and Webster, 1970–analysis), we will concentrate particularly on features deriving from the olfactory bulb studies on the rat, while supplying a skeleton of fundamental techniques used in this laboratory.

II. Normal material

A. Light microscopy; Golgi impregnation

The Golgi method and its various modifications are of the first importance in the study of individual neurons in the central nervous system; the peculiar features of Golgi-impregnation are that only about 1% of all neurons are stained, apparently in proportion to their relative frequencies, and are stained throughout their dendritic and axonal ramifications. This permits detailed study of the character and distribution of these processes and can serve as a basis for more detailed ultrastructural analysis of inter-neuronal connections; the Golgi method cannot however be used to determine these interneuronal connections at light microscopic level, since electron microscopy has clearly demonstrated that close proximity does not necessarily indicate synaptic contact. Axons have proved difficult to study by the Golgi methods, particularly if they are long, since they may not be stained throughout their length with some methods, and they are likely to pass out of the plane of section, being difficult to find in serial sections. Methods involving repeated impregnations appear to avoid the first problem, but naturally the technique is best suited to the determination of dendritic, short axonal or collateral processes (Cajal, 1911; Valverde, 1965; Price and Powell, 1970 a–d; Pinching and Powell, 1971b). There can be no doubt that the characteristic features of any neuron type can only be ascertained from the study of many examples in several brains, rather than by the selection of a single well-stained example (even though the latter may demonstrate many of the typical features). The extent to which the Golgi-impregnated cells are representative of the population under study is inevitably a problem, but studies with long series of sections of electron microscopic material, both normal and Golgi-stained (see below) strongly suggests that they are indeed representative.

To every laboratory its Golgi modification. Nevertheless, our satisfaction with the material obtained by the methods outlined in the appendix is sufficient to recommend its use, for the following reasons. It is basically simple and is easily modified for Golgi-E.M. studies on the same material (see next section); it gives repeatable, reliable impregnations, admirably suited to high magnification light microscopic studies. This latter is of considerable significance for correlation of electron microscopic data with Golgi impregnation light microscopy since the former demands much detail from the latter in terms of exact dendritic outlines and situations, etc., than is often provided by the descriptions in the literature (e.g. the deep dendrites of the granule cells, and the dendrites of the superficial short-axon cells). A further advantage is that material fixed for electron microscopy can be stained by these methods, they derive from the Golgi-Kopsch glutaraldehyde modification of Colonnier (1964).

B. Electron microscopy

The electron microscope is the only means of obtaining definitive morphological information regarding synaptic interconnections between nerve cell processes; this is because its high resolving power is sufficient to distinguish between simple proximity of neuronal profiles and synaptic specializations connecting them. The latter show clearly with the heavy metal stains used for electron microscopy as being quite distinct from simple membrane apposition. For this reason, ultrastructural studies now tend to dominate the field of anatomical studies on neuronal organization, and it is of the greatest importance to be able to correlate results obtained by other methods with those emerging from electron microscopic study.

Analysis of material

The study of electron microscopic material of the central nervous system is based on a series of assumptions, some of which derive from electron microscopy in general and some which are particular to the study of nerve cells and their connections; the validity of most of these assumptions can only be tested in physiological, biophysical or biochemical terms and such tests are at present regrettably not feasible. Perhaps first in importance is the assumption that the minute areas studied with the electron microscope are representative of the region; this is a direct corollary of the high magnification the technique offers. The observations that all the areas of the olfactory bulb that have been examined show essentially the same features lends some support to this assumption. Secondly, it is assumed that the morphological differences revealed by heavy metal staining and the other preparative procedures represent true functional differences. The cry of artefact is one that is commonly flung at the electron microscopist. Many of the criteria used in identification, and which are regularly fulfilled in many brains, rely entirely on these staining artefacts of electron microscopy; it is assumed that these bear the same constant relationship to the original living condition, and that differences in ultrastructure reflect a genuine difference in the unfixed state. Similarly, the tacit assumption is made that the things that are stained are the only things that are there; this is an indefensible view, but until electron histochemistry has developed further, it is not possible to deduce what the heavy metals presently in use do not show. Thus, although technically and physically at some remove from the original live state, the material used is, at least conceptually, close to actuality.

Cell soma identification

Many of these points are illustrated by the problem of identifying cell somata; this is based on the assumption that if a cell is functionally different, then its morphology will be different, especially its arrangement of nucleic acid material (Fig. 1). In the first place typing is based on the typing derived from Nissl-stained material and Golgi-impregnated material, so that each cell can be characterized according to two entirely different sets of criteria: the shape and size and contents of the cell soma on the one hand and the dendritic and axonal fields on the other. Such a correlation has been of paramount importance in studying the cells at glomerular level (Pinching and Powell, 1971b) since three neuron types lie in close proximity in the periglomerular region. The discovery of the third neuron type at this level, the superficial short-axon cell, started as the observation of a third type with the electron microscope, which was then characterized at light microscopic level, principally with the Golgi method; the evidence from this was then confirmed by electron microscopy, thus giving strong evidence of the identity of the types seen at the two ranges of magnification. The possibility of directly combining light and electron microscopic material of the same cell (Golgi-E.M.; see below) is the culmination of such a correlative approach. Somatic identification turned out to be especially valuable in the glomerular layer, since it was found that the external tufted and periglomerular cell somata contained characteristic vesicle types, sometimes associated with synaptic thickenings orientated away from the somata, suggesting a possible means of identifying their more distant processes.

Vesicles and membrane thickenings

In recent years, attention has been paid to the shapes and sizes of vesicles in the axon terminals of aldehyde-fixed central nervous systems. Observations in various regions showed the existence of certain classes of terminal showing "flattened" vesicles (Figs 4–6, 11–13), that is to say, populations of pleomorphic character, indicative of a cigar-shaped or disc-like nature. Other terminals always showed spherical type vesicles (Figs 3, 9–13), even when adjacent to terminals having markedly flattened ones. Uchizono (1965) suggested that the terminals containing flattened vesicles might be functionally

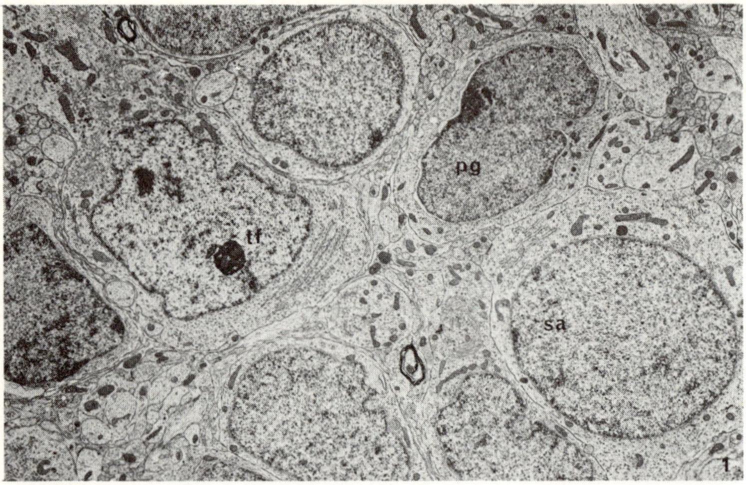

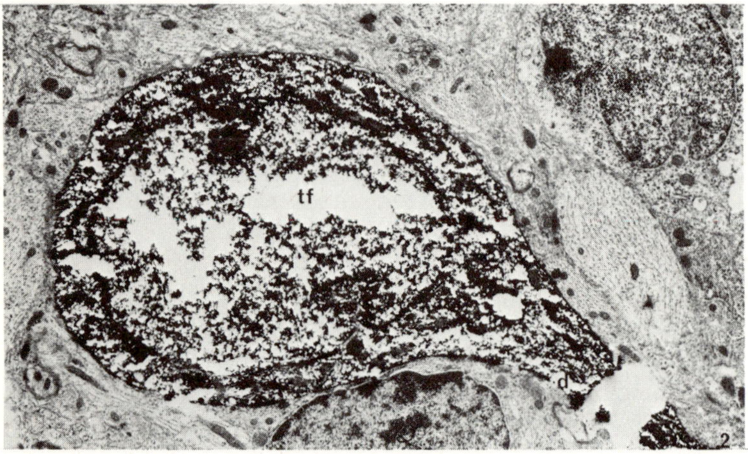

FIG. 1. Low magnification view of cell types in the periglomerular region, showing how they may be distinguished to some extent by the different characteristics of their nuclei and cytoplasm; note the different densities and arrangements of nuclear material and varying amounts of cytoplasm. x 8 100.

FIG. 2. Low magnification electron microscopic view of a tufted cell stained with the Golgi method; observe the very electron-dense, flocculated material within the cell and the outline of the nucleus. The break in the dendrite indicates a break in the original light microscopic section—one of the hazards of the combined method. x 10 800.

inhibitory and those with spherical vesicles excitatory. This idea has stimulated much work on the possibilities for the morphological identification of functional synaptic types (see Gray, 1969 for a review); although unproven, the concept has much to support it, even if the final synthesis is likely to be more complex. Colonnier (1968) demonstrated a correlation between flattened type vesicles and symmetrical type thickenings (Figs 4–6, 11–13), in which the post-synaptic density scarcely extends beyond the membrane, and between spherical vesicles and a marked post-synaptic web (Figs 3, 9–13). This correlation has been entirely borne out by the work on the olfactory bulb; it may be that the type of the post-synaptic thickening is the clearest indication of the functional nature of the synapse. The shape of the vesicles in electron microscopic material presumably represents the response of the equilibrium system of vesicle and cytoplasm to the fixative or preparative fluid, as there is little reason to suppose that they are flattened *in vivo*; that the contents of the vesicles, probably transmitter substances, play a dominant role in this response is not unlikely.

The observations of Price and Powell (1970b) on vesicle populations within different processes in the olfactory bulb, provided statistical evidence for two types of flattened vesicles–large (Figs 4, 5, 11, 13) and small (Figs 6, 12, 13); both are associated with symmetrical thickenings. The types may be distinguished visually and the distinction has been important in analysing the profile types of the bulb. The factors involved in producing this particularly useful artefact of flattening are not clear. The tonicity of the buffer solution seems to be of some importance (Valdivia, 1970), but other factors may be involved, causing considerable variation in the degree of flattening from brain to brain (e.g. Fig. 10), and a lesser degree even within a brain. Bodian (1970) has shown that the use of cacodylate buffer may cause one pópulation of vesicles to become flattened and, although this type has not been encountered in the olfactory bulb, it has been observed in the caudate nucleus (Kemp and Powell, 1971)– this type is also associated with symmetrical type thickenings.

Axons, dendrites and appendages: general criteria

Work on the olfactory bulb has provided unequivocal evidence of the existence of synapsing dendrites (Rall *et al.*, 1966; Price and Powell, 1970a, b, d; Pinching and Powell, 1971c) or Class B dendrites (Pinching and Powell, 1971d) and appendages, as well as evidence for the myelination of dendrites in certain sites and species (Pinching,

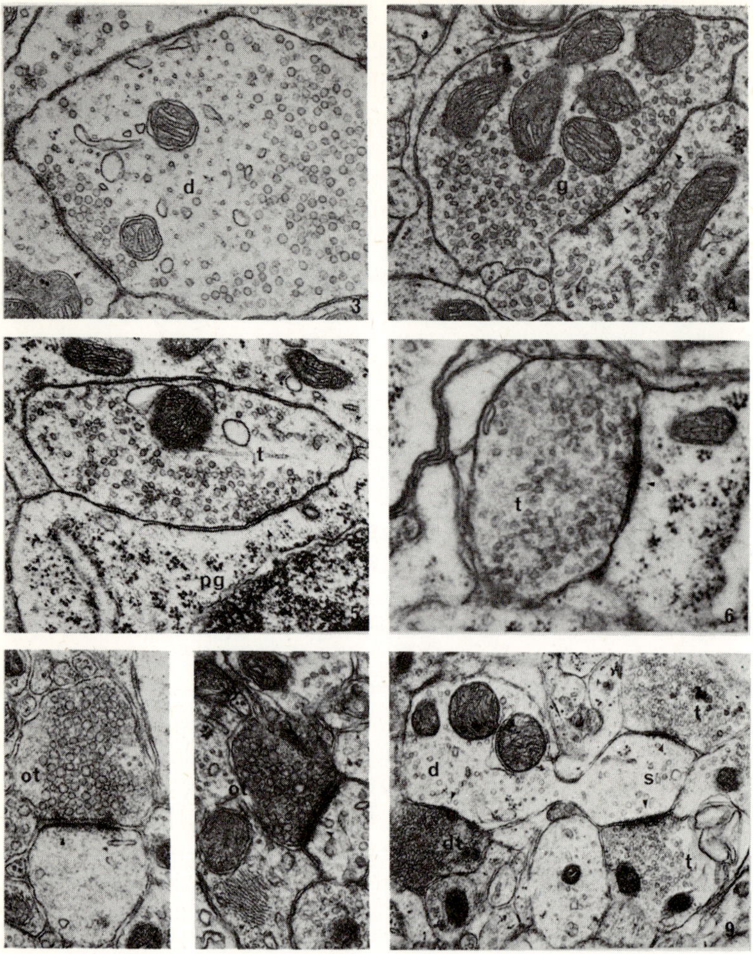

FIG. 3. Mitral or tufted cell dendrite making a synapse with a periglomerular cell profile; note the spherical vesicles and asymmetrical type synaptic thickening, as well as the characteristic neurotubules, cut in transverse section. x 53 400.

FIG. 4. Periglomerular cell gemmule containing large flattened vesicles and synapsing by means of a symmetrical type synaptic thickening. x 53 400.

FIG. 5. Periglomerular cell axon terminal synapsing with large flattened vesicles and symmetrical thickening onto a periglomerular cell soma; observe the alveolate vesicle. x 53 400.

FIG. 6. Axon terminal of a short-axon cell synapsing onto a periglomerular cell soma with small flattened vesicles and symmetrical membrane thickening; note the higher magnification from the preceding figures. x 60 000.

1971). Several factors have contributed to this, but perhaps the most important has been the fact that the distribution of axonal and dendritic processes within the bulb were well established from light microscopy. The external plexiform layer shows many unequivocal examples of reciprocal dendro-dendritic contracts between mitral cell dendrites and granule cell peripheral processes. The study on the glomerular layer, although confirming the presence of dendro-dendritic relationships in this layer, was less straightforward on account of the smallness of many of the profiles involved and their less obviously dendritic nature (Figs 3, 4, 10, 11); this fact made it necessary to examine carefully the criteria used for identification of isolated neuronal profiles in neuropil as axonal or dendritic, particularly with the help of serial sections.

What emerged from this study was that, in the primary analysis of a region, there is no substitute for the examination of *serial sections* to determine the identity of the neuropil profiles; continuity with a preterminal axon, a dendritic shaft, a cell soma, an axon initial segment or with a profile that contains some recognizable feature, is the most straightforward and certain way of identifying it. Naturally, information on the distribution of processes from light microscopy is the essential adjunct to such a study. Eventually, from these serial studies criteria will emerge for distinguishing between profiles, very often allowing their identification in single sections, although confirmation by serial examination is still valuable as a continuing check.

The most problematical differentiation is between a dendritic profile and an axon terminal (e.g. Figs 4 and 5); the problem is perhaps most crucial in the bulb in view of the large number of dendro-dendritic synaptic contacts made in this site. Nevertheless dendro-dendritic contacts have also been reported in the superior colliculus and thalamus; despite this, no universally applicable criteria for

---

FIG. 7. Early degeneration of a centrifugal fibre terminal 6 days after a lesion in the lateral olfactory tract. Note gliosis and glycogen granules. x 53 400.

FIG. 8. Degeneration of tufted cell recurrent collateral in the periglomerular region 5 days after a lesion in the superficial layers of the bulb about 5 glomeruli distant. Note the darkening and gliosis. x 53 400.

FIG. 9. A degenerating centrifugal fibre terminal, synapsing onto the shaft of a periglomerular cell dendrite; this dendrite gives off a spine, which receives asymmetrical synapses from two normal terminals which must be tufted cell collaterals in view of the complete lesion of the lateral olfactory tract. Experimental methods are the only ways of distinguishing between these otherwise similar types of axon terminal. x 39 900.

profile identification has yet emerged, although several are implicit in these studies. It is obviously dangerous to set out any lists of criteria, but it would seem useful to include here some possible indications of profile identity that have proved useful in the bulb, and which have been found to be accurate, if used with discretion. No single feature can be used to identify a profile, and despite many attempts to find one, no single set of features seems to be absolutely diagnostic, although they may give a strong indication.

Dendritic profiles commonly show ribosomal rosettes, many cisternae of agranular endoplasmic reticulum, often associated with mitochondria, and their shafts have a variable number and density of neurotubules (Figs 3, 13), that are fairly evenly spread through the profile; few, if any, neurofilaments are found. Ribosomal rosettes are frequent (Fig 11), and their presence is a strong indication of dendritic nature, although single ribosomes may be found in axons. Vesicles, when present, are unevenly distributed in dendrites, either closely associated with the presynaptic membranes of a synaptic specialization, or aggregated at branch points of the dendritic shaft (Figs 3, 11, 13); they are also found in small groups which are not associated with branching or synaptic contacts oriented away from the dendrites (e.g. in spines—Fig. 12). Most dendritic profiles, including appendages have a relatively inhomogeneous distribution of vesicles. The absence of neurotubules in a profile, and a rather flocculent appearance to the cytoplasm may be used as a criterion for an appendage (Figs 9, 12, 18), though the pedicle may contain neurotubules. The appendage may be a spine or a gemmule, the latter being defined as an appendage that gives off a synapse as well as receiving one; it may be difficult to distinguish between a gemmule and an axon terminal, and serial sections may be essential to make the distinction. Spine apparatus, although found in many other parts of the central nervous system, is rarely observed as a well-defined entity in the gemmules or spines of the bulb, but mitochondria are commonly found in appendages; vesicles may be found in both spines and gemmules. Axon terminals or en passant boutons, rarely show evidence of ribosomes, and when present, these are never aggregated into rosettes; few cisternae or mitochondria are seen and neurotubules are either absent (terminal) or restricted to one side of the profile (en passant); neurofilaments are commonly associated with the tubules. Vesicles are evenly spread through the profile in most cases tending to cover the whole, except in those regions containing neurotubules (Figs 5, 6, 9, 10, 12, 13). Despite the slightly unsatisfactory nature of these criteria, the majority of profiles in the bulb neuropil can be classified as axonal or dendritic—that this is so, is an extremely important background to cell type identification (see next section), and an essential one for understanding the function of the region. It has become increasingly difficult to substantiate a position of neuroanatomical identification that is based on the study of a single site and/or a single species; it is hoped however that these observations on the rat olfactory bulb, and which hold true for rabbit, cat and monkey olfactory bulbs, may lead to some clearer means of distinguishing between neuronal profiles in other sites. More general criteria are discussed by Peters *et al.* (1970).

Cell type identification of neuropil profiles

Once the profiles in the neuropil have been identified as axonal or dendritic, it becomes of interest, for organizational purposes, to identify them as belonging to a particular intrinsic cell type (whose somata have been identified—see above) or to a particular projection pathway. The essential background to this is again a knowledge from light microscopy, of the contributory processes to the site; this not only defines the limits of the problem, but will also indicate any special characteristics in morphology and distribution that may assist electron microscopic characterization. Dendritic processes are most easily characterized by the Golgi method in association with normal electron microscopic material, and axonal projections, both intrinsic and extrinsic, by the silver degeneration methods (see below and elsewhere in this volume) together with experimental electron microscopic material (see below).

Dendritic profiles may assume many different appearances in ultra thin sections, due to the varying planes of section, but with serial sections and favourable single sections, may be classed on the basis of outline as straight, varicose or irregular. Varicose dendrites (e.g. the deep dendrites of granule cells) typically show bulbous portions in which neurotubules are widely dispersed, and very thin and electron-dense portions into which the tubules funnel. Irregular dendrites (e.g. dendrites of some periglomerular cells) have a very variable outline and often follow a tortuous course; their neurotubules are similarly irregular in course. The presence or absence of appendages, determined from light microscopic typing, may generally be confirmed for any dendritic type using serial electron microscopic sections. The distribution and orientation of dendrites is another useful adjunct to identification of this sort, and may help to distinguish morphologically similar types; the use of Golgi-E.M. combined material is the most direct and sure way of confirming the correlations made between light

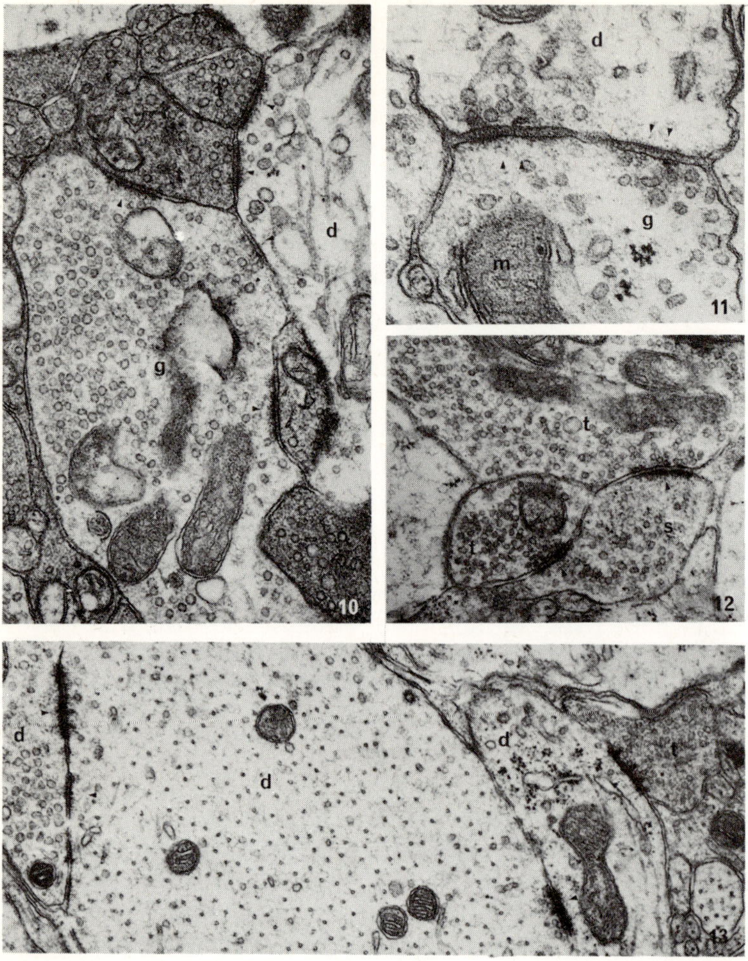

FIG. 10. Dark normal olfactory nerve terminals synapsing with spherical vesicles and asymmetrical thickenings onto two dendritic profiles, one of which is synapsing onto the other. Vesicles were not flattened in this brain, and this synapsing dendrite is probably a periglomerular cell dendrite synapsing with asymmetrical thickening onto a tufted or mitral cell dendrite. × 53 400.

FIG. 11. Reciprocal synapse between a tufted/mitral cell dendrite and a periglomerular cell gemmule; note the different characteristics of the two synapses. × 106 800.

FIG. 12. Axon terminal synapsing with spherical vesicles and asymmetrical thickening onto the head of a vesicle containing spine, that also receives a symmetrical synaptic contact from a terminal containing small flattened vesicles. × 66 900.

and electron microscopy. In the olfactory bulb, a further aid to identification is the presence of vesicles in the dendrites of several cell types and their related synaptic contacts. It has been suggested that those dendrites which make synapse as well as receiving them should be termed Class B dendrites and those that only receive them, Class A (Pinching and Powell, 1971d). Class B dendrites can be further subdivided on the basis of vesicles and membrane thickening type (see above).

Finally, from continuing study, it may emerge that dendrites either of similar or distinct morphology receive characteristic types of synapse, or receive them in a characteristic way, and this may be a help in identifying equivocal profiles. In the bulb, an additional method of identification is possible: the presence of somato-somatic and somato-dendritic synapses, or synapses orientated off the initial parts of dendrites, allows one to associate particular synaptic specialization types with given cell soma types. This is based on the assumption, termed the morphological corollary of Dale's principle, that a cell making synapses of one sort in one of its processes, will exhibit the same sort at any other synapses it may make; the same reasoning is used to associate Class B dendrites and axon terminals with the same cell type when they show the same synaptic specializations, and their distribution is consistent with such a view. As far as serial study of single cells allows one to confirm this, it has been seen to be a reasonable assumption.

III. Experimental material

A. Light microscopy; silver degeneration methods

The basis for and the application of these methods are discussed in the chapter by Lohman and Heimer, but mention can be made here of their value in correlative studies on the intrinsic organization and extrinsic connections of the olfactory bulb and the particular methods that have been used in this laboratory. The projection pathways to the bulb from central regions have been determined with these

---

FIG. 13. Serial arrangement: Large mitral cell dendrite in the periglomerular region–external plexiform layer boundary, making spherical vesicle/asymmetrical thickenings synapses with a granule cell gemmule (which makes a reciprocal synapse of opposite polarity) and with a periglomerular cell dendrite, that also receives a symmetrical synapse from an axon terminal containing small flattened vesicles. Compare the three vesicle types in this one field. x 42 600.

techniques (Cragg, 1962; Powell *et al.*, 1965; Heimer, 1968; Price and Powell, 1970e) and have formed the basis for studies of experimental material with the electron microscope (Price and Powell, 1970c; Pinching and Powell, unpublished observations). The main consideration in dealing with the pathways of central origin is in placing the lesions in such a way to avoid involving other pathways by damaging fibres of passage or by vascular necrosis. All lesions must be confirmed and their extent exactly mapped out; where it is not possible to achieve a selective interruption of only one pathway to the bulb, the results should be interpreted with caution. Recently, Land *et al.* (1970) have made an experimental study of the projection of the olfactory mucosa onto the bulb; their method shows cell somata well, which is of great value in dealing with the glomerular layer, and its findings correspond well with our own electron microscopic studies after more extensive lesions (Pinching, 1969).

In general, the experience in this laboratory has been that the most satisfactory results may be obtained by taking several series through a given brain and obtaining them with Nauta-Gygax, Fink-Heimer and Wiitanen methods or some combination of these, and analysing the results obtained by *all* methods. This may help to avoid the inevitable confusion that results from the use of widely differing techniques on the same basic material. Survival time is of the greatest importance in determining the results obtainable from the material and differs from pathway to pathway; it is imperative that the optimal survival time is found for each lesion. This is particularly important for finding terminal degeneration, which can be difficult to be certain of in the bulb. It may also be noted that the optimal survival times for light and electron microscopy may differ. From the technical point of view, it would seem worth noting that the ammoniacal silver is the critical solution with any of the methods and should be fresh and precisely made up.

For correlation with similar studies using electron microscopic material small lesions have been made in the superficial layers of the dorsal bulbar surface; this has been to determine the spread of degeneration of the intrinsic axons at glomerular level. The bulbs are cut either sagittally or coronally. A 1 in 5 series is taken and stained by the Wiitanen (1969) rapid method to locate the lesion and determine the broad extent of spread. Although this method was developed for the monkey, it works well for the rat and has the great advantage of staining cell somata rather better than the Fink-Heimer method, although terminal degeneration is often not as clear as in the latter. The region of degeneration is then stained in further series with the

Fink-Heimer and/or Nauta-Gygax methods, to give a clear indication of the inter-relationships of fibre and terminal degeneration and its site in relation to the bulbar layers; this has been a most useful background to the electron microscopic studies on similar material.

B. Electron microscopy; orthograde degeneration

The distribution of axonal degeneration after lesions of the afferent pathways to the olfactory bulb is determined by the silver degeneration methods at light microscopic level; further details on the mode and site of termination of these fibre pathways may be achieved by using experimental techniques for ultrastructural study. The lesions used are exactly equivalent to those made for the former (see below). The electron microscope can provide evidence to distinguish between regions of fibre passage and sites of termination; the experimental evidence can also allow one to identify the axonal profiles distinguished in normal electron microscopic studies. Many descriptions are to be found in the literature and each fibre pathway has its own characteristics (see, for example Price and Powell, 1970c; Pinching, 1969; Figs 7–9); we will therefore limit ourselves to a few basic points with regard to the principles of studying experimental electron microscopic material.

(1) The site and extent of the lesion must be confirmed by light microscopy.

(2) The distribution of fibre degeneration should be known in detail from light microscopy and used for selecting appropriate areas for ultrastructural study.

(3) A range of survival times should be used for each pathway, so as to find the optimal time for studying the mode of termination and to ensure that the degeneration being studied is indeed orthograde degeneration of the selected pathway. Optimal survival times vary widely and the kind of alteration seen in the degenerate profiles will also vary: swelling and pallor of the terminal and its vesicles are early changes in some pathways; darkening and loss of internal detail is usual at some stage for all pathways and is the most commonly studied type; filamentous degeneration may occur to varying extents in some pathways; glial engulfment is often an accompanying feature of degeneration and is of some help in confirming the degenerate nature of the profile, but its absence need not be significant; alterations in cytoplasmic inclusions are also a common feature.

(4) To ensure that terminals are being studied, clear synaptic membrane specializations must be evident; a piece of degenerate

material engulfed by glia means nothing unless serial sections show continuity with a region of synaptic contact with a dendritic profile. Where dendro-dendritic synapses are present, care must be taken to exclude the possibility of dark dendrites due to spontaneous or transneuronal degeneration (see below) or to poor fixation.

(5) The opposite side should be used as a control where feasible and be carefully examined.

(6) Findings should be correlated with typing derived from studies of normal material; where possible, specific localization of degeneration should be noted and confirmed and the post-synaptic profiles identified.

(7) If more than one fibre pathway has been involved, as in intrinsic lesions, attempts should be made to differentiate the types of axon terminal undergoing degeneration, on the basis of spatial distribution in relation to the lesion, optimal survival time, general appearance and synaptic morphology.

IV. Cell "marking" techniques

A. Combined Golgi impregnation and electron microscopy

This method seeks to combine the various advantages of the Golgi method and electron microscopy and to eradicate their disadvantages; naturally enough, the combination itself carries some of its own disadvantages, but there is no doubt that this approach is a most useful adjunct to the study of intrinsic organization in a region. It allows a single cell, stained and identified at light microscopic level by the Golgi method, to be prepared for electron microscopy and examined in serial sections, particularly for synaptic interrelationships; in this way a direct correlation can be made between the identities and characteristics of cell types at the two distinct levels of study, which provide quite different and complementary information about the cell's properties. The stained cell is easily picked out under the electron microscope enabling serial sections to be studied, without any doubt about whether one is still tracing the same profile. The method can be combined with experimental sectioning of axon pathways, as in the original study by Blackstad (1965).

The work of Blackstad (1965) and Stell (1965) on the use and techniques for combining Golgi impregnated material and electron microscopy in this direct manner, has been well reviewed recently (Blackstad, 1970). We will therefore only describe our own methods and experiences, which have derived from a study of the cells of the glomerular layers of the bulb and which have been based upon the techniques already in use in this laboratory; certain of our experiences indicate some of the problems of the technique in its present form, only some of which have been overcome. Combined Golgi-E.M. studies have also been described by Kolb (1970).

Preservation of material is poor (Figs 2, 14–17), probably primarily on account of the fact that the silver nitrate has to be made up in distilled water to avoid precipitation; however, with a knowledge of the normal electron microscopy of the region, including profile outlines and orientations, most processes can be recognized. Knowing the definitive light-microscopic type, its electron microscopic characteristics and relationships may be confirmed or discovered. Synaptic structures, both pre- and post-synaptic, are easily picked out and the layer of extra-cellular material in the synaptic cleft may often be resolved, giving a clear indication of the extent of the synaptic specialization (Figs 14, 15, 17). As reported by both Blackstad and Stell, the Golgi impregnation appears to be membrane-limited, occurring within the cytoplasm of cells, often sparing membrane-bound systems such as mitochondria or even neuro-tubules, or else being contained within them, such as within endoplasmic reticulum. The characteristic appearance of the cytoplasm of Golgi-impregnated cells under the electron microscope is of extremely electron-dense areas, of a granular nature, surrounded by slightly less dense material; in between these areas are empty regions of variable size (Figs 2, 14, 15, 17). The various manipulations suggested by Blackstad for preventing loss of stain were unsuccessful in this study, so it must be concluded that it is a part of, or an inevitable consequence of the staining process itself. The rounded edges of the dense material indicate that they may be aggregations of cytoplasmic material that have taken up the silver complex.

We have not seen any staining of the extracellular space as described by Blackstad, but on the other hand we have seen similar thin stained regions, that have been identified as belonging to the thin cytoplasmic lamellae of glia by their site, arrangement and continuity with glial profiles that are also stained (Fig. 16). Like him, we have observed the sudden cessation of staining in a large neuronal profile: this was the case of a dendrite of 2–3 μm diameter, cut longitudinally and passing in the superficial part of the external plexiform layer. The stain stopped abruptly in a straight line running diagonally across the dendrite; moreover, a mitochondrion and several neurotubules passed through the line, being unstained on both sides, but only surrounded by dense material on one. Such observations indicate that caution

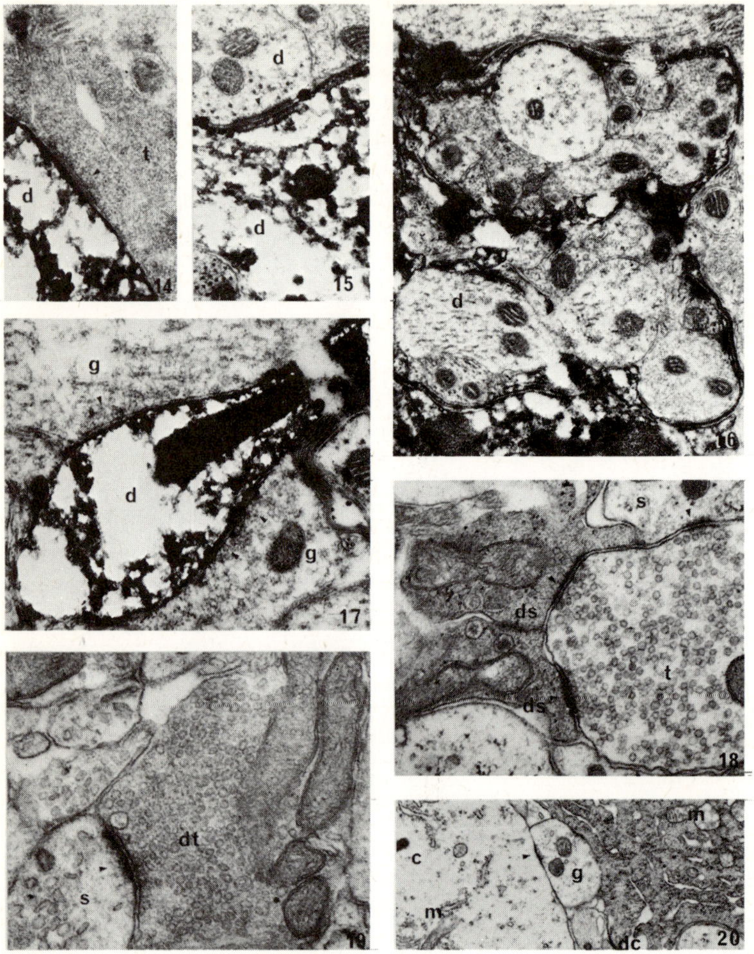

FIGS 14–17. Combined Golgi-E.M. material.

FIG. 14. Olfactory nerve terminal synapsing onto a varicose dendrite of the tufted cell in Fig. 2 in a glomerulus. × 53 400.

FIG. 15. Asymmetrical synapse off a dendrite of the same tufted cell as in Figs 2 and 14, situated at a branch point. × 53 400.

FIG. 16. Glial processes stained including thin glial lamellae surrounding a group of neuronal processes in the periglomerular region. × 32 000.

FIG. 17. Varicose dendrite of the same tufted cell as in Figs 2, 14, 15, receiving synapses from two processes. × 53 400.

must be used in identifying and characterizing cell processes in Golgi stained material. Finally, it may be noted that the technical procedures and examination needed to achieve satisfactory Golgi-E.M. material are very time-consuming and cannot at present be used for routine study; they are however of great value in confirming relationships and identities when the groundwork from normal material has been completed.

B. Transneuronal degeneration

The olfactory bulb is particularly sensitive to transneuronal cell degeneration and Matthews and Powell (1962) and Powell (1967) have given descriptions of the change at the light microscopic level. Recently these observations have been extended using the electron microscope (Pinching and Powell, 1971a) (Figs 18–20), and this paper should be consulted for a full description and discussion of the features of transneuronal degeneration at ultrastructural level. This type of degeneration has turned out to be a useful tool for "marking" cells and their processes in a similar way to Golgi-E.M. combination, in that whole cells are affected—at least in the olfactory bulb; this means that serial study of cells may be greatly facilitated and more internal detail may be seen than in the Golgi-E.M. technique. The dendrites, appendages (Fig. 18) and cell somata (Fig. 20) are characteristically darkened, with a relative disorganization of the nucleic acid components; membrane-bound systems are swollen, but the vesicles usually remain their normal size. Axon terminals undergo a characteristic granular change, with rather ill-defined vesicular profiles and their outlines are often irregular (Fig. 19).

With a background of light and electron microscopic knowledge,

---

FIGS 18–20. Transneuronal degeneration (rabbit).

FIG. 18. Transneuronal degeneration of granule cell spines (compare normal spine adjacent). Normal recurrent collateral terminal making asymmetrical synapses; note the dark granular cytoplasm of the degenerating profiles. × 53 400.

FIG. 19. Transneuronal degeneration affecting the axon terminal of a mitral cell recurrent collateral in the granule cell layer. × 66 900.

FIG. 20. Comparison of the cytoplasm of normal and transneuronally degenerating mitral cells. Note the dark granular cytoplasm, concentration of organelles in the cytoplasm and the swelling of mitochondria and granular endoplasmic reticulum. × 13 200.

this latter axonal change may be useful in determining intrinsic connections of the bulb, particularly the distribution of the recurrent collaterals of the tufted and mitral cells. These profiles have hitherto been identified as those terminals containing spherical vesicles and synapsing with asymmetrical thickenings that do not degenerate after complete isolation of the bulb from extrinsic sources, but which do degenerate after intrinsic lesions. This is rather unsatisfactory, and the occurrence of transneuronal change in some of these acons, without similar changes occurring in the terminals of central origin, provides a direct and complementary confirmation to the former definition. In order to use transneuronal degeneration as a marker technique, the cell types undergoing this change must be known and care must be taken in choosing a survival time since axonal change occurs rather later than the alterations in the corresponding cell soma and dendrites. Finally it should be noted that age and species are important factors in determining the optimal survival time, and indeed the occurrence of transneuronal change.

IV. Operative procedures

A. Olfactory mucosa removal

This relatively crude but selective operation has the advantage that a large number of fibres may be destroyed without opening the cranial cavity; its principal disadvantage is the relatively high mortality, due to the nature of the lesion and the young animals that have to be used to reduce the likelihood of "atrophic rhinitis". Anaesthetics within minimal or short-lasting respiratory depression are essential, and Avertin (tribromoethanol with amylene hydrate) gives good results; 0.5 ml Avertin is mixed with 25 ml normal saline and 2 ml 95% ethanol. A 200 g animal is given 1 ml mixture intraperitoneally and a further 0.2–0.5 ml after 5 min. if necessary. The skin over the nasal bones is divided anteroposteriorly and the pericranium separated; the dividing line between nasal and frontal bones is noted and the nasal bone of one side removed with small bone forceps (a drill must not be used, as it drives spicules of bone into the respiratory tract). The dorsal mucosa is removed with watchmakers' forceps and the whole area of olfactory epithelium is stripped off, especially on the septum and the undersurface of the cribriform plate. Cotton wool wrapped around forceps is a convenient tool. In this way most of the medial aspect of the olfactory bulb and the anterior part may be denervated. It is difficult to effect a complete posterior denervation and some of the conchae may also escape damage and this fact makes it difficult, if not impossible, to be certain of detecting reinnervation of the bulb were it to occur. There is considerable bleeding and measures should be taken to remove blood as quickly as possible until it abates. Penicillin powder may be sprayed into the wound and the skin incision closed; the animal is then allowed to survive for a variable period of time.

B. Lesions of central pathways

For lesions of the centrifugal pathways, the temporal approach of Powell *et al.* (1965) should be used, making the lesion far back in the white lateral olfactory tract to avoid damage to the anterior olfactory nucleus. More anterior lesions involve this nucleus and may be used to study the distribution of its fibres in the bulb, but it must be borne in mind that the centrifugal pathway has also been involved: damage to the anterior olfactory nucleus gives rise to degeneration in the fibres passing in the anterior commissure to the opposite bulb. Other central lesions may be affected using stereotactic apparatus (e.g. Price and Powell, 1970e). In all cases it is advisable to use young animals to avoid the complication of degenerative changes that occur in older animals.

C. Intrinsic lesions

The limits of the frontal bones covering the olfactory bulbs are noted after an incision has been made in the overlying skin. The bone covering one bulb is removed using a drill until a thin translucent layers is left, which may be removed carefully using watchmakers, forceps under a dissecting microscope; this avoids surface damage by the drill. The dura over the midpoint of the dorsal surface is carefully divided and a small hole made in the pia-arachnoid over the intended site of the lesion. A fine needle is then manipulated so as to make a small lesion in the underlying bulbar tissue of appropriate depth: this may be a point lesion or a short line running in coronal or sagittal planes. It should be emphasized that the lesion must be placed in such a way that its site can be located after fixation, in the case of electron microscopic material, as the small size of the lesion may mean that it is not evident on the bulbar surface. The wound is sprayed with penicillin and the skin incision closed; the animal is allowed to survive for a variable length of time.

V. Appendix: methods

A. Golgi impregnation

Fixation: Either the standard method for electron microscopy is used (see later section) or a fixative solution containing 5% glutaraldehyde and 2.4% potassium dichromate in phosphate buffer.

Staining. (a) Material of either sort is stained for 5 days in the glutaraldehyde-dichromate mixture above.

(b) The blocks are washed well in several changes of distilled water, rinsed with silver nitrate (0.75%) and then left for a further five days in the silver nitrate solution. They are then washed again in distilled water. Repeated impregnation may then be carried out by repeating (a) and (b) twice in the same order, washing carefully with distilled water between each fresh solution giving blocks two to three days in each stain. All solutions should be made up freshly and the pots kept in the dark during staining; it is advisable to use about 50 ml of each solution for a pair of olfactory bulbs, which are put through as single blocks for light microscopy.

Embedding is with Low-viscosity Nitrocellulose and mounting with XAM mountant and a cover slip. The method employed here appears to avoid fading of sections, over long periods of time.

B. Electron microscopy

Many of the problems and techniques involved in the electron microscopy of the central nervous system have been considered at length in recent articles (e.g. in Nauta and Ebbesson, 1970) and many excellent books are available on general principles for technique on electron microscopy (e.g. Pease, 1964; Kay, 1965; Sjostrand, 1967) and these should be consulted for the broader considerations of technique. For perfusion, it has been the practice to cool the animal under Nembutal anaesthesia to 15–20°C before perfusing it with fixative solution at about the same temperature; although this procedure may have little effect on the quality of the fixation, it allows time in case of any mishaps during preparation, tending to avoid the complication of asphyxic change. A very short preliminary wash with buffered saline (White's) is followed by a brisk perfusion with 150–200 ml of fixative (1% glutaraldehyde with 4% formaldehyde in phosphate buffer) after which a very slow perfusion rate is used for a further 200 ml of fixative; the formaldehyde is made up from paraformaldehyde with heating to 60–70°C. A prior injection of 1% sodium nitrite (0.1 ml) and heparin (0.2 ml) causes dilatation of cerebral blood vessels and prevents clotting. Perfusion is via the left ventricle and aorta, using a wide-bore glass cannula with bulb, and at a hydrostatic pressure at or just above the animal's normal blood pressure. The brain is left in the skull for some hours before being removed; care must be taken in removing the olfactory bulbs, as the olfactory nerves attach themselves firmly to the underlying bone – it is usually possible to sever the nerves close to the cribriform plate with fine watchmakers' forceps. The brain is immersed in fresh fixative solution for at least six hours.

Blocks of olfactory bulb are then cut (a minimum of six blocks from each bulb for the rat), given a very brief rinse with 10% sucrose in phosphate buffer and post-fixed with 2% osmium tetroxide, also in phosphate buffer; dehydration is in ascending alcohols, with block staining (1% uranyl acetate in 70% ethanol) at the 70% stage. After immersion in epoxypropane, followed by a 50:50 mixture of epoxypropane and embedding mixture, the blocks are embedded individually in plastic capsules using Araldite or Epon/araldite mixture. Beem capsules in which the lid acts as the bottom and the cone is cut off are useful for embedding, as they give a flat block face, whatever the size and shape of the block, and they may also be used many times since the block may be removed from the capsule without damaging the latter. The blocks are allowed to polymerize in 40°C (24 h) and 60°C (48 h) ovens. 2 μm sections are then cut off the trimmed block, mounted and stained with the method of Richardson *et al.* (1960)– 50:50, 1% methylene blue in 1% borax and 1% aqueous Azure II–for the orientation of the block. An area or "mesa" for study is marked out and trimmed; this should have a maximum size of 0.5 by 1 mm, but for routine work, 0.5 mm by 100 μm is a good size; with smaller mesas material is improved in reproducibility, and longer series of sections may be cut without interruption.

For ultrathin sectioning, the knife scoring angle we use is 45° and the clearance angle of the microtome 1–2°; cutting speed range is 1.6–2.0 mm per second for smaller blocks and slower speeds for larger ones. For manipulating the grids to pick up ribbons of sections from the knife water bath, a lip bent up at the edge of the grid improves manoeuvrability. Sections, once on the grids, may be dried with a commercial hair drier. Section staining is in 5% uranyl acetate in 50% ethanol at 40°C in a wet chamber, for 30 mins; the grids are then washed in warm ethanol (warmed in the same oven as the stain, to prevent rupturing of the sections) and dried lead citrate is used for further section staining (Reynolds, 1963); it is important after glutaraldehyde fixation to dilute this stain 1 in 5 with 0.01N Na OH to prevent the granular appearance caused by heavy lead deposition. Staining is at room temperature in a chamber containing N NaOH, to avoid precipitation with CO_2. The grids are rinsed in distilled water and dried. Stock solutions of both stains may be kept in the refrigerator, the lead citrate in a chamber with NaOH; it is not necessary to make fresh solutions for each staining routine.

For *serial sections,* small regular mesas must be used in order to obtain long series, and these are mounted on formvar-coated slit grids (slit–1 mm by 2 mm). A stock solution of polyvinyl formol formvar (1%) is made up in ethylene dichloride and stored in a flask at 0°C. The coating concentration determines the thickness of the film and it is necessary to achieve a balance between strength of the film and a minimal loss of resolving power; a 0.2% solution in ethylene dichloride is optimal in our experience. The solution is allowed to reach room temperature to avoid condensation on the slide which causes holes in the film; a 3″ by 1″ slide is cleaned in acid alcohol, wiped, rinsed in Teepol and wiped again and finally blown on to remove fluff. This is dipped into a coplin jar containing the formvar solution for 3–4 s and then rapidly withdrawn in the vertical position. The excess is drained and one side wiped clean; the film is allowed to dry, upside down under cover, and is then scored with a razor blade, just inside the edges and about one half inch inside the limits of the coating. The slide is breathed on, and inserted at about 45°, coated side up, into a dish of clean distilled water (surface dust may be removed with hard tissue paper). The film is thus released to float on the surface of the water and the grids may be carefully placed on the central portion of the film and pressed lightly onto it to ensure adhesion; a wire frame is brought under the film, so as to support it on all sides and the whole is removed from the water at an angle (this minimizes the pressure exerted on the film which tends to force it through the frame); it is allowed to dry under cover. The grids may be picked off with forceps, as needed. This technique has been discussed at length since its correct use is an absolute essential for good, repeatable serial sections; serial sections, in their turn, are an essential part of this kind of study.

C. Combined Golgi-E.M.

Golgi-impregnated material, preferably perfused with the primary staining fluid of the Golgi-Kopsch technique described above, and in blocks of typical electron microscopic dimensions (see above), is stained in 2% osmium tetroxide and embedded in Araldite; this is allowed to polymerize slowly at room temperature, or after 1 day in a 40°C oven only. When the blocks are hard enough to be gripped for cutting on a sledge microtome (several days), sections are cut at 20–25 μm with the knife edge steeply angled to the axis of travel. These sections are mounted in castor oil and a coverslip placed over them. They may then be examined and appropriate cells for study found. The dimensions of the sections are such that much of the dendritic fields of cells at the glomerular level is included, but not much extra stained material in the vicinity of the cell, which may confuse study with the electron microscope. The cells chosen are then photographed and/or drawn under high power, and their site in the section noted. The cover slip is then removed and the castor oil washed off with epoxypropane; the section may now be re-embedded by immersion in a drop of fresh Araldite in the lid of a Beem capsule; it is then firmly pressed down by a flat-surfaced, preformed and already polymerized Araldite block, which is in the remainder of the capsule. The whole is then polymerized at 60°C.

The composite block may then be examined under a dissecting microscope and a normal electron microscopic "mesa" made around the cell, avoiding as much other stained material as possible. This facilitates study and prevents excessive blunting of the microtome knife, allowing longish series of sections to be cut at a time. After cutting down the mesa with 2 μm sections to remove the fine surface layer of Araldite, ultra-thin sections are cut through the mesa and collected; these are examined so as to ascertain whether or not the cell has been reached, and more sections cut in serial order, until the cell or its processes appear. All sections are mounted on single-slit formvar-coated grids which allow easy serial study (see above); they are stained with uranyl acetate and lead citrate on the grid.

Acknowledgements

The authors would like to express their deep gratitude to Dr. T. P. S. Powell for his valuable advice and encouragement.

This work was supported by grants from the Medical and Science Research Councils.

Abbreviations

c = normal cell soma
d = dendrite
dc = transneuronally degenerating cell soma
ds = transneuronally degenerating spine
dt = transneuronally degenerating axon terminal
g = gemmule
m = mitochondrion
ot = orthograde degeneration of terminal
pg = periglomerular cell
s = spine
sa = short axon cell
t = axon terminal
tf = tufted cell

Arrow heads indicate synaptic thickenings.
Figs. 1–17 are from rat material, and Figs. 18–20 from rabbit.

References

Andres, K. H. (1965). Der Feinbau des Bulbus olfactorius der Ratte unter besonderer Berücksichtung der synaptischen Verbindungen. *Z. Zellforsch, Mikrosk. Anat.* **65**, 530–561.

Blackstad, T. W. (1965). Mapping of experimental axon degeneration by electron microscopy of Golgi preparations. *Z. Zellforsch. Mikrosk. Anat.* **67**, 819–834.

Blackstad, T. W. (1970). Electron microscopy of Golgi preparations for the study of neuronal relations. *In* "Contemporary Research Methods in Neuroanatomy". (Nauta, W. J. H. and Ebbesson, S. O. E., eds.), Springer-Verlag, Berlin.

Bodian, D. (1970). An electron microscopic characterization of classes of synaptic vesicles by means of controlled aldehyde fixation. *J. Cell Biol.* **44**, 115–124.

Cajal, S. R. (1911). "Histologie du système nerveux de l'homme et des Vertébrés". Maloine.

Colonnier, M. (1964). The tangential organization of the visual cortex. *J. Anat. (Lond.)* **98**, 327–344.

Colonnier, M. (1968). Synaptic patterns on different cell types in the different laminae of the cat visual cortex. An electron microscope study. *Brain Res.* **9**, 268–287.

Cragg, B. G. (1962). Centrifugal fibres to the retina and olfactory bulb, and composition of the supraoptic commissures in the rabbit. *Expl. Neurol.* **5**, 406–427.

Gray, E. G. (1969). Electron microscopy of excitatory and inhibitory synapses: a brief review. *Prog. Brain Res.* **31**, 141–155.

Heimer, L. (1968). Synaptic distribution of centripetal and centrifugal nerve fibres in the olfactory system of the rat. An experimental anatomical study. *J. Anat.* **103**, 413–432.

Hirata, Y. (1964). Some observations in the fine structure of the synapses in the olfactory bulb of the mouse, with particular reference to the atypical synaptic configurations. *Archvm. Histol. Jap.* **24**, 303–317.

Kay, D. H. (Ed.) (1965). "Techniques for Electron Microscopy". Blackwells, Oxford.

Kemp, J. M. and Powell, T. P. S. (1971). "The Synaptic Organization of the Caudate Nucleus".

Kolb, H. (1970). Organizations of the outer plexiform layer of the primate retina. Electron microscopy of Golgi-impregnated cells. *Phil. Trans. Roy. Soc. B.* **258**, 261–283.

Land, L. J., Eager, R. P. and Shepherd, G. M. (1970). Olfactory nerve projections to the olfactory bulb in the rabbit: demonstration by means of a simplified ammoniacal silver degeneration method. *Brain Res.* **23**, 250–254.

Matthews, M. R. and Powell, T. P. S. (1962). Some observations on transneuronal cell degeneration in the olfactory bulb of the rabbit. *J. Anat.* **96**, 89–102.

Nauta, W. J. H. and Ebbesson, S. O. E. (Eds.) (1970). "Contemporary Research Methods in Neuroanatomy". Springer-Verlag, Berlin.

Pease, D. C. (1964). "Histological Techniques for Electron Microscopy". Academic Press, New York and London.

Peters, A., Palay, S. L. and Webster, H. de F. (197). "The Fine Structure of the Nervous System". Harper-Row, New York.

Pinching, A. J. (1969). Persistence of post-synaptic membrane thickenings after degeneration of olfactory nerves. *Brain Res.* **16**, 277–281.

Pinching, A. J. (1971). Myelinated dendritic segments in the monkey olfactory bulb. *Brain Res.* **29**, 133–138.

Pinching, A. J. and Powell, T. P. S. (1971a). Ultrastructural features of transneuronal cell degeneration in the olfactory system. *J. Cell Sci.* 8, 253–287.

Pinching, A. J. and Powell, T. P. S. (1971b). The neuron types of the glomerular layer of the olfactory bulb. *J. Cell Sci.*

Pinching, A. J. and Powell, T. P. S. (1971c). The neuropil of the glomeruli of the olfactory bulb. *J. Cell Sci.*

Pinching, A. J. and Powell, T. P. S. (1971d). The neuropil of the periglomerular region of the olfactory bulb. *J. Cell Sci.*

Powell, T. P. S. (1967). Transneuronal cell degeneration in the olfactory bulb by the Golgi method. *Nature, Lond.* **215**, 425–426.

Powell, T. P. S., Cowan, W. M. and Raisman, G. (1965). The central olfactory connections. *J. Anat. Lond.* **99**, 791–813.

Price, J. L. and Powell, T. P. S. (1970a). The morphology of the granule cells of the olfactory bulb. *J. Cell Sci.* **7**, 91–123.

Price, J. L. and Powell, T. P. S. (1970b). The synaptology of the granule cells of the olfactory bulb. *J. Cell Sci.* **7**, 125–155.

Price, J. L. and Powell, T. P. S. (1970c). An electron microscopic study of the termination of the afferent fibres to the olfactory bulb from the cerebral hemispheres. *J. Cell Sci.* **7**, 157–187.

Price, J. L. and Powell, T. P. S. (1970d). The mitral and short-axon cells of the olfactory bulb. *J. Cell Sci.* **7**, 631–651.

Price, J. L. and Powell, T. P. S. (1970e). An experimental study of the origin and course of the centrifugal fibres to the olfactory bulb in the rat. *J. Anat., Lond.* **107**, 215–237.

Rall, W., Shepherd, G. M., Reese, T. S. and Brightman, M. W. (1966). Dendro-dendritic synaptic pathway for inhibition in the olfactory bulb. *Expl. Neurol.* **14**, 44–56.

Reynolds, E. S. (1963). The use of lead citrate at high pH as an electron-opaque stain in electron microscopy. *J. Cell. Biol.* **17**, 208–212.

Richardson, K. C., Jaret, L. and Finke, E. H. (1960). Embedding in epoxy resins for ultrathin sectioning in electron microscopy. *Stain Technol.* **35**, 313–322.

Sjöstrand, F. S. (1967). "Electron Microscopy of Cells and Tissues". Academic Press, New York and London.

Stell, W. K. (1965). Correlation of retinal cytoarchitecture and ultrastructure in Golgi preparations. *Anat. Rec.* **153**, 389–397.

Uchizono, K. (1965). Characteristics of excitatory and inhibitory synapses in the central nervous system of the cat. *Nature, Lond.* **207**, 642–643.

Valdivia, O. (1970). Methods of fixation and the morphology of synaptic vesicles. *Anat. Rec.* **166**, 392.

Valverde, F. (1965). "Studies on the Pyriform Lobe". Harvard University Press.

Wiitanen, J. T. (1970). Selective silver impregnation of degenerating axons and axon terminals in the central nervous system of the monkey. *Brain Research* **14**, 546–548.

7 Techniques for Investigating Single Unit Activity in the Vertebrate Olfactory Epithelium

ROBERT C. GESTELAND

Department of Biological Sciences
Northwestern University
Evanston, Illinois 60201

I. Introduction

This chapter is an attempt to commit to writing the techniques and the pitfalls in data interpretation which may be important to an investigator working on vertebrate olfactory receptor cells. Some of this material has been published. However, the literature is scattered and many of the published techniques neglect to discuss possible artifacts, alternative explanations and the small bits of laboratory practice upon which the success of an experiment and its correct interpretation rely.

There is a second reason for writing this chapter. Many of the techniques described here were developed or elaborated by Lettvin and have spread unevenly through the world of electrophysiology by word of mouth, project reports, and theses, and are thus inaccessible to most. Lettvin is responsible for three major advances in chemoreceptor physiology, all of which introduced new techniques. His method was the first which allowed recording from single units in contact chemoreceptors of insects (Hodgson *et al.*, 1955). The first use of metallic microelectrodes to record impulse activity from olfactory sensilla in insects occurred in his laboratory (Lettvin, 1960), as did the first records from vertebrate olfactory receptor cells (Gesteland *et al.*, 1963). Much of what is here is the distillate of his innovative techniques and profound insight.

How does an olfactory stimulus interact with the receptor cell membrane? What processes are involved in the generation of action potentials? How are the nature of the stimulus, its intensity, and duration represented in receptor cell action potential activity? Much remains to be learned before definitive answers to these questions can be given. Neurophysiologists rely primarily upon microelectrode techniques to obtain answers to such questions and the vertebrate olfactory receptor neuron has proved particularly refractory to microelectrode recording. Information about single cell responses became available *c.* 1960 from extracellular microelectrode studies, and before this, some of the response properties of single cells had been inferred from experiments in which larger electrodes were used to record the activity in small bundles of olfactory nerve fibres. At the present time, information about single units remains fragmentary and controversial, but enough has been learned so that we can say with assurance that olfactory receptors function in very complicated ways, even in the very simplest stimulus situations.

If a large reversible electrode is touched to the surface of the nasal mucosa, and an odorous stimulus is puffed into the nose, a voltage change will be recorded between this electrode and an indifferent electrode located elsewhere on the animal. Ottoson called this voltage the electro-olfactogram (EOG) and characterized many of its features in his elegant study published in 1956. He found that a short stimulus evokes a negative-going voltage with a rise time of less than a second. It decayed smoothly and approximately exponentially over a period lasting several seconds. The amplitude increased with the stimulus intensity (defined as the number of molecules arriving at the mucosa per second). It reached a maximum saturating amplitude of a few millivolts, which is different for different stimuli. He likened the voltage to receptor potentials recorded from other sense organs and suggested that this voltage was a measure of the excitatory effects of the stimulus on normally quiescent receptor cells. A longer stimulus produced a transient component like that from a short stimulus, followed by a maintained lower amplitude negative voltage which lasted as long as the stimulus and then decayed exponentially to zero. The rise times, fall times and amplitudes of EOGs evoked by different chemical substances varied regularly with the chemical nature of the stimulus. The differences in EOG shapes for different smells are not so divergent as to allow the identity of the stimulus to be inferred from the parameters of the response. EOGs to a particular substance at a particular intensity are quite alike from one animal to the next and quite alike from the nose of a single animal for many hours, as long as the tissue does not dry out. Ottoson investigated a small positive voltage which preceded the large negative wave and concluded that it was not due to the stimulus substance acting on the receptor cells. He has recently reviewed the state of knowledge about the EOG (Ottoson, 1971).

Recordings from filaments of the olfactory nerve in which the extracellularly recorded spikes from a large number of olfactory receptor cell axons are summated yields a similar description of receptor cell activity. The number of spikes evoked from a population of receptor cells increases smoothly as the intensity of the stimulus increases and the burst has a time course about like that of the EOG (Mozell, 1962). If the responses of the axons from the receptor cells localized in one area of the olfactory mucosa are compared with those from cells in another area, the results showed that the sensitivities of the cells were not homogeneously distributed, i.e. that there was spatial separation of receptor sensitivity (Mozell, 1964, 1971). This has since been shown to be true also for the EOG by Mustaparta (1971).

With the advent of extracellular recording of action potentials from single olfactory receptor cells, it became apparent that single cells respond in complex ways to different odours and that the cells are neither all alike nor can they be grouped into a few categories (Gesteland *et al.*, 1965). Studies on single chemoreceptor cells of the generalist type in insects showed that they have the same diversity of response properties as those in the frog (Boeckh *et al.*, 1965). Physiological studies on such a system are much more difficult than on a system composed of a homogeneous group of cells. There is no easy way to decide on adequate sample size. It is hard to assess the extent of distortions induced by the experimental technique.

The particular difficulties confronting the experimenter who works on olfactory receptors are as follows. The receptor cells in the noses of most animals are small and have long, unmyelinated, unbranching axons with an average diameter of 0.2 micron. Extracellular recordings in the region of the cell soma distort the cell's response properties, and activity in a single cell cannot be followed for more than fifteen minutes or so under these conditions. Recordings from the axon of a single cell do not seem to disturb the cell and often activity of the cell can be followed for two or three hours. However, extracellular axon recordings are difficult to accomplish because the voltages are small. More often than not, two or three axons will be equally close to the electrode tip, and the activity of one cell is difficult to distinguish from that of its neighbours. Individual nerve fibres have low average firing rates in the unstimulated state, and fire very irregularly with a maximum rate of the order of 30 per s when stimulated. Intervals between successive spikes may vary by a ratio as large as 100–1. Therefore, the search for a large amplitude extracellular spike is a slow one, since in order to find a cell, the electrode must be in very close proximity to the axon when a spike occurs. Some stimuli will excite a particular cell, some will inhibit it; some stimuli will cause complicated bursts of activity separated by intervals of silence, and some stimuli will have no effect. It often happens that when a large spike is found, none of the stimuli present in the stimulator are very effective in changing the cell's firing rate. In this situation, it is usually easier to look for another cell than to change stimulus bottles in the apparatus.

Many cells are sensitive to excessive stimulation. After a period of intense stimulation, cell excitability decreases, and responses to a particular stimulus at any intensity are not repeatable. For this reason, the strategy used in studying many sensory systems whereby a stimulus is repeatedly delivered in order to enhance cell firing rates, and hence speed the search for a single unit, is not so effective in the nose. Since different cells have widely differing thresholds for a particular stimulus, several different intensities of a large number of odours must be used on a cell in order to adequately define its response properties. For repeatable results, stimuli must be well separated in time and maximum intensities must be low.

If it were possible to reliably record for an extended period of time using intracellular micropipette electrodes, many of the questions about receptor mechanisms could be answered directly and the laborious extracellular techniques could be reserved for questions involving neural coding. However, this is very difficult. There is one brief report of successful intracellular recording from receptor cells in the frog's nose (Aoki and Takagi, 1968).

Currently the Mudpuppy, *Necturus,* appears to be an interesting experimental animal because the olfactory receptor cells are large

(Gesteland, 1971). Intracellular recording from those cells has also proven to be difficult. Much of the intracellular volume in these cells is occupied by the nucleus. There is relatively little cytoplasm and a small electrode-induced leak in the membrane quickly disrupts cell function.

II. Electrode properties and preparation

The ideal microelectrode would be small enough to penetrate and record internally from cell bodies of the size range commonly found. It would allow passage of polarizing currents through the electrode. It would have a reversible and stable d-c potential. It would have a time constant shorter than the fastest transients likely to be encountered (0.1 millisecond) and hence would induce no distortion in the amplitude or shape of biological signals over the frequency range from d-c to somewhere around 100 kHz. Its tip resistance would be low enough so that extracellular recordings of single unit activity would not be obscured by electrode noise. The properties of the tip in this use would be such that activity in neighbouring individual nerve axons could be easily distinguished. There is no such electrode. Instead, a judicious choice between several varieties of two basic types of microelectrodes must be made to suit the requirements of each experiment on nerve cells. This is particularly true in studies on olfactory receptor cells where all of the microelectrodes which have been developed so far must be pushed to their ultimate limits of performance in order to follow single cell activity.

The two categories of microelectrodes are the micropipette type, in which an electrolyte in a glass tube with a very fine tip makes contact with the biological tissue, and the metallic type, in which metal in the electrode tip is in contact with the biological preparation. The metallic electrode commonly has a tip diameter which varies between one and ten microns, does not record d-c potentials either reliably or stably, and can have a low impedance when used as an extracellular electrode. The impedance of the electrode-tissue interface decreases with increase in frequency and thus, this electrode is a high pass filter which accurately records transients and not steady d-c voltages. This liquid-junction type of electrode or micropipette, on the other hand, is d-c stable and reliably records resting potentials. Those commonly used have tip diameters less than 1 μm and usually cannot be resolved in an optical microscope. These are the only electrodes that work intracellularly. The smaller the tip, the higher is the tip resistance. Transient wave forms are distorted due to the fixed capacitance of the glass walls of the pipettes. Part of this capacitance can be compensated with amplifier circuitry. However, this compensation does not come cheaply. Decreases in response time of the electrode must be paid for by proportional increases in electrode noise. Thus, pipette electrodes are low pass filter and are a poor choice for recording low-amplitude, transient, rapidly changing voltages such as extracellular action potentials.

A. EOG electrodes

Fluid filled electrodes with very large tips are commonly used to measure the EOG responses in the vertebrate nose. The same kind of electrodes with much smaller tips are called micropipettes and have been occasionally successful for short periods in intracellular recording from olfactory receptors. Micropipettes are constructed by pulling heated glass tubing apart until the glass separates. The ends of the tubes will be open. The shape and size of the opening depend upon the properties of the glass and the properties of the machine which is used to pull the glass. There are as many variations in preparation techniques as there are experimenters using them.

For very large tipped electrodes ordinarily used to record EOGs, the simplest method of construction is to immerse a short length of the smallest diameter of capillary glass tubing in a hot solution of an electrolyte and a gelling medium and let it set. A chlorided silver wire is inserted in one end of the tube and the other end is touched to the surface of the olfactory mucosa. The electrolyte solution may be a Ringer's formula appropriate to the animal species or sodium chloride isotonic to the animal extracellular fluid (0.6% by weight for the frog, 0.9% for mammals). Most any other salt can be used as long as the gelling medium is stiff enough to prevent diffusion of the electrolyte solution onto the mucosa. Gelling media may be agar at a concentration between 0.3% and 0.5% by weight or gelatin mixed as if for dessert. A chlorided silver plate or coil of wire wrapped with saline moistened gauze or molded into saline in a gelling medium is used for the indifferent electrode. It is placed in contact with the animal, often in the mouth or under the head. If the electrode is placed too far caudad, electrocardiogram artefacts will intrude on the EOG record. It is useful to puff the odorous stimuli to be used at the indifferent electrode and at the recording electrode touching the animal any place except the olfactory mucosa to make sure that EOG-like voltages are not generated by electrochemical interaction between the stimulus substance and the electrodes. If the recording and indifferent electrodes are not of the same metal coupled to the animal with the same saline solution, there can be a significant d-c voltage between them which will cause a current to flow under certain recording conditions. This can result in an EOG-like artefact as stimulus-induced receptor cell conductance changes modulate the current generated by the electrodes. Other reversible electrode systems, such as calomel and zinc-zinc sulphate have been used for EOG recording. All are perfectly satisfactory when care is taken to eliminate artefactual stimulus-induced voltages.

Often it is more convenient to use small-tipped recording electrodes for EOG studies. In this case, the glass tubing is pulled down to an end with a tip diameter between 5 and 50 μm. Since this is difficult to do on most pipette pullers, fine tipped pipettes are pulled down and then broken to a convenient diameter by bumping their tips under microscopic observation against the edge of a glass slide. These electrodes are easily filled by immersing them in a hot solution of saline and gelling medium in a flask and evacuating it for a few minutes. The pipettes are filled when no bubbles can be observed while the flask is at reduced pressure. The filled electrodes are allowed to set. After this has taken place the spindle holding the electrodes is removed from the gelled medium and stored in a saline solution like that used to fill them with the gelling substance omitted. If the electrodes are to be stored for more than a day or two they should be cold to inhibit growth of micro-organisms.

Conventional ultra-fine tipped micropipettes also are satisfactory EOG electrodes. These pipettes have tip diameters of less than a micron and tip resistances greater than ten megohms when filled with 3 M potassium chloride. No gelling medium is needed because the tips are so small that there will be no appreciable diffusion of the electrolyte onto the nose. Small electrodes are used for studies of variations of EOG with depth below the mucosa surface because they penetrate with little deformation of the tissue. The resistance of these electrodes is high and therefore recording is somewhat noisy but adequate for most purposes.

Mucus is a low resistance medium compared to the resistance of the mucosa cell layer. Therefore current flow from any active cell will spread laterally across the surface for a distance, large compared to the cell diameter. Any electrode, however small, will thus measure the activity from cells over a relatively large area. As a result, EOG amplitudes and shapes are nearly independent of electrode tip diameters.

B. Intracellular micropipettes

Intracellular recording from single cells in the olfactory mucosa is a goal that has been pursued for many years. Since olfactory receptor organ cells are small, micropipette techniques must be pushed to their limits for intracellular studies. This restriction confronts the experimenter with a host of practical problems. These include inconsistency in making such fine-tipped pipettes, difficulty in filling the pipettes with electrolyte, degradation of the pipettes during storage, problems in inserting the pipette into a cell and physical limitations on the fidelity with which such an electrode can record both steady and transient voltages.

Methods and machines for making micropipettes have been discussed by Kennard (1958) and reviewed by Frank and Becker (1964). Tomita (1969) describes a technique for preparing very small-tipped micropipettes.

Since micropipette tips are too small to be resolved in an optical microscope, visual inspection of the tip provides only intuitive clues to the tip diameter. In practice what must be done is to make electrodes, fill them with an electrolyte, and measure the resistance. Relatively slight adjustments on the micropipette puller can have a large effect on tip diameter. It often takes extensive experimentation to reliably pull pipettes small enough to record from the cells of interest but not so small as to resist filling.

For conventional micropipettes, no particular care need be taken in the choice, size and kind of glass or in cleaning it. For very small tipped pipettes, these factors are important. Hard glass seems to work better than soft. The glass and the filling solutions must be clean. After the glass tubing is cut to the correct length for the electrode puller, it should be cleaned with a strong detergent or chromic acid solution, rinsed repeatedly in distilled water and then dried and stored in a capped bottle. The electrolyte solution should be filtered before use. I find that it is useful to use Millipore filters with submicron pore size for this procedure.

There are four principal ways of filling micropipettes with an electrolyte solution. The low pressure boiling method is commonly used and works well and quickly as long as the smallest pipette tips are not needed. The procedure occupies hours instead of minutes, however, for tips of resistance of 100 megohms and up. The advantage of the method is that the electrodes can be filled and used within an hour or so of the time that they are pulled. An interesting observation for very small micropipettes, presently not explained, is that they are harder to fill a half day or more after pulling. In fact, a large percentage of them will not fill at all. For small micropipettes it is best to pull, fill, and use them in as short a time period as is practical.

The low pressure boiling procedure is as follows. Several micropipettes are mounted, tip down, on a glass spindle and held to it by a rubber band. KCl solution (somewhat less than 3 M to allow for evaporation) is heated in a side-arm flask to about 80°C and the spindle with the electrodes is then inserted into the hot solution, pipette tips down, until the pipettes are totally immersed. The top of the flask is stoppered and suction is then applied to the solution until it boils vigorously. Since the boiling bubbles come off most easily at irregular surfaces, they appear at the tops of the pipettes and at the rubber band. If the flask of hot solution has been removed from the heat and made to sit on a metal surface at room temperature for a few minutes before suction is applied, then it is almost impossible to get boiling bubbles to form below the tips of the electrodes at the bottom of the flask. This reduces tip breakage, which occurs when pipettes are boiled over a flame. After about 20 s of moderate vacuum boiling, suction is cut off and air readmitted. Suction is then applied again, for several cycles of low-pressure boiling. The notion behind this procedure is that most of the air in the pipette will be exhausted in the first boiling and replaced with steam. On readmission of normal pressure, the steam will condense in the tip and electrolyte solution will flow into the large end of the electrode. There will be an entrapped air bubble near the tip. This bubble will be again diminished in following cycles because it dissolves in the hot solutions bounding it under normal pressure and is evacuated by the turbulence of reboiling. Addition of a drop of liquid detergent to the electrolyte solution speeds the filling process. For the smallest tips, it is often useful to gently boil the electrodes under reduced pressure for a half hour or so to completely fill them.

The second commonly used technique for filling micropipettes is by exchange diffusion. For this micropipettes are filled by vacuum boiling with warm distilled water, methanol, or a dilute electrolyte solution. Some hours or days, depending on the tip diameter, before the pipettes are to be used they are transferred into a solution of the desired electrolyte and allowed to stand. This electrolyte then enters the pipette by diffusion. For very small tips, exchange diffusion has severe limitations, in that the diffusion time is long and electrodes with small tips stored in solution for long periods are not so effective in recording intracellularly from very small cells. This is probably due to a partial dissolution of the glass at the tip. The procedure is improved by injecting the electrolyte solution into the electrode from the large end using a long fine hypodermic needle such as those used by gas chromatographers (0.13 to 0.25 mm o.d.). Diffusion into the tip then commences from both inside and outside.

Tasaki *et al.* (1968) published a new technique for filling micropipettes which is simpler and has some advantages over vacuum boiling or exchange diffusion. A few strands of fibreglass (glass wool, angel hair) are pulled through the glass tubing before it is formed into pipettes. The glass fibres pull down as the glass pulls and extend to or nearly to the tip in the final pipette. These now may be filled by injecting the electrolyte solution through a thin hypodermic needle inserted into the large end of the pipette. The needle must extend past the shoulder of the pipette to where the tip begins to pull out in its long thin taper. The filling solution should be filtered through a Millipore filter. The solution passes slowly along the fibreglass by capillary action to the tip. There will often be bubbles between the tip and the shank of the microelectrode, but enough fluid covers the inside surfaces of the glass to provide adequate electrical connection. Tasaki reported that tips of the same diameter had half the resistance when filled in this way than when filled by conventional boiling techniques. Miller and Dowling (1970) report that electrodes had the same resistance regardless of whether they were fibreglass-syringe-needle filled or vacuum-boil filled. For very small tips it may take 10 or 15 min. for the fluid ejected from the syringe needle to reach the tip, but a high percentage of the electrodes filled this way are usable.

There are two useful methods for introducing fibreglass into the glass tubes which are to be pulled into micropipettes. The glass tube can be connected to an aspirator and the fibreglass strands sucked through the tube and cut off at both ends. Alternatively, a fine wire with the end folded back to make a hook can be pushed through the glass tube until it protrudes about a centimetre beyond the end. Glass fibres can then be hooked, pulled through, and cut off at both ends. In general pipettes with many fibres fill more easily than those with only a few.

It is important with this technique, as with others, that after the micropipettes have been pulled to shape they be filled and used in a short period of time. Dry pipettes with fibreglass cores which stand overnight do not fill very well. For example, in a long series of pipettes which we made intending that the tip resistance should be between 100 and 200 megohms, about 80% filled the same day as they were pulled; less than 10% would fill the second day.

One serious limitation exists for the fibreglass core electrodes. The fibres function as a capillary surface to lead filling solution down into the tip. They also act in the same way to lead capillary-active substances from the biological preparation up into the electrode tip. In the nose and the tongue of the mudpuppy, the mucus secreted by these tissues rapidly fills the electrode tip and forms a high resistance plug (greater than $3 \times 10^9\ \Omega$) within a few seconds of being touched to the preparation. The electrodes are then useless for recording and passing electrical current. This plug is not removed by the usual techniques of feedback oscillation or current pulse injection. Whenever these electrodes are to be used, it is necessary to monitor electrode tip resistance continually as they are inserted into tissues.

Water can also be introduced into the tip of a micropipette without low pressure boiling or use of glass fibres. A syringe and needle is used to fill the barrel of the pipette. It is then very slowly advanced, tip first, through a warm heating coil. As the air-water interface nears the warm region, water evaporates and condenses in the cooler tip region. A stream of cooling air directed on the tip can speed the process. The entire filling process takes five to fifteen minutes depending upon tip size and apparatus parameters. The desired electrolyte is then introduced into the pipette by exchange diffusion. For extremely small tips, this filling process is tedious and diffusion times become very long. It is a convenient method for pipettes which will have tip resistances up to about 75 megohms when filled with potassium chloride.

It is convenient to discuss the properties of very small electrodes and their limitations in terms of a circuit model for the micropipette as it is used to measure a voltage. A micropipette can be thought of as a tapering volume of electrolyte that is connected to the measuring amplifier at the large end by a reversible metal electrode and at the other end to the biological preparation through a very small opening. Referring to Fig. 1 we see that part of the pipette which is immersed in the tissue is represented by the resistance of the electrolyte filling the pipette and capacitance across the glass wall between the electrolyte fluid in the pipette and the conductive extracellular fluid outside the pipette. The capacitance is proportional to the area of the immersed glass and inversely proportional to its thickness. The resistance of the electrolyte is proportional to the length of the column of fluid in the pipette and inversely proportional to its cross-sectional area. The resistance is greatest, therefore, near the tip where the cross-section is least. The metallic electrode inserted into the top of the pipette to connect it to the amplifier will have associated with it at the interface between the electrolyte and the metal an electrochemical voltage, a resistance which measures the energy expended in oxidation-reduction processes, and a capacitance due to the separation of charge at the interface double layer. When the metallic electrode has a surface area large compared to the micropipette tip area and is reversible, the impedance of the metal-electrolyte junction is negligible and we need only consider the electrochemical potential across the junction.

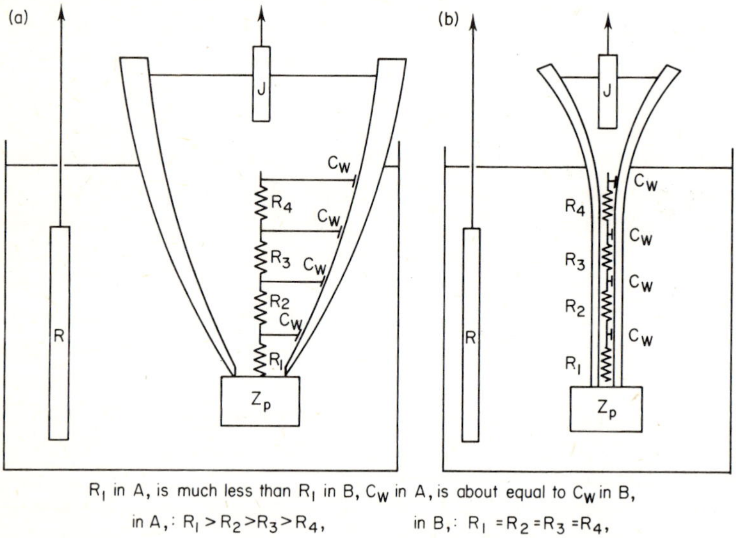

FIG. 1. A large-tipped micropipette can be visualized as equivalent to the electrode in drawing A. A small-tipped micropipette must be considered to be like that shown in B. Z_p represents the impedance of the biological preparation as seen by the electrode. The lumped resistances, R_t and capacitances C_w represent the resistance of the electrolyte filling the tube and the capacitance across the glass walls which must be present in any electrode are shown in C. The large tipped electrode has only a short length of small diameter tube and therefore R_t is low and concentrated at the tip. Only the glass capacitance close to the tip has any significant effect on the tip impedance. The small-tipped electrode, on the other hand has an extended length of high resistance electrolyte column and very thin walls. Therefore unit capacitance is larger and it is distributed all along the region with significant core resistance. R is the indifferent electrode and J is the electrode which connects to the fluid in the micropipette.

Immediately external to the tip of the micropipette there will be an impedance and a voltage due to the properties of the tissue being studied. The tissue immediately proximal to the micropipette tip can be considered as being immersed in a large volume of extracellular fluid. The resistance of this volume conductor is low enough to be neglected. A large reversible metal electrode is in contact with the extracellular fluid and is used to make the other connection to the recording amplifier. For convenience we will consider this electrode to be at ground potential. It is commonly referred to as the indifferent or common electrode in electrophysiology. Associated with the interface between this electrode and the extracellular fluid in which it is immersed, there will be an electrochemical voltage and an impedance. If this indifferent electrode is reversible and large, the impedance of the junction between it and the extracellular fluid can be neglected. There is a capacitance between the wire going to the micropipette and ground. There will be a capacitance across the input terminals of the recording amplifier also. It is desirable to keep these capacitances as small as possible for reasons which will appear subsequently. Therefore the wire from the amplifier to the micro-electrode should be as short as possible and it should be located physically far from any metallic conductors, i.e. it is best to avoid shielded wires.

The equivalent circuit in Fig. 2A represents the most desirable situation. It represents realistically that which occurs when micropipettes with resistances of the order of 10 megohms are used experimentally. In this situation, the micropipette resistance R_t is concentrated at the tip and the capacitance C_w across the micropipette wall can be thought of as lumped and connected between the inside of the tip and the extracellular fluid. The equivalent circuit includes the series combination of the voltage at the interface between the metal electrode and the micropipette electrolyte, E_j; the tip resistance of the micropipette, R_t; the voltage at the junction of the micropipettes and the biological preparation, E_t; the impedance of the biological preparation, Z_p; and the voltage between the extracellular fluid surrounding the biological preparation and the reference electrode, E_r. C_s represents the stray capacitance between the wires and across the amplifier input. C_w is lumped and occurs on the amplifier side of tip resistance. Generally, E_j and E_r are independent of time and current flow in the recording system. Their algebraic sum is balanced to zero by the d-c balance control of the recording amplifier. Any voltage change in the vicinity of the micropipette tip will be due to a biological phenomenon and will be the recorded signal. Furthermore, C_w and C_s can be lumped. The recording amplifier includes an electronic circuit which effectively places a negative capacitance across the input terminals of the amplifier. It is adjusted (see subsequent sections of this paper) so that the magnitude of the negative capacitance equals the sum of C_w and C_s. Since capacitances in parallel add, under this adjustment condition all of the capacitances disappear from

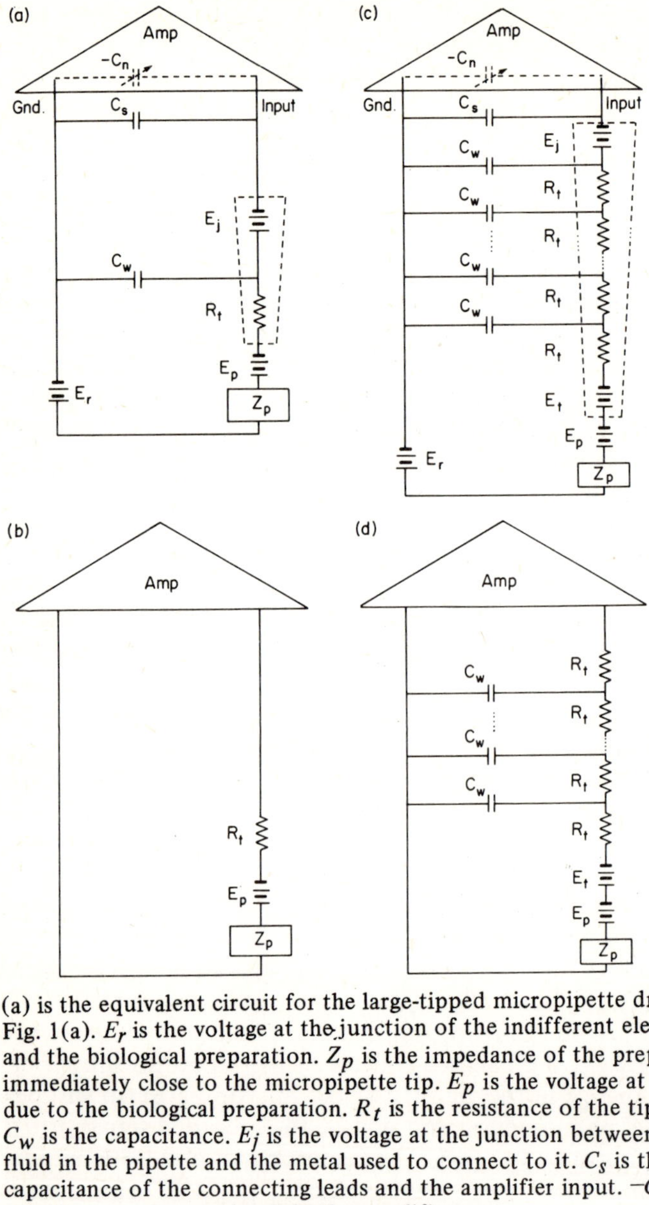

FIG. 2. (a) is the equivalent circuit for the large-tipped micropipette drawn in Fig. 1(a). E_r is the voltage at the junction of the indifferent electrode and the biological preparation. Z_p is the impedance of the preparation immediately close to the micropipette tip. E_p is the voltage at the tip due to the biological preparation. R_t is the resistance of the tip and C_w is the capacitance. E_j is the voltage at the junction between the fluid in the pipette and the metal used to connect to it. C_s is the capacitance of the connecting leads and the amplifier input. $-C_n$ is the negative capacitance circuit in the amplifier.
(b) is the result of the circuit shown in (a) under the conditions that the sum of E_r and E_j is constant and balanced to zero by the d-c balance controls of the amplifier and C_n is adjusted to be equal in magnitude to the sum of C_s and C_w.

the equivalent circuit. The resulting equivalent circuit is shown in Fig. 2B. If the recording amplifier has negligible input or leakage current and has an input impedance which is high compared to the sum of R_t and Z_p, any voltage change which occurs in the biological preparation will be accurately measured by the amplifier. There are no time constants in the measuring system which could distort the signal (time constants are products of resistances and capacitances and we have effectively removed all capacitances from our measuring system). If current is passed through tissue any voltage change which occurs with stimulation will be due to either stimulus induced changes in E_p or stimulus induced changes in Z_p. By measuring the stimulus induced voltage changes with and without applied current, effects of the stimulus upon biological potentials, E_p, and tissue impedances, Z_p, can be determined. Physically, this entire situation is represented in Fig. 1 by the convex pipette connected to the biological preparation by a small hole punctured in the rounded end.

When cells to be investigated are small, micropipette tips must be small and usually become long and thin. This situation is physically represented by the straight pipette in Fig. 1 and is more complex electrically. In this case, the resistance of the solution filling the pipette cannot be considered to be lumped at the tip orifice. The capacitance is distributed along the wall of the inserted glass and cannot be represented as a lumped capacitance. Furthermore, another voltage source appears in the system which is commonly called the tip potential of a micropipette. That it exists is easily demonstrated. A reversible electrode is immersed in a dish containing an electrolyte solution. The tip of a micropipette filled with the same electrolyte is also inserted into the dish. Into this micropipette is inserted a reversible metallic electrolyte identical to the one in the dish. The voltages developed at the metal-electrolyte junctions will be equal and connected in opposition. There are no concentration gradients in the system; therefore, no voltage difference should be measured between the two

---

(c) is the equivalent circuit for the small-tipped micropipette drawn in Fig. 1(b). Elements are the same as above. E_t is the tip-potential which occurs at very small tips due to the non-ideal behaviour of the electrolyte when it is in a very small channel.
(d) is the result of the same adjustment conditions as in (b). E_t cannot be assumed to be invariant and therefore cannot be omitted from the equivalent circuit. Due to the intervening $R_t s$, most of the $C_w s$ can not be compensated by C_n.
The dashed lines enclose those elements associated with the micropipette electrical properties.

metallic electrodes. Experimentally, it is found that if the glass micropipette tip becomes very small, a voltage appears in this system. If the tip of the pipette is broken off, the voltage vanishes; thus the tip potential is intimately related to the existence of a very small channel (Agin, 1969). The tip potential is likely due to the existence of fixed charges on the inside surface of the glass micropipette and the resultant interaction of ions in the electrolyte with these fixed charges. A significant fraction of the cross-sectional area of the small channel must be occupied by fluid in which the charged species cannot behave ideally.

The equivalent circuit of Fig. 2C approximates a small tip micropipette. The resistance of the electrolyte core is shown as a series of R_t's and a distributed capacitance across the glass wall is represented by the several C_w's. This is a lumped circuit approximation for a distributed delay-line network. Two voltage sources in series are shown at the tip. E_t represents the voltage due to the non-ideal behaviour of the electrolyte in the electrode tip. E_p is the voltage of the biological preparation which we would like to measure. The lag in the response time due to the distributed nature of capacitance at the electrode tip cannot be simply compensated. The negative capacitance circuit of the amplifier can only be used to eliminate the effects of stray capacitance far from the electrode tip. With optimal negative capacitance compensation, very small tipped electrodes will have rise times of several ms in response to a sudden voltage change at the tip. More elaborate compensating schemes (predictive filters) can be designed. Practically, these are cumbersome since the resistance at the tip of the electrode varies as it penetrates tissue and varies from tip to tip. It may be possible to find a geometry for the glass tip which minimizes the distributed capacitance. This has not been experimentally demonstrated. Therefore small tip micropipettes will be inherently slow to respond to rapid voltage changes.

As the electrode tip opening decreases its size, the resistance increases more than would be predicted from simple geometric considerations and assumption of ideal behaviour of solutions. Frank and Becker (1964) report that measurements on a series of electrodes filled with 3 M KCl with tip diameters of about 0.5 μm gave resistances of 10 to 30 megohms. Zettler (1970) reports that electrodes with tip outside diameters of 0.1 μm as measured with an electron microscope had a resistance of 8.8 megohms when filled with 3 M KCl. Snell (1969) reports a 10 megohm resistance for 0.1 μm o.d. tips. Tomita (1969) shows an electronmicrograph of a micropipette about 0.04 μm in diameter with a tip resistance of about 200 megohms when filled with 2 M KCl. Thus, as the tip diameters decrease by a factor of two, the resistance increases by a factor of 10–20. Since the tip resistance is proportional to the conductivity of the electrolyte, an electrode which has a resistance of 9 megohms when filled with 3 M KCl will be about 20 megohms when filled with physiological saline (0.9% NaCl) and between 40 and 80 megohms when filled with some of the dyes commonly used for depositing intracellular marks (Zettler, 1970; Davis, 1970).

In attempts to record intracellularly from the receptor cells of the frog, 30 mV amplitude intracellular spikes could be recorded occasionally when the tip resistance was greater than 200 megohms (Gesteland and Waloga, 1968). Similar results are obtained from the mudpuppy with 100 megohm electrodes. These electrodes were filled with 2.5 M KCl.

Figure 3 illustrates the response time for two different micropipette electrodes. The amplitude of the pulse is proportional to the electrode resistance. The rise time of the pulse measures the time delay due to the properties of the electrode. Trace A results when a 14 megohm micropipette is connected to a FET recording amplifier with a six-inch bare wire and the negative capacitance compensation is not used. The amplifier pass band is from d-c to 10 kHz. The rise time is 0.5 ms. Trace B shows the same electrode with the amplifier adjusted for optimum capacity compensation. The rise time is < 0.1 millisecond. Trace C shows a 250 megohm electrode with no capacity compensation. The rise time is 5 milliseconds. Trace D shows this electrode with optimum capacity compensation. The rise time is 1.5 milliseconds.

The tip potential E_t is diagnostic of another sort of trouble that occurs with micropipettes. It is common practice to select for use from a group of micropipettes of the same resistance those with the lowest tip potentials. Tip potentials in ordinary micropipettes of less than 5 mV are easily achieved.

Tip potentials, like resistance, increase rapidly as the tip diameter becomes smaller than 0.1 μm. The tip potential also increases as the ratio of potassium ion concentration in the electrode to potassium ion concentration in the fluid outside of the electrode tip increases. Snell (1969) suggests that the surface of the glass inside of the tip behaves like a potassium electrode. Small pipettes, therefore, measure some combination of the voltage that would be measured by an ideal large electrode and the voltage that would be measured by a potassium ion-selective electrode. Potassium ion concentrations are generally not known for tissues composed of small cells and restricted extracellular space. Therefore, there is no easy way to correct for the contribution of ion-selective processes to the voltage measured by the electrode.

It is often desired to pass d-c current through the microelectrode tip for the purpose of changing the membrane potential of the impaled cell or to eject a marker dye into the cell. At low current levels micropipettes behave like resistors, passing positive and negative current equally well. As the magnitude of the current increases, the tip begins to rectify, presenting a higher resistance to cathodal current than to anodal current. Micropipettes with large tip potentials show this non-linear behaviour at lower current levels than do those with small tip potentials.

Figure 2D shows the effective equivalent circuit for a very small micropipette connected by ideal metal electrodes to the recording amplifier which is optimally compensated. The tip resistance, tip capacitance and tip potential cannot be ignored and their effects

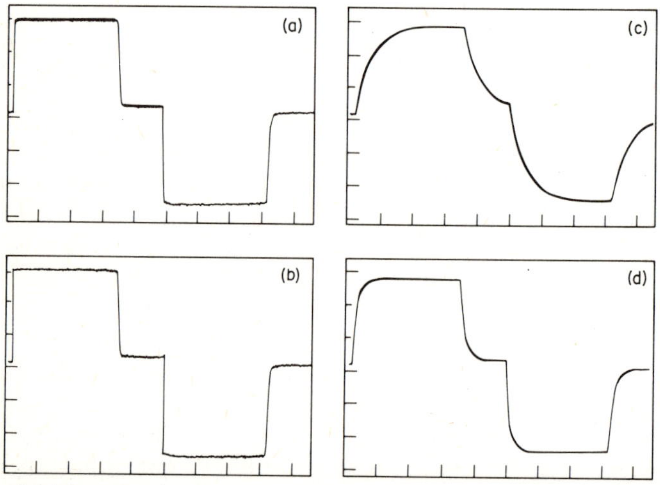

FIG. 3. (a) The response time of a 14 megohm micropipette to a square pulse with no negative capacitance compensation shows a rise time of about 0.5 ms.
(b) With optimum compensation, the same electrode will have a response time faster than 0.1 ms.
(c) A 250 megohm micropipette has a rise time of about 5 ms with no compensation.
(d) With optimum compensation, the rise time is no better than 1.5 ms. Thus, any voltage change measured with this electrode will appear to be longer lasting and of lower amplitude than it in fact is.
The horizontal axis calibration is 5 ms per large division. The vertical axis is 5 megohms per division in (a) and (b) and 100 megohms per division in (c) and (d).

must be considered when interpreting measurements on the biological preparation.

The properties of small electrode tips depend upon time. Tomita (1969) finds that electrodes with very small tip diameters work best when used the same day that they are filled. An electrode will commonly have a relatively high resistance shortly after it is filled. It will drop by a factor of two or three as it stands for a few minutes or hours. It will stay at this resistance for a period of hours or days and then may either gradually increase or decrease. There are at least three processes that contribute to these ageing effects. Glass dissolves in water, particularly in deionized water. Glass is etched more by basic solutions than by neutral or acidic ones. These processes will decrease tip resistance. Ability to record intracellularly from small cells decreases as the very thin glass at the tip etches away and the tip becomes more fragile and larger. Agin (1969) postulates that as glass hydrates it swells and that a very small micropipette is more like a leaky glass membrane than an open-tipped pipette. This process would increase tip resistance and make the electrode ion-selective in its response. The fixed charges present on the inside glass surface may also attract foreign materials in the filling solution which would tend to occlude the tip over a period of time. Lastly, if diffusion or a capillary fibre are involved in the filling procedure, time will be required for the process to near completion. This time may be very long (i.e. days) for ultra-small micropipettes with long tapers. The effects of silicates in water solution which have been studied in the controversy over "polywater" may also contribute to the anomalous properties of very small microelectrodes since the channel dimensions are like those in which "polywater" was discovered.

Silver wires and plates are commonly used as a connection to the micropipette and as the indifferent electrode making contact with the preparation. These must have a coating of silver chloride to make them reversible and therefore accurate indicators of electrochemical potentials. Non-reversible electrodes (commonly called polarizable electrodes) act like capacitances. Their voltage changes as charge accumulates and disperses at the electrode-electrolyte interface, and they are therefore not useful for measurement of d-c voltages. Silver chloride may be deposited on the silver surface electrolytically or by dipping it in molten salt. For the electrolytic method, two silver electrodes are immersed in a solution of sodium chloride. A battery and a variable resistance are connected between them. Current is adjusted so that a grey film is visible after a few minutes on the metal piece connected to the positive battery terminal. The light coloured coating of silver chloride gradually darkens upon exposure to light during the hours following the deposition. Alternately and somewhat simpler, silver chloride salt can be melted over a flame in a crucible. The silver wire or plate to be chlorided is dipped in the molten salt a few times to form the coating. In either case, the silver should be cleaned with abrasive paper or steel wool before it is chlorided.

More elaborate ways of preparing reversible silver chloride electrodes have been described and all are satisfactory.

C. Metal microelectrodes

Extracellular records of single olfactory receptor cell spikes have been obtained only with electrodes which are modified versions of the platinum black-tipped, metal-filled glass pipette described originally by Dowben and Rose (1953). Both a glass-coated, pointed, tungsten wire plated with platinum black and a metal alloy-fitted glass pipette plated with platinum black have been useful in recording from single frog olfactory receptors (Gesteland, 1961; Gesteland *et al.,* 1965; Altner and Boeckh, 1967; O'Connell and Mozell, 1969; Getchell, 1969). The latter form of the electrode has also proved to be successful in the tortoise (Matthews and Tucker, 1966), the catfish (Getchell, 1972), the vomero-nasal organ of the ground skink (Müller, 1972), and the olfactory receptors of the mudpuppy (Gesteland, 1971).

At any temperature above absolute zero, electrical noise will be present in any electrode system because of random motion of charged carriers. The greater the resistance and the greater the temperature, the greater the noise. There are additional sources of noise as well, but the resistance and temperature set a lower limit. The extracellular voltage change due to a nerve spike travelling along an unblocked axon is proportional to the second derivative of the membrane voltage change. This second derivative is tri-phasic and thus changes more rapidly than the action potential itself and is small. An electrode which has minimum noise in the upper frequency range of physiological signals and extended high frequency response will be most likely to record passing action potentials extracellularly. This is a unique property of platinum black metal microelectrodes.

Methods for constructing the glass-coated tungsten platinum black electrode and the alloy metal-filled pipette platinum black microelectrode in the form used for recording from olfactory receptor cells have been published (Baldwin *et al.,* 1965; Gesteland *et al.,* 1959). Since the latter electrode in the hands of the author is somewhat more effective for olfactory receptors, a detailed description of its preparation follows.

Glass micropipettes are pulled from hard glass on any microelectrode puller. Most often, commercially available tubing with a 2 mm o.d. and a 1 mm bore is used. The tip of the pipette is then broken off under microscopic observation by bumping it against the edge of a glass slide so that its outside diameter is 2–5 μm. It is a common failing when first working with these electrodes to make the tips too small. The glass tip is less likely to crack in the subsequent filling operation if the break is square and clean. Two extreme sorts of pipette tips can be realized. These are shown in Fig. 4. The uniformly

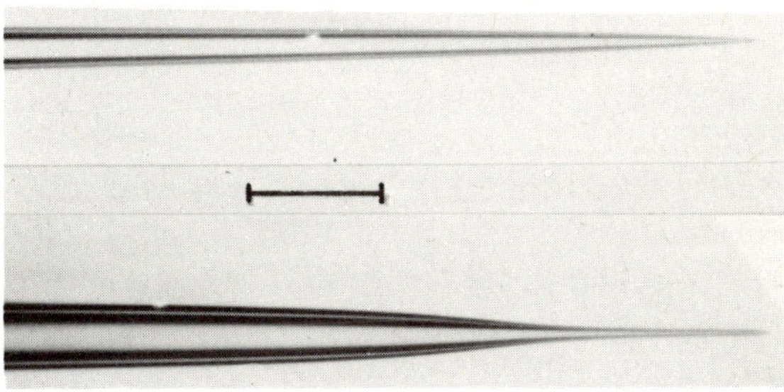

FIG. 4. Glass tubing can be pulled down into micropipettes with a wide variety of shapes. The upper photomicrograph is of a pipette with a nearly straight tapering form. The lower pipette is in the shape of a series of unduloids of decreasing periods and diameters. The straight taper shape is most useful for ultra-fine tipped micropipettes. The unduloid form is most useful for metal-filled microelectrodes because the tip can be broken off at a convenient diameter at one of the constrictions resulting in a clean break rather than a jagged end. The graticle in the photograph consists of lines spaced at 10 microns. Both types of tips appear to be of unduloid form at the very tip (which is not visible at this magnification). The appearance at the tip is due to diffraction patterns and the actual shape is more accurately inferred from the form of the broken end than from the properties of the taper.

tapering shape is most easily pulled but does not break as cleanly as does the unduloid. The unduloid shape is one of Plateau's volumes of minimal surface (Thompson, 1942), and the appearance of the tip of the glass micropipette is that of a series of constrictions or steps superimposed upon the continuous taper. When glass in such a shape is broken it tends to break at one of the constriction points. By suitable adjustment of the electrode puller, a constriction can be made to occur which will allow a clean break at a convenient diameter.

In order to fill the pipette with metal, an alloy must be used which will wet the glass surface when molten and which will not crack the glass because of volume changes upon cooling. A convenient material to use is Cerrelow 136, which is used in machine shops for die making. Its melting point is 58°C (136°F) and its composition is 49% bismuth, 21% indium, 18% lead and 12% tin. It is available from Cerro de Pasco Sales Corporation, 300 Park Ave., N.Y., N.Y. and from many machine tool supply houses. A metal slug with a diameter slightly less than the inside diameter of the glass tubing and about 1 cm long is needed. A big chunk of the alloy is melted in a beaker on a hot plate. Lengths of polyvinyl chloride tubing about 60 cm long are warmed by letting them lie on a hot plate surface at a temperature of about 60°C. Slugs which will fit into 1 mm i.d. glass tubing can be made in size 20 plastic spaghetti tubing which is sold by electronic supply houses. (Alphlex Tubing Division, Alpha Wire Corporation, Type PVC-105, size 20 plastic tubing). A piece of warmed tubing is quickly picked up, one end is inserted into the molten metal and the liquid metal is sucked up to nearly fill the tube. The tubing is quickly stretched and held firmly until the metal sets. The unfilled and distorted ends of the plastic and its metal core are cut off. The plastic then relaxes back to its unstretched larger diameter and the alloy wire is pulled out of the plastic sleeve and cut into short lengths. Since the surface of the alloy oxidizes, electrodes are most easily filled if the alloy places are no more than a week or two old.

After the glass pipette has been pulled to shape and the tip broken to the desired size, a piece of the alloy metal is dropped into the large end and a stiff copper wire is inserted behind it. The assembly is placed on a hot plate just warm enough to melt the alloy. The wire acts as a heat sink so that the alloy close to it remains hard and the alloy close to the electrode tip begins to melt. Gentle pressure is applied to the wire to extrude the alloy towards the pipette tip. Pressure is removed just as the alloy reaches the tip orifice. Alloy must not come out of the tip as this produces cracks in the glass resulting in noisy electrodes. A light can be adjusted so that it reflects off the alloy so that it can be watched without the aid of a microscope as it moves down the glass tube, just as one would watch mercury in a thermometer column. The hot plate temperature is critical. It must be warm enough so that the metal alloy begins to move towards the tip after about a minute of heating. If the plate is too hot, dimensional changes in the alloy upon cooling cause the glass near the tip to crack. For the filling process, the electrodes should be so positioned on the hot plate that by the time the extruded alloy reaches the electrode tip, the copper wire will have reached alloy melting temperature and will melt into the alloy. When the electrode is filled, it is gently removed from the hotplate. The alloy sets and solders to the copper wire. Sometimes, the metal does not quite reach the end, or vacuoles occur in the alloy column. Brief warming by passing the electrode near a hot light bulb will sometimes repair these troubles.

The copper push wires are made from a conveniently loose-fitting size of bus wire, also available from electronic supply houses. No. 20 or 22 U.S. wire gauge works well with 1 mm i.d. glass tubing. Lengths of this wire about one metre long are straightened by clamping one end in a vice and pulling on the other with pliers so that the wire stretches slightly beyond its yield point. It is then cut into convenient lengths, typically 5–8 cm.

Shortly before the electrode is to be used, platinum black is electrodeposited onto the tip as a small ball. Figure 5 is a scanning electron micrograph of the electrode tip. This is accomplished by passing a d-c current between the alloy-filled micropipette and a bright platinum wire immersed in a solution containing chloroplatinic acid. The plating current should be very small initially (about 0.1 μA) in order to form a low resistance surface. It can then be increased so that as the plating process is observed under a microscope, the platinum black ball forms in a few seconds. A three-volt battery with a resistor variable between 20 kilohms and 5 megohms can be used to provide the plating current. The negative terminal of the battery is connected to the electrode upon which the platinum black is to be deposited. The plating solution contains 1% chloroplatinic acid, a trace of lead acetate, and either gelatin or agar. If gelatin is used, 2 ml of a solution which is viscous but ungelled at room temperature is added to 100 ml of 1% chloroplatinic acid. If agar is used, it should be present at a concentration of about 0.05 per cent. Platinum black-plated microelectrodes will not record triphasic spikes from unmyelinated nerve fibres if gelatine or agar is omitted from the plating bath. Platinum black is an active catalyst for organic reactions and when it is placed in contact with biological tissue, it apparently causes a poorly conducting surface coat to form on the tip. Such electrodes record only the larger monophasic or diphasic spikes of blocked fibres. Almost any electrode can record these signals. However, it is hard to separate the activity in neighbouring cells when there is a conduction block. Even when it seems clear that

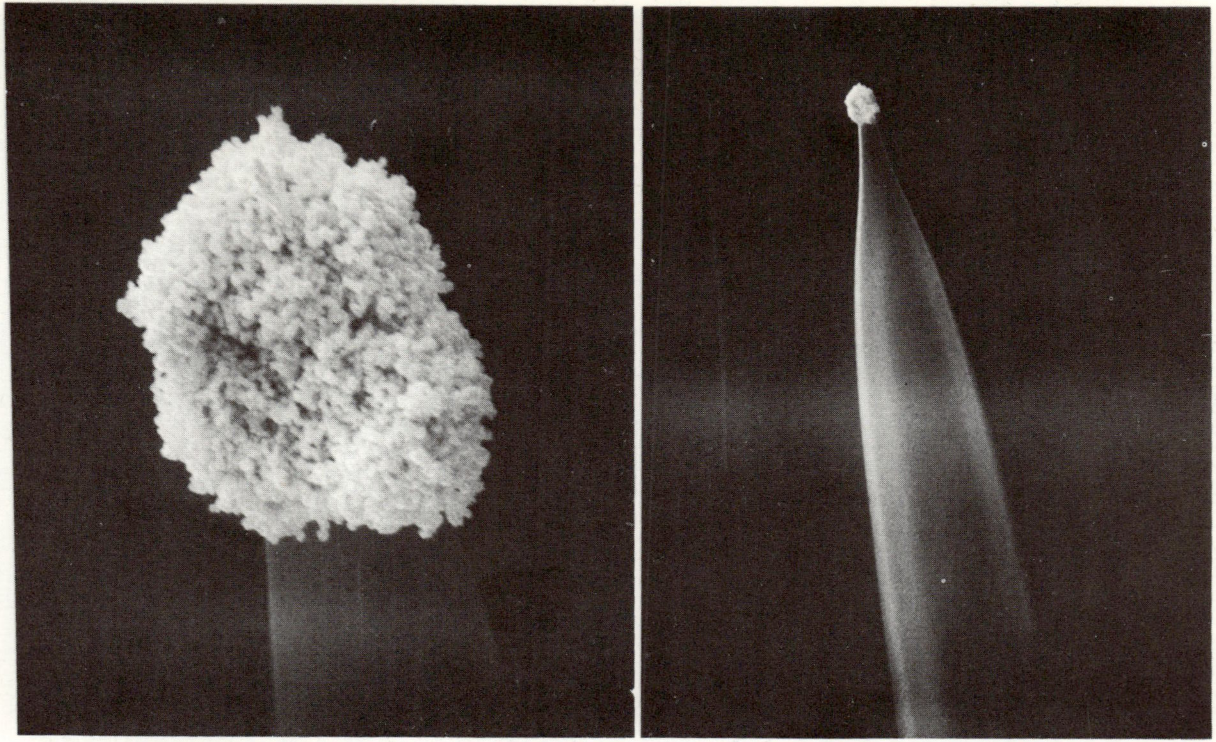

FIG. 5. The surface of a platinum-black tipped metal microelectrode is clearly seen in this photomicrograph taken with a scanning electron microscope. This electrode is about seven microns in diameter, slightly too large for useful studies of single axon spikes in olfactory receptor cell neurons.

the blocked spikes are from a single unit, they soon are lost. When agar or gelatine is present during the plating, the high surface area and resulting low impedance of the platinum black is retained, but its catalytic activity is lost. The electrode will record the triphasic voltage which characterizes a propagated action potential. Small movements of the microelectrode will then serve to make the signal recorded from one cell larger than those signals from neighbouring units.

It is best that the plating be done just before the electrode is to be used. This is because the tip resistance and, hence, the background noise of a plated tip increases with age. A previously plated electrode will work if a thin layer of platinum black is deposited just before it is used. If multiple platings make the tip too large, it may be removed by pulling the electrode between one's fingertips. A fresh ball of the desired size is then plated on. Some tissues, including frog noses, are sticky, and tend to pull the ball off the tip as the electrode is withdrawn from one recording location and removed to another one. The platinum black can be made more adherent if a layer of bright gold is electroplated onto the microelectrode before the platinum black is applied. The most adherent gold plate is formed from a plating solution called Dalic Process Alkaline Gold, Code 3020. It is available from Sifco Metachemical Co., 935 East 63rd St., Cleveland, Ohio, 44103. The gold is applied by passing a small current for a few seconds exactly as in the initial platinum black deposition process.

Sometimes, the electrodes are difficult to plate, probably because of contaminating films on their surfaces. This difficulty can be overcome by passing a brief pulse of current in the reverse direction from that used for plating. If an electrode will not plate, it means that the alloy has not filled the glass tube all the way to the tip, or that there is an air bubble in the electrode shank. After plating, the tip should be rinsed in distilled water and carefully examined as it is rotated on a microscope stage. Fuzzy regions on the glass surface near the tip indicate that the glass is cracked and the electrode will be noisy and useless for single cell studies. Such an electrode should be discarded.

The best results with these electrodes are achieved when they are constructed and plated as they are needed during the course of an experiment. If the electrodes are made before the experiment, slow phase changes in the alloy cause a significant fraction of those tips which would have worked when fresh to develop cracks in the glass which render them useless. I usually make a few electrodes before I begin the experiment, but expect that I will have to make additional ones during the course of a recording session. A good electrode can be rinsed and used another day if fresh platinum black is applied. Adherent mucus can be removed with a solution consisting of equal parts concentrated sulphuric acid and 30% hydrogen peroxide and a trace of silver nitrate.

Both the resistance and capacitance of these tips are functions of frequency, as well as of the tip surface area. Therefore, no simple measuring method is available which can be used routinely to separate good electrodes from bad ones. The most useful procedure makes use of wide band noise generated by the electrode. The electrode is capacitively coupled to a high impedance input amplifier which has a 20 megohm resistor connected between the input terminal and ground. The amplifier high frequency cut off should be set at 10 kHz and the low frequency cut off at a convenient value between 20 and 100 Hz. With the electrode in air the noise from the 20 megohm resistor will produce about 100 microvolts peak-to-peak signal on the oscilloscope. When the electrode is advanced so that the tip contacts extracellular fluid, the peak-to-peak noise should drop dramatically. For electrodes which work well in the frog olfactory nerve, the noise will look about like that which would be seen if the input resistance to the amplifier had suddenly decreased to a value near 100 kilohms, i.e. about 20 microvolts peak to peak. Large noise spikes or oscillations are due to cracks in the electrode tip and the electrode must be rejected. This testing procedure will work, of course, only when the self-noise from the amplifier is well below that to be expected from a good electrode.

To understand why metal electrodes work for extracellular spikes but not for intracellular resting potentials, it is useful to consider their impedance properties. From physical chemistry we find two extreme kinds of electrodes. The first is the reversible type, in which ions from the solution are charged and discharged, so that a steady current is possible and the d-c potential of the electrode has a well-defined value depending on the current and the composition of the solution. The second type is the ideal polarized electrode, in which no reduction and oxidation of ions takes place, no steady current can pass, and the transient currents which flow represent charging and discharging of a double layer made up of the electrode and ions very close to its surface. These form a structure which acts like a capacitance whose value is dependent on the voltage across it. This electrode has no well-defined d-c potential. It may vary widly under apparently identical circumstances, and is enormously influenced by traces of impurities. Actual electrodes always combine both ideal types, and their impedance as a function of frequency shows the extent to which one or the other mechanism dominates their behaviour.

A bright platinum electrode in the spinal cord or in spinal fluid, both rich in absorbable compounds which presumable retard oxidations and reductions at the interface, behaves like a capacity even at low frequencies. Its impedance is proportional to ω^{-1}, over a very wide range (ω is angular frequency). If the electrochemical reaction is itself quite rapid, and the current is limited by the diffusion of the reacting ions and the products between the surface of the electrode and the bulk of the solution, the impedance for high frequencies will

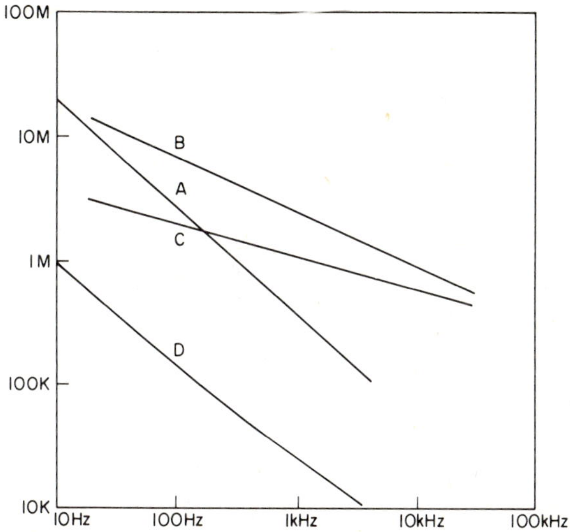

FIG. 6. Curves showing the variation of resistive component of metal electrode impedance with frequency. These are normalized for apparent surface areas of approximately 50 square microns. Curve A is bright platinum in spinal fluid. The slope is slightly less than proportional to ω^{-1}. Curve B is for bright platinum in saline solution. The slope is nearly proportional to $\omega^{-1/2}$. Curve C is for chlorided silver in saline solution. The slope is approximately proportional to $\omega^{-1/4}$. Curve D is for a platinum black ball in saline. The slope is proportional to $\omega^{-0 \cdot 8}$. The capacitive components of impedance vary in approximately the same way with frequency as do the resistive components. For bright platinum in spinal fluid (A) X_c is about 10 times R. For bright platinum in saline (B) X_c is 0.9 times R. For chlorided silver in saline (C), X_c is 0.3 times R. For platinum black in saline, X_c is about 5 times R. The relatively low values of impedance components and steep slope of the platinum black curve show why it is so effective in recording transient events of very low amplitude. The contributions of the electrolyte solution to the measured resistances have been subtracted to yield the data plotted.

decline as $\omega^{-1/2}$. If the electrochemical reaction is slow, the impedance will be lower for very small ω and higher for very large ω than in cases where diffusion is predominant. A bright platinum electrode in physiological saline behaves this way.

In practical cases, it happens that in homogeneities in the electrode material will spread out the band of frequencies for which the impedance declines only slowly, since the rate of the reaction is strongly dependent on the d-c potential of the electrode, and this will vary over the surface. In cases where an insoluble reaction product covers the electrode, the multiplicity of diffusion paths of different length may have a similar effect. Such a case is the Ag-AgCl electrode whose impedance is approximately proportional to $\omega^{-1/4}$ over a wide range.

In Fig. 6 the resistive and reactive components of the impedance of several kinds of metal electrodes are plotted as functions of frequency. The platinum black ball is clearly superior to the others. Its impedance components are least, even though its size is smaller by a large factor.

Signals can be corrupted in two ways by electrodes such as these. Because the impedance properties of the tip are frequency-dependent, amplitude distortion of a signal will occur. Because the impedance has a resistive component, noise will be introduced. The circuit schematic in Fig. 7 shows an electrode with a tip resistance Z_e connected to an amplifier with an input impedance of Z_l. At any particular

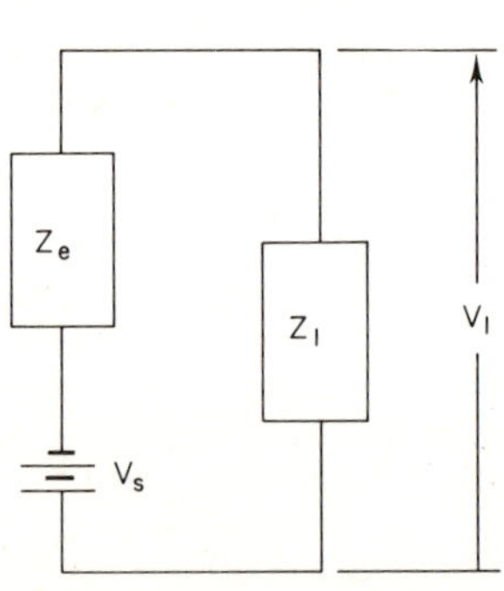

FIG. 7. The equivalent circuit of any voltage measuring scheme is shown here. V_s is the voltage to be measured. Z_e is the impedance of the measuring electrode. Z_1 is the input impedance of the measuring device (amplifier). V_1 is the measured voltage. The measured voltage is less than the true source voltage and can be calculated for any frequency from the equation:

$$V_1 = V_s \frac{Z_1}{Z_1 + Z_e}.$$

frequency the voltage input to the amplifier will be related to the source voltage by the equation for a voltage divider:

$$V_l = V_s \frac{Z_l}{Z_e + Z_l}.$$

Z_e is a function of frequency and will be large at low frequencies. Thus low frequency components in a signal will be attenuated more than high frequency components. The steeper the slope of the electrode impedance versus frequency curve, the more extreme the distortion. If there were no noise in the system, it would be possible to compensate for this distortion by passing the signal through a filter which emphasizes the higher frequencies to the same extent that a particular recording electrode diminishes them. However, all physical systems at temperatures above absolute zero inherently produce electrical noise. The rms noise voltage generated by a metal microelectrode in a given narrow frequency band is most easily specified in terms of the equivalent noise resistance at room temperature, R_n. The relation between these quantities is

$$E_{\text{rms noise}} = \sqrt{4\text{kTRn}\Delta F}$$

Here k is Boltzmann's constant and T is the absolute temperature. Nyquist's Thermal Noise Theorem states that R_n should be equal to the real part of the electrode impedance at the same frequency. The proof of this theorem assumes that the system is in thermal equilibrium, which would in this case require that the current through the electrode be zero, and that there be no chemical reaction taking place at the surface of the electrode. These conditions are difficult to satisfy in practice but the equation does set a minimum value for the noise that can be expected. When the resistive component of a platinum electrode is measured by conventional a-c bridge techniques the value is found to agree with high accuracy to the resistance determined by measuring the narrow band noise voltage as a function of frequency generated by the electrode. The noise voltage is high at low frequencies and is larger for smaller tips. Low frequency components of the signal to be measured will be obscured by noise. Any technique which reduces the resistive component of impedance without increasing the tip size increases the likelihood of detecting small, slowly-changing voltages. The superiority of the platinum black electrode at frequencies above 300 Hz with respect to noise is evident from the curve showing the resistive component of its impedance in Fig. 6.

None of the classical reversible electrode systems has been demonstrated to reliably measure d–c voltages in living tissue as a microelectrode. These electrodes are not useful for intracellular measurements, even if they are fabricated with sufficiently small tips. The Ag-AgCl electrode fails to measure (Cl-), apparently for a number of reasons. First, the solid AgCl diffuses away from the tip of a microelectrode with great speed and can hardly be continuously formed with an imposed current. This will be the difficulty generally for very small electrodes requiring saturation of the neighbouring solution with a sparingly soluble salt, since the time of diffusion varies with the square of the linear dimensions of similar structures. Second, silver forms extremely stable complexes with organic molecules having attached amino and sulfhydryl groups. The stability of almost all of these is greater than the insolubility of AgCl. This difficulty is more important than the other, for it applies to almost all heavy metals one might hope to use: Hg, Sb, Bu, Pb, Dc, Cu, etc. Finally, the reduction-oxidation potential of axoplasm is low enough to reduce methylene blue, which places it below hydrogen. At this potential AgCl is reduced to metallic Ag and Hg_2Cl_2 to Hg, the insolubility not being adequate to the nobility of the metals. This difficulty also applies to other heavy metals and to oxide electrodes (Sb and Bi) used to measure pH.

The inert or noble metal electrode (the polarizable type) should be free of these difficulties and measure the reduction-oxidation potential of the solution. The difficulty here is that the oxidations and reductions take place only through enzyme-substrate complexes of high molecular weight, activation energies being too high to permit any equilibrium to be attained otherwise. The consequence is an impedance far too high to make the small electrodes usable at d-c or very low frequencies even with the best control of the leakage current inevitably imposed on the electrode by the measuring system. If there were a rapidly reversible system present to buffer the potential, such as $Fe^{2+} - Fe^{3+}$, this could be avoided, but no inorganic system seems usable, and organic ones seem to be all of the quinoid type which is absorbed to proteins with great avidity. Thus, no metal microelectrodes appear to able to measure d-c voltages in small cells.

We conclude then, that metallic microelectrodes will not be useful for measuring resting potentials and slowly changing voltages. They are high pass filters. Beceause of their dramatically lower resistance at high frequencies they will be much superior to micropipettes for recording extracellular transient voltages. So it is that the platinum black pre-poisoned electrode is the only one that has proven to be useful for studying extracellularly action-potentials in single olfactory nerve axons.

III. Input amplifiers and electrode tests

Fluid-filled micropipettes are d-c stable low pass filters and inherently noisy due to their high resistance, while metal microelectrodes are high pass filters which are not d-c stable and can have very low wide band noise. Physiologists commonly use separate amplifiers for the two electrode types. The amplifier for fluid-filled micropipettes must be direct-coupled, needs negative capacitance compensation circuitry and will sacrifice signal-to-noise ratio in order to achieve the highest possible input impedance and lowest input leakage current. The combination of electrode offset voltage and membrane resting potential can be relatively large, and therefore the amplifier has to be able to handle input signals of about one volt. The amplifier for metal micropipettes can be a-c coupled. In fact, the microelectrode must be connected to the amplifier through a capacitor so that whatever d-c leakage current occurs at the amplifier input cannot flow through the microelectrode. This amplifier must be optimized for lowest possible noise. This is often accomplished by sacrificing very high input impedance and d-c stability. The amplifier should have good high frequency response but need not have a large dynamic range since the input signals will never exceed one hundred millivolts and are more usually less than one millivolt peak-to-peak.

An admirable alternative to the two amplifier requirement is an input coupler designed by Lettvin (1969). This amplifier has nearly optimal noise performance for use with metal microelectrodes, and accomplishes this without the sacrifice of features required for micropipette studies. The circuit and a few notes regarding its construction are included in an appendix to this chapter. It will be useful here to give its performance specifications in some detail as a guide in selecting amplifiers for those who chose to use commercially available units. The specifications for the version of the circuit which has been in regular use in my laboratory are as follows.

When the circuit is constructed to have a gain of ten, it has a dynamic range of plus or minus one volt at the input. The output level is adjustable over a range of plus or minus 0.5 V with respect to ground referred to the input. The noise with the input grounded over a bandwidth from d-c to 100 kHz is less than 25 microvolts peak-to-peak referred to the input and the drift for constant supply voltages and room temperature is less than 20 microvolts/hour. From a low impedance source with a 20 megohm impedance and optimum adjustment of the negative capacitance, the rise time is between three and five microseconds. The noise is 500 microvolts peak to peak referred to the input for a 100 kHz bandwidth. The leakage current at the input can be adjusted to be less than 10^{-12}A and will be stable at its adjusted value over a long period of time (weeks). This current is independent of the voltage applied to the amplifier input, i.e. it does not vary as the signal being measured changes value. It is possible to work with the current adjusted to less than 10^{-14} amperes for shorter periods of time.

If only metal electrodes are going to be used, any low noise a-c coupled source follower, emitter follower, or cathode follower amplifier is adequate. The noise with the input terminal short-circuited must be less than 30 microvolts peak-to-peak referred to the amplifier input. The overall gain of the entire recording system must be high enough so that a 50 microvolt peak-to-peak signal can easily be discriminated. It is convenient to have adjustable low and high frequency filters on the oscilloscope or the pre-amplifier.

When working with metal microelectrodes, they should be connected to the amplifier as shown in Fig. 8. There is a resistive path to

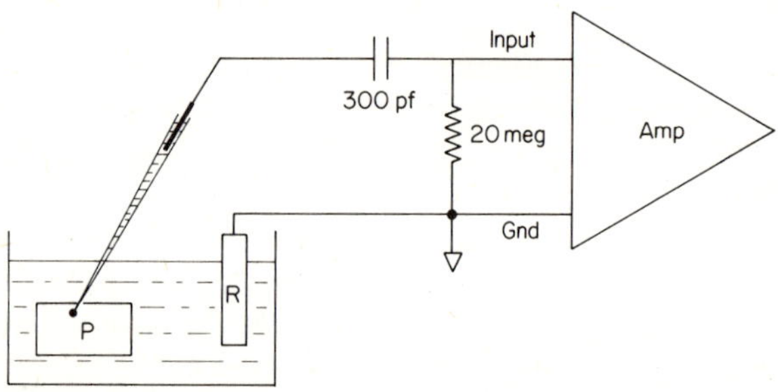

FIG. 8. Schematic diagram shows components used to couple a metal microelectrode to the measuring amplifier for recording extracellular action potential activity. *P* is the biological preparation. *R* is the indifferent electrode.

ground through which amplifier input leakage current can flow and a capacitor connecting the electrode to the amplifier input. Typically, a 20 megohm resistor is convenient with a capacitor chosen for a low frequency cut-off between 10 and 60 Hz, i.e. about 300 pf. Shielded wires in high impedance situations degenerate high frequency components in a signal and cause noise because of the inner wire moving with respect to the shield. An optimum recording situation is one where the amplifier is located very close to the electrode and connected to it by an unshielded wire. To avoid ground loops, each piece of apparatus in the experimental set up should be connected to the ground terminal on the pre-amplifier, and there should be only one ground path from the pre-amplifier to the oscilloscope and other equipment. I find it most convenient to use a battery powered pre-amplifier and to establish connection to the building ground through the three wire power cord on the oscilloscope which the preamplifier feeds. The ground connections between this oscilloscope and all other line operated instruments can be established by the shields on the signal leads interconnecting them. No other instruments should be connected to the building ground through their three-wire power cords (a three-wire to-two-wire adaptor should be used on each of these power cords with the green pigtail on the adaptor floating). To test for ground loops, the input terminal on the pre-amplifier should be connected directly to the ground terminal on it. No trace of 60 Hz or 120 Hz signal should be seen on the oscilloscope. No trace means less than 20 microvolts peak-to-peak.

For work with micropipettes of extremely small tip size most any commercially available negative capacitance d-c coupled amplifier is adequate. The lower the input or leakage current and the higher the input impedance, the better. Negative capacitance must always be used with very high impedance electrodes. Since negative capacitance circuitry adds noise in proportion to the magnitude of the capacitance to be compensated, the experimental apparatus should be arranged to keep stray capacitance around the electrode and amplifier to an absolute minimum. This means a very short connecting lead from the amplifier to the electrode and no shield around the electrode and its connecting leads. It is useful to connect a one pf capacitor and a 10 or 20 megohm resistor to the amplifier input terminal, keeping the leads as physically short as possible. The one pf capacitor is used to introduce a signal for the purpose of measuring the electrode resistance. The resistor is for calibration of the measuring circuit. For calibration, the microelectrode is raised so that it does not touch the preparation, and a clip lead is used to connect the open end of the resistor to ground. For recording, this clip lead is removed and the microelectrode is advanced to touch the preparation.

It is most convenient to measure micropipette resistance by a method which will not compromise amplifier input impedance and will not provide a d-c path to ground shunting the electrode. Such a measuring apparatus need not be disconnected from the amplifier during the course of an experiment and therefore can be used to continuously monitor electrode resistance. This measurement is accomplished by coupling a triangular or trapezoidal voltage source to the electrode through a small capacitor (Lettvin *et al.*, 1958). This measurement scheme has the further advantage of providing a method for optimum adjustment of the negative capacitance circuit.

Figure 9 shows the arrangement for measuring micropipette resistance. An adjustable amplitude trapezoid wave form is connected

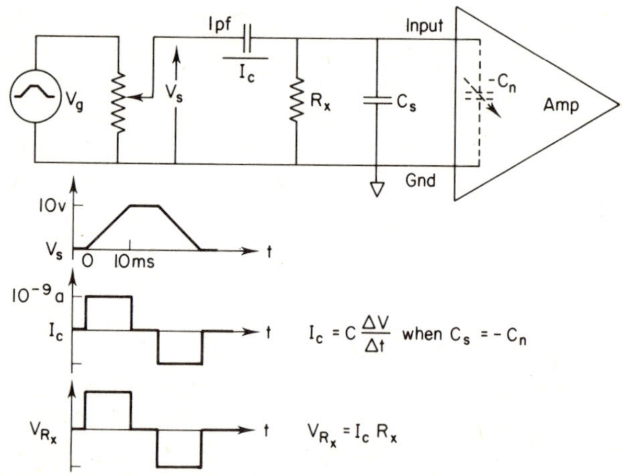

FIG. 9. A. Schematic diagram shows method for connecting trapezoid waveform generator and negative capacitance amplifier to measure the resistance of fluid-filler micropipette electrodes. V_g is the trapezoid or triangle waveform generator (for instance, see Appendix II). R_x is the unknown resistance of the electrode. C_s is the capacitance of the connecting wires and the electrode tip. $-C_n$ is the negative capacitance circuit of the amplifier. The current flowing through the 1 pf coupling capacitor, i_c, also flows through the electrode, R_x, when C_n is adjusted to be equal in magnitude to C_s. The potentiometer, R_s, is used to adjust the source voltage to calibrate the system for a known value of resistance, R_x.
B. The expected values of i_c and the voltage across R_x are shown as a function of time for a 10 volt peak amplitude trapezoid source and an electrode resistance R_x of 10 megohms.
The oscillograms in Fig. 3 were made using this circuit (the numerical values were different, however).

through the one pf capacitor to the input terminal of the electrometer amplifier. R_x represents the resistance to ground which is to be measured. C_s represents the stray capacitance from the input terminal of the amplifier to ground. C_n is the adjustable negative capacitance circuit in the amplifier. When C_n is adjusted so that it equals C_s, their algebraic sum is zero and they vanish from the equivalent circuit. Under these conditions, the one pf capacitor and R_x form a differentiator. The current through the capacitor is the derivative of the source voltage and the voltage across R_x is the product of this current and R_x. Under the condition that the voltage across R_x is small compared to the source voltage (this is equivalent to requiring that the impedance of the coupling capacitor be very much larger than R_x), the electrode resistance is measured by the amplitude of the rectangular pulses which are the derivative of the trapezoid or triangle source voltage. To use this measuring scheme, a known resistance, say 10 megohms, is substituted for the microelectrode, R_x. Assume that trapezoid takes 10 ms to go from zero to maximum amplitude. From the relation $i = C\frac{dV}{dt}$, the current through the coupling capacitor will be 10^{-9} amps during the ramp portion of a trapezoid with a peak amplitude of 10 volts. From Ohm's law the voltage across a 10 megohm resistor will consist of alternating positive and negative 10 millivolt amplitude pulses. In effect, we have calibrated the amplifier so that it reads one millivolt per megohm. The 10 megohm known resistance is then removed; a microelectrode is put in its place and is lowered to touch the preparation. The negative capacitance control of the amplifier is adjusted to give the best possible rectangular pulse. The amplitude of the pulse will measure the electrode resistance and the rise time of the pulse will indicate the slowness of the response of the microelectrodes to a fast transient. The voltage source for this measurement can be generated by integrating the square calibration voltage available from most oscilloscopes to produce a triangular wave form. Alternatively, commercially available triangle or trapezoid wave form generator can be used. Appendix II shows a simple circuit which will produce a trapezoidal wave form from a pulse generator such as the Tektronix Type 161. If the gate voltage from the oscilloscope is used to trigger the trapezoid generator, a stable pulse display is obtained.

Conventional d-c bridge techniques or ohmmeter circuits are not useful for measuring micropipette resistances because they must be disconnected from the system while the micropipette is in use. Furthermore, most commercially available devices for measuring high resistances apply a large voltage to the unknown. Because a micropipette is an electrochemical system, the resistance is only defined under conditions where the system is operating reversibly, i.e. when the voltages are small compared to any electrochemical potentials of the system. Alternating current methods can of course be used to measure micropipettes resistance as long as the measuring frequency is low compared to the high frequency fall off of the pipette.

Metal microelectrodes are characterized by an impedance which is an inverse and variable function of frequency. Therefore, the trapezoid technique for evaluation electrode impedance is useless. It is important to have some convenient measure of this property, however, because for a given tip diameter, the lower the impedance, the better the possibility of picking up extracellular action potentials. A-C bridge techniques work well (Gesteland and Howland, 1959). However they are tedious because measurements must be made at several different frequencies. The easiest and quickest method for evaluating metal microelectrodes is to watch the noise as the electrode is brought in contact with the extracellular fluid of the preparation. Capacitance measuring methods such as the commonly used L-C meter technique are not such useful predictors of metal microelectrode effectiveness in recording small extracellular voltages because the resistive component of electrode impedance is ignored in these measurements.

There is an easy way to measure input current of the amplifier and this should be done periodically, since, in most equipment this will change as battery voltages change. The current through a capacitor equals the product of its capacitance and the rate of change with time of the voltage across it. Therefore, if a known value of capacitance is connected between the amplifier input and ground, the rate at which the voltage across it charges is the measure of the current flowing. The voltage across a one pf capacitor (10^{-12}f) will change at the rate of one V/s for a steady current of one pa (10^{-12}A). To make the measurement of the amplifier one need only temporarily connect a convenient size capacitance, say 100 pf, from the input terminal to ground. Then, using a clip lead, the capacitor is momentarily short circuited. The short circuit is then removed, and using a slow oscilloscope sweep rate, the charging rate in V/s can be directly observed. Of course, there must be no path to ground from the input terminal other than the 100 pf capacitor. Amplifier leakage currents larger than 10^{-11}A make work with very small tipped micropipettes inaccurate. If this leakage current varies with signal voltage, inaccuracies are compounded.

IV. Searching for single units

A. Intracellular recording

The search with intracellular micropipette electrodes for an olfactory receptor cell is straightforward because the signals are large, but very difficult because the cells are small and easily damaged when the electrode penetrates the cell membrane. As the electrode is advanced sudden changes in the d-c voltage occur. In most tissues, the shift will be in the negative direction when the electrode enters a cell. The magnitude of the shift is variable depending upon the amount of damage done by the entering electrode. When the electrode is in the cell and the cell is adequately stimulated, a depolarization or hyperpolarization will occur. The former, if it is large enough, will be followed by evoked action potentials or an increase in the firing rate in the case of a cell in which there is spontaneous spike activity. Hyperpolarization, on the other hand, will tend to suppress spontaneous activity. When the electrode is recording action potentials intracellularly, they are seen as positive-going monophasic transients of constant amplitude. Spike amplitudes recorded from mudpuppy olfactory cells can be as large as 50 millivolts and as small as 5 millivolts. There is no doubt that cell damage causes a reduction of recorded spike amplitude. As a micropipette is inserted into the olfactory epithelium, many d-c voltage shifts occur which are not accompanied by action potentials. Delivery of a stimulus often evokes a depolarization or hyperpolarization with a time course like that seen in receptor cells. However, no spike response is seen. It is a reasonable assumption that the electrode is in a cell other than an olfactory receptor cell in these cases. However, it is also clear from attempts at marking electrode locations by an ejected dye, that stable resting potentials and apparent generator potentials can be recorded from mucus layers and diffusion-restricted extracellular spaces. The only acceptable criteria which will establish that a micropipette is recording responses of olfactory receptor cells are therefore the occurrence of intracellular spikes or deposition of a dye which is subsequently histologically established to be localized in the interior of an olfactory neuron.

The sequence of events which usually occurs when a receptor cell is impaled is as follows (see Fig. 10). There is a sudden negative shift in d–c potential, typically 50 millivolts, followed by a burst of action potentials. These are regularly spaced and occur at a high rate, as much as 300 spikes per second. The first spike is large, 50 or 60 millivolts. The amplitude of successive spikes decreases rapidly, reaching an amplitude of a few millivolts in less than ten seconds. As the spikes decrease in amplitude, the firing rate decreases regularly and the spike duration increases. The first spike has a duration of two or three milliseconds. By the time the amplitude has declined to five millivolts, the duration has increased to about ten milliseconds. At this point, either of two things can occur. The amplitude of the spikes can continue to decrease as they fire in a regular fashion until they are lost in the noise. Alternatively, the spike amplitude may remain constant at around five millivolts and the spike firing pattern change

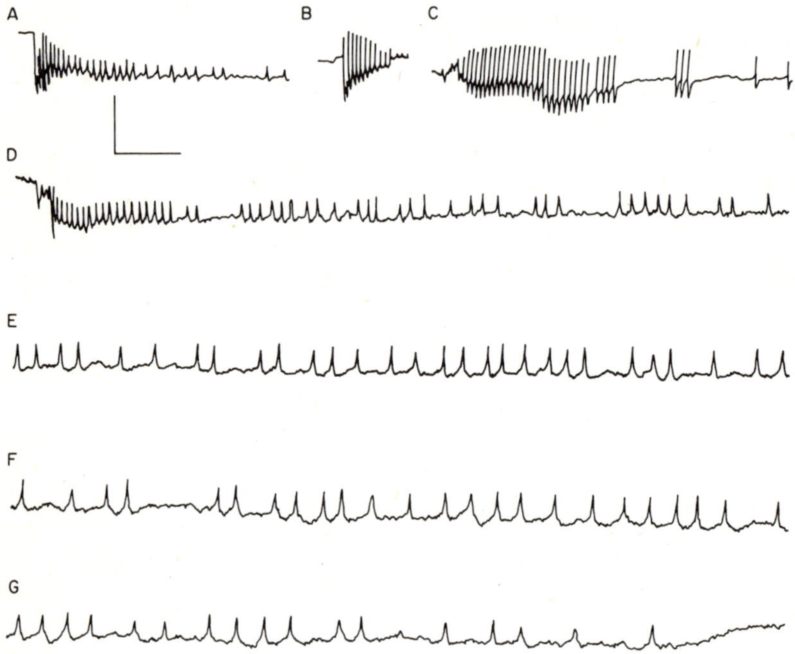

FIG. 10. The electrical events recorded when a 200 megohm micropipette electrode penetrates olfactory receptor cells in the nose of a mudpuppy are shown here. The top traces (A, B, C) show responses of cells upon penetration which do not lead to stable intracellular recordings. There is a burst of injury-evoked spikes which decline in amplitude accompanied by a decline in resting potential. The lower four traces (D, E, F, G) are segments of the record from a successful penetration. The regular burst which follows penetration (D) subsides, followed by the irregular interval firing pattern which is characteristic of the unstimulated activity of these receptors. After about 20 seconds the spike amplitude increases slightly (E), presumably due to cell membrane sealing around the electrode tip. The cell in this instance was lost after about a minute as the electrode came out of the cell (G). Variations in the interval between impulses do not correlate with the membrane potential recorded in the vicinity of the electrode. The calibration marks indicate 0.5 sec on the horizontal axis and 20 mv positive upward on the vertical axis.

from very regular interspike intervals to very irregular ones. These action potentials change from monophasic shape to diphasic as the intervals change from regular to irregular. The average rate will be about one spike per second. The intervals between successive spikes can differ by a large factor from one spike to the next. This pattern is like that seen with an extracellular electrode when it records from a single axon of a cell in the unstimulated state. A stimulus delivered to the nose produces a depolarization and an increase in spike rate if it is of appropriate kind and intensity. These observations are consistent with the hypothesis that regularly occurring spikes originate near the electrode tip because of local damage while the small irregularly occurring diphasic spikes originate in membrane far from the electrode location. They do not invade the cell membrane near the electrode. The electrode records events at the distant location electrotonically. If the electrode has penetrated the dendrite of the receptor cell, which is long and thin, and the spike originates in the axon on the other side of the large soma region, it ought to produce spikes of the sort observed.

B. Extracellular recording

The search for a single receptor cell with an extracellular metal electrode is tedious because the signals are very small and the cells are closely packed so that most often the microelectrode picks up activity from several neighbouring cells. The search must be done by watching the electrical record on an oscilloscope as the electrode is advanced into tissue. The noise becomes more spike-like when the electrode is near a cell. This is easily seen with a little practice. I find a display sensitivity of about 50 microvolts per centimetre and a sweep rate of 10 milliseconds per centimetre most convenient for searching. The amplifier is set for a low frequency cut off of 30 Hz and a high frequency cut off of 10 kHz. The more usual method of searching for cell spikes using an audio amplifier rather than visual observation of an oscilloscope does not work at all because the amplitude and frequency spectrum of the extracellular spikes as the electrode comes into the vicinity of the cell are very much like the amplitude and frequency spectrum of the noise of a metal microelectrode in extracellular fluid.

Monophasic, diphasic and triphasic spikes of all possible polarities are recorded by extracellular metal electrodes. Where an electrode tip just touches the mucosa surface, small, long, positive monophasic spikes are seen. These are shown in Fig. 11A. They come from many

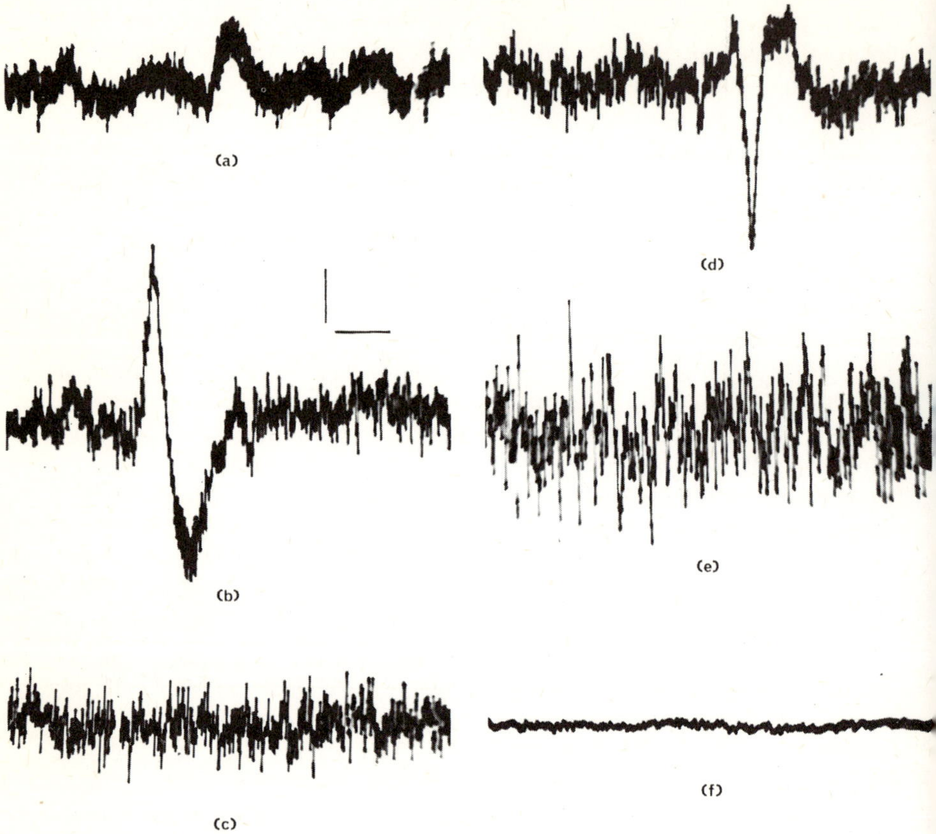

FIG. 11. Typical records obtained with an extracellular electrode in the nose of a frog are shown here. Trace (a) shows monophasic, low amplitude positive spikes recorded as the electrode makes contact with the surface of the epithelium. Trace (b) shows a diphasic spike recorded when the electrode tip is advanced so that it is in the region of the soma of the receptor cell. Trace (c) shows noise of slightly increased amplitude with no sign of neural activity which occurs as the electrode is advanced further with the tip in the region of the basement membrane. Trace (d) shows a triphasic spike recorded with the electrode tip deeper, below the basement membrane in the region of the receptor cell axons. Trace (e) is the noise due to an 18 megohm input resistor on the amplifier input with the electrode in air. Trace (f) is the amplifier noise with the input connected to ground. In all cases the recording system bandpass was 30 Hz to 10 kHz. The calibration marks indicate 50 microvolts positive upward on the vertical axis and 5 milliseconds on the horizontal axis.

receptor cells near the electrode tip, and no amount of manoeuvring of electrode position or change in size or shape of the electrode tip permits separation of the spikes of one cell from those of its neighbours.

As the electrode penetrates deeper, larger, often diphasic spikes are seen which are illustrated in Fig. 11b. There are often two or three clearly different spike amplitudes. If a stimulus is delivered, spikes of one amplitude can respond differently from spikes of another amplitude. The largest signals come from a receptor cell electrically most closely coupled to the electrode. If the electrode is moved slightly, the relative amplitude of the different spikes changes. Sometimes it is possible to manipulate the electrode so that the spikes from one cell are clearly larger than the spikes from all of the other cells nearby. This is not an effective way to study a single cell, however, because the cell lasts for only a short time. Seldom more than 10 or 15 minutes is available before the spike amplitude decreases and the cries are lost in the background of its neighbours. It is likely that the pressure of the large electrode tip on the soma membrane slowly damages it. I have the impression, but no proof, that when very regular spike intervals are recorded, they are diphasic spikes from an electrode tip located in the soma layer of the mucosa. I think that regular intervals are a sign of a pathological state of olfactory receptor neurons.

As the electrode is further advanced, signs of spike activity diminish and a region is entered characterized by a doubling or tripling of random noise with no sign of spikes imbedded in it (Fig. 11c). Here the electrode is apparently near the basement membrane. Upon further advances with a fortunately constructed electrode, the noise level abruptly drops and triphasic spikes, often as large as 500 microvolts, can be seen (see Fig. 11d). As in the soma region, spikes of several different amplitudes which respond differently to a stimulus often occur. It is easier here to separate one cell from its neighbours by slight movements of the electrode. Unlike the others, triphasic spikes (which we assume are axon action potentials propagating past the electrode tip) from a single cell can be followed for many hours. As long as stimulus intensities are kept low, a particular stimulus will evoke nearly the same response anytime it is delivered during a two or three hour recording period. As the electrode sits in its recording position the action potential amplitude often spontaneously grows for a few minutes. After this time it is hard to again "lose" the cell. Slight movements of the electrode don't greatly affect the recorded signal. It is as if the receptor cell axon had glued itself on to the electrode tip.

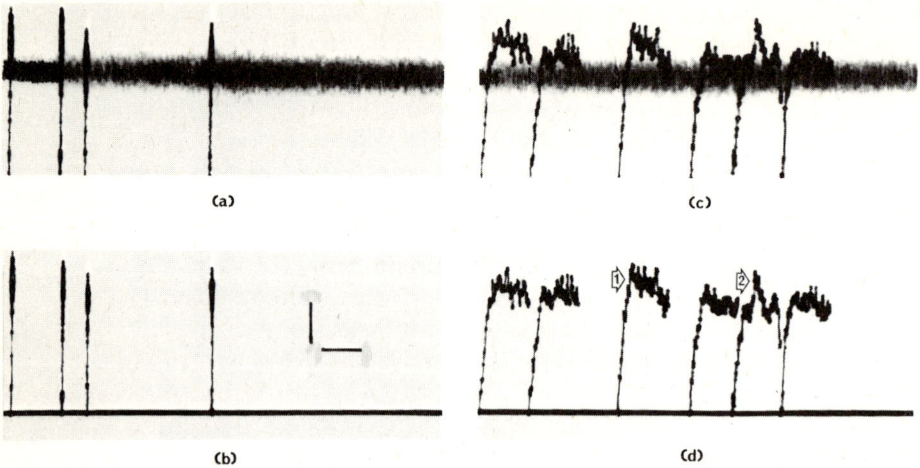

FIG. 12. Two axons close to an extracellular microelectrode may produce action potentials of nearly identical amplitudes. They usually have discriminable shapes. This figure shows such a situation. Trace A shows the unprocessed record with the spikes intensified. Trace B shows the action of an a-c coupled amplitude selector gate which is turned on for a period of a few milliseconds following the negative peak of the action potential. (Appendix III describes such a device). Trace (c) is like trace (a) except that the sweep is accelerated for 3 milliseconds following the negative spike peak. Trace (d) is the corresponding expanded sweep of the amplitude selector gate output. In (d) two clearly discriminable spike shapes can be seen (arrows 1 and 2). The spikes were triphasic with a peak to peak amplitude of about 200 microvolts. The recording system bandwidth was 100 Hz to 3 kHz. The calibration marks indicate 50 microvolts positive upward on the vertical axis and 10 seconds on the horizontal axis. During sweep expansion, the horizontal axis calibration line represents 3 milliseconds.

The single most difficult problem encountered in extracellular recording is separation of activity in neighbouring cells. The noise level is always large enough to contribute to variations in the amplitude from one spike to the next. Because metal electrode noise is not white but instead an inverse frequency function, simple statistics are not adequate to prove that spike amplitude variation is random. There are two practical techniques to help ascertain whether the action potentials observed are coming from one cell or several cells. First the electrode can be manipulated to maximize the amplitude of the largest spikes. Then several different stimuli should be delivered to test for the presence of an axon close to the electrode which is quiescent in the absence of stimulation. If the variation in amplitude of the largest spikes is about like that of the background noise, then it is likely that the activity being observed is due to one cell only. The second criteria can then be applied. For this a fast sweep rate is used and the detailed shape of each successive large spike is observed. It is convenient to use a storage oscilloscope with a sweep rate from 2–5 ms/cm, an oscillographic recorder with a bandwidth of at least 3 kHz and a chart speed adequate for studying the spike shape, or Polaroid photographs from the face of a conventional oscilloscope. Since the average firing rate of olfactory receptor cells is low, the process is tedious and consumes a considerable amount of recording material. A convenient alternative is to adjust an amplitude selector so that it triggers on the large spikes and no others. This can be used to trigger the sweep on an oscilloscope so that the selected spikes appear superimposed with a fast time base. If the variations in spike amplitude are not because of random noise but are in fact because of activity from two cells equally close to the electrode, clear differences in shapes of individual spikes will be observed. This effect is shown in Fig. 12. When this occurs, it is necessary to reposition the electrode such that spikes of one shape are clearly distinguishable in amplitude from spikes of a different shape. During the course of an experiment, every now and then, the singularity criteria should be rechecked. This is because it occurs more often than one would like that a cell close to the electrode is quiet until an appropriate stimulus for it is used and it responds with a spike amplitude close to that of the cell which is being followed.

Appendix III shows a simple modification of a Tektronix Type 161 Pulse Generator which allows it to be used as an a-c coupled amplitude selector. The time constants are optimized for olfactory receptor spikes. It contains a gated amplifier which is turned on for an adjustable number of milliseconds following a spike exceeding a certain amplitude and turned off otherwise. Commercial amplitude discriminator circuits are available, some of which will select for several different spike amplitudes simultaneously. These are useful provided that appropriate filtering is used so that low frequency noise does not distort the selection criteria and that spike amplitude differences are well above the high frequency noise amplitude in the signal. It is unusual to have signal amplitudes so large that the activity of several different cells can be reliably tracked for an extended period of time.

V. Interpretation of single unit responses

Surprisingly little can be said with certainty about the activity of olfactory receptor cells. It is as if the nose has evolved in such a way as to confound all sensible electrophysiologists. We will consider each of the major difficulties of interpretation in turn.

First, it is crucial to know with certainty that the spikes being observed as the stimulus is varied are, in fact, from the axon of one cell and not from the axons of several cells. Many unmyelinated axons are enclosed within each invagination of a supporting cell. The extracellular electrode has a tip which is ten times the diameter of an axon. Thus it must be close to many axons. This problem has earlier been considered in the case of the optic nerve of the frog (Lettvin *et al.*, 1959; Maturana *et al.*, 1960). In this case, the number of axons in the same bundle were fewer than in the case of the frog olfactory nerve. That low-impedance metallic microelectrodes can in fact separate the activity of neighbouring axons is shown by positioning the electrode so that spikes of a particular shape are clearly larger than the rest. In this situation, we never see two spikes of the same shape partially superimposed. If the electrode were recording spikes from two cells, we would expect that now and then the two would fire at the same or nearly the same time. Anatomically, the region in which single unit axon spikes is best recorded is immediately below the basement membrane. This is where axons are farthest separated from each other. No successful single unit recordings have been reported from the highly compact more central portion of the olfactory nerve which is the region where the anatomical relationships between first cranial nerve axons and sheath cells have been studied.

Second, there is a significant problem in choosing an appropriate measure of response beyond the trivial statement that a stimulus affects a cell when it causes a change in impulse activity of the cell. We have several alternatives. Generally, experimenters have described a stimulus as being excitatory when it causes a cell to begin firing soon after the stimulus is delivered or when it increases the firing rate of an active cell. An inhibitory stimulus is one which depresses the quiescent impulse discharge rate. Stimuli can be rank ordered in terms of their effectiveness in modifying spike activity. Inhibitory effects turn out to be difficult to catalogue. A typical discharge of an unstimulated cell shows an average firing rate of about one impulse per second. If the discharge pattern is monitored for a long period of time during which no stimuli are delivered, this average rate can change by an order of magnitude in either direction. It is not clear whether this is because of inadvertent stimulation by changes in the atmospheric background or whether it is because of changes in the body chemistry of the frog. It is much harder to detect inhibition in a cell if its quiescent discharge rate averages one spike in ten seconds than it is in a cell averaging ten spikes per second. Furthermore, neither the evoked spike response nor the resting discharge is regular. Adjacent intervals between impulses commonly differ by a factor of 30 in duration, and it is unusual to see several spikes occurring sequentially with approximately equal intervals between them.

The time at which the stimulus arrives at the receptor cell cannot be measured accurately. If a second electrode is used to monitor the EOG, this electrode shows a voltage change occurring between 200 and 400 ms after butanol vapour reaches the mucus surface. Since the EOG is a measure of spike generating current flow for the average of the cell population, there is no reason to expect a correlation between EOG onset and change in evoked spike activity for a particular cell unless all of the cells responding to butanol respond alike. In fact, however, the impulse pattern evoked from different cells by any particular stimulus varies greatly in terms of onset time, number of spikes evoked, and duration of the effect. It is often observed that a longer than average interval occurs between the time that the stimulus vapour is turned on and the marked increase in firing rate of an excited cell. Should this be interpreted as a response that is excitatory, or one which is initially inhibitory and then excitatory? Or is this long interval the result of random firing rates of the unstimulated cell and biased experimental observation coupled with a long diffusion time for the stimulus. The long interval could also be due to a primary effect of the stimulus interacting with the mucus. That is, increased stimulus concentration in the mucus could initially cause a change in the fluid near the receptor cell due to ionic and osmotic effects. The stimulus would then exert its direct effect later because of its low diffusion constant. At the present time these alternatives cannot be resolved. The importance of inhibition as a measure of cell response is widely debated. All that can be said with certainty is that experimenters who look for it, find it, some of them rarely, others nearly as often as they find excitation. Since the origin of positive waves in the EOG is debated, EOG evidence for inhibitory processes is equivocal.

Another commonly observed response is a rapid burst of spikes, often lasting for seconds, occurring as much as ten seconds after the delivery of a brief stimulus and after termination of an extended stimulus. Occasionally, stimuli evoke alternate periods of spikes and quiet periods during and following cessation of the stimulus without changing the average firing rate from that observed in the unstimulated state. To an observer, there is no question that the cell responded to the stimulus. However, by the usual criteria in neurophysiology, the number of spikes occurring during the response period was not different than the number of spikes in the same period of time preceding the stimulus. The most reasonable attitude one might assume in analysing the present state of olfactory physiology might well be that an appropriate measure of a response has not been established and that as a result it might well be wise to adopt a sceptical nature with respect to those experiments which pretend to show how the information from olfactory receptors is encoded.

Third, it is important to establish clearly that there are no acceptable established relationships between stimulus parameters and receptor cell responses. If we consider one particular stimulus we find that different cells which respond to it have different sensitivities. An intensity which evokes a just observable change in impulse activity in one cell may be orders of magnitude higher than the intensity of that substance required to produce an observable change in another cell. Furthermore, there is no simple correlation between the amplitude of the EOG evoked by a stimulus of a particular intensity and the probability of observing a vigorous response in a single cell. Further, for most stimuli, cells do not respond in a monotonic way as the intensity is varied. For instance, when a cell has been located, it can be stimulated with a variety of substances at convenient concentrations and one of these to which the cell responds vigorously can then be explored with respect to intensity. First, it is easy to find with a suitably dilute solution of this substance which does not affect the resting discharge of the cell. As the intensity is raised, in say half-log or third-log steps, a concentration is found which evokes a noticeable change in impulse activity, say, excitation. As the intensity is further raised, the number of spikes evoked increases. If the nose is allowed suitable time to recover from each stimulus, repeated stimulation at one intensity produces about the same pattern of response and about the same or slightly fewer evoked spikes. There is a concentration above which the number of spikes evoked decreases as the intensity increases. The pattern of activity tends to be preserved. If, for instance, the stimulus tends to evoke bursts of spikes separated by silent periods, the bursting pattern is maintained at higher stimulus intensities, but the number of spikes per burst decreases. With still further increase in stimulus intensity, the discharge rate of spikes in the unstimulated state decreases and the rate of spikes evoked by the stimulus is actually lower than what was the firing rate of the cell in the unstimulated state before this sequence of stimuli was started. Even in this state, the pattern of activity in response to the stimulus appears to be related to the pattern of activity evoked by low intensity stimulation. Even though single unit firing rate increases with stimulus intensity to some maximum and then declines, EOG amplitudes over this intensity range increase monotonically with stimulus intensity and reach a limiting maximum value. The experiment has not been done to determine whether a cell which has been rendered nearly unresponsive to a stimulus by high intensity stimulation with that substance will respond in a more normal fashion to other stimuli. Since a particular cell changes in its response with stimulus intensity, stimulus duration, stimulus repetition rate and stimulus substance, it requires a quite complex experiment to separate these effects.

Fourth, and finally, the nose apparently represents the ultimate complexity in the way it analyses stimuli which differ in kind. None of the proposed ordering schemes which attempt to group the thousands of different odorous substances into categories of similarities appear to reflect the ordering properties which describe responses of frog olfactory receptors. Unlike photoreceptors in the eye, experimenters have yet to discover primes, that is, the dimensions of the space into which the receptor cells map the nature of the stimulus. If several stimuli are tried successively on one cell, it is possible to rank order them with the most excitatory at the top and the most inhibitory at the bottom. If the same substances are tried on another cell, the rank order is different. Of two substances which have nearly identical effects on one cell, one may strongly excite another cell, and the other strongly inhibit it. There are several possible explanations for this apparent chaos. First, it may be that a normalization process needs to be discovered which will have the effect of removing stimulus intensity information from the spike burst evoked and that the information remaining in the impulse interval patterns might then more clearly show its relationship to the nature of the stimulus. It is also possible that the differences seen from one cell to the next are because of the stimulus history of the nose (i.e. that each stimulus molecule denatures a receptor molecule which must then be slowly returned to its nascent state). If the response of a cell reflects the recent history of what has happened to the various receptor sites on it, experiments in which stimuli are delivered in a jumbled sequence of kinds and intensities might be expected to yield the result that no two cells appear to be alike.

Another possibility is that each receptor cell in fact is different, that is, each looks at the world from a different point of view. This is called holistic coding and it is the most nihilistic of the possibilities from the point of view of a physiologist. If this is the way the nose is organized, no amount of work on the responses of single cells will serve to reveal the mechanisms involved in the olfactory sense.

These four kinds of problems of interpretation are important for the experimenter to have in mind. To distinguish between the various possibilities will require experiments which are much more carefully thought out and executed than any of those which have been reported to date. Some recent attempts to by-pass electrophysiological studies of single unit responses by going directly to determination of the chemistry of the interaction between stimulus and receptor may provide fresh insights. However, not until the physiologist can listen to the language of the cells of the brain when the nose is presented with an odour and guess the nature and intensity of the stimulus will it be possible to claim an understanding of the physical basis for olfactory sensation. We are far from this goal, not only for lack of good theoretical models, but also because the experimental art for studying this sense is primitive and the experiments most difficult.

Acknowledgements

J. Y. Lettvin conceived of much that is reported here. His collaboration has been a continuous pleasure. The late W. H. Pitts wondered how noses work and provided the enthusiasm to keep me at single cell studies when the path became tortuous. G. Waloga, T. V. Getchell, S. Wilson, M. L. Getchell, J. Y. Stick and W. Muller have participated in various aspects of this work in my laboratory. S. Sonnet and M. Lanus helped with the manuscript.

Financial support came from the National Science Foundation Grant No. GB-30520, the United States Air Force Office of Scientific Research Grant No. AFOSR 71-2124, and the Air Force Systems Command, Aerospace Medical Division Contract No. 33615-67-C-1497.

Appendix I

A versatile input amplifier suitable for use with both ultra-fine tipped micropipettes and extracellular metal microelectrodes is shown in Fig. 13. Some notes on its construction and adjustment follow.

The FET input transistor, T_1, can be a Type 2N3686 or 2N3088. Other low noise types with $V_{PO} < 1$ V, $I_{GSS} < 0.1$ na and I_{DDS} between 0.5 and 1.0 ma will work. T_2 is a Type 2N5163 or 2N3086 with $I_{DSS} = 3$ ma. All other transistors are low noise, high β silicon junction types of the low cost variety. *N*-type 2N5133 and *p*-type 2N5138 work well. The diode is a silicon type, for instance FD300.

Potentiometers should be high-quality carbon types such as Ohmite Type AB. Connecting leads should be short with all component leads cut to as short a length as is convenient for wiring. The trimmer capacitor C_i is a glass tubular type with a range of 0.5 to 5.0 pf. This sets the range of the negative capacitance compensation available. Lowest noise results when the trimmer is set at as low a value as possible.

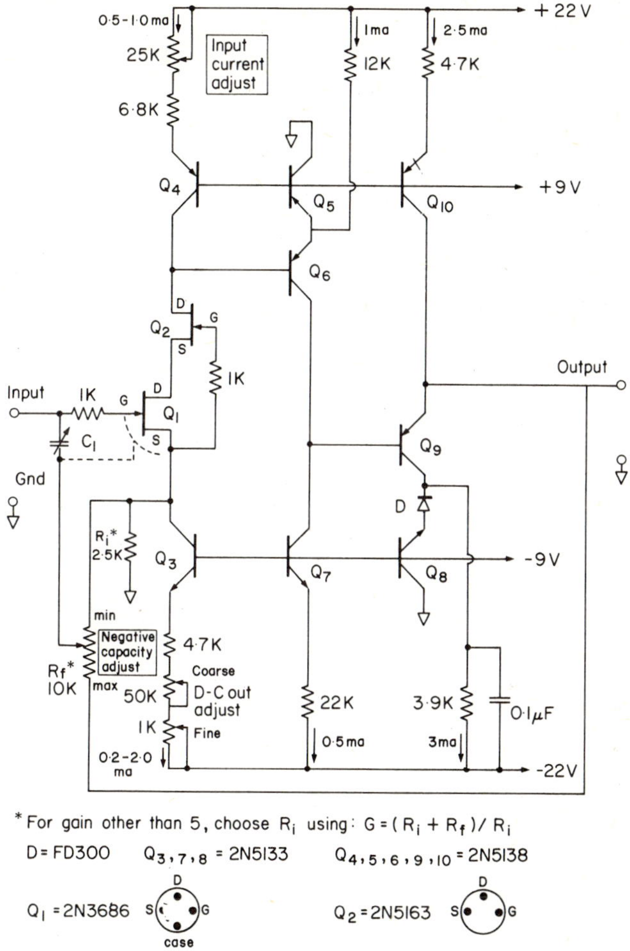

FIG. 13. An input amplifier suitable for use with microelectrodes. Details concerning construction and testing are in the text.

The potentiometers which adjust input current and d-c output level interact somewhat. The input current control should be initially set in about the centre of its range and, with the input grounded, the d-c output level set to zero. The input current is then set to zero or for slight tendency for the input to drift positive using the method described in the section on Input Amplifiers and Electrode Tests. It may be necessary to repeat this process several times to converge on the correct setting for the input current.

If the input FET has a separate lead to the case, it should be connected as shown by the dotted line.

Appendix II

The simple circuit shown in Fig. 14 will produce a trapezoidal waveform from a rectangular pulse input. As shown here, it can be added to a Tektronix Type 161

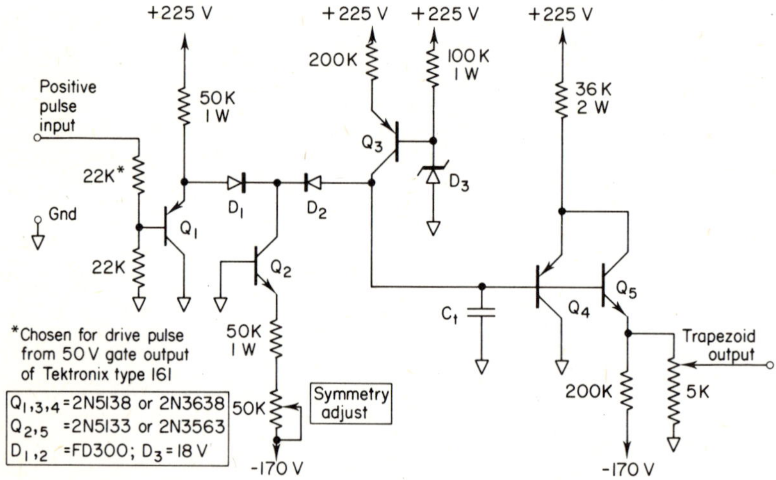

FIG. 14. This circuit produces a trapezoidal output voltage in response to a positive input pulse.

Pulse Generator without modification of the pulse generator circuitry. The duration of the rising phase and falling phase of the trapezoid is determined by C_t. Some useful values are:

C	Rise time
1.0 μf	15 ms
0.3	5.0
0.1	2.2
0.01	0.16

Appendix III

The amplitude selector used to display the extracellular action potentials in Fig. 12 was constructed by modifying a Tektronix Type 161 Pulse Generator in a fairly simple way. The necessary changes in the circuits associated with vacuum tubes V_1 and V_4 are shown in Fig. 15. The positive pulse output can be

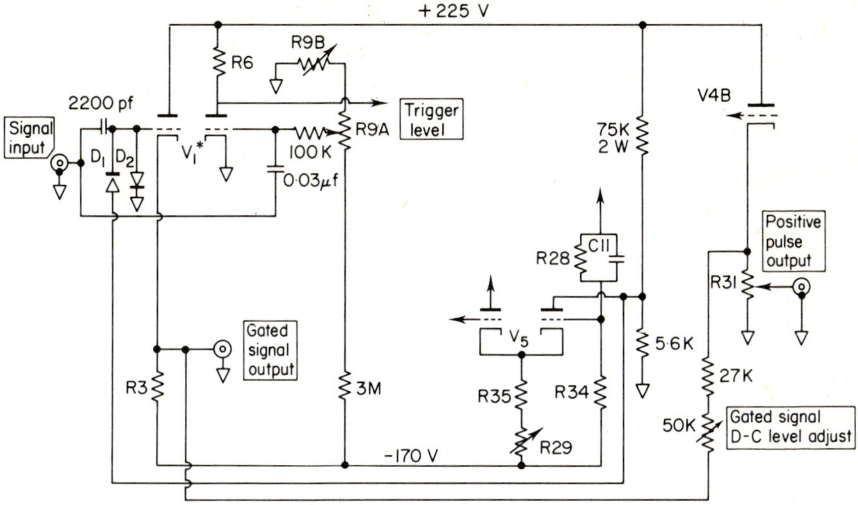

FIG. 15. Modified portions of the schematic diagram of a Textronix Type 161 Pulse generator which allows it to be used as an amplitude selector.

used for brightening the displayed spikes. R_1 is selected to produce zero voltage offset at the onset of the selected spike. R9A adjusts the level of positive voltage at which the gate amplifier turns on to pass the selected portion of the signal. The original pulse width controls of the pulse generator determine the duration of the period that the gated amplifier is turned on.

References

Agin, D. P. (1969). Electrochemical properties of glass microelectrodes. *In* "Glass Microelectrodes" (Lavallee, M., Schanne, O. and Hebert, N., eds.), Chapter 4. John Wiley and Sons, New York.

Altner, H. and Boeckh, J. (1967). Über das Reaktionsspecktrum von Rezeptoren aus den Riechschleimhaut von Wasserfröschen (*Rana esculenta*). *Z. Vergl. Physiol.* **55**, 299–306.

Aoki, K. and Takagi, S. F. (1968). Intracellular recording of the olfactory cell activity. *Proc. Jap. Acad.* **44**, 856–858.

Baldwin, H. A., Frenk, S. and Lettvin, J. Y. (1965). Glass-coated tungsten microelectrodes. *Science, N.Y.* **148**, 1462–1463.

Boeckh, J., Kaissling, K. E. and Schneider, D. (1965). Insect olfactory receptors. *Cold Spring Harbor Symp. Quant. Biol.* **30**, 263–280.

Davis, W. J. (1970). Intracellular dye injection technology. Notes from the Neurophysiological Underground, No. 1. Stanford. Personal communication.

Dowben, R. M. and Rose, J. E. (1953). A metal-filled microelectrode. *Science, N.Y.* **118**, 22.

Frank, K. and Becker, M. C. (1964). Microelectrodes for recording and stimulation. *In* "Physical Techniques in Biological Research" (Nastuk, W. L., ed.), **5**, 23–85. Academic Press, New York, and London.

Gesteland, R. C. (1961). Action potentials recorded from olfactory receptor neurons. Ph.D. Thesis, M.I.T., Cambridge.

Gesteland, R. C. (1971). Chemical mechanisms and neural coding in vertebrate chemoreceptors. Final report to Aerospace Medical Research Laboratory, Wright-Patterson Air Force Base, Ohio.

Gesteland, R. C. and Howland, B. (1959). Bridge for measuring the impedance of metal microelectrodes. *Rev. Sci. Instr.* **50**, 262–264.

Gesteland, R. C., Howland, B., Lettvin, J. Y. and Pitts, W. H. (1959). Comments on microelectrodes. *Proc. Inst. Radio Engr.* **47**, 1856–1862.

Gesteland, R. C., Lettvin, J. Y., Pitts, W. and Rojas, A. (1963). Odor specificities of the frog's olfactory receptors. *In* "Olfaction and Taste" (Zotterman, Y., ed.). Pergamon Press, New York.

Gesteland, R. C., Lettvin, J. Y. and Pitts, W. H. (1965). Chemical transmission in the nose of the frog. *J. Physiol.* **181**, 525–559.

Gesteland, R. C. and Waloga, G. (1968). Unpublished contract progress report.

Getchell, T. V. (1969). A functional characterization of the olfactory receptor membrane. Ph.D. Thesis, Northwestern University, Evanston.

Getchell, T. V. (1972). Personal communication.

Hodgson, E. S., Lettvin, J. Y. and Roeder, K. D. (1955). Physiology of a primary chemoreceptor unit. *Science, N.Y.* **122**, 417–418.

Kennard, D. W. (1958). Glass microcapillary electrodes. *In* "Electronic Apparatus for Biological Research" (Donaldson, P. E. K., ed.), Chapter 35. Butterworths, London.

Lettvin, J. Y. (1960). Quarterly Progress Report, M.I.T. Research Laboratory of Electronics. No. 58, page 225.

Lettvin, J. Y. (1969). Quarterly Progress Report, M.I.T. Research Laboratory of Electronics. No. 93, pages 279–286.

Lettvin, J. Y., Howland, B. and Gesteland, R. C. (1958). Footnotes on a headstage. *Inst. Radio Engr. Trans. Med. Elec.,* ME-**10**, 26–28.

Lettvin, J. Y., Maturana, H. R., McCulloch, W. S. and Pitts, W. H. (1959). What the frog's eye tells the frog's brain. *Proc. Inst. Radio Engr.,* **47**, 1940–1951.

Matthews, D. P. and Tucker, D. (1966). Single unit activity in the tortoise olfactory mucosa. *Fed. Proc.* **25**, 329.

Maturana, H. R., Lettvin, J. Y., McCulloch, W. S. and Pitts, W. H. (1960). Anatomy and physiology of vision in the frog. *J. Gen. Physiol.* **43**, 129–175.

Miller, R. F. and Dowling, J. E. (1970). Intracellular responses of the Müller (glial) cells of mudpuppy retina: their relation to the b-wave of the electroretinogram. *J. Neurophysiol.* **33**, 323–341.

Mozell, M. M. (1962). Olfactory and neural responses in the frog. *Amer. J. Physiol.* **203**, 353–358.

Mozell, M. M. (1964). Evidence for sorption as a mechanism of the olfactory analysis of vapours. *Nature, Lond.* **203**, 1181–1182.

Mozell, M. M. (1971). Spatial and temporal patterning. *In* "Handbook of Sensory Physiology, Chemical Senses I" (Beidler, L. M., ed.), **4**, 205–215. Olfaction. Springer-Verlag, Berlin.

Müller, W. (1972). Personal communication.

Mustaparta, H. (1971). Spatial distribution of receptor responses to stimulation with different odours. *Acta Physiol. Scand.* **82**, 154–166.

O'Connell, R. J. and Mozell, M. M. (1969). Quantative stimulation of frog olfactory receptors. *J. Neurophysiol.* **32**, 51–63.

Ottoson, D. (1956). Analysis of the electrical activity of the olfactory epithelium. *Acta Physiol. Scand., 35 Suppl.* **122**, 1–83.

Ottoson, D. (1971). The electro-olfactogram. *In* "Handbook of Sensory Physiology, Chemical Senses I" (Beidler, L. M., ed.), **4**, 95–131. Olfaction. Springer-Verlag, Berlin.

Snell, F. M. (1969). Some electrical properties of fine-tipped pipette microelectrodes. *In* "Glass Microelectrodes" (Lavallee, M., Schanne, O. and Herbert, N., eds.), Chapter 7. John Wiley and Sons, New York.

Tasaki, K., Tsukahara, Y., Ito, S., Wayner, M. J. and Yu, W. Y. (1968). A simple, direct and rapid method for filling microelectrodes. *Physiol. and Behav.* **3**, 1009–1010.

Thompson, D. W. (1942). "On Growth and Form", Chapter V. Cambridge University Press, Cambridge.

Tomita, T. (1969). Single and coaxial microelectrodes in the study of the retina. *In* "Glass Microelectrodes". (Lavallee, M., Schanne, O. and Herbert, N., eds.) Chapter 8. John Wiley and Sons, New York.

Zettler, F. (1970). Kalt gefüllte Mikroglaskapillaren und ihre electrophysiologisch relevanten Eigenschaften. *Z. Vergl. Physiol.* **67**, 423–435.

8 Methods in Electrophysiological Studies of the Olfactory System

KJELL DØVING

The Institute of Zoophysiology
University of Oslo,
Blindern, Oslo 3,
Norway.

Advances in our knowledge of the function of the olfactory system have been hampered mainly because of the multitude of technical difficulties encountered by the student of this discipline of physiology. Some of the problems that arise are general to studies of sensory physiology and neurophysiology. Many problems, however, are particular to the olfactory system. These difficulties are mainly connected with the technical administration of chemicals to the olfactory epithelium, and with obtaining sufficient purity and known concentrations of the odours used.

The neurophysiological problems arise with the inaccessibility of the olfactory epithelium and the olfactory bulb, especially in birds and mammals. In the teleosts, amphibians and reptiles the olfactory system may be reached by simple surgical procedures. The main problem has been, and still is, the task of recording from single olfactory receptors. The small physical size of the olfactory receptor cell somas (Andres, 1965) and their still smaller axons (Gasser, 1958; Gemne and Døving, 1969) has been the main obstacle in obtaining crucial information about the transmission of chemical messages to nerve impulses. The problem of recording the activity from single olfactory receptors will be discussed elsewhere in this volume by Gesteland (for vertebrates).

The present description of the techniques involved in stimulation with odorous substances and recording with electrophysiological techniques reflects my own interest and research activity. This has been concerned with the activity of the neurones in the olfactory bulb during stimulation with chemical substances. The methods which have been used and are described here are in no way unique, and are compiled from various articles and monographs treating these subjects; references to these articles will be supplied in the text.

I. Stimulation

The main problem in olfaction research is determining the exact nature of the stimulant. Contrary to general belief, purity of odorous substances is not only of concern to the chemist working on flavour, but also to the olfactory physiologist. As it is, the latter is usually forced to choose between doing no experiments at all or doing experiments with the available odours and their impurities. This situation has led to much dissatisfaction, and even though the chemical industry shows interest in providing pure substances for olfactory research, the actual materials delivered often contain a surprisingly high level of impurities.

One of the worst problems is that some impurities present in only small amounts may have a high vapour pressure and thus dominate the head space and hence the odorous stimulation. GLC analysis of the head space is therefore an essential test in olfactory stimulation procedure.

In attempts to overcome these difficulties several precautions should be taken. The first would be to clean the carrier gas of the stimulating system. As the carrier gas for odour stimulation one can use air, nitrogen or other nontoxic gases. Air is the cheapest agent, and with sufficiently good filters one can use an air pump or the compressed air normally supplied in laboratories. It is advisable to get rid of acids and oil droplets in the air from laboratory compressors by washing and drying as indicated in the diagram of Fig. 1. Further cleaning with charcoal and paper filters (to remove the fine dust from the charcoal) will suffice in normal cases. The oxygen present in air and other gas mixtures is often undesirable because it may oxidize the stimulating substance. In many cases, of course, the substances are so impure that the additional impurities brought about by oxidation will not appreciably alter the stimulating conditions. However, the use of N_2 is highly recommended. If one uses a high-purity quality of nitrogen, the traces of oxygen present can be removed by pyrogallol.

In Fig. 1 the use of pyrogallol, charcoal and sterile filters is shown schematically. The charcoal filter can be made from glass tubing

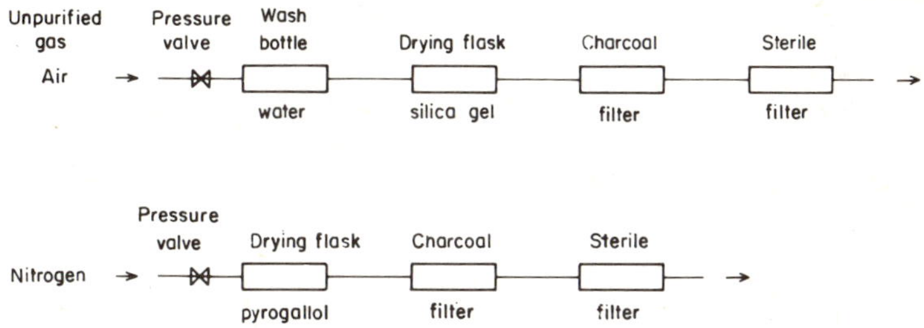

FIG. 1. Schematical drawing of filter devices for air and nitrogen.

with cylinder joints. The size should be adjusted to the air flow wanted. A filter volume of a size giving a filtering time of over 2 min. would suffice in most cases. The charcoal filters usually give off dust particles, which are very efficiently removed by paper filters (MS-sterile filter, Stora Kopparberg), or by other means, e.g. glass fibre.

A. Stimulation with an open system

In terrestrial animals where the olfactory epithelium is easily uncovered and exposed with little or no trauma for the animals, an open delivery system for the different odours is most convenient.

The simplest of these consists of a glass syringe, the inside of which is blotted with an odorous substance. The opening tube is directed onto the epithelium, and the plunger is pushed either mechanically or by hand to give an odorous flow of air. This was the principle of the olfactometer first described by Zwaardemaker and used by him in studies of human olfactory thresholds. This method can also be applied in electrophysiological studies where the concentration is of little importance or the stimulating effectiveness of the odours can be controlled, e.g. by monitoring the electro-olfactogram (EOG) (Ottoson, 1956; Døving, 1964, 1965, 1966a, b). Naturally the concentrations are not easily controlled under these circumstances, and this method of stimulus delivery is unsuitable for work where the absolute concentration of the stimulant is of importance.

The use of syringes may be refined to give exact mixtures of odorous gases by an injection apparatus in which one can adjust the rate of mixing by varying the speed of piston movement and/or size of the syringes.

Another way of varying the concentration of the odorous gas flow is by letting the gas pass over a solution of the wanted odour. A good solvent for all kinds of organic substances is diethyl-phthalate. Unfortunately the solvent may influence the vapour pressure, and thereby the activity of the odorous substance, by influencing the concentration over the solution.

The kind of vessel shown in Fig. 2 has been used in connection with the application of odours to the olfactory epithelium of frogs and rats. (See Drake *et al.*, 1969.) The vessel holds 12 ml and is filled with 2 ml of solution. At equilibration there will be 10 ml of gas at known concentration above the solution. This can then be driven onto the olfactory epithelium by letting gas through the vessel. If it is necessary to hold a uniform concentration throughout the stimulation period, the initial part of the gas which has been in the outlet tube should be removed by a screen with suction. This screen is then removed during the actual stimulus delivery.

In the application procedure used by Drake *et al.* (1969), 25 of the vessels drawn in Fig. 2 were mounted on a turn-table with a central delivery shaft. For a closer description see the paper by Mustaparta (1971). The arrangement sketched below made it possible to have the outlet in the same position relative to the preparation for each stimulus.

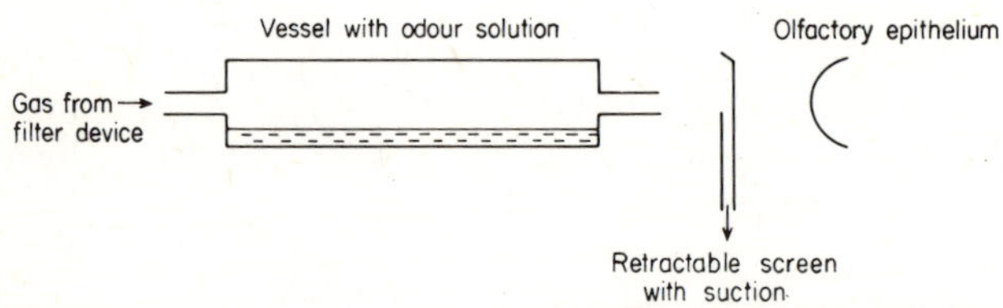

FIG. 2. Principle of an open delivery system for odour stimulation.

When an open stimulus arrangement is used for application of odours to the rat or similar mammal, it is necessary to perform sniffing action for the animal. This is because the resistance to the flow of air in the nasal passages is too high in a chronic experimental condition. When the tracheal cannula is inserted during the operative procedure, a nasal cannula is also inserted from the trachea up into the nasal pharynges (see below). One can apply suction via this tube through the nose simultaneously with the flow of air through the odorous vessel, to ensure that the stimulant does reach the olfactory epithelium. It is definitely necessary to control the flow of gas so that one does not create excessive sniffing action for the animal during stimulation. Otherwise a large amount of "false" air may be drawn into the nose. Gas flow measurements can easily be made using burettes with soap water and a stop-watch.

B. Stimulation with a closed system

A system whereby the stimulus is transported to the olfactory epithelium sealed from the surroundings is of great advantage in olfactory experiments. In a closed system it is not only possible to determine the exact amount of the test substance, but also to control the rate of gas flow and the timing of the stimulus. If one is going to the trouble of constructing such a stimulus device, however, it is of utmost importance that extremely pure samples of odour substances can be obtained.

The diagram in Fig. 3 gives the essential principles in such a device. It consists of a stimulus delivery system, an equilibration device, a concentration detector and a four-way-valve. The carrier gas should preferably be nitrogen, and with the use of nitrogen it is necessary to pass air through the nose when not stimulating.

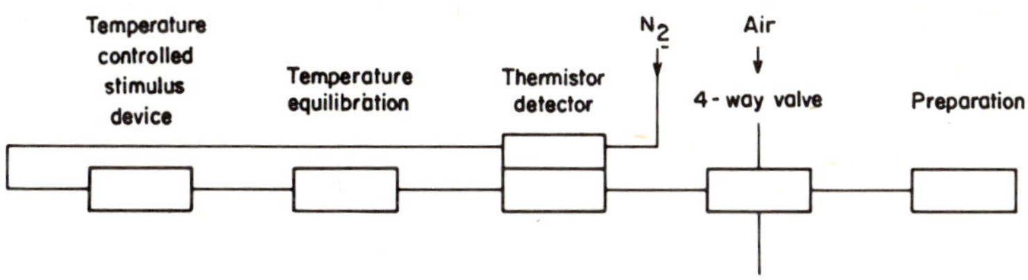

FIG. 3. Diagram of a closed delivery system for odour stimulation in nitrogen atmosphere.

The use of a four-way magnetically driven valve makes possible both a continuous flow of air through the nose and allows convenient control of the stimulation time.

There are several solutions to the many problems that arise in making this stimulating apparatus. The description given below is only one; a great many more may be equally efficient.

The essential part of the stimulating device is the stimulus delivery system shown in Fig. 4. Small amounts of the pure substance, 0.5 ml, are placed in a glass beaker which fits a spherical joint. The gas stream is directed on to the sample of the chemical. The temperature at the surface of the sample is measured by a thermistor detector. The beaker is fitted into an oven the temperature of which can be raised or lowered by means of heating and cooling coils. The temperature is controlled by another thermistor detector. The current through the heating coil is controlled by a linear temperature programmer (Perkin Elmer Co.).

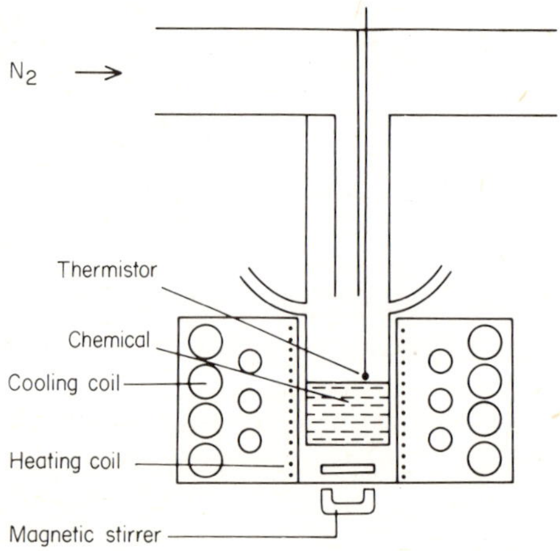

FIG. 4. Schematical drawing of a temperature-controlled sample chamber for delivery of known concentrations of odorous atmosphere.

By this device the concentration of the stimulant can be calculated when the rate of gas flow, the vapour pressure of the substance and the temperature at the surface are known. When the gas flow is higher than that necessary to provide saturation at the surface, the concentration must be monitored. In the system drawn in Fig. 3 a thermistor detector has been used. This is favourable because it does not interfere with the substance, it is not necessary to split the beam and its function is not extinguished by sudden changes in the pressure as is the case with a flame ionization detector. The thermistor detector is also more sensitive than a hot wire detector at room temperature.

The stimuli can be given through the four-way-valve as indicated in Fig. 3. By using a magnetic valve the exact timing of the stimulus can cal vibrations, and to prevent these being transmitted to the preparation where they are highly undesirable, it is necessary to have a flexible joint between the valve and the preparation.

II. Recording techniques

A. d.c. recording of slow potentials

The recording of very slow potentials from the surface or interior of nervous tissues is beset with certain difficulties. A rather detailed description of some of the pitfalls and ways of overcoming them will therefore be given.

The slow potential recorded from the surface of the olfactory epithelium upon stimulation with odorous gas has become a useful parameter in olfactory research. This slow potential, named the electro-olfactogram or EOG (Ottoson, 1956), can be easily recorded with proper techniques, but inadequate methods may result in highly "original" results. (For review see Ottoson, 1970.)

The crucial point in any d.c. recording is the connection between the lead and the solution, or between the lead and the tissue. Direct contact as in the latter case is unfavourable because of the unstable redox potential at the junction. Anyone can assure himself of the phenomenon by doing this little experiment. Solder platinum wire to two leads and dip them in Ringer solution. Observe the potential difference with a d.c. recording apparatus at a band width from d.c. up to 1 kHz at an amplification of 0.1 mV/cm. Slow potential drifts as well as spontaneous discrete potentials of varying amplitudes will be seen. This obvious control should not be omitted even when more adequate electrode systems, some of which will be described below, are used.

The electrode most suitable for d.c. recording is the calomel half-cell. In the original version it has the form:

$$Pt \mid Hg \mid Hg_2Cl_2 \mid KCl\text{, saturated}$$

The salt can be saturated or be at any other known concentration of KCl. The purity of the different components should be as high as possible.

The electrode is most easily made by melting platinum wire in a sodium-glass tube into which mercury is poured. Above the mercury the paste of mercury, colomel and potassium chloride solution is added. Frequently tubing is used for the salt bridge to the tissue, until the Ringer in agar-agar solution makes contact with the tissue.

The highly toxic effect of potassium makes this conventional electrode used in physical chemistry unsuitable, because some potassium will diffuse out to the tip of the recording electrode with fatal effects, even when extreme precautions are taken. However, the electrode can be made with approximately similar properties by replacing the potassium with sodium. Sodium is less toxic and will not impair the function of the nervous tissue as will potassium.

The design of the electrode can be suited to particular purposes, and several possible systems are described by Carter and Silver (1961). In other chapters of the same monograph, theories behind electrode systems are described and discussed in detail.

A widely used electrode system in neurophysiological research is the silver, silver chloride, salt bridge. A simple way to make this is by anodic coating of silver wire in a dilute hydrochloric solution. These electrodes may be used directly on nervous tissue, but it is not advisable to apply them directly to the olfactory epithelium. They can be stable for long periods of time, but may become unstable for no obvious reason. The electrodes are photosensitive and may change potential when struck by light. In addition they are mechanically unstable and susceptible to break-down by accidental scratching of the silver chloride coating.

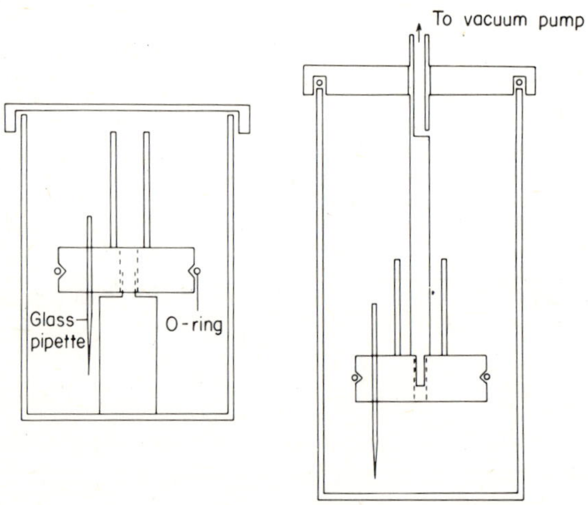

FIG. 5. Holder for glass microelectrodes fitting into a storage cylinder (left) and to a glass beaker for boiling (right).

Lately the Al Chem Company has produced a silver, silver chloride electrode where the coating is made porous by a special sintering process. These electrodes are made in threads or "pellets" which are mechanically and electrically more stable than the traditional home-made electrodes. An application of one of these electrodes is shown in the design in Fig. 6. These electrode holders may be purchased from WP Instruments, Inc.

The connection from the electrode to the tissue is very important. It is best to use a physiological saline solution, preferentially in agar-agar in glass capillaries. When the tip is small, on the order of 1 μm one can use isotonic or higher concentrations of sodium chloride. Higher concentrations increase the conductivity (and thereby decrease the resistance) of the electrodes. Simple metal electrodes are quite unstable and thereby unsuitable for d.c. recording.

Preparing electrodes for d.c. recording of the slow potentials is comparatively easy. The glass electrodes are drawn out to a thin taper. Over a small flame the glass capillary is bent just at the shaft. The electrode tip is cut and the end carefully melted to blunt the sharp edges. Filling the electrodes with Agar-Ringer can present some difficulty, but careful boiling under vacuum is usually sufficient to ensure a proper filling.

B. Microelectrodes

One of the difficulties in studies of the function of nervous structures arises from the specialization of nerve cells. Even within morphologically homogeneous structures like the olfactory bulb and the olfactory epithelium, the cells are specialized. Even cells which are morphologically similar might have quite different properties and react to special stimuli. This is characteristic of the nervous system, and presumably applies also to the olfactory structures. It is a result of the intricate connections which the cells make with each other. The system is therefore inaccessible to any kind of study which does not reveal the activity of single cells. Recording of "multi-spike activity" can only remain inadequate for revealing how the system functions.

In the last two decades there has been a tremendous advance in knowledge of the function of nervous structures because techniques have become available for recording from single cells. In the following will be described two methods available for recording the extracellular activity from single nerve cells in the central nervous tissue.

C. The metal microelectrode

Steel (Fe), platinum (Pt), wolfram or tungsten (W), copper (Cu), silver (Ag) and other metals have been used in attempts to make microelectrodes. Copper, platinum and silver usually become too soft and are not advisable since they bend and are easily blunted. The wolfram tungsten electrode described by Hubel (1957, 1959) is an excellent tool for recording single unit activity. The fact that a single electrode can be used over and over again fully compensates for the time it takes and the difficulties encountered in making them.

The electrode is prepared from 100 μm or thicker wire of tungsten conveniently obtained from a local light bulb factory. The wire should preferably be stretched in a flame, whereby it is straightened out, and cut in about 7-cm-long pieces. The sharpening is performed in saturated KNO_2 (potassium nitrate solution). The electrode is dipped repeatedly in the solution to about 5 mm either by a mechanical device or by hand while applying about 5 VAC. The other electrode can be the central carbon rod from a 1.5 V flashlight battery. When the electrode is sharp it will usually become too thin and fragile at the tip to be useful in electrophysiological work. It must therefore be blunted. This is done by slowly lowering the electrode tip into the solution while applying 1.5–2 V. When the first bubbles of etching appear the electrode is kept still until no more etching can be seen. Too long and slender electrodes are difficult to insulate and bend easily. The etching solution must be removed from the electrode by distilled water, and the electrode may be kept in a small reagent glass. When dry it is ready for coating. The trickiest part in preparing tungsten electrodes is to cover the electrode with insulation all the way except the tip. The lacquer proposed by Hubel is a clear vinyl lacquer S 9865 obtained from Stoner-Mudge CO, or the European equivalent.

One gets two cans: the lacquer and the corresponding thinner. The lacquer is too thin and should be heated or poured in a beaker and allowed to thicken for two or three days with frequent refills. When it has reached the consistency of "soft porridge" it can be used for coating.

The lacquer must be poured in a beaker to form a top meniscus, in order to let the thinner flow off, or be blown off, the lacquer surface so that a coating is formed. The electrode is dipped, tip up, into the solution until it is barely seen in the binocular microscope. A film is formed by gently blowing over the surface and then the electrode is lifted up into the air. A tent is thereby formed, and by continuously raising the electrode this tent should narrow and follow smoothly around the electrode shaft. Frequently a series of drops are formed which makes the electrode unsuited for deep recordings in nervous tissue because it will push the tissue instead of sliding through. When a tip of 15–35 μm is free from coating the electrode is suitable for intracellular work in the vertebrate CNS.

D. The glass microelectrode

Glass electrodes have found wide use in electrophysiological disciplines. They are readily made, filled and used, and can be formed to suit different applications. Several aspects of glass electrodes are covered in the monograph by Lavallée *et al.* (1969). It seems superfluous to present a thorough discussion of them here. The following is a short description of the making of glass microelectrodes and some practical hints on their use.

Several models of microelectrode pullers are commercially available and a few of them can be made in a standard equipped workshop. Pyrex glass is far superior to ordinary sodium glass for making microelectrodes for intra- or extra-cellular work. The Pyrex electrode can be bent and, for example, when recording intracellularly in muscles, the glass microelectrode with long shaft can follow the movement exerted by contraction of the muscle without jumping out of the cell.

Corning Pyrex glass with an outer and inner diameter of 1.4 and 0.8 mm, respectively, is suitable for most purposes. The glass should be rinsed in sulphuric acid and washed repeatedly in distilled water, whereafter it should be kept in a dry and dust-free environment until used. To avoid excessive handling of the electrodes after pulling I have found it convenient to use the form of electrode holder shown in the drawing of Fig. 6. A batch of up to 30 electrodes can be

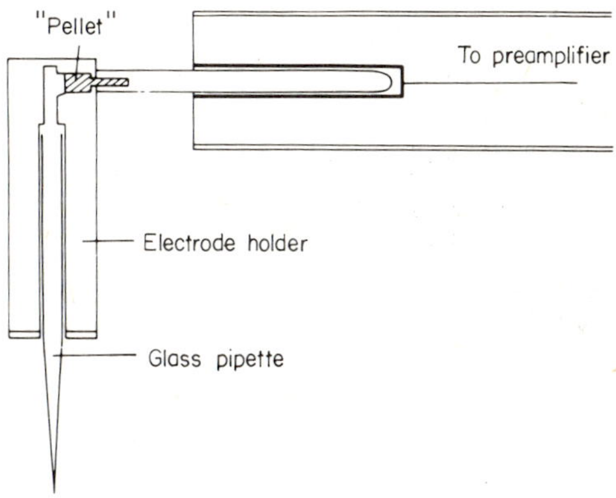

FIG. 6. Microelectrode holder in electronic preamplifier probe with silver/silver-chloride pellet.

fastened to the holder at a time. The electrodes can be kept dry in the perspex jar shown to the left in Fig. 6 and filled by boiling in the glass beaker device shown to the right. The beaker is filled two-thirds and the electrode holder fastened to the lid. The beaker is then heated to 95°C, water suction is applied and the water is allowed to boil for 10 minutes. The pressure should then be gradually decreased to normal, whereupon the electrodes are filled with the desired solution. Zettler (1970) has described a method for filling glass microelectrodes at room temperature. For intracellular work a standard solution is 2.5 N KCl. When using extremely fine pipettes with more than 100 Mohm resistance, it is desirable to have a 2.0 N solution to reduce the probability of crystals forming in the tip. These can plug the electrode or break the tip.

For extra-cellular work with an electrode resistance of less than 10 Mohm and a considerably larger tip diameter, from which the internal solution can leak out, it is necessary to use a salt other than KCl. A 4.0 or 2.5 N NaCl solution is frequently chosen.

The fine microelectrodes made in a puller usually have too high a resistance for extra-cellular work. It is then necessary to break the tip. This can be done before filling or after. I have found it convenient to break the tip just before use by lowering the electrode onto any metal surface connected to ground. One can then hear in the audio monitor where the electrode hits the hard surface. If this is done carefully, just the fine tapered tip breaks off.

The microelectrodes can also be filled mechanically in the following way. After pulling, the tip of the electrode is broken under the microscope. In this procedure only the fine tip will break off, if it is done carefully. The electrode is then filled from behind up to the shaft, avoiding the formation of bubbles. The electrode is fastened to a syringe with a clamping device which assures a water-tight connection. On pushing the plunger the solution will be pressed forward up to the tip and the electrode will be filled. By this method electrodes are easily filled at room temperature, but their resistance will not be high of course, probably less than 10 Mohm. The method permits filling of electrodes with any kind of solution. It is especially useful with Procion Yellow fluorescent dye, which does not tolerate heating above 40°C (see below).

In making recordings of single unit activity in the olfactory bulb it is of importance to know the timing of the stimulus to be able to describe the pattern of the unit activity in relation to the stimulus. As shown by Ottoson (1956) a short blast of odorous air onto the epithelium produces long-lasting excitation of the olfactory receptors which quite outlasts the actual blast of air onto the olfactory mucosa. In some studies of unit activity in the bulb the EOG from the sensory epithelium has been recorded simultaneously (Døving, 1964). This is shown schematically in Fig. 7. In the rat the peripheral response is not that easily obtained as in the frog, and because of the intricate

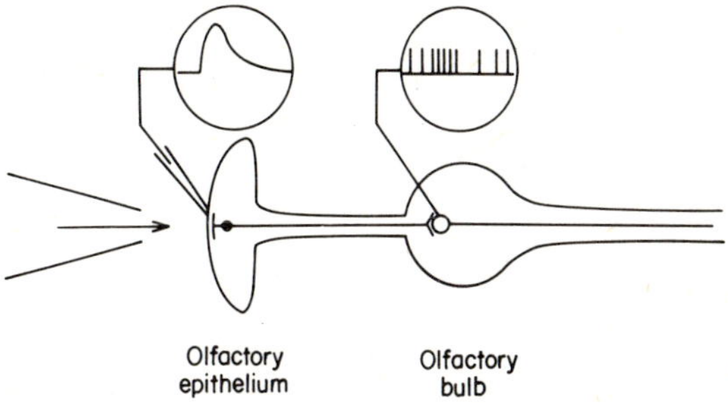

FIG. 7. Schematical drawing of the principle for recording simultaneously the activity of single units in the vertebrate olfactory bulb and the EOG from the olfactory epithelium.

system of conchae the timing may differ from one place in the turbinales to another. Ottoson (1959) has demonstrated the close correlation between the EOG and the slow potentials at the surface of the bulb. Instead of recording the peripheral response, one can therefore record potential changes on the surface of the bulb and use that for coordinating the activity of the single unit with the time course of the stimulus.

III. Material

A. Frogs

Frogs are widely used in neurophysiological studies because they are readily available and easily handled. The olfactory system is easily exposed which makes them convenient subjects. *Rana temporaria* should be kept in clean plastic tanks with access both to dry places and clean water. At a temperature of about 5°C they may eat some liver.

The anaesthesia can be given by placing a few crystals of urethane (ethyl carbamate) on the back of the animal. Other anaesthetics are nembutal and MS 222 (Sandoz, tricain-*methansulphonat*). Injections are best made into the dorsal lymph sac of the animals.

D-tubocurarine chloride can be used to release muscle tension in a dose of less than 0.3 ml of 0.3% solution. The respiratory activity is blocked by this agent, and to ensure proper oxygenation an air supply should be provided. This can be done by letting air into the lungs via a plastic tube in the trachea. The outlet is provided in the space between the tube and the trachea. A continuous flow of humidified air can thereby be supplied for the animal which will be sufficient together with the oxygen taken up by the skin. The ventral olfactory epithelium and the brain are easily exposed by cutting away the skin and removing the dorsal cartilage of the nasal capsule. Care should be taken to avoid cutting the supraorbital arteries and veins. The general anatomy of the frog is described by Gaupp (1899) and the olfactory bulb has been investigated by Ramón y Cajal (1922).

The frogs are fastened to the cork plate support by a few pins and kept moist by a covering of soft tissue paper. A small hole is made in the skull above the olfactory bulb and the cavity filled with oil. Pulsations due to heart beats are easily reduced by a clear plastic sheet with a hole for the microelectrode.

B. Rats

The use of rats in acute preparations for olfactory research requires a few precautions. Ordinary laboratory rats frequently acquire rhinitis which reduces and also sometimes obliterates the sense of smell due to partial or total infection of the olfactory epithelium. This can be prevented by rearing sterile animals in an isolater. Less drastic measures may also be fruitful, but care should be taken to control the condition of the rats in this respect.

A variety of anaesthetics may be used. Urethane is favourable because one dose of exactly 7.48 ml/kg of 20% solution will keep the animal in good operative condition, although it is advisable to keep it warm by a thermostatic heating device. Intra-peritoneal injections will last for 12 h without need for artificial respiration. As the animals seldom survive this anaesthesia, urethane should not be given when it is necessary to use the rat after the experiment.

When the animal is anaesthetized the throat is cut open and the trachea is dissected free. To avoid bleeding, one should seek to tear the muscles and connective tissue apart instead of cutting with scissor or knife. Two tracheal cannulas are inserted, one into the lungs, the other up into the choana. The latter should not be pushed too far since its purpose is to "sniff" for the animal, and advancing it past the turbinates will prevent odorous air from coming in contact with the olfactory epithelium. This can be seen in Fig. 8. The lung tubing, which should be comparatively short, can have an outer diameter of

2.2 mm; the choana tubing should be smaller, e.g. 1.9 mm. The tubes should be fastened by a suture around the trachea. The wound is sewn together and the tubes secured in their positions.

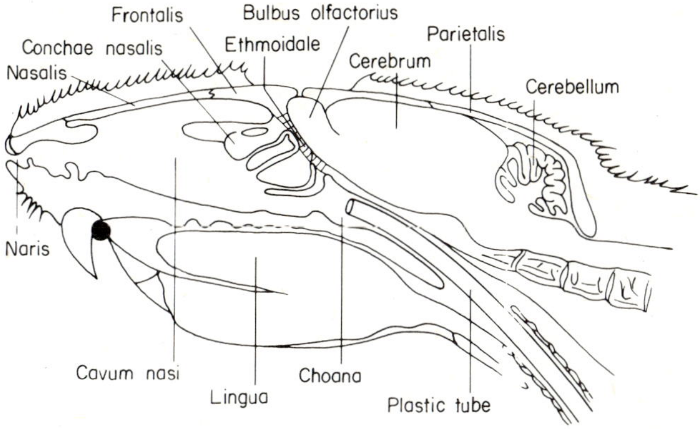

FIG. 8. Drawing of a medial section of a rat head with a plastic tube fitting into the choana.

The animal is now ready to be placed in the head holder (Rat holder, La Precision Cinematographique). The supporting points are the ear canals (meatus) and the incisors (Fig. 8). Because the standard holder supplied for the muzzle is not acceptable for use in our studies, we have constructed a slightly different holder which does not cover the nose and is especially designed to leave the nostrils open.

The fixation points are shown in Figs 8 and 9. Origo in the co-ordinate system created by these supporting points is the section between the vertical and horizontal lines. (The anatomy of the rat

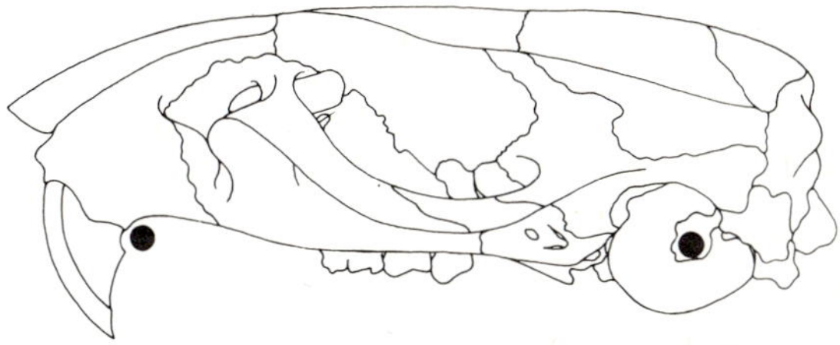

FIG. 9. Lateral view of the rat skull. The supporting points behind the incisors and in the meatus are indicated in black.

brain is described in several monographs and atlases (Green, 1959; Zeman and Innes, 1963; Pellegrino and Cushman, 1967).)

The olfactory bulb can now be exposed by removing the skin above the *Os frontalis* as shown in Fig. 8. Care should be taken to avoid cutting the inferior cerebral vein which drains the olfactory bulb. Two branches of this vein are located in the groove between the bulbs and the cerebellum underneath the skull (Fig. 10).

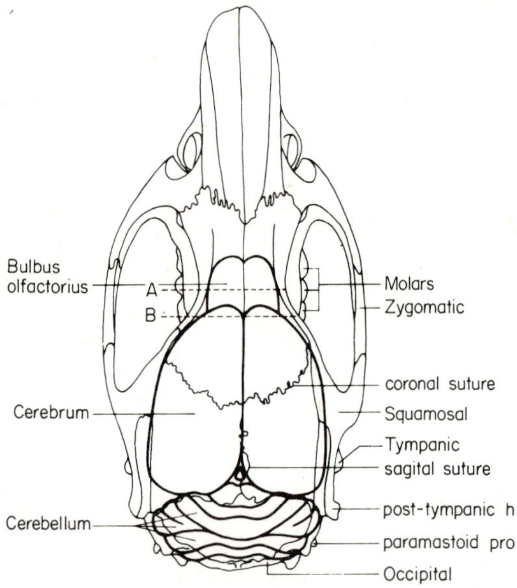

FIG. 10. Dorsal view of the rat skull with the outline of the olfactory bulb, cerebrum and cerebellum indicated with heavy lines.

For recording with microelectrodes it is unnecessary and undesirable to make a large hole in the *Os frontalis.* A small hole just sufficient for the electrode to enter prevents obtaining too great an amplitude of pulsations due to heart beats. Further reduction can be obtained by applying the plastic sheet described for the frog.

C. Determining electrode coordinates

In most neurophysiological work it is desirable to know from which cells in the brain structure one is recording. For this purpose there are several alternative or simultaneous methods available. The commercially available sterotaxic holders and atlases are not very useful since none to my knowledge include the olfactory bulb. It is possible, however, to make an atlas for the bulb and the anterior olfactory nucleus by standard histological procedures.

The brain is fixated *in situ* by perfusion with saline and fixative (96% ethanol and 10% formalin, 1:1 or 2:1). Serial sections normal to the dorsal brain surface are cut at 15 μm thickness and every fifth section is used for mapping the structures. Each section is then drawn on an acetate sheet from a projection of the section 100 times magnified. The drawings are stacked with the appropriate spacers to make a three-dimensional model of the bulb. Sections 18 and 45 of such a series are shown in Figs 11 and 12. With the aid of these sections one can approximately determine the location of the electrode tip in the different brain structures.

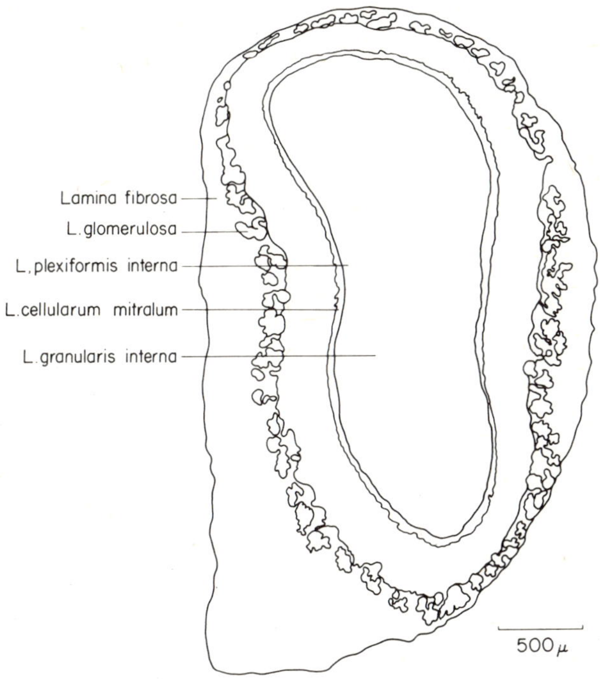

FIG. 11. Transverse section (nr. 18) of the olfactory bulb at the level A indicated in Fig. 10.

A convenient origo for the electrode coordinates at the surface of the olfactory bulb is the branching point of the inferior cerebral vein posterior to the bulb. The depth is located by adjusting the vertical readings for bulb surface contact and the final recording position of the electrode. In this procedure one is greatly aided by the anatomical peculiarities of the olfactory bulb. As seen from Fig. 11 the olfactory bulb is a nicely layered organ. The olfactory nerves come from the ventral aspect of the bulb and encircle the structure. The synapses with the various secondary neurones, the glomeruli, are distributed around the periphery. The mitral cell bodies are located about 300 μm from the dorsal surface of the bulb. These cells are the largest in the olfactory pathway and the cells most easily recorded by microelectrode techniques.

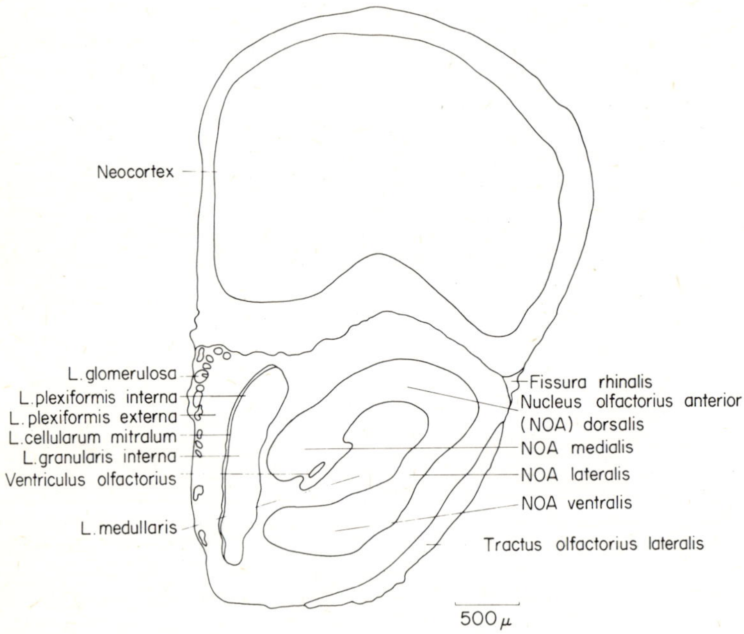

FIG. 12. Transverse section (nr. 45) through the anterior olfactory nucleus at the level B indicated in Fig. 10.

D. Marking

Knowing the electrode tip position in the structure is an insufficient criterion for identification of the cell from which one has been recording. In the highly complex matrix of cells in the central nervous system it is not even sufficient to locate the tip exactly within a certain layer by electrolytic lesion or by depositing stains from the tip of the electrode. Such localization techniques give a general indication of from which cells one has been recording, but this evidence is probably not better than that obtained by stereotaxic methods.

A new and highly promising method for identification of cell structures has been developed lately by Stretton and Kravitz (1968). These authors filled micropipettes with Procion Yellow M-4R (ICI) and, after intracellular recording, injected the fluorescent dye into the cell by electrophoresis. The results of this method for the lobster neuropil are fascinating. The same method has successfully been applied in the vertebrate spinal cord (Jankowska and Lindstrøm, 1970). In this way it is possible to determine the ramification and connections of a particular cell, the function of which has been explored previously by electrophysiological means. This method would combine neurophysiological and neuroanatomical efforts in exploring the properties of the nervous system. The possibility of using this technique on the olfactory bulb with its conspicuous stratification and relatively simple structure offers a unique challenge for neurophysiologists to gain insight into the function of the intracellular action essential for odour discrimination.

Appendix

Addresses

Al Chem. Co.
413 W. Imperial
Le Habra, California 90631, USA

ICI Ltd.
Dyestuff Division, P.O. Box 42
Hexagon House, Blackley, Manchester
M9 3DA England

La Précision Cinématographique
19 Rue des Parisiens
92 Asniéres-sur-Seine, France

Perkin Elmer Corporation
Norwalk, Connecticut 06852, USA

Stoner-Mugde Company
Pittsburg 33, Pa., USA

Stora Kopparberg
Fack 79101 Falun 1, Sverige

Tandberg Radiofabrikk A/S
Kjelsåsvn. 161
Oslo 4, Norge

WP Instruments, Inc.
2822 State Street
Hamden, Connecticut 06517, USA

References

Andres, K. H. (1965). Differenzierung und Regeneration von Sinneszellen in der Regio olfactoria. *Naturwissenschaften* **52**, 500.

Carter, D. B. and Silver, I. A. (1961). Microelectrodes and electrodes used in biology. *In* "Reference electrodes" (Ives, D. J. G. and Janz, G. J., eds.), 464–523. Academic Press, New York and London.

Drake, B., Johansson, B., von Sydow, E. and Døving, K. B. (1969). Quantitative psychophysical and electrophysiological data on some odorous compounds. *Scand. J. Psychol.* **10**, 89–96.

Døving, K. B. (1964). Studies of the relation between the frog's electro-olfactogram (EOG) and single unit activity in the olfactory bulb. *Acta Physiol. Scand.* **60**, 150–163.

Døving, K. B. (1965). Studies on the responses of bulbar neurons of frog to different odour stimuli. *Rev. Laryng. (Bordeaux)*, **86**, 845–854.

Døving, K. B. (1966a). An electrophysiological study of odour similarities of homologous substances. *J. Physiol. (Lond.)*, **186**, 97–109.

Døving, K. B. (1966b). Analysis of odour similarities from electrophysiological data. *Acta Physiol. Scand.* **68**, 404–418.

Gasser, H. S. (1958). Comparison of the structure as revealed with the electron-microscope and the physiology of the unmyelinated fibres in the skin nerves and in the olfactory nerves. *Exp. Cell Res. Suppl.* **5**, 3–17.

Gaupp, E. (1899). Anatomie des Frosches. *In* "A. Ecker's and R. Wiedersheim's Anatomie des Frosches" 2, Publishers aufl. 124.

Gemne, G. and Døving, K. B. (1969). Ultrastructural properties of primary olfactory neurons in fish (*Lota lota* L.) *Am. J. Anat.* **126**, 457–475.

Green, E. C. (1959). Anatomy of the rat. *Trans. Am. Phil. Soc.* **27**, 370 pp.

Hubel, D. H. (1957). Tungsten microelectrode for recording from single units. *Science, N.Y.* **125**, 549–550.

Hubel, D. H. (1959). Single unit activity in striate cortex of unrestrained cats. *J. Physiol.* **147**, 226–238.

Jankowska, E. and Lindström, S. (1970). Morphological identification of physiologically defined neurones in the cat spinal cord. *Brain Res.* **20**, 323–326.

Lavallée, M., Schanne, O. F. and Hébert, N. C. (1969). "Glass Microelectrodes". John Wiley and Sons, New York.

Mustaparta, H. (1971). Spatial distribution of receptor responses to stimulation with different odours. *Acta physiol. scand.* **82**, 154–166.

Ottoson, D. (1956). Analysis of the electrical activity of the olfactory epithelium. *Acta Physiol. Scand.* **35**, suppl. 122.

Ottoson, D. (1959). Comparison of slow potentials evoked in the frog's nasal mucosa and olfactory bulb by natural stimulation. *Acta Physiol. Scand.* **47**, 149–159.

Ottoson, D. (1970). The electro-olfactogram. A review of studies on the receptor potential of the olfactory organ. *In* "Handbook of Physiology: Chemical senses. Part I, Olfaction", Beidler, L. M. (Ed.). Springer-Verlag. In press.

Pellegrino, L. J. and Cushman, A. J. (1967). "A Stereotaxic Atlas of the Rat Brain". Appleton-Century-Crofts, New York.

Ramón Y Cajal, P. (1922). El cerebro de los batracios. "Libro en Honor de D. S. Ramón y Cajal. Jiménes y Molina". Madrid, 13–59.

Stretton, A. O. W. and Kravitz, E. A. (1968). Neuronal geometry: Determination with a technique of intracellular dye injection. *Science, N.Y.* **162**, 132–134.

Zeman, W. and Innes, J. R. M. (1963). "Craigie's Neuroanatomy of the Rat", 230 pp. Academic Press, New York and London.

Zettler, F. (1970). Kalt gefüllte Mikroglaskapillaren und ihre electrophysiologisch relevanten Eigenschaften. *Z. Vergl. Physiol.* **67**, 423–435.

9 Human Psychophysics in Olfaction

E. P. KÖSTER

Utrecht University,
The Netherlands

I. Introduction

Since 1859, when Fechner (1859) defined psychophysics as "the exact science of the functional relationships between the physical and the psychical (mental) world", psychophysicists have tried to establish the laws relating the physical stimuli from the environment to the sensations of the human observer.

In those fields of investigation in which the nature of the physical stimulus could be defined clearly, as in vision and audition, the contributions of the psychophysicists have led to the formulation of major theories on the functioning of the sensory system and to an understanding of complicated phenomena such as colour mixing. In other fields of research, like olfaction, where the exact nature of the physical stimulus is still unknown and where as a consequence the functional properties of the stimulus can not be varied at will by the experimenter, the results have been rather limited. Since the only functional property of the olfactory stimulus that can be varied systematically is the concentration of the odorous substance which reaches the nose (not even the olfactory epithelium) and since this property appears to be related to the intensity of the olfactory sensation, most experimenters have turned to psychophysical methods for assessing odorous intensity. These methods will be discussed in some detail in this chapter.

A few investigators have also used psychophysical methods for the assessment of odorous qualities. Although the value of their results remains limited by their arbitrary choice of odorous substances, their methods are interesting and will be discussed briefly.

It can not be the aim of this chapter to treat all technical procedures used in psychophysics in full detail. It seems more important to give a survey of the existing methods, to discuss the advantages and disadvantages of their use in olfaction and to indicate what kinds of questions can be answered with them.

II. Psychophysical methods for the assessment of odorous intensity

A. General

In psychophysics, a discussion of intensity traditionally centres around three questions:

(1) How strong should a stimulus be to become perceptible (detection problem)?
(2) How much should the physical intensities of two stimuli differ in order to be perceived as distinct (discrimination problem)?
(3) How is the perceived intensity of a stimulus related to its physical intensity (scaling problem)?

These three questions will be treated in the following sections.

B. How strong should a stimulus be to become perceptible?

This question leads directly to the concept of the "absolute" or "detection" threshold and to the problems involved in threshold measurement. It is generally agreed upon by all psychophysicists that the "absolute" threshold is a statistical rather than a truly absolute concept and that for any given sense modality there is no one physical intensity which constitutes the breaking point below which no stimulus is noticed and above which all stimuli are perceived. In fact, the chance that a stimulus will be perceived increases gradually over a certain range of intensities. Figure 1 gives the theoretical curves expressing the relationship between detection probability (percentage of positive response) and concentration of an odorous stimulus for two subjects (S and S') of different sensitivity (S being more sensitive than S'). By general agreement the detection threshold is defined as the concentration at which a subject gives a positive response in 50% of the cases in which the stimulus is presented to him. Thus, in Figure 1 the concentrations I and I' are the threshold concentrations for S and S' respectively.

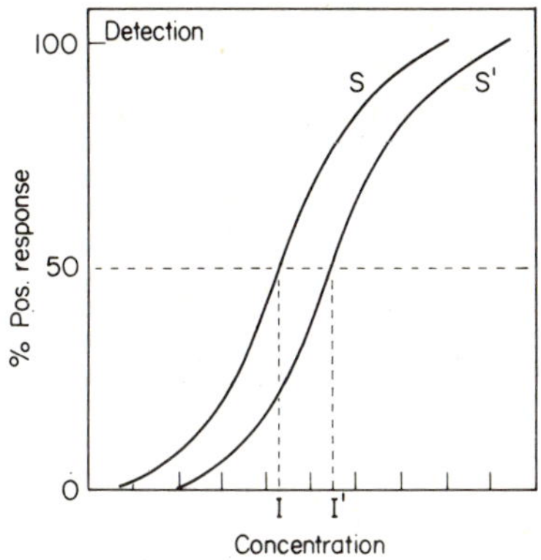

FIG. 1.

There are a number of classical methods for threshold measurement. They differ both in the order in which the stimuli are presented to the subject and in the nature of the responses asked from the subjects. Not all of these methods are suitable for use in the study of olfaction and with most of them special precautions should be taken.

The method of limits

With the method of limits a number of stimuli differing in small discrete steps on a logarithmic concentration scale (usually ^{2}log), chosen such that the lowest one is not perceptible and the highest one is clearly perceived, are presented alternatingly in ascending and descending series. An ascending series is continued until the subjects' response changes from "no" to "yes" and a descending series is stopped after a change in the response from "yes" to "no". In order to avoid that the subject builds up an expectation about the position of the transition point in the series merely on the basis of the number of trials or the time elapsed since the beginning of the series, the intensity which serves as the starting point is varied at random in successive ascending and descending series. The physical intensities of the stimuli at the transition points are averaged over a number of at least 10 ascending and 10 descending series and this average is considered to be the threshold for a given subject. Normally, the transition points in the ascending series lie at a somewhat higher intensity than those in the descending series. This phenomenon is understandable, if one realizes that in the descending series the subject knows better what he is looking for in the next trial because he has experienced the stimulus in the last one, whereas in the ascending series his memory of the stimulus quality may fall short, because he has not perceived it for some trials. Also, the subject might have a tendency to perseveration in his answers, which might lead him to continue to say "no" in the ascending series and to say "yes" in the descending series.

In olfaction, however, this overlap between the transition points in the ascending and the descending series is very seldomly, if ever, found. On the contrary, one usually finds a gap between the transition points, those of the descending series being higher than those found in the ascending series. This phenomenon can be illustrated by the results of Pangborn *et al.* (1964) showing the percentages of positive response to different concentrations of 2-heptanone obtained with three forms of stimulus presentation: sequential-up, sequential-down and randomized order. As can be seen from Fig. 2, the lowest threshold (equivalent to lowest transition point) is obtained with the sequential-up method and the highest one with the sequential-down method, the threshold obtained with the randomized order taking an intermediate position.

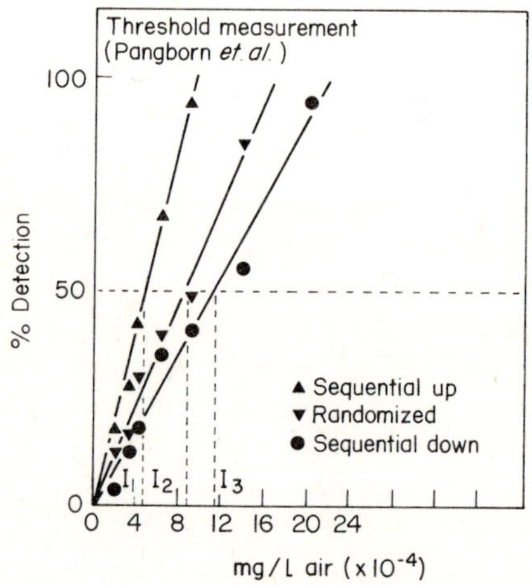

FIG. 2.

This phenomenon can best be explained in terms of olfactory adaptation. As was shown by Köster (1971) strong stimuli have a larger adapting effect than weak stimuli, but even weak stimuli of short duration reduce the olfactory sensitivity for quite some time (up to 2 min.). In Pangborn's study the interval between stimuli was only 15 s and since with the sequential-down method the preceding stimulus is always stronger than the one following it, whereas with the sequential-up method the preceding stimulus is weaker than the one following it, differences in the adapting effects exerted by the preceding stimuli may very well explain the differences in threshold found with the two methods. Since, unless sufficient time is allowed for full recovery of the olfactory sensitivity between stimuli, adaptation affects above all the thresholds obtained with the sequential-down method and to some extent also those obtained with the randomized order method, the sequential-up method is preferable. It is both more economical (shorter inter-stimulus intervals permitted) and more sensitive (lower threshold) than the sequential-down method, which is usually avoided.

The method of constant stimuli

Compared with the randomized presentation method, the sequential-up method has the disadvantage that quite a number of extra stimuli may be needed to avoid the expectation effects mentioned earlier. Since these extra stimuli are always too low to be perceived, they do not contribute to the actual measurement. For this reason and because it is easier to automate experiments in which the order of presentation of the stimuli is independent of the subjects' response (especially when one administers the same stimulus to a group of subjects at the same time), many investigators have chosen for the method of randomized presentation or, as it is often called, the method of constant stimuli. Five or seven stimuli of different concentrations are selected on the basis of a very quick survey with the ascending method in such a way that the weakest one is perceived in about 10% and the strongest one in about 90% of the cases. These stimuli are then presented a large number of times (at least 50) to the subjects in a randomized order or in an order which is systematically varied in such a way that each stimulus follows each other in an equal number of times, the order of the pairs of successive stimuli being randomized as much as possible. The interval between two successive stimuli should be at least one minute to reduce adapting effects and, in order to avoid general fatigue on behalf of the subjects, the experiment is usually split up in separate sessions in each of which maximally 35 stimuli are presented. At the end of the experiment the percentages of positive response to each of the five or seven stimuli are calculated for each subject and plotted against the concentration (Fig. 3a). When the percentages of positive response are expressed in *z*-scores (or standard deviation units) (Fig. 3b) a straight line can be fitted to them and the threshold can be determined by interpolation at the 50% (*z*-score = 0) point. Thus, all measurements have contributed to the determination of the threshold.

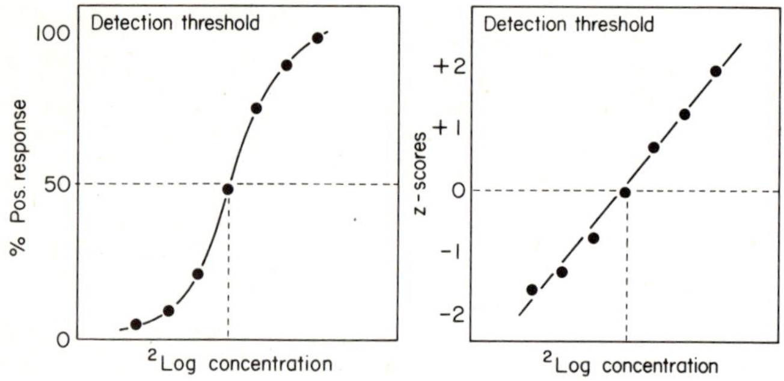

FIG. 3.

The use of blank stimuli and the forced choice method

In both methods described so far the subjects were asked to indicate whether they did or did not perceive the stimulus presented to them by simply saying yes or no. This procedure has a number of disadvantages. Subjects may be biased to saying either yes or no on a number of grounds which are quite unrelated to their actual sensitivity. If, for instance, the threshold measurement is carried out as part of a selection procedure for panel members in sensory evaluation research and if being a panel member is a very desirable position, subjects could cheat and say yes much more often than they actually perceived the stimulus. But even if subjects have no intention to cheat they may be biased in one way or another without being aware of it.

In order to be able to check the influences of such response biases, experimenters have introduced blank stimuli in their series, i.e. stimuli not containing the odorous substance, but consisting of the solvent only.

If the subject often answers positively to these stimuli, he may be told that he is not performing well, because he perceives stimuli that do not exist.

Telling the subject that he is "imagining things" has the disadvantage that it may make the subject over-cautious and may lead him to adjust the criterion of what he accepts as a proper stimulus in such a rigorous way, that this threshold becomes artificially high. He is not prepared to do guesswork anymore for fear of being trapped by the blank stimuli.

Correcting the responses of the subject after the experiment may be better if one does not want to bias the subject in the direction of saying no whenever he is in doubt, but it has certain disadvantages too. It presupposes that the subject keeps his criterion and therefore the amount of guessing constant during the whole experiment. However, the experimenter has no direct way of controlling the truth of this presupposition. Therefore, a number of experimenters have used a method which forces the subject to make guesses whenever he is in doubt. With this method, the "forced choice" method, the subject receives a number of stimuli (usually 4 or 5) in each presentation. These stimuli may be presented either simultaneously in separate places (vision) or in temporal order (audition, olfaction). All but one of the stimuli are blanks and the subject is instructed to find the stimulus which is different from the others. Even if he can not tell the difference at all he has to make a choice by guessing. After the experiment his results are corrected by the following formula:

$$P_{(corr)} = \frac{P_{(obs)} - P_{(chance)}}{100 - P_{(chance)}} \times 100$$

$P_{(corr)}$ = Percentage of positive response to a stimulus after correction.

$P_{(obs)}$ = Percentage of positive response given by the subject to that stimulus.

$P_{(chance)}$ = Percentage of positive response expected on the basis of chance (guessing) alone (i.e. 25% if 4 stimuli are given in each presentation).

Figure 4b illustrates the results of such a correction procedure. The

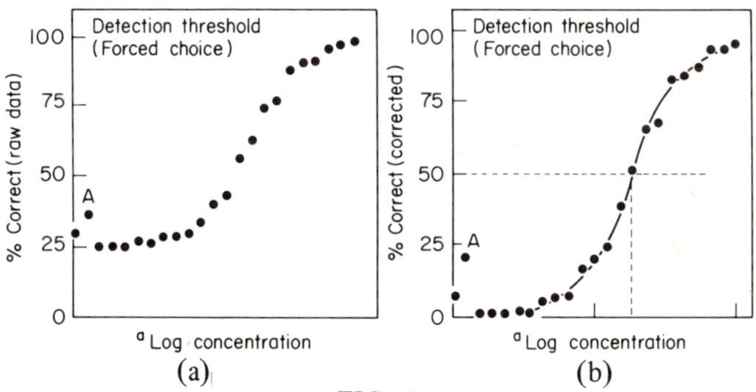

FIG. 4

percentages in Fig. 4a may of course be transformed into z-scores in order to obtain a straight line as in Fig. 3b.

In olfaction, the forced choice technique has often been used in detection threshold measurement. It is usually combined with the method of constant stimuli. The blank stimuli which are presented with the odorous stimulus in each presentation contain only the diluent (usually benzyl benzoate, diethyl phthalate or water) which is used to reduce the concentration of the odorant stimulus. Great care must be taken not to provide the subject with other clues to the detection of the correct response than the olfactory ones. Since the subject has the blanks at hand for direct comparison very small differences in colour or amount of substance in the test bottles may provide such clues, which might remain unnoticed in successive presentation of single stimuli.

It is also very important to use the same batch of diluent for both the preparation of the odorous concentrations and the blank stimuli. Very slight differences in purity of different batches may spoil the experiment by providing extra clues. In some instances, when a diluent is used which is almost, but not completely odourless, the subject may do very well on concentrations which lie clearly below the concentration level at which he has already reached chance performance (point A in Fig. 4a). The explanation of this strange phenomenon lies in the fact that the minute quantities of odorous substance which in themselves are no longer perceptible block the odour of the diluent by the well-known mechanism of odour compensation (suppression of the perceived odorous intensity) which tends to occur in mixtures with uneven mixing ratios (Köster, 1968; MacLeod, 1968). In these cases the subject detects the right stimulus because its smell is *weaker* than that of the surrounding blank stimuli.

Signal detection theory

In 1961, Swets, Tanner and Birdsall demonstrated that mathematical decision theory provides an elegant theoretical framework for the formal description of the influences of response biases on psychophysical measurements.

Although psychophysicists had been aware of the influence of response bias on their measurements, not much had been done to evaluate the influence systematically up to that time. In recent years the influences of response biases have been investigated thoroughly. An extensive account of this research has been given by Green and Swets (1966) and more recently by Engen (1971). Here only a simplified synopsis of the theory will be given for the sake of brevity. According to signal detection theory, signals always appear on a background of "noise", from which they must be distinguished. The "noise" (N) may arise from a variety of sources including external events on the one hand and spontaneous neural discharges on the other. Its intensity is supposed to be normally distributed and if one adds a signal (S) of constant intensity to it, the resultant intensities of signal plus noise ($S + N$) are supposed to be normally distributed too. This is illustrated in Fig. 5.

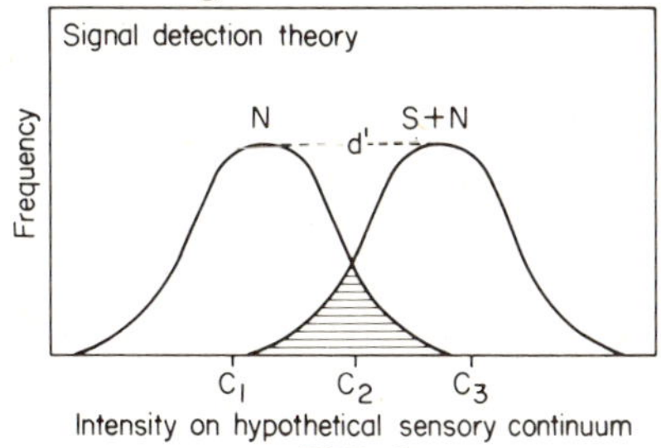

FIG. 5.

From this figure it may be seen that the N and $S + N$ distributions may have a considerable amount of overlap (shaded area). This means that a subject, confronted with a value on the sensory continuum under these overlapping parts, is faced with the problem to decide whether this value belongs to the N distribution or to the $S + N$ distribution. This holds of course for all values on the continuum, but it becomes particularly difficult for the subject for values under the shaded area. Since he has no clue to guide him, a certain amount of guesswork will have to be done and the theory supposes that he will set up and maintain a certain criterion value (c) of his own. Values above this criterion value he will ascribe to the $S + N$ distribution (i.e. he will say "yes" to them in a psychophysical experiment) and values below the criterion value he will consider to be noise only (he will say "no"). The value at which the subject sets his criterion seems to depend mainly on two kinds of response bias, his expectations about the frequency of occurrence of the signal and his motivation to give or avoid particular responses (i.e. "yes" or "no"). Each time the subject in a psychophysical experiment gives a response and depending on the presence of the signal at the moment of the response one of the following four signal-response combinations will occur:

		Signal	
		On	Off
Response	Yes	Hit	False Alarm
	No	Miss	Correct Rejection

Now if the N and $S + N$ distributions did not overlap and if the subject were a perfect observer, only Hits and Correct Rejections would be found. However, since the two distributions do overlap, the subject is caught in an ambiguous situation in which each criterion which he sets will, along with Hits and Correct Rejections, inevitably result in a number of Misses or False Alarms or both. It is this fundamental ambiguity of the situation which renders the subject sensitive to the influences of response biases while he is setting his criterion.

Thus, if a high positive reward is attached to scoring Hits or a form of punishment is given whenever he produces a Miss, whereas no reward or punishment is given for False Alarms and Correct Rejection, the subject will tend to shift his criterion in the direction of c_1 and, by doing so, will indeed receive more reward or avoid more punishment.

Also, telling the subject that he is "imagining things" (see the example in the preceding section) will have the opposite effect and will make the subject shift his criterion in the direction of c_3. Thus, he avoids giving False Alarms (and being called a liar) at the cost of making a larger number of Misses (higher threshold). In the forced choice situation the criterion of the subject is again forced in the direction of c_1 and a correction by means of the formula mentioned above is applied afterwards on the data to reduce the large number of False Alarms which necessarily accompany such a criterion-value. It is easy to see why signal detection theorists object to this crude treatment.

Expectations based on knowledge about the frequency of occurrence of the signal may also influence the criterion-values set by the subject. If the subject knows that the frequency of occurrence of the signal is high, he tends to shift his criterion in the direction of c_1 thereby producing both more Hits and more False Alarms, but if the subject knows that the frequency of occurrence of the signal is low he will shift his criterion towards or even beyond c_3 and by doing so, he will increase the number of Correct Rejections and Misses.

The influence of expectation has been investigated in experiments in which the subject is told in advance what the frequency of occurrence of the signal will be. In a number of successive experiments with the same stimulus intensity this frequency is varied from 1 signal in every ten presentations (0.10) to 9 signals in every ten presentations (0.90). By plotting the proportion of Hits against the proportion of False Alarms for each of the separate experiments an isosensitivity curve can be obtained (Fig. 6b) which illustrates the influence of the response bias based on expectation.

The two different curves in Fig. 6a show the results found for two stimulus intensities. The distance of these curves to the diagonal line is indicative of the strength of the stimulus. The larger this distance is, the stronger is the stimulus. If the two curves in Fig. 6a are the results of two different subjects both receiving the same stimulus intensity, the subject producing the upper curve is more sensitive than the other one. The distances to the diagonal in Fig. 6b are equal to d' in Fig. 5. In olfaction, a number of investigators have used signal detection methods (Semb, 1968; Corbit and Engen, 1971; Janssen, 1970) and have shown that the general principles of signal detection theory are true also for this sensory modality.

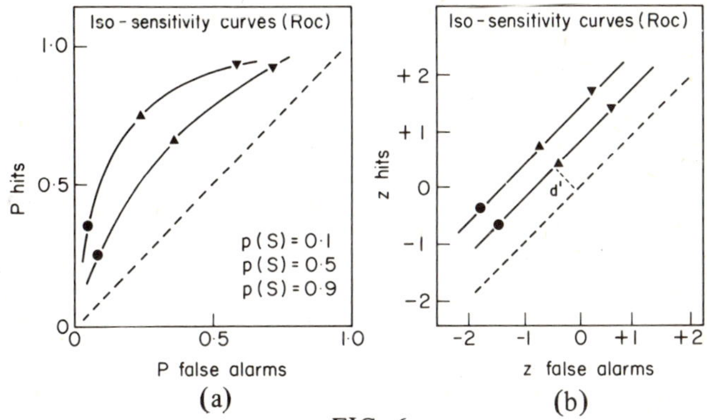

FIG. 6.

Although signal detection theory has made a very important contribution to the study of response bias phenomena and has made psychophysicists much more aware of the kind of decisions they ask of their subjects, its practical value in psychophysical measurement of intensity remains limited due to the very large number of presentations needed to obtain an isosensitivity curve as shown in Fig. 6b. This number of presentations can be reduced by applying a special procedure called confidence rating, in which the subject is asked to rate the degree of certainty with which he judges that the stimulus belongs to the N or the $S + N$ distribution. But even if the number of presentations is decreased with this method, it still remains large and the conversion of the certainty data into Hit and False Alarm proportions is based on a number of suppositions about the consistency of the subjects behaviour which are still open to some questioning. Thus, quite recently, Janssen (1972) demonstrated that the criterion used by the subject does not often remain constant during the course of an experiment.

The lessons about the influences of response bias taught by signal detection theory can of course also be applied in the work with the less demanding methods described earlier. Experimenters should try to keep the subjects' expectations about signal occurrence as constant as possible by carefully balancing their series for signal frequency and they should try to keep the level of motivation to give a particular response constant by giving the subjects so-called pay-off functions, i.e. a system of rewards and punishments (bonuses and financial penalties) for Hits, False Alarms, Misses and Correct Rejections. It should be remembered that a change in the signal frequency may completely change the effective value of such pay-off functions.

C. How much should the physical intensities of stimuli differ to perceive them as distinct?

Fundamentally, there is no difference between the detection problem described above and the discrimination problem to be treated here. In both cases the subject is asked to tell the difference between a stimulus and something else, whether it is another stimulus or a more or less "noisy" background. In fact, the detection problem can be considered as a special case of the more general discrimination problem.

Discrimination threshold determination

Just as for detection one can determine a threshold for discrimination. This threshold, which is often called the differential threshold or the just-noticeable-difference (jnd) is defined as the difference between the physical intensity which is just detected by the subject in 50% of the cases.

Usually the jnd is determined by a paired comparison method in which a standard stimulus (*S*) is presented together with a comparison stimulus (*C*). The subject is asked to indicate whether he perceives the comparison stimulus to be stronger than the standard or not. A series of comparison stimuli (usually 7 or 9) is selected on the basis of preliminary tests in such a way that the intensities of the *C*s are about equally distributed around the intensity of the *S*, the middle intensity of the *C* series being equal to *S*.

The distances on the intensity scale between the successive *C*s in the series are equal. The highest *C* should be perceived in about 90% of the cases to be stronger than *S* and the lowest one should be perceived in about 90% of the cases to be weaker than *S*. Each *C* is paired with *S* a large number of times (at least 50) and the pairs are presented in a random order to the subject. After the experiment the percentages of "C stronger than S" judgements are calculated for each pair and these percentages are plotted against the intensity scale as in Fig. 7. The same can be done for the percentages "weaker".

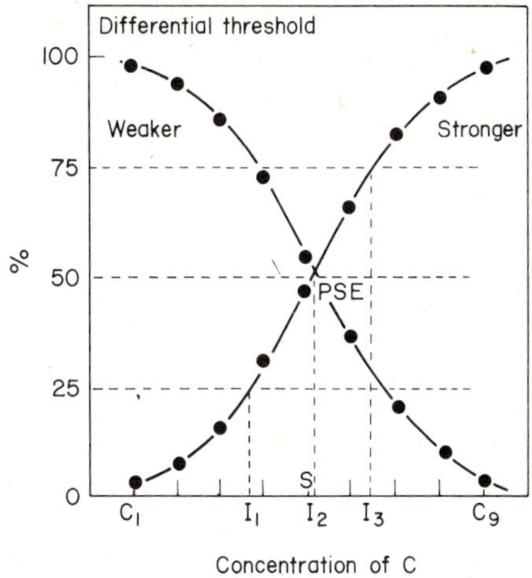

FIG. 7.

Before discussing the way in which the differential threshold (jnd) can be obtained from these data some general points should be raised. Since S and C_5 have exactly the same intensity, the point of subjective equality (PSE), i.e. the point at which the C is judged to be stronger than S in 50% of the cases (and therefore also weaker in 50%) ought to be found right above this intensity. However, this is often not the case. The PSE may be shifted to the higher or to the lower side of the intensity ranges as a result of a number of factors. In the first place the subject may be biased to saying "stronger" more easily than saying "weaker" or vice versa. All that has been said above about response bias applies here too. Expectation and motivation may have a large influence.

In order to check on the effects of expectation it is very important that the intensities of the Cs are indeed equally distributed around the intensity of the S. Only in this way will the subject build up the expectation that "stronger" and "weaker" ought to apply about equally often.

Another phenomenon that may influence the responses is adaptation. If C and S are equal in intensity and are presented shortly after each other, the second one will usually appear to be weaker. This effect can be counteracted by presenting the pairs of stimuli equally often in the C–S order as in the S–C order to the subject. In olfaction this precaution is particularly important because adaptation may have a very large influence.

The influence of adaptation is somewhat counteracted by a tendency, which most subjects seem to have, to overestimate the intensity of the last stimulus of a pair of stimuli. This general phenomenon, which is often called "time error", is usually attributed to the fact that the memory trace of the first stimulus may already be fading while the second stimulus is presented. There exists a vast and somewhat confusing literature on time error effects, which need not concern us here, because the effects of time errors can largely be reduced by counterbalancing the stimulus orders as was shown for adaptation.

Usually, when the PSE is shifted, all the points of the curve are shifted with it. This is especially true if response bias or time error cause the shift. In these cases the form of the curve may not be distorted at all or only to a very limited degree. Adaptation, which may affect the perceived intensities at the lower end of the scale somewhat more than those at the higher end of the scale, may cause some distortion. Nevertheless, if all the precautions mentioned earlier are taken, the effects on the shape of the curve remain small enough to use the curve for the calculation of the differential threshold even if the PSE is shifted considerably. This is possible because the jnd is in principle a difference measurement between two intensities and because shifting both intensities in the same direction by an equal amount does not alter the difference between them. Thus, in calculating the difference threshold (jnd) we should depart from the observed intensity which belongs to the PSE rather than from the physical intensity of *S*. The problem of finding the jnd can then be restated as follows: How much does one have to add or to subtract from the intensity corresponding with the PSE to reach an intensity which the subject perceives as being different from it in 50% of the cases? Since a paired comparison experiment is typically a forced choice experiment with a 50% chance level, the percentages "stronger" or "weaker" have to be corrected with the formula described above under "forced choice method". This means that the corrected percentages "stronger" and "weaker" at the PSE become zero and that the 75% "stronger" and 75% "weaker" (i.e. 25% "stronger") points of the original data are transformed to the corrected 50% points. The jnd is the difference in the values of the intensity scale corresponding with these corrected 0% and 50% points. Converting the percentages to *z*-scores makes it possible to fit a straight line to the data and to let all points contribute to the threshold determination (see Fig. 3b).

An easier and more direct, although less precise, method of determining the jnd around a given PSE is to take the difference on the intensity scale between the 25% and 75% points of the original data and divide it by two (Fig. 7).

Weber's law

If one varies the value of *S* one soon finds out that the jnd is not independent from the intensity of *S*. The higher the intensity of *S* is, the larger is the difference threshold. This relationship has been formalized in Weber's law, which reads: jnd = $c.I$, when I is the intensity of *S* at which the jnd is measured and c is a constant, called the Weber fraction.

Weber's law holds well for most sense modalities in the middle range of the intensity scale. Stone and Bosley (1965) have shown that this is also true for olfaction. The Weber fractions found for different substances varied from 0.20 to 0.36.

Matching of odorous intensities

The paired comparison method can also be used to match the perceived intensities of different odorous substances. Thus, in experiments on the odorous intensity of odour mixtures (Köster, (1968, 1969) used a fixed standard stimulus (a molar concentration of *m*-xylene) and a number of concentration series of different pure odorous substances and of their mixtures in a number of fixed mixing ratios as the comparison stimuli. For each pure substance and each mixing ratio the concentration corresponding with the point of subjective equality to the standard stimulus was determined. In this way it could be shown that odorous substances when mixed in certain ratios will suppress each other's odorous intensity.

Matching of odorous intensities is also important if one wants to prepare stimuli for quality assessment. Intensity differences should be avoided whenever the odorous qualities of odorous substances are compared, because otherwise the subjects are often unable to single out the quality aspects.

D. How is the perceived intensity of a stimulus related to its physical intensity?

So far, only methods for the measurement of the subjects' absolute or differential sensitivity have been described. By these methods the smallest differences in physical intensity that lead to a change in the sensation experienced by the subject have been determined. Although this is useful information for a number of practical and theoretical purposes, it is clear that the sensations experienced by subjects may vary over much larger ranges than the jnd or the absolute threshold. Thus the question arises as to how these larger differences in sensation or perceived intensity are related to the differences in the physical intensities of the stimuli provoking them. Does this mean that if one doubles the physical intensity of a stimulus the sensation caused by it is also doubled? Or may one assume (as is still often done in aroma research) that by setting up a scale of physical intensities in which the absolute threshold is chosen as the scale unit, one ensures that equal intensities on this scale will give rise to equal intensities of sensation, regardless of qualitative or even cross-modal differences in the stimuli employed?

In other words, is there a direct linear relationship between the physical intensity (I) of the stimuli and the strength of the sensation (S) experienced by the subject? Although there is still some discussion about the exact nature of the relationship, everybody seems to agree that it is not a linear one.

Fechner assumed that Weber's law was true and that the jnd, being the minimal amount of change that could be detected, should be considered to be the scale unit on the physical scale corresponding with one unit of sensation. On this basis he proposed a logarithmic relationship of the following form: $S = k \log I$, in which k is a constant which may differ for different stimulus qualities of different sense modalities. For the last twenty years this law has been challenged by S. S. Stevens, who proposes a different law reading: $S = I^n$, in which n is again a constant which may vary with the stimulus quality or the sense modality investigated. This power law, which can be transformed onto $\log S = n \log I$, states that there is a linear relationship between the logarithm of the physical intensity and the logarithm of the sensation or perceived intensity. The constant n gives the slope of the straight line in a double logarithmic plot. Since it can not be the aim of this chapter to discuss the theoretical merits of these laws and of other variants proposed, we will limit ourselves to a short description of the methods most commonly used in olfaction for the determination of sensation values of physical intensities. For a full description of the theoretical and practical implications of the use of scaling methods we strongly urge the reader to consult the excellent review by Engen (1971), which uses many examples from the field of olfaction.

It should be remembered that even if some of the scaling methods discussed below do not give us a direct insight into the exact nature of the sensation scale, they may provide a means for matching stimuli at fixed levels of sensation provided the subject can handle them in a reliable way.

Category scaling

One of the first methods which comes to mind if one is faced with the problem of determining the strength of sensation in relation to the physical intensity of the stimulus, is to let the subject rate the strength of his sensation on a given rating scale. Such a scale might have the following form (adapted after Engen, 1971):

Description	Scale Value
Very very weak	1
Very weak	2
Weak	3
Medium-weak	4
Medium	5
Medium-strong	6
Strong	7
Very strong	8
Very very strong	9

The subject is asked to assign each of the stimuli presented to him to one of the scale categories by mentioning the appropriate scale value. Usually, at the beginning of the experiment, he receives a "very strong" stimulus and a "very weak" one, in order to provide him with some indication of the stimulus range. Often the scale is reduced to seven or even five categories, because many experimenters feel that the subject can not handle too many categories adequately. In these cases the descriptions "very very weak (strong)" and "medium-weak (strong)" are usually omitted.

For each of the presented stimuli (presented at least 10 times) the mean of the category judgements is calculated and plotted against the physical intensity. Figure 8 shows the results of two experiments with the same stimuli, but with different numbers of categories on the sensation scale.

As can be seen from this figure, the two curves are similar in shape, both levelling off at their extremes. This levelling off is one of the disadvantages of the use of category scaling. The subject is confronted with a set of categories which is forced upon him and he seems to use it more freely in the middle range than at its extremes. His reluctance to mention the extreme categories may also explain that the curve found with the nine categories is somewhat steeper than the curve found with the five categories. In both cases the extremes are avoided, but this affects the five category scale more than the nine category one.

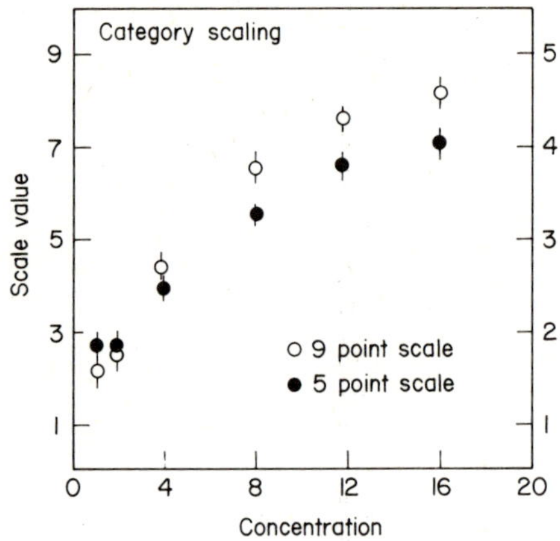

FIG. 8.

The fact that the subject behaves in this way has theoretical importance. It means that the categories do not represent equal steps of sensation to him and in fact there does not seem to be a good reason why his judgement "very strong" should differ by an equal amount of sensation from his judgement "strong" as the "strong" judgement differs from the "medium-strong" judgement. But if the distances in sensation between the categories are not equally large and if the relationship between the category judgements and the physical intensity is by no means a simple one, the category scale does not seem to provide a very satisfactory answer to the question, which of the two laws mentioned above does apply best to the relationship between sensation and physical intensity. To settle that matter, a method is needed which permits the subject to choose his own sensation scale and does not force him to think in predetermined categories. Methods like magnitude estimation and cross-modality matching (see below) seem to meet these criteria in a better way. Nevertheless, category scaling has also some advantages. The subject has little difficulty in performing the task, because he is usually familiar with this type of rating procedure. As a result he uses the categories in a rather reliable way as can be seen from the standard deviations given at the different points of Fig. 8. Therefore, category scaling may be a very good method for matching stimuli. In olfaction it has been used on several occasions (e.g. Mitchell, 1971).

Magnitude estimation

Magnitude estimation is the most direct of all scaling methods, because it leaves the subject completely free to set up his own sensation scale. The subject receives a series of intensities in random order and he is instructed to assign numbers to each of these intensities in such a way that he feels that the ratios between the assigned numbers correspond with the ratios between the observed intensities. He is told that he may use any numbers he wants, including decimal numbers. The method has been devised by S. S. Stevens and is generally believed to provide the best evidence for Stevens power law.

Two variations of the method can be used. In the first one (fixed modulus method) a standard intensity (usually the middle one from the series) is given to the subject and he is asked to call the sensation value corresponding to that intensity "10". In the second variation (free modulus method) no standard is given and the subject is given complete freedom to choose his own scale.

For each stimulus intensity the geometric mean of the numbers assigned by the subject in the different presentations is calculated by converting the numbers into logarithms, averaging the logarithms and reconverting the average into its antilog. This last step is often omitted and the averages of the logarithms are plotted directly against the logarithms of the concentration. Figure 9 gives an example of the

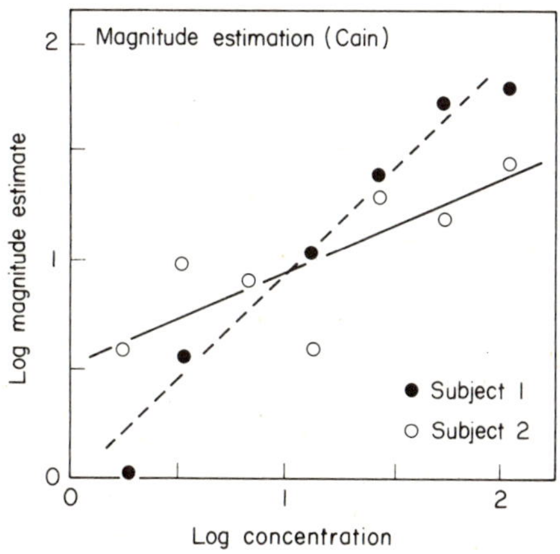

FIG. 9.

results of two different subjects who received each of nine intensities twice (Cain 1968; in Engen, 1971).

As can be seen from this figure, the results of the subjects differ considerably in two important aspects: the steepness of the straight line that can be drawn through the points of the individual subject and the goodness of fit of the lines to the points through which they are drawn.

Subject 1 needs about a full log step of concentration to let his magnitude estimate grow by one log step, whereas Subject 2 needs two log steps of concentration to reach the same increase in magnitude estimate. Does the difference in exponent of the Stevens' power function, which is directly reflected in the difference in slope of the two lines, mean that the two subjects have indeed a very different range of sensations when they judge the same stimuli or does it mean that the subjects simply use the numbers in a different way? The question is hard to answer. It is known that many inexperienced subjects have difficulties in using a scale that is based on ratios. Some tend to produce a category scale and some start using numbers in an almost haphazard way. The large dispersion of the points around the line drawn for Subject 2 makes one wonder whether the subject has used the ratio scale effectively. Even an adequate ordering of the stimuli according to their intensity is lacking. This casts some doubt on the precision of magnitude estimation, as an indication of sensation strength. Of the 15 subjects in Cain's study who received each of the nine intensities twice only one (Subject 1 in Fig. 9) produced magnitude estimates in an order which correlated perfectly with the order of the stimulus intensities, and Subject 2 in the figure does by no means represent the only or even the worst case of non-linearity in the set of data.

The results might perhaps be improved considerably by taking more measurements per point, but even then the large differences in the slopes of the lines remain. In his discussion of the individual results of Cain's experiment, Engen gives 0.56 as the mean slope for the whole group of subjects. However the standard deviation is 0.22 and the variation is therefore rather large.

Although individual functions, based on enough measurements per point and obtained with well-trained subjects, usually support Stevens' law, the large variability round the individual points and the large variability of the slopes of the lines obtained for the individual subjects, cast serious doubts on the meaningfulness and the practical value of the method. Simple matching of intensities on the basis of equal sensation values appears to be more reliable when carried out with either category scaling or paired comparison methods than with magnitude estimation.

Cross-modality matching

With magnitude estimation, the subject is asked to indicate the strength of his sensation by giving a numerical value. With cross-modality matching, the subject is asked to express the strength of his sensation by matching it with an equally strong sensation in another sense modality. This method has two advantages: it is independent of the subjects' "number behaviour" and it is based on comparative rather than on absolute judgement. The method may also be used with different stimuli from the same modality as is shown by Engen (1971).

Normally, a stimulus of one of the two modalities is presented to the subject at a fixed intensity and he can vary the other one until he feels that the strength of sensation provoked by it equals that of the fixed one. This procedure is then repeated at another intensity level of the fixed stimulus and so on. However, it is difficult to perform such a measurement with the olfactory stimulus as the variable one because even if one uses an olfactometer which allows one to vary the odorous intensity continuously over a large range, the adaptation caused during the long lasting matching procedure will disturb the results. Therefore, true cross-modality matching with a variable stimulus from a modality that is easier to handle and which does not adapt so rapidly, is preferred. Usually auditory stimuli (1000 Hz or white noise) are used for this purpose. The odorous

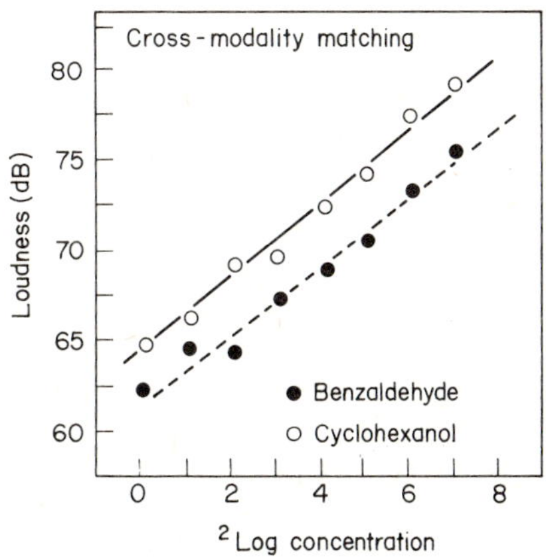

FIG. 10.

concentrations are each presented a number of times in random order and the sound level at which the subject starts his matching is varied systematically to make sure that such variables as the time or the number of turns on the potentiometer, which is used to vary the sound intensity, do not influence the measurement in a systematic way. After the experiment, the mean of the sound intensities (in dB) judged to be equal to a given odour intensity is calculated for each of the odour intensities. Since the dB scale is a logarithmic scale no conversion to logarithms is needed as with the numbers in magnitude estimation. The results of such an experiment with two different odorous substances are given in Fig. 10.

The results obtained for both substances are in good agreement with Stevens' power law and the slopes of the straight lines drawn through the points hardly differ. This resemblance in slope is, however, a mere coincidence. Usually the slopes differ more. In general, it appears that cross-modality matching is a much easier and more natural task for the subject than magnitude estimation. As a result both the intra- and the inter-individual variability in performance are smaller.

III. Psychophysical methods for the assessment of odorous quality

As was pointed out in the introduction, in the history of psychophysics far more attention has been devoted to the systematic measurement of odorous intensity than to the assessment of odorous quality. Early investigators like Linne, Zwaardemaker and Henning have tried to classify odours on a purely intuitive basis and until recent years only occasionally research has been carried out to verify their conclusions. In much of this research in recent years, methods have been used which do not measure the similarity between odours in a direct way, but which rely on some form of intensity measurement like cross-adaptation (Ohma, 1922; Cheesman *et al.*, 1953, 1956; Köster, 1971) or anosmia (Guillot, 1948; Amoore and Venstrom, 1967) to draw indirect conclusions about odour similarity. These methods have not been very successful.

The question arises as to what methods are available to attack the problem of odorous quality and qualitative similarity more directly. Perhaps the question should be divided into two separate ones:

(1) How can similarities between odours be measured?
(2) How can existing similarities or differences in odorous quality be specified?

These two questions will be treated briefly here and a number of the most important methods used will be described.

A. *How can similarities between odours be measured?*

A number of methods have been employed to measure the relative similarities in a given set of odours. The results of some of these methods have been used to investigate the concept of similarity itself. What attributes must odours have in common to be called similar by the subject or, stated otherwise, what attributes or dimensions does the subject use to judge the similarity of odours? To answer this question, multidimensional scaling methods have been developed by Torgerson (1958) and by Kruskal (1964a, b). In these methods the similarity measures obtained by one of the methods to be described here are considered to be indications of the distances of the stimuli in a multidimensional similarity space. The method then provides a means to find the minimal number of dimensions on which the data can be fitted with a reasonable goodness of fit between the data and the model imposed on them. Here we will not go into the intricacies of this method, but we will simply describe the methods by which the original similarity measures are obtained.

Scaling with paired comparison

The simplest thing one can do to find out how similar two things are to a subject is to give them both to him and to ask him to rate their similarity on a scale. Usually a category scale is employed and all possible pairs of stimuli in a given set of stimuli are presented (in both possible orders of pair members) to the subject for judgement. Varying the order of the pair members is important, because, since the subject is free to choose the dimension which he thinks is most important for judging similarity and since the one pair member may induce him to choose a dimension different from the other one, the results of the two presentation orders may differ. If, for example, one of the odours is an unpleasant odour of rotten fruit and the other is a typical fresh fruit odour, the subject may use pleasantness-unpleasantness as the most important dimension for judging the similarity in the one order of presentation and fruitiness in the other order. The result may be the same, but does not have to be so. After the experiment, the mean of the category scale values is calculated and these mean scale values are converted into relative distances of the stimuli in a multidimensional similarity space. Woskow (1968) used this method to scale the similarities between 25 odorous substances and found that his data could well be fitted in a three dimensional space. Therefore, it appeared that his subjects used only three dimensions of any importance to judge the similarities. The first and most prominent dimension found by Woskow correlated very highly with a separate rating of the substances on a pleasant-unpleasant scale, suggesting that this hedonic dimension played a very important role in judging the similarity between the odours.

Ordinal scaling methods

Although similarities may be expressed directly in scale values, some experimenters feel that it is better to ask the subject to give relative instead of absolute judgements of similarity. In such cases ordinal scaling methods are often used in which the subject simply indicates which of two or more different pairs of stimuli shows the greatest similarity between its pair members. A number of these ordinal methods have been developed by Coombs (1964). The simplest one of them is the "method of propellors".

Out of a given set of stimuli (A, B . . . . . N) all possible combinations of three stimuli are presented to the subject in all possible orders like A-BC, A-CB, B-AC, B-CA etc., and the subject is asked to indicate which of the two stimuli which are presented last to him resembles the first one most. Thus, the subject compares two similarities (between A and B and between A and C, for instance) and makes a relative judgement about them. From the data obtained in this way, information about the relative distances between the stimuli can be deducted and further analysed by multidimensional analysis.

Since presenting the triplets of the stimuli in all three orders is rather time consuming, some experimenters prefer to give the three stimuli in a random order and to ask the subject which two stimuli in the triplet are most alike and which two are most different. With this so-called "complete method of triads", the subject has to compare and rank-order three similarities (A-B, A-C and C-B) per presentation. Obviously, this task is more difficult for the subject than the task in the method of propellors where only two similarities are compared. The complete method of triads has been tried in olfaction by the author, but the task seems too difficult for the subjects, who tend to have forgotten their judgement on the first similarity by the time they are judging the third one. If, on the other hand, one leaves the subjects free in the number of times they want to sniff at all three stimuli, thus providing a better possibility for remembering the three similarities, the subjects cross-adapt themselves too much and this may distort the results to a considerable degree.

The other methods of ordinal scaling described by Coombs demand even larger sets of stimuli (4 or 5) or more similarity judgements per presentation. Since the complete method of triads already seems to be unsatisfactory in olfaction, these methods will not be discussed here.

Confusion matrix

A third and more indirect way of judging the similarities between stimuli is based on measuring the degree in which subjects confuse stimuli. The more readily two stimuli are confused, the more they must be alike.

In a typical experiment based on this principle, the subject would first receive each stimulus of the set together with a special code by which he will have to name the stimulus throughout the rest of the experiment. After this first series, the stimuli are presented to the subject in random order an equal number of times and each time a stimulus is presented the subject is asked to identify it by giving the code. If he gives a wrong code, he is corrected and the next stimulus is presented to him. Thus, the subject is confronted with a paired associate learning task. He is requested to learn until he has reached a certain criterion (for instance: no mistakes in two successive presentations of the whole set of stimuli). In the meanwhile all his answers are recorded and a confusion matrix is drawn up giving for each stimulus the number of times each different code was mentioned. Separate confusion matrices may be made for each block of 5 or 10 presentations of the whole set of stimuli in order to check on the speed of learning.

The data in these confusion matrices can be converted into distances between the stimuli and a multidimensional analysis can be carried out on them.

In order to use this method correctly one must take great care in selecting the code names which the subject has to learn and one must be sure that the subject is equally well acquainted with all of these code names before the actual experiment starts in which they are paired with the odours. If the codes are not well chosen the subjects may be led to respond according to similarities between the code names rather than according to the similarities in their olfactory sensations or they may have a preference for naming a specific code name whenever they are in doubt.

Therefore, such simple codes as the numbers from 1 to n or the first letters of the alphabet should be avoided. Two- or three-digit numbers or two-letter codes are perhaps best, provided one makes sure that there are no striking resemblances between two or more of the individual codes. Very special codes like multiples of 100 or 111 should also be avoided.

Before starting the actual experiments with the odours, the subjects should know the code names by heart, but not in any fixed order. To achieve this, the subjects are trained in a recognition experiment in which the codes are presented randomly among other codes of the same type. The subject is asked to indicate whether or not the codes presented to him belong to his set of codes and he is trained until he recognizes all of his codes correctly at least five times in a row and makes no mistakes with the other codes.

After this criterion has been reached, the actual experiment in which the odours are paired with the odorous stimuli can begin. The pairing of the codes with the odours is varied systematically over the subjects so that each code is paired with each odour an equal number of times. This means that at least as many subjects are needed in the experiment as there are odours.

As can be shown (Köster, 1972) the results obtained with this method do agree with those of experiments in which paired comparison combined with category scaling is used but the confusion method is much more time consuming.

Descriptive methods

A still less direct method for assessing qualitative similarities is to let the subjects scale their impressions of each of the stimuli on a number of well defined descriptive attributes such as camphoraceous, minty, etc. The extent to which these dimensions are judged to be relevant in the description of the quality of two odorous stimuli is considered to be an indication of their similarity. This method has the disadvantage of being very indirect insofar as the similarity measurement itself is concerned, but it has the advantage that it provides a more direct indication of the nature of the qualitative similarity. Thus, extra experiments to specify the nature of these similarities can be avoided.

The danger of this method is that it relies to a large extent on semantics. There may exist—and this is perhaps especially true for olfaction where there seems to be a fundamental lack of good descriptive terminology—dimensions of similarity for which the subjects or the experimenters have no adequate terms and which may remain unnoticed in the experiment as a result of this lack of descriptive terms.

B. How can existing similarities or differences in odorous quality be specified?

There are two approaches to the problem of specification of stimulus quality. The first one, which has already been mentioned in the preceding section, tries to specify the quality directly in descriptive terms and relies therefore heavily on semantics. The second one tries to avoid semantic problems by using standard qualities as references. In the latter case odorous qualities are specified in terms of similarity to a set of well-known odorous compounds.

Both methods have been used extensively and with varying success. Harper, Bate-Smith and Land (1968) have given a very good survey of the existing methods in their book "Odour description and odour classification". Rather than to go into a lengthy description of all the possible techniques and the many difficulties encountered in using them, we would recommend this book to the reader.

IV. The use of psychophysical data

Before concluding this chapter some remarks should be made about the use of psychophysical data for theoretical or practical purposes.

In psychophysics the subject is used as a measuring instrument. As in all problems of measurement a thorough knowledge of the properties of the measuring instrument is needed to evaluate the validity of the results obtained with it. This is the reason that a large number of psychophysicists are mainly interested in the form of psychophysical functions and in the possible sources of experimental bias such as order-, practice-, expectancy- and motivational-effects. Thus, it often seems that psychophysics has become a goal in itself. Its objective appears to be to unravel the intricacies of the measuring instrument. The external stimuli only provide an adequate and simple means to study the decision processes, which are supposed to take place in the subject, in a reduced and well controlled experimental situation.

To those who simply wish to use psychophysical methods to evaluate the external stimuli in a practical setting, much of what is presented as psychophysics may therefore seem unnecessarily complicated.

To a certain extent this is true. If one wishes to assess the intensity difference between two stimuli, one need not go into all the details of each individual's response bias or personal psychophysical function. In such a case it is enough to obtain the judgements from a large group of subjects, *provided* one takes into account such factors as order effects by carefully balancing the presentation orders, position effects by systematically varying the position of the stimuli, coding effects by choosing codes that are of equal attractiveness or by varying the codes systematically over the stimuli for different subjects, and expectancy and motivational effects by keeping the stimulus probability constant in all phases of the experiment and by giving a fixed pay-off function which does not favour saying either "yes" or "no" too strongly.

Thus, although it is by no means necessary to study all these effects for their own sake, one should recognize them as possible influences on the experimental outcome and adequately control them. In this context another problem should be discussed. Some psychophysicists are vehemently opposed to the use of group data, because in group data two sources of variability have been mixed. The truth of this objection can not be denied, but its validity depends very much on the use one wants to make of the data. This may be illustrated by a simple example.

If one wants to determine the average threshold for a given substance in a large population, one may determine the psychophysical function for each of a small number of subjects in a rigorous way by presenting all the stimuli in a constant stimuli experiment a large number of times to each subject. From each of these psychophysical functions one may obtain the individual threshold and one may average these thresholds to get the group threshold.

All of the subjects may have the same psychophysical function as for instance the steepness and form of the sigmoid curve (as is suggested for two different subjects in Fig. 1). In that case one may generalize and conclude that the psychophysical function for that substance is characterized by that steepness and form. This is a good method to obtain insight in the properties of the measuring instrument.

At the same time, however, the subjects which all function in the same way, may differ considerably in their sensitivity as could be seen from the variability around the mean threshold. Moreover, this variability measure might not be very reliable because only a small number of subjects took part and it may well be that the sample was not truly representative of the population. If on the other hand one would present the same stimuli used in the threshold determination only once (or a small number of times for reasons of economy) to a large group of subjects and plotted the percentage of positive responses given by the subjects to each of these stimuli against their concentration, one would also obtain a sigmoid curve. This curve, from which the average threshold of the group may be determined in the same way as the individual thresholds of the subjects from their psychophysical functions in the previous experiment, does not represent the psychophysical function, however. Its form reflects the variability in sensitivity between subjects rather than the growth of the detection probability with increasing concentration within each of the subjects. In fact, the individual psychophysical functions may be much steeper than the average group function obtained in this way. But, although the group function does not give us any insight in the functioning of each of the individual subjects, it is of considerable practical value, because it makes a prediction of the percentage of the population that will detect a stimulus of a certain concentration possible, provided the sample of subjects tested was large enough.

Thus, both methods have their own right, but they answer different questions. The decision on which one is to be used depends on the kind of question which is to be answered and on the necessity to economize in the experimental procedure, for it always remains possible to combine them at the expense of more experimental effort.

References

Amoore, J. E. and Venstrom, D. (1966). Correlations between stereochemical assessments and organoleptic analysis of organic compounds. *In* "Olfaction and Taste II" (Hayashi, O. O., ed.), Pergamon Press, Oxford.

Cain, W. S. (1968). Olfactory adaptation and direct scaling of odour intensity. Thesis, Brown University, 1968 (cited in Engen, 1971).

Cheesman, G. H. and Mayne, S. (1953). The influence of adaptation on absolute threshold measurements for olfactory stimuli. *Q. J. Exp. Psychol.* **5**, 22–29.

Cheesman, G. H. and Townsend, M. J. (1956). Further experiments on the olfactory thresholds of pure chemical substances, using the "sniff-bottle method". *Q. J. Exp. Psychol.* **8**, 8–14.

Coombs, C. H. (1964). "A Theory of Data". Wiley, New York.

Corbit, T. E. and Engen, T. (1971). Facilitation of olfactory detection. *Percept. Psychophys.* **10**, 433–436, 1971.

Engen, T. (1971). Psychophysics. *In* "Experimental Psychology" (Kling, J. W. and Riggs, L. A., eds.). Holt, Rinehart and Winston, New York.

Fechner, G. (1859). "Elemente der Psychophysik". Leipzig.

Green, D. M. and Swets, J. A. (1966). "Signal Detection Theory and Psychophysics". Wiley, New York.

Guillot, M. (1948a). Anosmies partielles et odeurs fondamentales. *C. R. Hebd. Acad. Sci. Paris* **226**, 1307–1309.

Guillot, M. (1948b). Sur quelques caractères des phénomènes d'anosmie partielle. *C. R. Soc. Biol. Paris* **142**, 161–162.

Harper, R., Bate Smith, E. L. and Land, D. G. (1968). "Odour Description and Odour Classification". Churchill, London.

Janssen, W. (1970). Paper given at the Nuffic Summer School: "Odour Perception: multidisciplinary research methods", Utrecht, August-September, 1970.

Janssen, W. (1972). Criterion variability and the number of response categories in a movement detection task.

Köster, E. P. (1968). Relative intensity of odour mixtures at supra-threshold level. *Olfactologia* **1**, 29–41 (*Suppl. Cahier d'Oto Rhino-Laryngologie* **3**, No. 5).

Köster, E. P. (1969). Intensity in mixtures of odorous substances. *In* "Olfaction and Taste III" (Pfaffmann, C., ed.). Rockefeller University Press, New York.

Köster, E. P. (1971). Adaptation and cross-adaptation in olfaction. Thesis, Utrecht, The Netherlands.

Köster, E. P. (1972). Odour similarities in nine odorous compounds. A methodological study. Report of the Psychological Laboratory of the University of Utrecht, 1972.

Köster, E. P. and MacLeod, P. (1974). See Chapter 13, in this volume.

Kruskal, (1964a). Multidimensional scaling by optimizing goodness of fit to a non-metric hypothesis. *Psychometrika* **29**, 1–27.

Kruskal, (1964b). Non-metric multidimensional scaling: A numerical method. *Psychometrika* **29**, 115–129.

MacLeod, P. (1968). Interactions dans un mélange d'odeurs-étude électrophysiologique. *Olfactologia* **1**, 25–27.

Mitchell, M. J. (1971). Investigations of olfactory similarity scaling. Thesis, University of Canterbury, New Zealand, 1971.

Ohma, S. (1922). La classification des odeurs aromatiques en sous-classes. *Arch. Neerl. de Physiol.* **6**, 567–590.

Pangborn, R. M., Berg, H. W., Roessler, E. B., Webb, A. D. (1964). Influence of methodology on olfactory response. *Percept. Mot. Skills* **18**, 91–103.

Semb, G. (1968). The detectability of the odour of butanol. *Percept. Psychophys.* **4**, 335–340.

Stone, H. and Bosley, J. (1965). Olfactory discrimination and Weber's law. *Percept. Mot. Skills* **20**, 657–665.

Swets, J. A., Tanner, W. P. and Birdsall, T. G. (1961). Decision processes in perception. *Psychol. Rev.* **68**, 301–340.

Torgerson, W. S. (1958). "Theory and Methods of Scaling" (7th ed., 1967). Wiley, New York.

Woskow, M. H. (1968). Multidimensional scaling of odours. *In* "Theories of Odour and Odour Measurement" (Tanyolac, N., ed.). Robert College, Bebek, Istanbul.

10 Laboratory Methods for Obtaining Olfactory Discrimination in Rodents

DAVID A. STEVENS
Clark University

Discrimination learning is a method widely incorporated into experimental investigations of olfaction and the role of odour stimuli in the mediation of behaviour. Since there can be no discrimination without detection, the method provides a means of assessing the detectability of an odorant. Such investigations are done to determine the concentrations of odorants at absolute thresholds, to determine the ability to detect odorants following surgical or chemical intervention, and to test hypotheses about conditions which may produce pheromones. Studies of discrimination ability are also a means of investigating odour classification. If two odorants are detected but not discriminated it can be argued that they are perceived as more similar than odorants readily discriminated. And some workers have used discrimination techniques not so much to study olfaction as to study learning processes as they may vary with the modality of the stimuli presented.

Regardless of the purpose of studies requiring the acquisition of an olfactory discrimination, there are problems of control common to many of the various discrimination methods employed. It is a truism that it is essential to control all potential stimuli available to the subject. While that principle need not be restated, it might be worthwhile to review some of the common sources of possible confounding stimuli, both olfactory and within other sensory modalities.

The usual kind of confounding is the presence of a cue which correlates with the presence of an experimental stimulus. In situations where there are two response locations, control of confounding spatial cues is achieved by having the two locations equally often correct. Gellerman (1933) and Fellows (1967) have published specially prepared sequences for this purpose. Confounding visual cues are most often a problem when odorants are presented passively, as with cups or capsules of odorant accessible to the subject, or when an odorant such as urine is applied to a particular surface of the apparatus. In the former example, new containers can be used on each trial so that any differential visual characteristics will not be consistently associated with the reinforced response. In the latter example, another substance not containing the experimental odorant (e.g. distilled water, other urine) could be used to remove the confounding between the visual and olfactory components of the stimulus.

When odorants are presented actively in air streams there is usually no problem with confounding visual cues, but there is a possibility of confounding by differential auditory and pressure cues. The olfactometer must be constructed so as to minimize the possibility of such cues being produced. Air travelling through gas washing bottles containing liquids of different densities, or through stopcocks or needle valves set to pass different volumes could have differential auditory characteristics. Since rodents hear ultrasonic sounds, the use of wide-spectrum white noise to mask such sounds would seem to be desirable. In addition to careful physical control of the odour delivery system, it is also necessary to perform control trials in the course of the experiment to demonstrate that only olfactory stimuli are determining the animal's discriminative behaviour.

Perhaps less obvious are the sources of confounding by olfactory stimuli. These can come from at least four sources. First, in the course of the experiment, the *subjects* themselves can be a source of discriminable odours. For example, rats, mice, and gerbils emit discriminable odours associated with the omission of reinforcement (frustration) (Ludvigson and Sytsma, 1967; Davis, 1970; Topping and Cole, 1969). The ability of rats and mice to discriminate odours emitted by rats and mice subjected to aversive stimuli has also been demonstrated (Carr, Martorana and Krames, 1970; Stevens and Köster, 1972; Valenta and Rigby, 1968). Since reinforcement and frustration are typically a part of experimental designs involving discrimination, such confounding must be expected. These cues should not introduce a bias, however, for these cues are correlated with the location of the reinforcer which would ordinarily be controlled by positional sequences.

The second source of confounding odours is the *experimenter.* McCall *et al.* (1969) have shown that rats spend more time on that side of an open field closest to their caretaker, who could be smelled but not seen, than on that side closest to a stranger. Extensive direct contact with the caretaker was not a necessary condition to produce this preference; the animals were handled for only 20 s once a week over a period of 11 weeks, plus 3 min. right before testing. Obviously it is important that the experimenter stand in a neutral location when testing subjects in situations where his odour could be detected.

The third potential source of confounding odours is the *experimental treatment.* For example, electric shock is often used to induce stress in studies of reactions by rodents to odours produced by stressed conspecifics. It is possible that ozone, or the odours of burnt skin or hair produced by the shock could confound the stress-mediated odours produced by the subject. For this reason Stevens and Köster (1972) chose to stress their rat subjects by a combination of proprioceptive, visual and auditory stimuli.

The fourth source of confounding odours in discrimination is the *reinforcer.* Rats are able to discriminate the odour of a single 45 mg food pellet from clean air in a T-maze (Southall and Long, 1969). Clearly it is necessary to withhold food or any other reinforcer with an odour until its odour could not bias the observed behaviour, i.e. after a choice response has been made, or after the critical physiological records have been taken.

Artifactual odorants can not only bias choice behaviour, but can also introduce error in experiments where purity of the experimental odorant is critical. This error, the mixing of uncontrolled odorants with the experimental ones, would be an important one in psychophysical studies, and those where electrophysiological responses are being recorded. This would be especially true when subject-emitted odorants are present, since such odorants are pheromonal and thus likely to be more salient than the experimental stimuli. Situations in which there is constant evacuation of air, such as wind tunnels, or where the experimental odorants are presented in relatively confined areas may be preferred where purity is essential.

I. Classification of discrimination methods

Discriminations can be classified into two groups on the basis of the nature of stimulus presentation.

A. Simultaneous discrimination

This term refers to situations where the correct stimulus (the stimulus responses to which are reinforced) and the incorrect stimulus (the stimulus responses to which are not reinforced) are presented at the same time. This method can be used only when different locations are provided for correct and incorrect responses, and thus is limited to T- and Y-mazes, and discrimination boxes. Simultaneous discrimination has the advantage of allowing direct comparison of the two stimuli by the subject, but has the possible disadvantage of requiring the discrimination where there can be an interaction of the two stimuli. The presence of one of the stimuli could affect sensory thresholds, the perceived intensity, or the quality of the other stimulus. While this should not be a problem if one of the stimuli was clean air, it could be a critical factor if two odorants were used.

B. Successive discrimination

This term refers to situations where only one of the stimuli to be discriminated is presented on any one trial. Direct comparison of stimuli by the subject is not possible, but also the effect of any direct interaction of odorants on their perception is avoided.

Within the general classification of successive discrimination the methods can be further broken down into those in which the subject's task is either to respond or not in the presence of one of the stimuli—a go-no go situation—and those in which one response, a left turn for example, is made to one stimulus and another response, a right turn, is made to the other stimulus. This latter type, sometimes called a sign-differentiated spatial discrimination, in fact involves both successive and simultaneous discrimination; successive discrimination of the olfactory stimuli, and simultaneous discrimination of position. That is, odours, presented successively, are cues or signs for the appropriate spatial response. An important difference between these types of successive discrimination is the level of response differentiation required. While the animal need only differentiate responding and not responding in the go-no go type, it must differentiate between the two active spatial responses, turning right and left, in the latter type.

While simultaneous discrimination cannot be obtained in all situations, that is not true of successive discrimination. If there is but a single response location, the go-no go method is used. If more than a single active response is possible, then the sign-differentiated spatial discrimination (also called two-choice successive discrimination) is used.

Discrimination test situations can also be divided into two general types on the basis of the nature of the response required by the subject. In the first group, which includes mazes, discrimination boxes, and runways, the subject is required to move (or suppress movement) to one or more particular locations as a function of the stimuli presented. In the second group, the operant chambers, the subject is required to differentially perform some instrumental act, such as lever pressing, which does not require any change in location. Examples of the former class of experimental situations will be presented first,

C. T- and Y-mazes

These terms are used for situations where the subject is required to travel from a starting point to a choice point, and thence to one of two goal areas, either 180° apart (T-maze) or less than 180° apart (Y-maze). A typical example of a simultaneous discrimination using a Y-maze is an experiment by Bowers and Alexander (1967). The apparatus is shown in Fig. 1. Odorized air was blown into the rear of

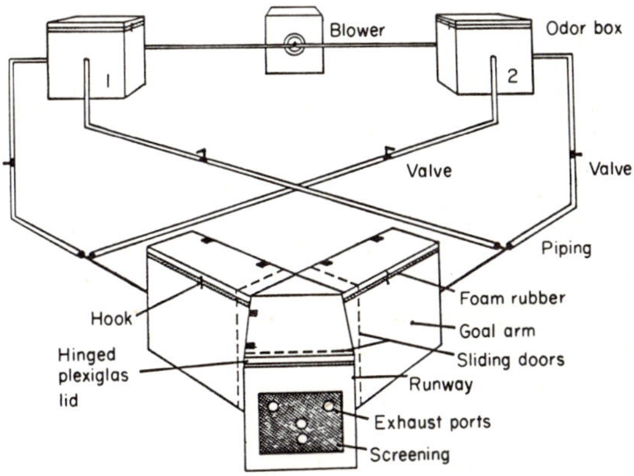

FIG. 1. A typical Y-maze for obtaining olfactory discrimination (Bowers and Alexander, 1967: Reproduced by permission of American Association for the Advancement of Science).

each arm and allowed to travel through the apparatus and allowed to escape through ports in the end of the stem. A door was placed in the stem to provide a start box, and there was a door in each arm just past the bifurcation to prevent retracing by the subject after an arm was entered. Water was the reinforcer. In that study, in which water-deprived mice (*Mus musculus*) were subjects, relatively rapid discrimination of four pairs of odorants was obtained. The pairs of odorants, in order of testing were (1) cinnamon oil *v.* juniper oil, (2) odour of male mice of two species (*M. musculus v. Peromyscus maniculator*), (3) odour of male *v.* female *M. musculus,* and (4) odours of two individual male *M. musculus.* A fifth, control condition, in which empty odorant boxes were used yielded a chance level of performance. In this study all subjects were tested with the same stimulus of each pair of odours as the correct one. A better procedure would have been to use half of the mice with one odorant of each pair correct and test the remaining subjects with the other odorant designated as correct. Then, any pre-experimental preferences could not confound the assessment of discrimination ability. Another refinement would have been to counterbalance the order of presentation of the different tests so that any transfer effects attributable to sequence of testing would balance out. Latin square designs are available for this purpose.

Another example of olfactory discrimination using a maze is Long and Tapp's (1970) study of the effects of olfactory tract and bulb lesions on odour discrimination. A T-maze was used with a stem and arms of about equal length. The air was drawn into the arms through slots in the rear walls by a blower mounted at the end of the start box. For each trial, the air drawn into one of the arms was odorized by a cup of moist powdered food, located behind a barrier at the end of the arms. One hundred trials were given to rats with bilateral olfactory tract lesions, bilateral bulb ablations, or sham operations. All of the sham control subjects learned to a criterion of ten consecutive correct responses in about 50 trials or less, while none of the experimental subjects did as well, with eight rats failing to meet the criterion.

As noted earlier, it is normally unwise to use food rewards in odour discrimination tasks, for the food odour could mask the test stimuli, alter their quality through cross adaptation, or serve as a cue itself. In this experiment, however the use of a food reward was permissible since food served as the test odorant. The odour of the food reward could not mask or alter the experimental odour, nor could it provide a spurious cue since the food odour was the correct stimulus for all subjects.

Simple mazes can be used in successive discriminations. Two methods are available. One technique is to present on half of the trials one of the stimuli in both arms and reinforce the subject's choice of the right-hand arm, and on the other trials present the other stimulus in both arms and reinforce choice of the left-hand arm. The second technique is to present one or the other of the stimuli at the bifurcation, rather than in the arms, and make reinforcement contingent upon right-hand turns in one stimulus condition and upon left-hand turns in the other stimulus condition. The major advantage of the first variation is that the subject is in the presence of the odorant when reinforcement is delivered or withheld. Since the stimuli are presented only at the choice point in the second variation, this contiguity of stimulus and reinforcement is not possible. The major advantage of the second variation is that since the stimuli are introduced only at a single point in the apparatus the olfactometry may be less complicated and the clearing of the apparatus of odours between trials is facilitated.

An interesting example of a successive discrimination in a T-maze is a study done by Morrison and Ludvigson (1970) to test the hypothesis that reward and nonreward produce discriminable odours in rats. A standard T-maze with doors at the start box and in each alley was used. Replaceable paper covered the walls and floor. The odours of reward (R) and nonreward (N) were obtained by placing one of a group of hungry rats at the choice point in the maze for 30 s and either feeding it or not. A third stimulus condition (C) in which no animal was placed in the apparatus was used. Four experimental groups were formed. Group RN was reinforced for entering the right-hand goal box (for example) after a stimulus animal had been rewarded at the choice point, and reinforced for entering the left-hand goal box when a stimulus animal had been unrewarded at the choice point. The other groups were RC, requiring a discrimination between the previous presence of a rewarded stimulus animal and no predecessor; NC, requiring a discrimination between a nonrewarded stimulus animal and no predecessor; and CC, where no predecessors were used and no discrimination should have been possible.

A total of 84 trials were given. The results showed that over the last 18 trials, Groups RN and NC were responding at a level of about 62% correct, while Groups RC and CC were responding at a chance level. While the level of accuracy shown by Groups RN and NC was far from perfect, the discriminations were sufficient to show that there is a discriminable residue left by nonrewarded rats.

D. Discrimination boxes

Discrimination boxes differ from T- and Y-mazes only in that the choice point is not at the intersection of relatively narrow alleys but rather within a single chamber. This distinction is not a clear one, and a T-maze with a relatively wide stem and relatively short alleys could be called a discrimination box as well as a maze.

In its simplest form a discrimination box is a chamber with two response locations. An example is the apparatus described by Moulton (1960) in a study using aliphatic acetates as odorants. A three-sided box with a metal floor was used. Protruding from the middle wall were two metal water founts. Disposable nylon capsules which held a small wad of cotton saturated with an odorant could be attached to the founts. The fount with the odorant designated as incorrect, and the metal floor were connected to the outputs of an electric shock generator. The subjects for this study were rats which had been pretrained in the circular choice apparatus described below. Testing involved simply placing a water-deprived animal in the box, observing its response, and removing it until the next trial. Drinking at the correct fount was permitted while drinking at the incorrect fount was punished by electric shock. The position of the correct fount was determined by Gellerman sequences. Learning was quite rapid in this study; subjects not having had previous discrimination training except that given during pretraining (which involved discrimination of the experimental odorant) learned to a criterion in 38 trials or less. The pretraining was probably largely responsible for this rapid acquisition. Since it also involved avoidance of an electrified fount on the basis of the experimental odorant, positive transfer of the discrimination would be expected. The pretraining also served to reduce the disruptive effects of the shock. In a previous attempt to train rats to discriminate odorants in a similar box but without the pretraining, performance was severely disrupted by the punishment given for incorrect responses.

Simple discrimination boxes have disadvantages. One problem is that there is usually poor control over the behaviour of the subjects following their choice responses. The most troublesome behaviour is the subject's correction of incorrect responses, i.e. withdrawing from the incorrect location and moving to the correct one before removal from the box by the experimenter. The use of goal boxes eliminates this by providing for doors which can be lowered behind the subject to prevent retracing. An example of a simple box with goal compartments is shown in Figure 2. Miller and Erickson (1966) used this to demonstrate that solutions of chloride salts can be discriminated by the rat on the basis of odour. Two swinging doors were located at one end of the main chamber. Holes in the doors allowed the subject to sniff at tubes containing the stimulus material. The animals were trained to push open a door which gave access to a goal box. A water reward was provided in the box behind the correct door. Discrimination of chloride salt solution of 0.1 M concentration *v.* water was found at levels of 75% correct or better after 80 trials. Since the odorants were located at the goal box door, and reinforcement was given at the rear of the goal box there was not contiguity between stimulus and reinforcement with this apparatus. Perhaps a higher level of discrimination would have been obtained if such contiguity were the case.

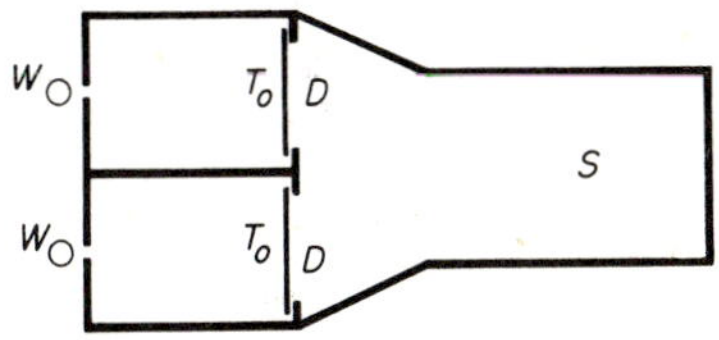

FIG. 2. Discrimination box with goal compartments; S–starting place, D–doors, T–odorant tubes, W–water (Miller and Erickson, 1966: Reproduced by permission of *Physiology and Behaviour*).

One problem with simple discrimination boxes is the possibility of bias since the trial is initiated by the experimenter placing the subject in the apparatus. It is possible, and indeed likely, that the location and body orientation of the subject will affect its behaviour. Thus there is the possibility that the experimenter, having knowledge of the location of the correct stimulus, might unintentionally influence the animal's choice by the manner in which it is placed in the chamber and the trial begun. This can be overcome by the use of a start box equipped with a door, and having an opaque cover. An appropriate procedure then would be to place the subject in the start box facing away from the door, close the cover, wait a fixed period of time, perhaps 5 s and then open the door. No experimenter induced bias is possible since the opaque cover prevents the experimenter from seeing the orientation of the animal when the trial is begun.

A discrimination box which features both a start box and goal boxes which provide contiguity between the stimulus and reinforcement was used by Jennings and Keefer (1969). In this apparatus, shown in Fig. 3, three goal boxes are used. These could be shifted left or right as a unit so that either the right-hand and centre boxes, or centre and left-hand boxes were accessible from the start box. By providing the correct odorant in the centre box, and the incorrect odorant in the two side boxes, the relative position of the two stimuli can be shifted quickly and easily. Air flow was obtained by producing negative pressure in the start box by a fan, which pulled odorized air through the two goal boxes in line with the start box. An expanded-metal guillotine door was located between the start and goal boxes. This type of apparatus has the advantage of isolating each odour in separate goal boxes, thus eliminating the need for long intertrial intervals to clear the odours. However, that advantage is overshadowed by the fact that there is confounding between all

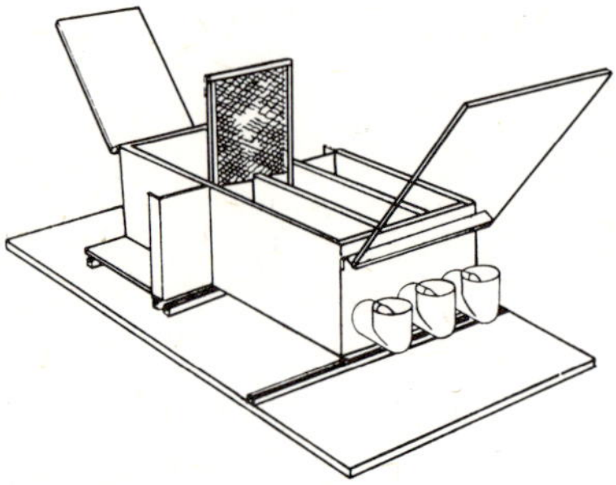

FIG. 3. Discrimination box using movable alleys and a start box (Jennings and Keefer, 1969: Reproduced by permission of *Psychological Reports*).

discriminable nonolfactory aspects of the goal boxes and the odours presented through the boxes. Furthermore, since the two side boxes are always incorrect the possibility of confounding by subject-emitted odours resulting from nonreward is present. Since it is most difficult to be certain that all three boxes are identical, this confounding between particular surfaces and experimental stimuli must be avoided. Another disadvantage with this apparatus is that the start box door is located at the choice point. Thus, any biases that might result from start box orientation would be magnified. Such a bias can be minimized by locating the start box some distance from the choice point.

A good resolution of these problems is found in an automated discrimination box designed by Moulton (Moulton, Celebi and Fink, 1970). Bias attributable to the presence of an experimenter is eliminated by having the subject initiate each trial. Response, stimulus and reinforcement are contigious in time and place, and with proper olfactometry there is no possibility of nonolfactory factors confounding the discrimination.

The apparatus consists of a rectangular box. At one end there is a glass treadle which when pressed begins a trial. At the opposite end are two stimulus compartments which are separated from the main chamber by glass guillotine doors. Air, with or without an odorant, passes from the floor of each compartment up through exhaust ducts in each compartment's ceiling. The stimulus compartment floors are hinged and contain water cups. Photo-cell beams are directed through each of the stimulus compartments. The criterion for choice is the breaking of a beam for 9 seconds. Thus a subject can insert its nose briefly into the air streams to sample the flow without registering a choice. The sequence of events from the resting state are as follows: Depression of the treadle initiates the programmed assignment of correct and incorrect odorants to the left and right stimulus compartments according to Gellerman sequences. After a 5 s delay the stimulus doors open and remain so until a choice is made. If that choice is correct, the incorrect stimulus door closes, water is dispersed in the correct compartment and 5 s drinking time is given. Then the rear of the compartment floor raises, which drains any remaining water and forces the subject's head out of the compartment. The stimulus door closes and the floor returns to the horizontal position. If the choice was incorrect, the correct door closes, the incorrect compartment floor raises without water having been delivered, the door closes and the floor returns to its normal position. The cycle begins again when the treadle is depressed. The subjects can be shaped by the experimenter to perform, or as Moulton reports, simply left in the box for a day or two on a water deprivation schedule.

Another type of discrimination box is the circular choice apparatus (Moulton and Eayrs 1960). This has proven to be an effective means of training rats to discriminate odours and can be used for critical testing of odour detection. It differs from traditional discrimination boxes in that confounding by spatial location is avoided by having stimulus and response locations slowly move around the circumference of the apparatus rather than having fixed stimulus and response locations and controlling for spatial cues by randomly switching the location of the stimuli. One version of this apparatus is shown in Fig. 4. Two metal water founts protrude into the chamber from the wall. Nylon capsules containing a plug of cotton are attatched to the founts. One of these is impregnated with an odorant. The founts are electrically isolated from the metal wall, floor and ceiling. By connecting one lead from an electric shock source to one of the founts and the other lead to the floor, a conditioned avoidance of drinking from that fount can be established. The two founts slowly rotate around the apparatus to eliminate spatial confounding. Visual cues are eliminated by placing the apparatus in total darkness and using electronic sensors to determine the subject's responses to the two founts. When water deprived rats are placed in the apparatus for two or three nights a reliable discrimination is obtained. While not well suited to studying the course of acquisition of discrimination, this apparatus has the advantage that up to three rats can be trained at once. These can then be tested individually using discrete trial procedures in the same apparatus (Furchtgott and Kimbrell, 1967; Moulton and Eayrs, 1960) or in another discrimination box (Moulton, 1960). In the former case the position of the founts is varied randomly from trial to trial. The subject is introduced, allowed one choice, and then removed.

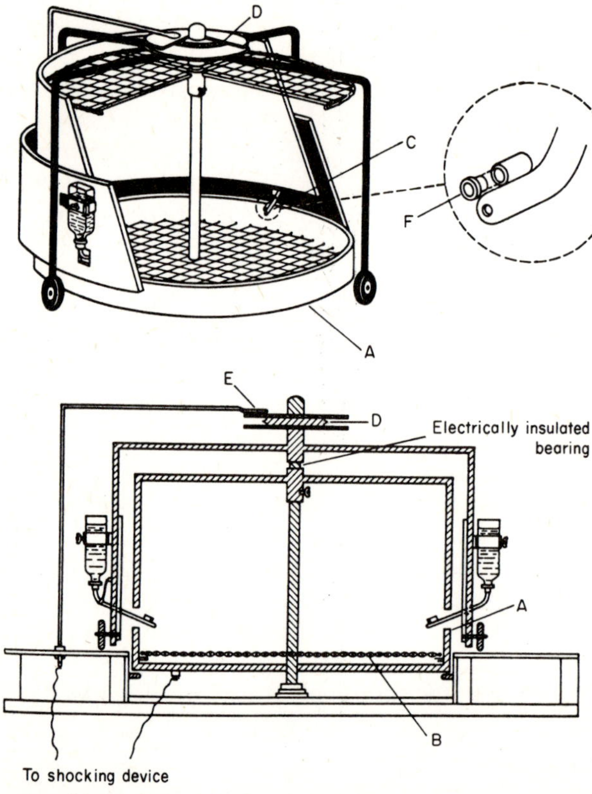

FIG. 4. Circular discrimination box; A–inner wall; B–wire mesh floor; C–metal water fount; D–pulley; E–metal brush; F–nylon odorant capsule. (Eayrs and Moulton, 1960: Reproduced by permission of *The Quarterly Journal of Experimental Psychology*.)

E. Runways

Successive discrimination can be obtained in a straight alley or runway equipped with photo-cell operated clocks, or other means of measuring the speed of travel. The procedure is simple; reinforcement is given with one stimulus present and not given with another. While this is an example of the "go-no go" type of successive discrimination, in fact subjects are rarely trained to the point of not running in the unreinforced stimulus condition, but are usually trained until the response times on all unreinforced trials are longer than those found on all reinforced trials within a block of trials.

This method was used in an ingeneous study by Ludvigson and Systma (1967) who were also testing the hypothesis that rats receiving both reinforcement and nonreinforcement emit one or more discriminable odours associated with the reinforcement conditions. A 4 ft enclosed straight alley was used with a start box and goal box at either end. All subjects were rewarded with food on a double-alternation schedule, i.e. a series of trials in which two reinforced trials are followed by two nonreinforced trials, etc. Eight trials a day were given for 13 days. All subjects within an experimental group were run on a particular ordinally numbered trial before given their next trial. That is, all subjects within a group received trial 1, then all were given trial 2, etc. until all eight daily trials were given.

The groups differed in that for one group (Group P) the particular double-alternation pattern was the same for all subjects (RRNNRRNN), while for the other (Group NP) each subject was run on a different double-alternation sequence than that given the subject just previously run. For example, the sequence to the first subject was RRNNRRNN, for the second subject it was RNNRRNNR, for the third it was NNRRNNRR, etc. Thus within Group P there was perfect correlation between the reinforcement condition on any trial except the first, and whether or not reinforcement was given to the previously run animal. But for Group NP there was no such correlation between the reinforcement conditions given to successively run animals. The rationale for this procedure was that it is generally accepted that rats are unable to learn double-alternation sequences *per se.* If double-alternation patterning of responses is found, then it must be due to some cue present at the time of the response. If the hypothesis that rats produce differential odours following reinforcement and non-reinforcement is correct, then the animals in Group P would have stimuli available for patterned responding, while those in Group NP would not.

The results confirmed the hypothesis. On the last day of training the subjects in Group P were running about three times faster in the goal box on reinforced trials than on nonreinforced trials, while there was no difference in goal box speeds in Group NP. A final day of testing was given in which no reinforcement was given on any trials to either group. That condition produced a decrease in speed in

Group P but not in Group NP. These findings showed that there are differential cues left by rats which can serve as conditioned stimuli. It is very likely that these cues are olfactory since Seago, Ludvigson and Rembley (1970) have shown that the patterning described is not found in anosmic subjects.

Runways have also been used to show disruptive effects of conspecific odours on behaviour. While this method does not involve a discrimination in the strictest sense—there is no discriminative training involved—it is similar enough to discriminative methods to warrant presentation here.

The general procedure is to train the subjects to travel down the runway until running times are stable, and then on critical trials introduce the experimental stimuli. An example of this method was reported by Stevens and Saplikowski (1973). Rats were trained to run down a paper-covered glass runway 2 m, long. On the critical day six trials were given. On trials 1–3 and 5, distilled water was applied to a section of the runway. On trials 4 and 6 blood, homogenized leg muscle, or homogenized brain from rats was applied. On encountering the blood or muscle one or more fright reactions, such as freezing, elimination, and stopping and turning around were seen. Upon being removed from the apparatus some subjects bit the experimenter, or showed other emotional responses. These reactions were not seen prior to the application of the blood and muscle, and were not found when the brain homogenate was used. In a subsequent study tissue from guinea pigs was used as well as that from rats. Fright reactions were again found with rat blood and muscle, but no disruption was found with rat brain, or with the blood, muscle or brain from guinea pigs. As well as suggesting that the observed alarm reactions are made to conspecific tissue only, these latter results show that the responses were not to visual stimuli, for the rat and guinea pig preparations were identical in appearance.

F. Operant chambers

The term *operant chambers* is used here for apparatus in which the response required of the subject is not to move to different locations, but rather to differentially respond by lever-pressing or performing some other response not related to different distinct locations in the apparatus. Since the subject remains more or less stationary it is necessary to bring the odorant to the subject by way of a moving air stream.

One of the earliest olfactory operant chambers was the wind

tunnel designed by Pfaffman *et al.* (1958) for rats. It consisted of a cylinder 4 in. in diameter and 12 in. long. Near one end was a lever and a water-dipper. A stream of air, which was either odourless or odorized was introduced into the tunnel at the end nearest the lever and exhausted out the other end. The subjects were trained to press the level for water reinforcement on a variable-ratio schedule when odourless air (S^D) was delivered. Reinforcement was never given when the odorant (S^Δ) was delivered. A variable-ratio reinforcement schedule is one where reinforcement is given only occasionally such that a predetermined proportion of responses are, on the average, reinforced. This schedule is useful for it produces a relatively high, stable rate of responding. Thus, when the discrimination is established a reliable rate of responding is found with S^D, the "correct" stimulus, and there are no, or very few responses made with S^Δ, the "incorrect" stimulus.

Using a similar apparatus, Goff (1961) found that it was possible to withhold reinforcement for 10 s after a brief presentation of the odorant (S^Δ) and obtain excellent discrimination. The advantages of using a puff of odorant instead of a prolonged presentation are great; the possibility of sensory adaptation is reduced, and the clearing of the chamber of odorant is more easily obtained.

Another efficient operant chamber is shown in Fig. 5 (Braun *et al.*, 1967). Although different in many details from the Pfaffman *et al.*

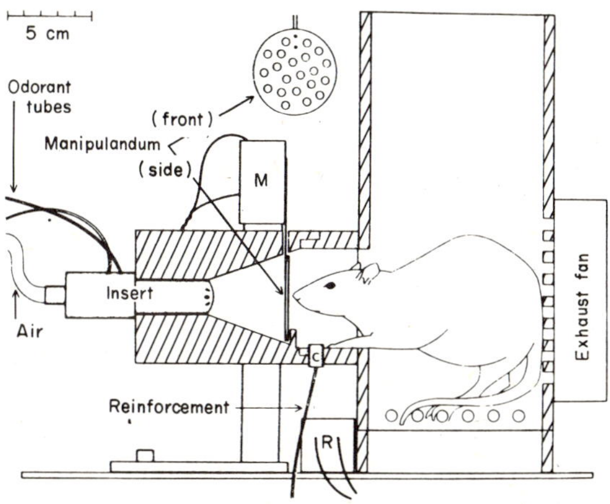

FIG. 5. Operant chamber employing a nose-plate manipulandum (Braun *et al.*, 1967: Reproduced by permission of The Psychonomic Society, Inc.).

wind tunnel, the major difference is that here the required response is the pressing with the nose of a perforated Teflon plate through which the odorant is flowing rather than the pressing of a lever with a paw. This modification, together with the relatively small size of the chamber, assures that the animals will be oriented to the air stream during training.

The reinforcement (water, or 0.1 ml of a 0.05% saccharin solution) is delivered into a Teflon cup. An audible click was provided to cue the delivery of the reinforcer so that the unreinforced responses made during the presentation of S^D need not be interrupted by the subject examining the cup. Using this apparatus, Braun and Marcus (1969) obtained stable discriminative responding in different rats given acetone *v* geranial, propanol *v.* cineol, or camphor *v.* benzyl acetate within only six half-hour training sessions.

The chambers used by Pfaffman *et al.* and Braun *et al.* use a single manipulandum (lever or nose-plate) and thus require a go-no go paradigm. If active responding to both of the stimulus conditions is desired, two levers are necessary. A two-lever operant chamber is described by Bennett (1968). He placed a small chamber equipped with two levers and a water cup in a bell jar into which clean air continuously flowed. First the subjects were trained to press either of the levers each time a buzzer sounded. Then discrimination training began. Pressing one of the levers was designated as the correct response when one odorant (cineol) was added to the air stream and pressing the other lever was the correct response when the other odorant (octaneol) was delivered.

The subjects were given 20 trials a day. For each trial the buzzer sounded and simultaneously one of the two odorants was delivered into the chamber. As soon as either of the levers was pressed the buzzer and odorant delivery were stopped, and if the response was correct a water reinforcer was given. If no response was made within 15 s the trial terminated (buzzer and odorant delivery stopped). After a 1 min. interval, the next trial automatically began. With this procedure discrimination to a criterion of 17 correct responses in 20 trials for three consecutive days was obtained in most subjects in a mean of 420 trials. Animals were also trained in an odour detection paradigm which was identical to the above one except that clean air replaced one of the odours. The rate of learning in this case was about the same as when two odours were discriminated; about 390 trials were required.

The rate of acquisition found by Bennett was considerably lower than that obtained by Braun and Marcus. Although it is impossible to compare the stimulus intensities used in the two studies, they were well above threshold in both and it is improbable that differences in odorants or concentrations produced the large differences in learning rates found. It is more likely that differences in the nature of the required responses were critical. Activation of Braun's nose-plate manipulandum required head movements while Bennett's lever was designed to be pressed by a paw. The rat may be better "prepared" to associate head movements, which are an important part of olfactory orientation, with odours than to associate paw movements, which are not instrumental in olfaction. Another difference in the nature of the responses concerns the degree of response differentiation required. The two classes of responses in Braun's apparatus was pressing *v.* not pressing the plate, while in Bennett's apparatus they were pressing lever A *v.* pressing lever B. Discrimination of levers was prerequisite to odour discrimination in the latter case but in Braun's situation the only response discrimination needed was between moving the head and not. It is safe to assume that head movement control is acquired by adult subjects long before entering experimental situations, while discrimination of two identical, spatially close levers must be learned in the test situation.

The conditioned suppression method is another technique using a fixed response location. With this method the subject is trained on an intermittent reinforcement schedule to emit a response such as lever pressing. Then, when a stable, relatively high rate of responding is established, a warning stimulus is presented followed by punishment by electric shock. The administration of the shock results in suppression of the previously ongoing response for a short time. After repeated presentations of the warning stimulus followed by shock response suppression is found upon the onset of the warning stimulus. Suppression of response thus indicates detection of the stimulus.

While this technique has been used extensively with other modalities, its use in olfactory studies has been largely in demonstrating odour discrimination in the pigeon (Henton, 1969; Henton *et al.*, 1966, for example). However, two studies have been done with rats (Davis, 1971; Davis and Tapp, 1968). Those studies showed absolute thresholds of 5.78×10^{-8}g/l for cyclohexanone and 2.7×10^{-11}g/l concentration for amyl acetate. These levels of concentration are far below those found for the rat for amyl acetate and other substances determined by other methods, both behavioural and electrophysiological, suggesting that conditioned suppression may be a particularly sensitive method. A more detailed discussion of conditioned suppression as a psychophysical method is found in Smith (1970).

II. Concluding comments

A review of the literature concerning the acquisition of olfactory discrimination in rodents shows a great deal of variance in experimental results. With the rat, relatively rapid discrimination has been obtained in some situations, but in others acquisition is surprisingly slow. Many workers have found that acquisition of olfactory discrimination is far more difficult to obtain than, say, visual discrimination, despite ample anatomical and behavioural evidence indicating that rats are macrosmatic.

The majority of techniques for olfactory discrimination have been derived from techniques used for obtaining discrimination in other modalities. This use of similar techniques across modalities is based on the widely held principle that the laws governing classical and instrumental conditioning are general, i.e. the specific stimuli and responses used are not critical. This premise, however, may be false.

A recent article by Seligman (1970) has as its theses that as a result of evolutionary history animals are more or less "prepared" to associate particular stimuli with particular responses, and that the laws of learning may vary with the degree of preparedness. Seligman notes, for example, that rats readily learn to avoid gustitory stimuli that are followed by illness, even when the illness occurs as long as four hours after the sampling of the gustitory stimuli, but they cannot learn to avoid auditory or visual stimuli that precede illness (Garcia and Koelling, 1966).

Rats readily learn to run or jump to avoid electric shock, but learning to press a lever to avoid shock is acquired at a slow rate. Dogs easily learn to go to the left or right when auditory cues are presented above or below the animal, but they show slow learning if the tones differ in pitch but not location. The opposite results are found if a go-no go response is required. Then, the discrimination of spatially different tones is learned slowly, and the pitch discrimination is learned rapidly (Lawicka, 1964). The implications are clear. The types of conditioned stimuli presented, responses required, and unconditioned stimuli used as reinforcers may be of critical importance. One should not expect that experimental conditions optimal for one modality would be optimal, or even appropriate, for another. All of this suggests that as workers in this field continue to experiment with different methods, they be alert to the increasing amount of data emerging on the role of pheromones and other odours in mammalian behaviour. Hopefully as more is learned about how responses to olfactory stimuli have been shaped by evolutionary history the conditions defining appropriate laboratory situations for studying olfactory discrimination will become increasingly apparent.

References

Bennett, M. H. (1968). The role of the anterior limb of the anterior commissure in olfaction. *Physiol. Behav.* **3**, 507–515.

Bowers, J. M. and Alexander, B. K. (1967). Mice: Individual recognition by olfactory cues. *Science, N.Y.* **158**, 1208–1210.

Braun, J. J. and Marcus, J. (1969). Stimulus generalization among odorants by rats. *Physiol. Behav.* **4**, 245–248.

Braun, J. J., Wermuth, B. and Haberly, L. (1967). An olfactory discrimination apparatus: Modification of the basic "wind-tunnel" design. *Psychon. Sci.* **9**, 515–516.

Carr, W. J., Martorano, R. D. and Krames, L. (1970). Responses of mice to odours associated with stress. *J. Comp. Physiol. Psychol.* **71**, 223–228.

Davis, R. G. (1971). Olfactory psychophysical measurements in the rat with a conditioned suppression behaviour. Unpublished doctoral dissertation. Vanderbilt University.

Davis, R. G. and Tapp, J. T. (1968). A reliable method for determining olfactory thresholds in animals. Paper presented at the meeting of the Psychonomic Society, St. Louis, November 1968.

Davis, S. F. (1970). Conspecific odors as cues for runway behavior in mice. *Psychon. Sci.* **19**, 169–170.

Fellows, B. J. (1967). Chance stimulus sequences for discrimination tasks. *Psychol. Bull.* **67**, 87–92.

Furchtgott, E. and Kimbrell, G. McA. (1967). Olfactory discrimination in prenatally X-irradiated rats. *Radiat. Res.* **30**, 217–220.

Garcia, J. and Koelling, R. (1966). Relation of cue to consequence in avoidance learning. *Psychon. Sci.* **4**, 123–124.

Gellerman, L. W. (1933). Chance orders of alternating stimuli in visual discrimination experiments. *J. Genet Physiol.* **42**, 207–208.

Goff, W. R. (1961). Measurement of absolute olfactory sensitivity in rats. *Amer. J. Psychol.* **74**, 384–393.

Henton, W. W. (1969). Conditioned suppression to odorous stimuli in pigeons. *J. Exp. Anal. Behav.* **12**, 175–185.

Henton, W. W., Smith, J. C. and Tucker, D. (1966). Odor discrimination in pigeons. *Science, N.Y.* **153**, 1138–1139.

Jennings, J. W. and Keefer, L. H. (1969). Olfactory learning set in two varieties of domestic rat. *Psychol. Rep.* **24**, 3–15.

Lawicka, W. (1964). The role of stimuli modality in successive discrimination and differentiation learning. *Bull. Pol. Acad. Sci.* **12**, 35–38.

Long, C. J. and Tapp, J. T. (1970). Significance of olfactory tracts in mediating response to odours in the rat. *J. Comp. Physiol. Psychol.* **72**, 435–443.

Ludvigson, H. W. and Sytsma, D. (1967). The sweet smell of success: Apparent double alternation. *Psychonom. Science* **9**, 283–284.

McCall, R. B., Lester, M. L. and Corter, C. M. (1969). Caretaker effect in rats. *Develop. Psychol.* **1**, 771.

Miller, S. D. and Erickson, R. P. (1966). The odor of taste solutions. *Physiol. Behav.* **1**, 145–146.

Morrison, R. R. and Ludvigson, H. W. (1970). Discrimination by rats of conspecific odors of reward and nonreward. *Science, N.Y.* **167**, 904–905.

Moulton, D. G. (1960). Studies in olfactory acuity III. Relative detectability of *n*-aliphatic acetates by the rat. *Q. J. Exp. Psychol.* **12**, 203–213.

Moulton, D. G. and Eayrs, J. T. (1960). Studies in olfactory acuity II. Relative detectability of *n*-aliphatic alcohols by the rat. *Q. J. Exp. Psychol.* **12**, 99–109.

Moulton, D. G., Celebi, G. and Fink, R. P. (1970). Olfaction in Mammals–Two aspects: Proliferation of cells in the olfactory epithelium and sensitivity to odours. *In* "Ciba Foundation Symposium on Taste and Smell in Vertebrates". (Wolstenholme, G. E. W. and Knight, J., eds.). Churchill, London.

Pfaffman, C., Goff, W. R. and Bare, J. K. (1958). An olfactometer for the rat *Science, N.Y.* **128**, 1007–1008.

Seago, J. D., Ludvigson, H. W. and Remley, N. R. (1970). Effects of anosmia on apparent double alternation in the rat. *J. Comp. Physiol. Psychol.* **71**, 435–442.

Seligman, M. E. P. (1970). On the generality of the laws of learning. *Psychol. Rev.* **77**, 406–418.

Smith, J. (1970). Conditioned suppression as an animal psychophysical technique. *In* "Animal Psychophysics: The Design and Conduct of Sensory Experiments" (Stebbins, W. C., ed.). Appleton-Century-Crofts, New York.

Southall, P. F. and Long, C. J. (1969). Odor cues in a maze discrimination. *Psychonom. Sci.* **16**, 126–127.

Stevens, D. A. and Köster, E. P. (1972). Open-field responses of rats to odours from stressed and nonstressed predecessors. *Behav. Biol.* **7**, 519–525.

Stevens, D. A. and Saplikoski, N. J. (1973). Rats' reactions to conspecific muscle and blood: Evidence for an alarm substance. *Behav. Biol.* **8**, 75–82.

Topping, J. S. and Cole, J. M. (1969). A test of the odor hypothesis using Mongolian gerbils and a random trials procedure. *Psychonom. Sci.* **17**, 183–184.

Valenta, J. G. and Rigby, M. K. (1968). Discrimination of the odor of stressed rats. *Science, N.Y.* **161**, 599–601.

11 Determination of Odour Preferences in Rodents: A Methodological Review*

RICHARD L. DOTY
Monell Chemical Senses Center,
University of Pennsylvania,
Philadelphia, Pennsylvania

In studies using nonhuman organisms, one stimulus is generally assumed to be positively preferred to another stimulus if the organism behaviourally increases the probability of being stimulated by that stimulus over the probability of being stimulated by the other. The present review critically examines a number of basic methodologies commonly used in the determination of odour preferences of small rodents. Paradigms related to the determination of olfactory discrimination, *per se,* are not included (e.g. Bowers and Alexander, 1967; Braun *et al.*, 1967; Dagg and Windsor, 1971; Eayrs and Moulton, 1960; Goff, 1961; Hahn and Simmel, 1968; Moulton and Eayrs, 1960; Pfaffman, Goff, Bare, 1958).† Results of specific studies are discussed only if they add to the description of a particular procedure.

* The major portion of this paper was written during the author's tenure as a graduate student at Michigan State University and as a postdoctoral fellow at the University of California, Berkeley. The author wishes to thank Dr. J. A. King for his constructive comments on a previous version of the manuscript. This work was supported, in part, by USPHS Traineeship 5-TO1-MH-10611 and NIMH Predoctoral Research Fellowship 1-FO1-MH-49081.

† It is usually assumed that if a preference occurs, the stimuli in question are discriminable. However, a lack of preference does not necessarily imply a lack of discrimination.

The present compendium appears warrented in light of the recent surge of interest in mammalian olfactory communication (e.g. Gleason and Reynierse, 1969; Johnston *et al.*, 1970; Ralls, 1971).

Five types of preference paradigms are conceptualized in the present paper: (a) approach; (b) forced approach-avoidance; (c) bar-press stimulus presentation; (d) odour trail; and (e) sniff-rate analysis. Although these categories are not necessarily mutually exclusive, they provide a useful taxonomy for the classification and comparison of odour preference test situations.*

I. Approach paradigm

The most common odour preference test situation reported in the literature is one in which two or more odorants are placed at different locations within a test chamber or home cage. The shapes of the test chambers used vary from simple squares to interconnected rectangular compartments, Y-mazes, and round arenas. Duration and/or frequency of a subject's investigation of an area near the odorant usually serves as the dependent measure, although other variables are used. Several studies have used the amount of food or water consumed near the odorant as their dependent variable (e.g. Barnett and Spencer, 1953; Drickamer, 1972; Millman, 1968). One study required a subject to dig through wood-chip bedding toward an odour source before its presence was recorded electronically (Howard *et al.*, 1969). Studies utilizing discrete one-trial choices between portions of a test apparatus containing experimental odours (e.g. Y-maze arms) frequently express the preference measure in terms of the proportion of subjects choosing the appropriate section (e.g. Godfrey, 1958; Leone and Moltz, 1971; Thiessen *et al.*, 1971).

When duration of investigation is the dependent measure, most studies have recorded such times manually on stopwatches (e.g. Bronson and Caroom, 1971; Carr *et al.*, 1965; Carr *et al.*, 1970; Carr *et al.*, 1970; Doty *et al.*, 1971; Doty and Anisko, 1973; Johnson, 1972; Pfaff and Pfaffman, 1969; Quadagno and Banks, 1970; Scott

* A number of terms used in this chapter (e.g. "preference", "aversion", "attraction") have varying degrees of semantic overtones which smack of anthropromorphism. For example, definitions of preference commonly include terms such as "liking", which only can be inferred for nonhuman organisms. With full realization that a more explicit vocabulary of behavioral definitions is needed, I have elected to use some of these terms, albeit sparingly, since (a) they are widely used by modern behaviorists, (b) they appear to function relatively well in communicating essentials, and (c) this chapter is not the most appropriate forum for semantic rectification.

and Pfaff, 1970; Stern, 1970; Thiessen *et al.,* 1971). Others have minimized the potential recording bias on the part of the experimenter by monitoring the investigatory behaviour with photoelectric cells or mechanical tripping devices connected to electrical recording systems (e.g. Baran and Clickman, 1970; Doty, 1972; Doty, 1973; Landauer and Banks, 1972; Lydell and Doty, 1972; Mainardi *et al.,* 1965; Marr and Gardner, 1965; Moore, 1965; Smith, 1965).

There are a number of advantages to the use of the approach paradigm. First, it is easy to set up and can be placed, in some cases, within the subject's home cage or home range, making it amenable to field as well as laboratory situations. Furthermore, many of the problems associated with the testing of an animal in a strange environment can be eliminated. Second, various numbers of odorants can be tested simultaneously. Third, its reliance upon simple exploratory or foraging behaviours eliminates the necessity of elaborate training or shaping procedures to produce the desired behaviour.

Unfortunately, a number of potential disadvantages of this paradigm are also apparent. First, depending upon the size and construction of the test situation, the possibility of odour diffusion and mixing cannot be ruled out, particularly where exhaust systems are absent or poorly planned. Second, this paradigm makes the assumption that a positively preferred odour elicits marked approach behaviour in the subject. It is conceivable that an animal can smell an odour from a distance and not approach directly towards its source, depending upon such factors as the odorant's concentration and volatility. I have noticed such long range behaviour in rats towards urine odours, although usually approach was elicited as well (Lydell and Doty, 1972). Third, position preferences, particularly in male subjects, often occur in approach test situations (e.g. Barnett and Spencer, 1953; Doty, 1971). Such preferences potentially limit the use of repeated measures with the same subjects, require that care be taken in counterbalancing the positions of the odorant presentations, and frequently necessitate the use of a relatively large number of subjects to decrease error variance. Fourth, test situations which use the amount of food or water consumed as their dependent measure may confound olfactory and gustatory factors (e.g. some of the "odour" stimuli may become ingested and tasted), limiting the degree of generalization of their results to other situations. Furthermore, this confounding may vary with the deprivation level of the organism. As a hypothetical example, the odour of roses may elicit more investigation than that of food pellets in a satiated rat, whereas rose odour associated with food may be avoided by a hungry rat in favour of food pellets not so odorized, or food pellets associated with a different odour. Imagine the odour of roses in your breakfast cereal at 7:00 a.m. Fifth, this procedure does not necessarily allow accurate measurement of negative odour preferences or aversions. For example, both a neutral and a highly aversive odour may receive an initial exploratory sniffing bout by a rodent, and then receive no more attention during a test session. A literal interpretation of the data would suggest that both odorants are equally attractive, contrary to reality. A sixth limitation of this paradigm is related to the potential interaction between a subject's activity level and the preference measure. During periods of high circadian running activity, for example, a preference observed during other periods may wash out, depending upon the type of test apparatus (Doty, 1971).

II. Forced approach-avoidance paradigm

A test situation which theoretically measures both the attractive and aversive aspects of an odorant can be devised, assuming that the organism is strongly motivated to approach the attractive odorant and move away from the aversive one. To measure both attraction and aversion, a relatively small two-chamber apparatus is used. Air or some other presumed neutral stimulus is present or contingent upon the subject's presence in one chamber, whereas the experimental odorant is present or contingent upon the subject's presence in the other chamber. Attraction of the subject to the odorant is operationally defined as the subject's spending more than 50% of a test period on the odour side, and avoidance is defined as spending less than 50% of the test period on that side. A preference measure between two odorants can be inferred indirectly by comparing the relative time spent with each odour when they are individually paired against a control alternative in separate tests. Experimental test situations which can be conceivably categorized as forced approach-avoidance include those of Carter (1972), Doty (1973), and Rottman and Snowdown (1972).

The major advantage of the forced approach-avoidance paradigm is that both aversive and attractive qualities of odorants can be theoretically determined. However, this procedure has many shortcomings. First, developing preference measures for a number of odorants becomes quite time consuming. Only one odorant can be tested at a time, and large numbers of subjects must be used to decrease the size of the error variance resulting from large individual differences in the preference measure. Since the animal must be on one side of the apparatus or the other, his position is constantly being recorded during the test session, even though it is unlikely that he is attending to the environmental odours throughout the entire period. Second, it is conceivable that a preference measure inferred from this paradigm is not the same as a preference measure obtained from paradigms in which more direct comparisons of two odour stimuli are possible (Doty, 1973). Although no data are available which directly test this point, results from a study using a bar-press stimulus presentation paradigm suggest this possibility. Tapp and Long (1971) found that a hierarchy of odour preferences obtained by giving a rat a choice between a lever producing a puff of air and one producing a puff of odorized air correlated only 0.02 with a hierarchy obtained from a paired-comparison two-lever situation. Research is needed to determine if similar preferences are obtained from the approach and forced approach-avoidance paradigms. Third, this paradigm requires the testing of subjects in a test chamber independent of the home cage. Finally, a number of the shortcomings of the approach paradigm are also present in the forced approach-avoidance situation. For example, circadian activity and position preferences within the test apparatus conceivably influence the dependent measure.

III. Bar-press stimulus presentation paradigm

A number of studies have used a test paradigm in which a puff of odorized or nonodorized air is made contingent upon a bar press in a two-lever modified Skinner box (e.g. Long and Tapp, 1967; Tapp and Long, 1968; Long and Stein, 1969; Long and Tapp, 1970; Tapp and Long, 1971). Odours are subsequently exhausted from the test apparatus by means of a flow-through exhaust system. This test situation differs from the approach paradigm primarily in its brief presentation of the odorant (usually 1-s puff/bar press) and in its use of an extremely objective and measurable dependent variable.

The bar-press stimulus presentation paradigm has a number of positive features. First, the use of bar presses as a dependent variable allows the reinforcing properties of various odorants to be determined using standard operant conditioning equipment. Second, various parameters of the odorant presentation can be independently manipulated with maximal precision and minimal difficulty (e.g. duration of odour puff, concentration of odorant, schedule of reinforcement, magnitude of energy expenditure required to produce an odour puff, etc.). Third, response rates elicited in food deprived rats for odours of powdered food (e.g. Long and Tapp, 1967) are quite high in comparison to response rates elicited in nondeprived ones, suggesting the sensitivity of this measure to biological need states of the organism. Fourth, since the subject must perform a discrete behaviour in order to smell an odorant, bypassing of the recording system is not possible, as in the approach paradigm. Fifth, it is conceivable that a number of odorants could be presented simultaneously in this test situation by simply adding additional levers. Sixth, a measure of odour aversion could be determined, at least theoretically, by requiring a subject to push a lever to turn *off* an odorant

Major drawbacks of this paradigm include: (a) its reliance upon complex mechanical and electrical components; (b) difficulties involved in adequately cleaning odorants from the valves and tubes after their use; (c) potential influences of position preferences and activity levels upon the subjects' responses; and (d) the necessity of shaping animals to press the bars in some test situations.

IV. Odour trail paradigm

Anecdotal and experimental evidence suggests many rodents are capable of following and detecting subtle differences in odour trails laid by conspecifics (e.g. Ewer, 1968; Eibl-Eibesfeldt, 1970). If a rodent consistently chooses to follow one trail rather than another it can be assumed operationally that he prefers the former to the latter, at least at the initial and/or subsequent choice points. Such a conceptualization differs from the alternative odour preference paradigms in terms of the unique presentation of the stimuli and the types of odorants typically tested.

Douglas (1966) systematically examined various cues used by rats in producing "spontaneous alternation" in the T-maze and found that the tendency of the rat to avoid its own recently laid odour trail was a major determinant of the alternation. Klein and Brown (1969) found that rats made either anosmic or both anosmic and blind did not alternate at rates significantly above chance levels, whereas control and osmatic blinded rats did alternate at such levels, supporting Douglas' findings. These studies suggest that the rats preferred earlier laid trails or clean flooring to recently laid ones in this test situation.

A series of studies using straight alley runways indicate rats deposit a distinctive odour trail when their expectation of food reward has been recently frustrated (e.g. Ludvigson and Sytsma, 1967; Moorison and Ludvigson, 1970; Seago *et al.*, 1970). This odour can be used as a discriminable stimulus in learning situations and apparently produces a decrement in the running speed of animals passing over it.

Although the tracking of one odour trail rather than another clearly meets our general definition of preference, attempting to define a preference in terms of an increase in running speed is a more difficult task unless running speed is related to the amount of odorant reaching the olfactory mucosa. Unfortunately, no data are available which test this possibility directly. A common sense analysis would suggest an odour or lack of odour which does not signify "frustration" would be preferred to an odour which does. If it is assumed that running speed has no significant influence on olfactory or neurological inputs, the finding that rats actually spend more time in the presence of "nonreward" than "reward" odour in the runway test situation is in direct opposition to our initial definition of preference. It may be useful to circumvent this paradox by formulating a different operational definition of preference for this test situation, although some basic assumptions must first be met. If (a) it can be demonstrated that the "nonreward" odour increases the level of some bodily state responsible for the decrement in running speed (e.g. "anxiety"?) over the level produced by the "reward" odour, and (b) if it can be independently demonstrated that the organism prefers the former over the latter level (of arousal?), we have a seemingly conceptually-valid operational situation from which to infer that an increase in running speed denotes a preference. At the present time, however, such an analysis lacks direct confirmation and appears nonparisimonious. Furthermore, activity level possibly influences olfactory stimulation in rats (e.g. Welker, 1964).

The use of the odour trail paradigm for determining odour preferences is probably most appropriate in ethological studies of homo- and hetero-specific social and sexual attraction. This paradigm may also be useful in determining subtle aspects of predator-prey relations in small mammals such as shrews. However, this test situation is likely less amenable to accurate testing of general odour preferences than some of the other paradigms.

V. Sniff-rate analysis paradigm

A novel approach to the measurement of odor preferences was developed by Teichner (1966) and Teichner *et al.* (1967). By means of an elaborate electronic system these workers monitored the number, duration, and intensity of sniffs by unrestrained rats in an odour-controlled environment. Output from microphones located in the test arena was passed through appropriate circuits to an ink writing recorder and a pair of stereophonic earphones worn by one of the experimenters. The "repellency" and "attractiveness" of various concentrations of a number of different odorants were determined by comparing the amount of sniffing during a control period with the amount during a period when an odorant was present. Decreased sniffing was the measure of aversiveness and increased sniffing the measure of attractiveness. These authors report that the less restriction to movement imposed upon their subjects, the more they sniffed. This finding is reminiscent of common observations of sniffing in dogs following an odour trail, and suggests some interesting hypotheses for future studies. For example, approach tests in large arenas, as well as odour trail preference tests in long runways, may be optimally suited for animals whose sniff rates are markedly effected by locomotion.

The sniff-rate analysis paradigm appears to be one of the more sophisticated preference test situations reported in the literature. Care seems advisable, however, in interpreting what increases and decreases in sniffing rates signify. Welker (1964) reports, for example, that "mildly novel" visual, auditory, tactile, or olfactory stimuli are capable of arousing a characteristic sniffing pattern in the rat, and that removal of the olfactory bulbs has no marked effect upon the average rate of sniffing or upon the integration of various behaviours related to sniffing. Furthermore, strong stimulation of any of the aforementioned modalities tended to inhibit sniffing in a number of subjects. Keeping these potential problems in mind, the sniff-rate analysis paradigm still appears to have many merits and potential applications in the development of odour preference hierarchies.

VI. Conclusion

The present paper has classified odour preference tests into five basic types. Although it is not a comprehensive review of all olfactory preference studies reported in the literature, it does present a framework which includes most, if not all, odour preference test situations devised to date. More research is required, however, before definitive comparisons of odour paradigms can be made with optimal precision.

An investigator interested in developing valid odour preference hierarchies must concern himself not only with appropriate test situations, but with parameters as basic as odorant concentration. Surprisingly, few investigators studying rodent olfactory preferences have carefully addressed this problem. As a hypothetical example, a determination of whether or not rats prefer lemon odour to orange odour probably would depend upon the concentrations of the odorants used. Thus, a mild concentration of lemon odour may be preferred to a strong concentration of orange odour, whereas a reversal of the concentrations might likely lead to a reversal of the relative preferences. One way of determining which of these odorants is most preferable would be to compare the most strongly preferred concentration of orange odour to the most strongly preferred concentration of lemon odour. Other techniques to answer such a question might be developed which are less time consuming and more appropriate to a given concentration range of specific interest. It is probable that valid hierarchies of odour preferences will depend upon careful consideration of such details.

References

Baran, D. and Glickman, S. E. (1970). "Territorial marking" in the Mongolian gerbil: A study of sensory control and function. *J. Comp. Physiol. Psychol.* **71**, 237–245.

Barnett, S. A. and Spencer, M. M. (1953). Responses of wild rats to offensive smells and tastes. *Brit. J. Animal Behav.* **1**, 32–37.

Bowers, J. M. and Alexander, B. K. (1967). Mice: Individual recognition by olfactory cues. *Science, N.Y.* **158**, 1208–1210.

Braun, J., Wermuth, B. and Haberly, L. (1967). An olfactory discrimination apparatus: Modification of the basic "wind-tunnel" design. *Psychonom. Science,* **9**, 515–516.

Bronson, F. H. and Caroom, D. (1971). Preputial gland of the male mouse: Attractant function. *J. Reproduction and Fertility,* **25**, 279–282.

Carr, W. J., Krames, L. and Costanzo, D. J. (1970). Previous sexual experience and olfactory preference for novel versus original sex partners in rats. *J. Comp. Physiol. Psychol.* **71**, 216–222.

Carr, W. J., Loeb, L. S. and Dissinger, M. E.(1965). Responses of rats to sex odors. *J. Comp. Physiol. Psychol.* **55**, 415–417.

Carr, W. J., Wylie, N. R. and Loeb, L. S. (1970). Responses of adult and immature rats to sexual odors. *J. Comp. Physiol. Psychol.* **72**, 51–59.

Carter, C. S. (1972). Effects of olfactory experience on the behaviour of the guinea-pig (*Cavia porcellus*). *Animal Behav.* **20**, 54–60.

Dagg, A. I. and Windsor, D. E. (1971). Olfactory discrimination limits in geoils. *Can. J. Zoo.* **49**, 283–285.

Doty, R. L. (1971). Homospecific and heterospecific odour preferences in sexually-naive *Peromyscus maniculatus bairdi* and *Peromyscus leucopus noveboracensis.* Unpublished doctoral dissertation, Michigan State University, 1971.

Doty, R. L. (1972). Odor preferences of female *Peromyscus maniculatus bairdi* for male mouse odors of *P. m. bairdi* and *P. leucopus noveboracensis* as a function of estrous state. *J. Comp. Physiol. Psychol.* **81**, 191–197.

Doty, R. L. (1973). Reactions of deermice (*Peromyscus maniculatus*) and white-footed mice (*Peromyscus leucopus*) to homospecific and heterospecific urine odors. *J. Comp. Physiol. Psychol.* 296–303.

Doty, R. L. and Anisko, J. J. (1973). Procaine hydrochloride olfactory block eliminates mating behavior in the male golden hamster. *Physiol. Behav.* 395–397.

Doty, R. L., Carter, C. S. and Clemens, L. G. (1971). Olfactory control of sexual behavior in the male and early-androgenized female hamster. *Hormones Behav.* 2, 325–335.

Douglas, R. J. (1966). Cues for spontaneous alternation. *J. Comp. Physiol. Psychol.* **62**, 171–183.

Drinkamer, L. C. (1972). Experience and selection behaviour in the food habits of *Peromyscus:* Use of olfaction. *Behaviour,* **41**, 269–287.

Eayrs, J. T. and Moulton, D. G. (1960). Studies in olfactory acuity. I. Measurement of olfactory thresholds in the rat. *Q. J. Exp. Psychol.* **12**, 90–98.

Eibl-Eibesfeldt, I. (1970). "Ethology: The Biology of Behavior". Holt, Rinehart and Winston, New York.

Ewer, R. F. (1968). "Ethology of Mammals". Plenum, New York.

Gleason, K. K. and Reynierse, J. H. (1969). The behavioural significance of pheromones in vertebrates. *Psychol. Bull.* **71**, 58–73.

Godfrey, J. (1958). The origin of sexual isolation between bank voles. *Proc. R. Phys. Soc. Edinburgh,* **27**, 47–55.

Goff, W. R. (1961). Measurement of absolute olfactory sensitivity in rats. *Amer. J. Psychol.* **74**, 384–393.

Hahn, M. E. and Simmel, E. C. (1968). Individual recognition by natural concentrations of olfactory cues in mice. *Psychonom. Sci.* **12**, 183–184.

Howard, W. E., Palmateer, S. D. and Marsh, R. E. (1969). A body capacitor-olfactometer for squirrels and rats. *J. Mammalogy* **50**, 771–776.

Johnson, R. E. (1972). Sex pheromones of the golden hamster. *Amer. Zool.* **12**, 662 (Abstract).

Johnston, J. W., Moulton, D. G. and Turk, A. (Eds.) (1970). "Communication by Chemical Signals". Meredith, New York.

Klein, D. and Brown, T. S. (1969). Exploratory behavior and spontaneous alternation in blind and anosmic rats. *J. Comp. Physiol. Psychol.* **68**, 107–110.

Landauer, M. R. and Banks, E. M. (1972). The influence of early olfactory experience on social preferences in mice (*Mus musculus*). *Amer. Zool.* **12**, 652 (Abstract).

Leone, M. and Moltz, H. (1971). Maternal pheromone: Discrimination by pre-weaning albino rats. *Physiol. Behav.* **7**, 265–267.

Long, C. J. and Tapp, J. T. (1967). Reinforcing properties of odors for the albino rat. *Psychonom. Sci.* **7**, 17–18.

Long, C. J. and Tapp, J. T. (1949). An apparatus for the assessment of the reinforcing properties of odors in small animals. *J. Exp. Anal. Behav.* **11**, 49–51.

Long, C. J. and Tapp, J. T. (1970). Significance of olfactory tracts in mediating response to odors in the rat. *J. Comp. Physiol. Psychol.* **72**, 435–443.

Long, C. J. and Stein, G. W. (1969). An analysis of the reinforcing properties of food odour. *Can. J. Psychol.* **23**, 212–218.

Ludvigson, H. W. and Sytsma, D. (1967). The sweet smell of success: Apparent double alternation in the rat. *Psychonom. Sci.* **9**, 283–284.

Lydell, K. and Doty, R. L. (1972). Male rat odour preferences for female urine as a function of sexual experience, urine age, and urine source. *Hormones and Behav.* **3**, 205–212.

Ludvigson, H. W. and Sytsma, D. (1967). The sweet smell of success: Apparent double alternation in the rat. *Psychonom. Sci.* **9**, 283–284.

Mainardi, D., Marsan, M. and Pasquali, A. (1965). Causation of sexual preferences of the house mouse. The behaviour of mice reared by parents whose odour was artificially altered. *Atti Soc. Ital. Sci. Naturali e del Museo Civico di Storia Naturale di Milano,* **104**, 325–338.

Marr, J. N. and Gardner, L. E. (1965). Early olfactory experience and later social behaviour in the rat: Preference, sexual responsiveness, and care of young. *J. Genet. Psychol.* **107**, 167–174.

Millman, B. (1968). Odour preferences and aversions in the rat. Unpublished mast thesis, University of Calgary (Canada).

Moore, R. E. (1965). Olfactory discrimination as an isolating mechanism between *Peromyscus maniculatus* and *Peromyscus polionotus. Amer. Midland Naturalist* **73**, 85–100.

Morrison, R. R. and Ludvigson, H. W. (1970). Discrimination by rats of conspecific odors of reward and nonreward. *Science, N.Y.* **167**, 904–905.

Moulton, D. G. and Eayrs, J. T. (1960). Studies in olfactory acuity. II. Relative detectability of *n*-Aliphatic alcohols by the rat. *Q. J. Exp. Psychol.* **12**, 99–109.

Pfaff, D. and Pfaffmann, C. (1969). Behavioral and electrophysiological response of male rats to female rat urine odors. *In* "Olfaction and Taste: Proceedings of the Third International Synposium, N.Y." (Pfaffman, C., Ed.), pp. 258–267. Rockefeller University Press.

Pfaffman, C., Goff, W. R. and Bare, J. K. (1958). An olfactometer for the rat. *Science, N.Y.* **128**, 1007–1008.

Quadagno, D. M. and Banks, E. M. (1970). The effect of reciprocal cross fostering on the behaviour of two species of rodents, *Mus musculus* and *Baiomys taylori ater. Animal Behav.* **18**, 379–390.

Ralls, K. (1971). Mammalian scent marking. *Science, N.Y.* **171**, 443–449.

Rottman, S. J. and Snowdon, C. T. (1972). Demonstration and analysis of an alarm pheromone in mice. *J. Comp. Physiol. Psychol.* **81**, 483–490.

Scott, J. W. and Pfaff, D. W. (1970). Behavioral and electrophysiological response of female mice to male urine odors. *Physiol. Behav.* **5**, 405–411.

Seago, J. D., Ludvigson, H. W. and Remley, N. R. (1970). Effects of anosmia on apparent double alternation in the rat. *J. Comp. Physiol. Psychol.* **71**, 435–442.

Smith, M. H. (1965). Behavioral discrimination shown by allopatric and sympatric males of *Peromyscus eremicus* and *Peromyscus californicus* between females of the same two species. *Evolution,* **19**, 430–435.

Stern, J. J. (1970). Responses of male rats to sex odors. *Physiol. Behav.* **5**, 519–524.

Tapp, J. T. and Long, C. J. (1968). A comparison of the reinforcing properties of stimulus onset for several sense modalities. *Can. J. Psychol.* **22**, 449–455.

Tapp, J. T. and Long, C. J. (1971). Olfactory preferences in rats. Unpublished manuscript, Vanderbilt University.

Teichner, W. H. (1966). A method for studying olfaction in the unrestrained rat. *J. Psychol.* **63**, 291–297.

Teichner, W. H., Price, L. M. and Nalwalk, T. (1967). Suprathreshold olfactory responses of the rat measured by sniffing. *J. Psychol.* **66**, 63–65.

Thiessen, D. D., Lindzey, G., Blum, S. L. and Wallace, P. (1971). Social interaction and scent marking in the Mongolian gerbil (*Meriones unguiculatus*). *Animal Behav.* **19**, 505–513.

Welker, W. I. (1964). Analysis of sniffing of the albino rat. *Behav.* **22**.223–244.

12 The Simultaneous Measurement of Neural Response and Performance in Rats and Rabbits involved in an Odour Detection Task

DAVID G. MOULTON
Monell Chemical Senses Center,
The University of Pennsylvania

I. Introduction

A central problem in sensory physiology is to identify the peripheral and neural events essential to sensory performance and establish the principles that dictate the transformation of peripheral into central neural events. In some instances the analysis must begin by determining which receptor systems are involved. (Are the responses of certain mammals to pheromones controlled by the vomeronasal, or olfactory organ, or both?) But even when this is established we can seldom predict performance from a knowledge of receptor properties alone. At each successive relay station between the receptor and the highest receiving centres, afferent signals are potentially vulnerable to distortions that may progressively erode fidelity. Most significantly, only certain features may be selected for transmission. In other words, there are no *a priori* reasons for assuming a linear relation between receptor output and that of the corresponding higher receiving systems.

We might nevertheless argue that, for all practical purposes, this assumption is essentially valid over most sensory systems. Thus, there are numerous reports of a close correspondence between neural and behavioural events. Particularly significant are the extensive studies of Mountcastle and his associates relating neural and psychophysical response functions in somethesis. They conclude that, in the somesthetic sphere, the nervous system operates on its input in a linear manner and that the relation of the organism to its environment is set by the functional properties of the first order fibres (Mountcastle, 1968; Mountcastle *et al.*, 1972; Talbot *et al.*, 1968; Merzenich and Harrington, 1969; Harrington and Merzenich, 1970). Close correspondences have also been noted between subjective magnitude estimates and responses of the chorda tympani nerve to different intensities of taste solutions (Borg *et al.*, 1967, 1970) and between subjective visual perception in man and extracellular recordings from the cat visual cortex (Jung, 1961).

As against this evidence, however, there are examples of striking incongruities between neural events and subjective responses. Thus if sound intensity is increased over a certain range, subjective loudness rises. But in the cat auditory nerve, response magnitude reaches a maximum at medium intensity levels and then declines (Stevens and Davis, 1938; Kiang, 1965). Furthermore, how can we account for our ability to perceive a noise pressure range of 1–1 000 000 when the discharge rate of neurones in the auditory pathway may be only 1–100—and that in the presence of a continuous barrage of spontaneous activity (von Békésy, 1968)? Finally, for certain somatic intensity functions, Rosner (1961) provides evidence that subjective and physiological variables are non-linearly related.

When we turn to olfaction similar contrasts are evident. Stimulus-response curves for amyl acetate show remarkable correspondences, whether derived behaviourally in the rat (Moulton, 1960), or electrophysiologically from either the rat primary olfactory neurones (Mathews, 1972) or the rabbit olfactory bulb (Mozell, 1958) (Fig. 1). Even the marked reversal in slope of the curve appears in each case. On the other hand, the well-known decline in the subjective intensity of certain odours when presented over prolonged periods is not matched by any corresponding decline in the electrical activity of the olfactory bulb of the rabbit when the olfactory organ is repeatedly stimulated with butanol vapour (Ottoson, 1959).

Thus despite growing evidence of a linear relation between afferent events and behavioural response for a variety of sensory functions, exceptions are sufficiently frequent to urge caution. In particular, we can seldom predict with any confidence either the nature of sensory performance, knowing the associated neural response function for a given set of stimuli, or the reverse. Both sets of information are necessary.

In practice, the neural information is generally obtained from animals immobilized by anaesthetics, surgery or neuromuscular blocking agents. For some purposes this approach may be entirely adequate. Indeed, it has frequently been used in studies that show correlations between behavioural and neural events. But on the other hand this has also been the approach used by workers who have failed to show close correspondences between these events. Thus the possibility exists that discrepancies may in some cases arise as a result of distortions in the neural data imposed by the condition of the animal.

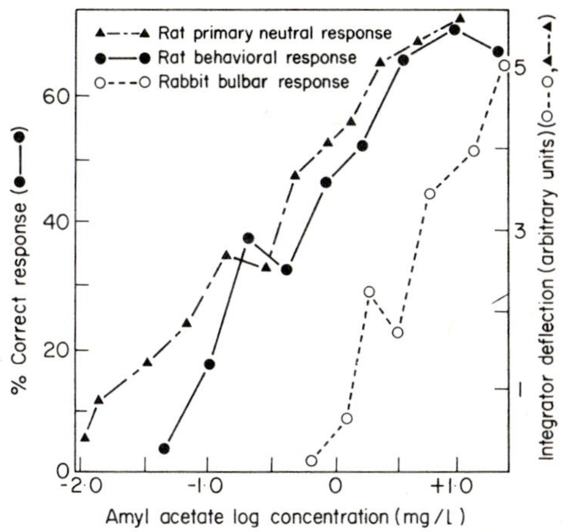

FIG. 1. Comparison of stimulus-response curves for *n*-amyl acetate derived by electrophysiological and behavioural techniques. Behavioural data for the rat are expressed on a scale in which 0 = chance score. (The integrator deflection magnitudes have been converted from the arbitrary units originally published to a common scale.) It has been suggested that the reversals in these curves may reflect the presence of two different populations of receptors or receptor sites one of which saturates at a lower concentfation than does the other (Moulton, (1960).

In particular, changes occur in the amplitude of odour-evoked responses and background activity in the olfactory bulb with changing levels of anaesthetics (Adrian, 1950; see also Moulton and Tucker, 1964). Among specific effects reported in other systems, pentobarbital can block synaptic transmission (Löyning *et al.*, 1964), chloralose reduces recurrent inhibition from Renshaw cells to motoneurones (Shimamura *et al.*, 1968), and sodium pentothal alters receptive fields in the cat's visual cortex as compared to those defined in the cerveau isolé preparation (Robertson, 1965). In addition, anaesthetics and other immobilizing agents have well-known effects on respiratory, circulatory and other body functions (e.g. Korner *et al.*, 1968). Restraint alone more than doubled corticosterone secretion in rats and caused a marked increase in acid concentration of the gastric juice (Knigge *et al.*, 1959; Brodie *et al.*, 1962).

Clearly, then, for the most valid comparison between behavioural and neural events, these events should be measured simultaneously in the unrestrained behaving animal. And there are other advantages in the use of this approach. In particular, it can obviate the need for the large numbers of subjects and replications involved in establishing the existence of events with long latencies having a time course in the order of days or weeks. Such events include alterations in sensitivity to odours as a function of circadian rhythms; of changing hormonal, nutritional, age, stress, or disease factors; or during learning of an odour detection task. The chronic preparation also makes more feasible simultaneous recordings of several variables, such as activity from multiple sites in the olfactory bulb. And it allows investigations of the influence of the normal patterns of sniffing and of neural "focusing" including the influence of the centrifugal fibre system on bulbar response. A disadvantage is that it has not yet been demonstrated that intracellular recordings are feasible in moving animals. There are also species limitations in the use of unrestrained animals. For example, the monkey will tug at any accessible lead or microdrive.

In the remainder of this chapter we will consider selected techniques for deriving comparable records of electrical and behavioural response to odour stimulation with emphasis on relating bulbar unit and multiunit activity to the performance of mammals simultaneously engaged in an odour detection or discrimination task.

However, the problems of technique, design and interpretation which this approach imposes can be formidable and practical solutions are not always optimal for some of the difficulties encountered. What follows is an attempt to outline selected approaches to these problems, without necessarily implying that they are the only solutions possible. Furthermore, it gives little attention to methods of recording and analysis equally appropriate to the study of epithelial or bulbar activity in acute preparations or to the range of behavioural and olfactometric methods available, since these are described in other chapters of this book.

II. Approaches to the simultaneous recording of neural and behavioural response to odours

The investigator who wishes to record, simultaneously, neural and behavioural events elicited by the same stimuli has the choice of using restrained, partially restrained or freely moving animals. In general, the less restraint he imposes, the more difficulties he will encounter in controlling stimulus variables and in maintaining and transmitting the signal from the animal to the recording apparatus. The more restraint he imposes the more likely he is to exclude the normal physiology and behaviour of the species under investigation.

A key question in stimulus control is mode of delivery of the stimulus to the receptor surface. It may be necessary or sufficient to allow the animal to breathe normally. For example, Macrides and Chorover (1972) have shown that single units in the olfactory bulb of rodents show inhalation-dependent patterns of firing. The extent to which normal sniffing is necessary to allow the full development of these patterns is not clear. But studies with animals that sniff normally are clearly needed. If the animal controls its own breathing, however, the rate of delivery of odorous molecules to the receptor surface cannot be controlled or accurately measured. Unless the animal is trained to detect odour, or to sniff when a stimulus is presented, it is not even certain that it will sniff the odour.

An alternative approach, which allows control of duration of the stimulant, is to inject odours by cannula directly into the nasal chamber. However, this may still allow the animal to sniff and so dilute the injected odour with air or vary the flow rate of the stimulus. In addition, connections have to be made to the odour source which restricts the animal's freedom of movement. Nevertheless, for certain purposes this could be an effective technique.

Theoretically, at least, recording of neural activity from freely moving animals is best achieved by employing telemetry devices. But if some degree of restraint is acceptable, signals can be transmitted—with less practical difficulties—by direct wire. In either case two methods of recording are possible: with electrodes that can be repositioned during the course of the experiment or with electrodes chronically implanted in one position. The first method is suited to the extracellular recording of single unit activity and even intracellular recordings are possible. However, it has often the same limitations as conventional single unit recording techniques in that units can be held for only limited periods. On the other hand a number of units can be sampled during a recording session. A basic component is a microdrive for positioning the electrodes.

Numerous microdrive assemblies have been described.* An example is illustrated in Fig. 2. In brief, a tubular steel or teflon implant with a central cavity is fitted into a trephine hole in the skull over the exposed surface of the olfactory bulbs. Fluid is introduced into this well and a hydraulic micropositioner is attached to the tube to form a closed system. The microelectrode (tungsten, stainless steel, or glass-insulated platinum-iridium) is then lowered into the bulb until a unit is identified. Full details for the construction of such a microdrive assembly are given by Evarts (1966). A cheaper system, claimed to be less cumbersome, has been developed around a disposable hospital syringe (Bridgeman, 1972). Associated with this microdrive is an impedance reduction unit (in the form of a removable plug) also secured on the animal's head.†

Major difficulties with this technique are the loss of a unit when the animal moves abruptly and the general problems of backlash and drift in the microdrive. The first difficulty can be minimized or eliminated by restraining the animal, and it is probable that both difficulties are reduced by the use of remotely controlled microdrives in combination with telemetry devices (Edge *et al.*, 1969).

The main disadvantage of most techniques involving microdrives, however, is that long-term studies of the same unit or group of units are not possible. This has been overcome by John and Morgades (1969a) who have designed a microdrive technique to allow repositioning of chronically implanted multiple microelectrodes (glass-insulated Pt-Ir.). Humphrey (1970), achieves a similar result by using micrometer screws to drive the electrodes.

However, the investigator may wish to record simultaneously both neural and behavioural responses from each member of a group of animals on a daily basis. For this and certain other purposes immovable chronically implanted electrodes may offer a less expensive and simpler solution. The following section outlines an example of such an approach.

* For example: Hubel, 1960; Fuster, 1961; Baust *et al.*, 1961; Evarts, 1966; Oomura *et al.*, 1967; John and Morgades, 1969a; Barmack and Hayes, 1970; Bridgeman, 1972.

† Microdrives with direct manual or hydraulic control are available from: Frederick Haer and Co., Inc., Ann Arbor, Michigan 48106 and Neuro Probe Inc., 7400 Arden Road, Bethesda, Maryland 20034.

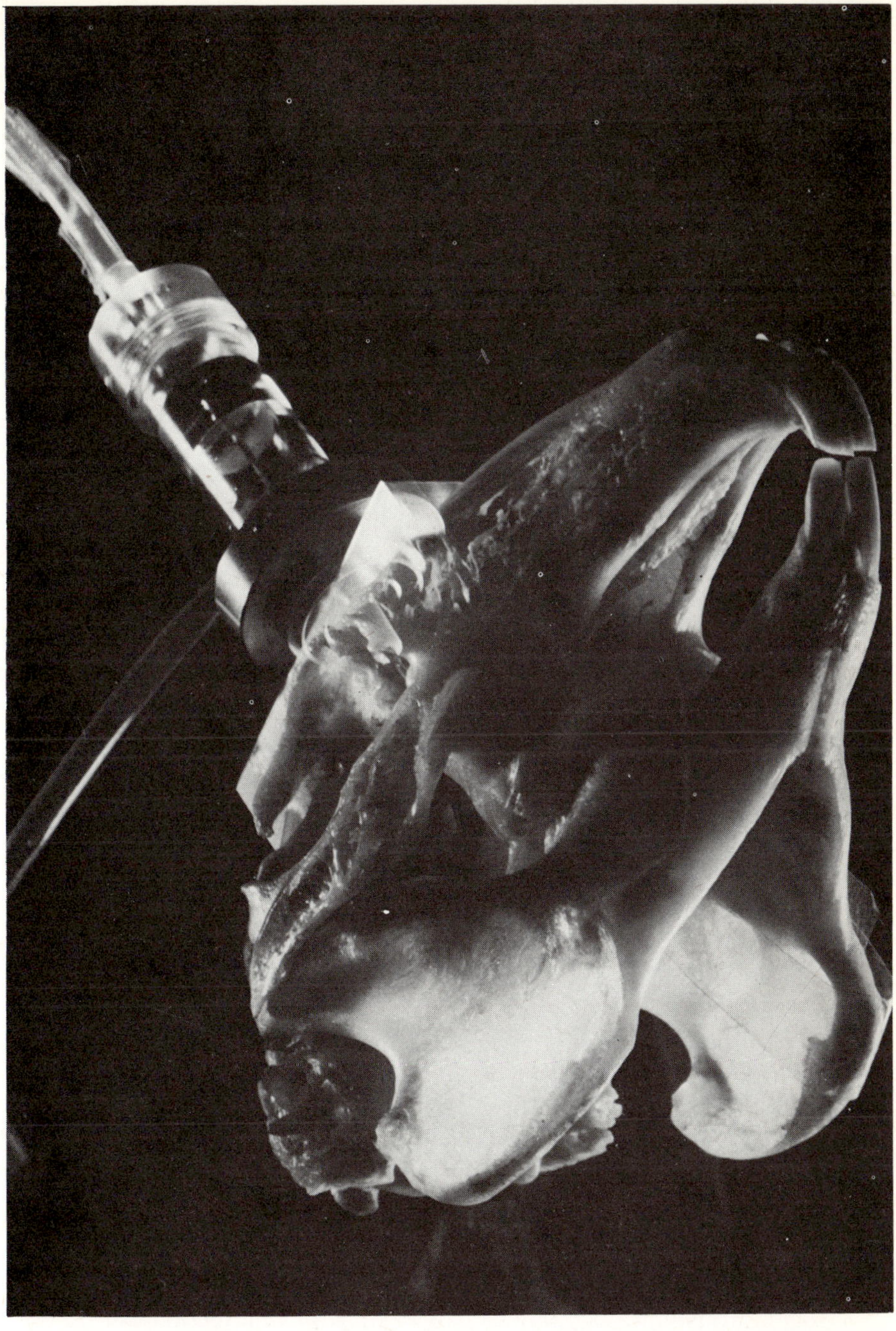

FIG. 2. Hydraulic microelectrode positioner in place on rabbit's skull (modified from a design described by Hubel, 1960). The unit normally rests on a hollow teflon ring screwed into a trephine hole in the skull. However for purposes of illustration this has been replaced with a lucite block in this figure. The piston in which the shaft of the tungsten microelectrode is mounted is moved downwards by the introduction of water above the piston.

III. Techniques for chronically implanting electrodes in the olfactory bulb of rats and rabbits

The range of techniques necessary to record from freely moving animals covers electrode implantation and associated procedures; signal transmission, monitoring, recording, processing and analysis; odour production and delivery, and measurement of sensory performance. This section concerns implantation procedures in the rat and rabbit as examples of small and larger mammals.

A. Electrode implantation in the rat

The individual steps in the construction of an electrode and its implantation in the brain of a chronic rat preparation (as well as procedures for histological examination of the brain after experimentation) have been outlined in detail by Skinner (1971), while Olds *et al.* (1969) give information for the implantation of multiple microelectrodes. We have used essentially comparable techniques to prepare and implant electrodes in rat olfactory bulbs. But in addition to the direct wire recording methods, one can also transmit multiunit and occasionally single unit activity by telemetry devices while the moving rat is involved in an odour detection task (Moulton, 1968; Goldberg and Moulton, 1969; Goldberg, 1970). What follows is a summary of the implantation technique. [See Skinner (1971) for fuller details and for the addresses of manufacturers of the materials required.]

If electrodes are to be chronically implanted they should be of a material that is tolerated by cortical tissue and is sufficiently rigid to penetrate to the target site without buckling. Nichrome or stainless steel meet these requirements. It might be argued that the flexibility of these metals would also be an advantage in so far as it allows the electrodes to move with the brain rather than resist and destroy the tissue. However, the fact that both glass-insulated Pt-Ir and tungsten microelectrodes are also reported to be effective in chronic single unit studies (John and Morgades, 1969a, 1969b; Naka and Kido, 1967) suggests that this may not be a significant factor.

Bipolar electrodes used with differential recording techniques are superior to unipolar electrodes because they reduce movement artifacts. Suitable bipolar stainless steel electrodes can be made from 0.0025 in. (63.5 mμ) diameter wire insulated with teflon or quadruple enamelled. (Twist two lengths of wire together using a dental drill if necessary, and cut the tips with scissors.) For single unit recordings it is helpful to sharpen the tips electrolytically and reinsulate leaving the tips exposed (cf. Green, 1958).

Rats (250g–350g) can be anaesthetized with intraperitoneal injections of chloralose; pentobarbital or Equithesin (a combination of pentobarbitol, chloral hydrate and magnesium sulphate, 0.15–0.18 ml/100g body weight; Jensen-Salsbery, Kansas City, Mo.) but until it is shown that inhalation of ether does no damage to the olfactory epithelium it should not be used as a method to supplement the anaesthesia.

Anaesthetize the rat, shave its head and place it in a stereotoxic instrument. Make an incision about 2 cm long extending anteriorly from a point about 1 cm posterior to bregma. Retract the skin and remove the exposed periosteal connective tissue by blunt dissection. Using a # 80 bit, drill a hole at a point about 9.0 mm anterior to bregma and 1.0 mm lateral to the midline to allow entry of a single electrode (Goldberg, 1970). It is important to slice the dura before inserting the electrode or the tip will bend and direct the electrode in a plane parallel to the surface of the skull. To provide support for dental cement, drill four holes around the perimeter of the exposed area away from the electrode hole (avoiding the midline longitudinal suture). Screw stainless steel machine screws (0–80 x 1/8 in.) into these holes. They should be of a diameter that allows the screws to self-thread. It is not necessary to pre-tap the bone. Cement usually adheres well to the combination of dry bone and screws. If difficulties arise a primer may be useful in ensuring an efficient bond (Caulk Cavity primer; Horovitz and Chow, 1963; Rose, 1966).

The sterilized electrodes should be straight and precisely aligned above the hole. Although some workers (e.g. Blum, 1962) have used more elaborate techniques to ensure the correct angle of entry, a simple solution is to secure a hypodermic needle on the arm of a micromanipulator and thread the electrode through the lumen until the tip protrudes about 15 mm. To secure the electrode in position, use a plastic wedge. It rests inside the hub of the needle and tightens by a screw sunk horizontally through the wall. It is also helpful to work under a Zeiss or other dissecting microscope with a 12.5 cm working distance since this allows unimpaired manipulation under stereoscopic observation (16 or 25 x suffices). Activity is monitored on an oscilloscope and audio monitor as the electrode is slowly lowered into the olfactory bulb. With practice the drive can be halted at about the level associated with maximum amplitude of unit activity.

Perhaps the most critical stage in implantation is securing the electrode shaft in the drill hole. Unit activity will cease or decline markedly in amplitude with small movements of the electrode tip and cortical tissue can easily be damaged. The skull surface immediately around the shaft should be free of any cerebrospinal fluid that exudes through the drill hole. Fine spindles of cotton or tissue paper are adequate for this purpose. A strip of Gelfoam folded around the shank will reduce access of toxic substance in the cement to the brain. This strip is overlain with dental cement. The cement is sufficiently hard in about five minutes (if initially at optimal consistency) to allow removal of the encasing hypodermic needle. First, release the electrode shaft by cautiously loosening the screw of the wedge. Record the unit responses to odour stimulation at that site before cutting the electrode shaft. Then slowly raise the needle until clear of the electrode.

Whether direct wire or telemetry is used to transmit activity from the electrodes to the amplifier or receiver, it is generally necessary to terminate the free ends of the electrodes in an accessible outlet. One or more pairs of contact pins from a miniature connector or wire form contacts (Frommer, 1970) serve well and can be anchored to the skull with dental cement. Scrape off the insulation from a few mm of the exposed ends of the electrodes and crimp, spot well or solder them into the contacts of the miniature connector (cf. Skinner, 1971; Annau and Weinstein, 1968; Milstein, 1969; Ray and Yates, 1970). Crimping has the advantage in that it does not involve heat which might damage the tissue at the tip of the electrodes. After insulating the contact base and all exposed uninsulated electrode surfaces with dental acrylic, cement the connector onto the skull. Pull the skin around the margins of the cement dome. A two-week post-operative period allows time for the development of a glial capsule around the implant. The skin heals tightly around the dome.

B. Electrode implantation in the rabbit

The small size of the rat limits implants in the olfactory bulb to a few pairs of electrodes. However, the investigator may wish to record activity from multiple channels including, for example, activity from non-olfactory structures and from different sites in the olfactory epithelium. For this purpose a larger animal is more suitable. The rabbit, in particular, offers certain advantages (Moulton, 1973). Thus the olfactory bulbs are well developed, have been extensively studied both electrophysiologically and anatomically (see Shepherd, 1972), and are easily accessible without the need to penetrate overlying cortical tissue; intravenous injection of anaesthetics is favoured by the large marginal vein of the ear; the social behaviour of wild rabbits is known to be extensively controlled by odours (Mykytowycz, 1965); while the domestic rabbit is docile and an excellent subject for both instrumental (Swadlow, 1970; Oakley, 1971) and classical conditioning (e.g. Baru, 1954) studies.

With minor modifications, the techniques outlined for the rat are also applicable to the rabbit. However, as the number of implants increases so does the need for an effective support structure. Thus frequent plugging and unplugging of connectors and the torque imposed by cables can place stress on the electrode shafts and adjacent areas of the skull and may lead to displacement of electrode tips with resulting loss of activity. For some purposes, a simple socket holder, such as Reeves and Hagamen (1968) describe, overcomes these problems. But if multiple telemetry units are used their correct orientation is important in reducing electromagnetic coupling between neighbouring units. It is also convenient, in some cases, to begin implanting electrodes on one day and complete it on another. In the interim, as well as in the long term, it is necessary to protect unconnected electrode shafts and exposed skull areas.

One solution to these problems is to design a scaffold roughly in the form of a hollow open dome, the base of which can be separated from the upper shell (Fig. 3). Both can be made from a self-setting resin.* Contacts embedded in the walls of the roof allow connection with six or more telemetry units or pins for direct coupling (Çelebi, 1971). The base is anchored to the skull first so that an area overlying the olfactory bulbs remains exposed. Electrodes are implanted within this area and the upper shell attached later. While we have not investigated the possibility, this design should also lend itself to the inclusion of miniaturized preamplifiers (cf. Oomura *et al.*, 1967; Rokushima *et al.*, 1968; Spinelli *et al.*, 1970).

The walls of the basal section of the scaffold are predrilled to accommodate the machine screws which attach it to the skull.

After anaesthetizing the animal with pentobarbital, position it in a stereotoxic apparatus head-holder, and shave the hair overlying the skull with animal clippers. Make an incision 25 mm anterior to the frontal-nasal suture at the midline and extended about 70 mm posteriorly. Retract the skin and bluntly dissect the subcutaneous tissue and periosteum from a flattened oval region 30 x 65 mm. The

* For example: "Quickmount" (Fulton Metalurgical Products Corp., Pittsburgh, Pa.). Welker (1967) has outlined techniques for making casts.

underlying skull surface is dried (a dry bone surface is essential) and roughened with a dental drill. Arrest any bleeding from intra-medullary space in the bone with bone wax, and coagulate any severed vessels.

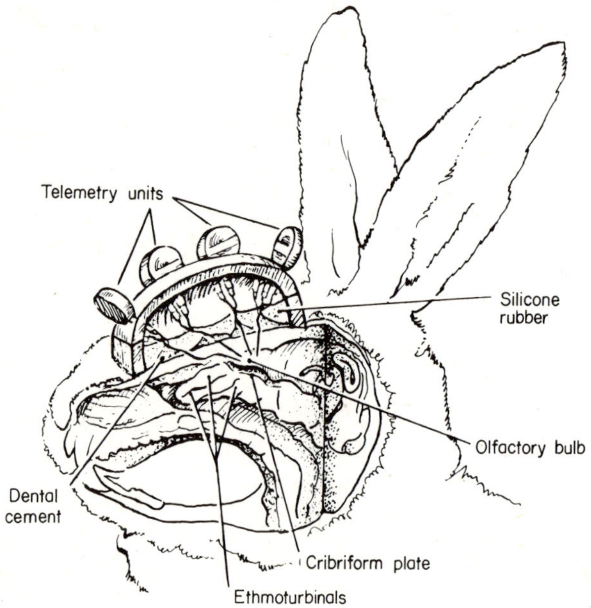

FIG. 3. Section through head of rabbit and support scaffold to show implanted electrodes and telemetry units. (From Moulton, 1973.)

The anterior pole of each bulb in a 6–8 month old New Zealand rabbit lies about 7–9 mm posterior to the fronto-nasal suture at the midline and extends about 22 mm posteriorly. The maximum width of each bulb is about 10 mm and has a depth (from the dorsal to the ventral surface) of about 9 mm. The centre of the bulb surface lies 3 mm below the skull surface at the midline. The cribriform plate, which has several larger foramen suitable for implanting electrodes in the primary neurons, is 1 mm thick.

After holes have been drilled in the skull for electrode placement the base of the scaffold is positioned on the skull so that these drill holes lie within the exposed centre well of the support. The base is attached by five screws staggered around its walls. (This number is necessary because the cranial bone of the rabbit does not provide good stability for threaded screws.) Use the shafts for these screws to align the bit used to drill holes in the skull. Care should be taken to avoid penetrating through the bone into the cranial cavity. Secure the base with screws directed at right angles to the plane of the skull. No screws should impinge on the cortex.

Implantation of sterilized electrodes differs from the techniques outlined for the rat, only in that the walls of the basal scaffold provide support for the dental cement. When the cement anchoring the electrode shafts has hardened to form an array of domes overlay the surface with dental acrylic or silastic rubber to provide additional support and protection, colour-code the electrodes, and map their locations.

To prepare the free ends of the electrodes for attachment, strip the insulation off a few centimetres of the tip. (Support the electrodes with tweezers to block transfer of stress to the lower shaft.) Crimping the free exposed tips of the leads to contacts is best done under magnification. If necessary it can be completed when the animal is unanaesthetized but restrained. Note the position of electrode pair before connecting, and check the insulation of their shafts. The contacts, which protrude from the domed shell portion of the scaffold, are taken from miniature connectors. Screw the shell down on to the base portion of the scaffold and allow the animal to recover. Electrodes, and associated scaffold, remain in place two years or more. Only occasionally have we noticed skin infection and there has been no evidence of pressure necrosis in so far as this is reflected in the loosening of screws retaining the scaffold.

Another approach to electrode implantation involves constructing an assembly of macro- or microelectrodes which is then lowered into the bulb in a single insertion (Johnson *et al.*, 1972; Marg and Adams, 1967).

IV. Signal transmission

A. Direct wire coupling

The simplest solution to transmitting neural events to the recording apparatus is a direct cable connection. But this has two disadvantages: the animal's movements are restrained, and they induce noise. The first problem is inherent in the method but the second can be virtually eliminated. Sources of noise include static noise due to cable distortion and capacity modulation of the electrode tip potential (Oomura *et al.*, 1967). Swivel devices that confer full rotational freedom on the cables, a preamplifier insensitive to such artifacts, and short low-noise shielded cables* minimize distortion (Sutton and Miller, 1963; Mundl, 1971; Edelson and Holdstock, 1969). Miniaturized preamplifiers that can be mounted directly on the animal's head, seem to offer the best solution to remaining noise problems (e.g. Oomura *et al.*, 1967; Rokoshima *et al.*, 1968; Spinelli *et al.*, 1970).

B. Telemetry

Despite the limitations on freedom of movement that direct wire methods involve, many will prefer them to the additional difficulties imposed by telemetry. Yet ultimately, telemetry seems likely to become the best practical as well as theoretical approach to the coupling problem. Even now it offers the only way of transmitting information from animals moving more than a limited distance from the recording apparatus. Its attractiveness is growing with the development of more reliable circuits allowing miniaturization of units; the advent of field effect transistors; the appearance of descriptive texts (e.g. Mackay, 1970); the commercial availability of telemetry units, and the growing interest in monitoring neural events in freely moving animals (e.g. Black-Cleworth and Verberne, 1972). It also has the advantage of minimizing stress on the implant on the animal's head.

Several circuits have been reported which are suitable for transmitting a single or multi-unit activity (Skutt *et al.*, 1967; Beechey and Lincoln, 1969; McElligott *et al.*, 1969; Marshall and Çelebi, 1970; Brown *et al.*, 1971). Units constructed from these circuits range in size from a disc 11 mm in diameter and 6 mm thick (Skutt *et al.*, 1967) to a unit 1 x 1 x 2.5 cm (Beechey and Lincoln, 1969). The disc unit is shown diagrammatically in Fig. 3 and weighs about 1.5 g, including battery. It has been used in this laboratory to record both single and multi-unit activity from the rat and rabbit (Skutt *et al.*, 1967; Moulton, 1968; Goldberg and Moulton, 1969; Goldberg, 1970; Çelebi, 1971). Up to two units can be mounted on the rat head and at least six on the rabbit. The range of the unit is about 100 feet. It transmits at a fixed carrier frequency which can be set within a few MHZ when constructed. Unless this frequency coincides

* For example: Microdot. 250–3838, Microdot Inc., South Pasadena, California 91030.

with that of a strong FM radio station, this limitation will not normally present difficulties. However, if a number of units are stacked together mutual interference between fundamental and in some cases, harmonic frequencies, may prevent effective transmission. Optimal spacing of telemetry frequencies minimizes this problem and avoids conflict with strong FM radio signals. This can be achieved with a modification outlined by Marshall and Çelebi (1970) that allows the transmitter to be tuned, over a wide range. (These authors outline in detail the procedure for constructing the unit, list commercial sources for the components and point out some of the difficulties of multichannel telemetry.)

A further difficulty of telemetry is the disappearance or reduction in amplitude of the signal when the transmitting and receiving antennas are in certain orientations relative to one another. There are several solutions to this problem including the use of a stronger transmitter and a transmitter that sets up a spinning or circularly polarized field (Mackay, 1970), or–ideally–a circuit which virtually eliminates this problem (Brown *et al.*, 1971). For some purposes, however, it is sufficient to ensure that critical information is obtained only when the animal's head is optimally oriented relative to an omnidirectional or other receiving antenna (Çelebi, 1971).

Finally, movement artefacts occur when the animal bumps the transmitter unit against a wall of the test apparatus or licks a water reward. Bouts of teeth "chattering", chewing, or other muscle artefacts also disrupt the trace (Moulton, 1968; Goldberg, 1970). However, such artefacts are usually obvious.

For most purposes standard FM radio receivers with drift control are adequate both for reception and auditory monitoring of the signal. Additional monitoring and recording can be done by conventional techniques.

V. Odour production, measurement, delivery and control

Simultaneous recording of neural and behavioural responses to odours imposes few requirements not met separately in behavioural or electrophysiological studies (see Moulton, 1973). There is one area, however, where the problems are more specific. To identify neural correlates of behavioural responses one may need to detect events occurring in the order of ms. Several factors control both the probability of detecting the neural event and the accuracy with which its magnitude, duration and time of occurrence are defined. These include molecular species of the odour, its pulse form, the degree of background odour contamination, and the animal's mode of sampling the odour.

The physico-chemical properties of the odorant determines its degree of adsorption to the walls of the olfactometer and delivery system and thus the shape and duration of the odour pulse. They also determine the time required to rid the delivery system of the odorant. The more volatile and less polar the odorant the more readily will it desorb. Musks and ionones are examples of odorants with unfavourable characteristics.

Where a contrast between an odour-induced and a control ("blank" air) response is required and subjects are involved in an odour detection task, a two-choice situation involving simultaneous but separate delivery of odour and air has certain advantages. These are more appropriately considered in the next section.

VI. Measurement of sensory performance

Quantitative studies of odour detection or discrimination performance have drawn on a variety of training procedures including classical and operant conditioning, and conditioned suppression. Any of these methods can be combined with simultaneous recordings of neural response, provided that certain precautions are taken. For example, transients generated by shocking devices, solenoids, movement, etc. must be identified and either suppressed or taken into account during data analysis.

Before beginning training the animal should be habituated not only to the behavioural situation but also to the restraint imposed by direct wire coupling or the weight of the telemeter units. Heart rate

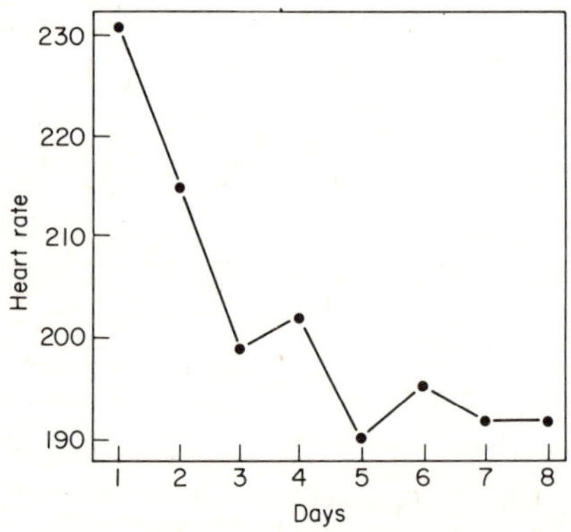

FIG. 4. Habituation to test environment. Mean heart rate of seven rabbits placed in test chamber for two hours daily for eight days.

provides a simple indicator of the animal's degree of habituation. It is sufficient to place the animal in the test situation for two hours daily until heart rate declines to a relatively stable plateau. In the case of the rabbit this can take about 7–10 days (Moulton, unpublished. See also Fig. 4). Rabbits may require up to 7 days of water deprivation before they will work for a water reward. Their total daily consumption is in the order of 30 ml water per animal (Swadlow, 1970).

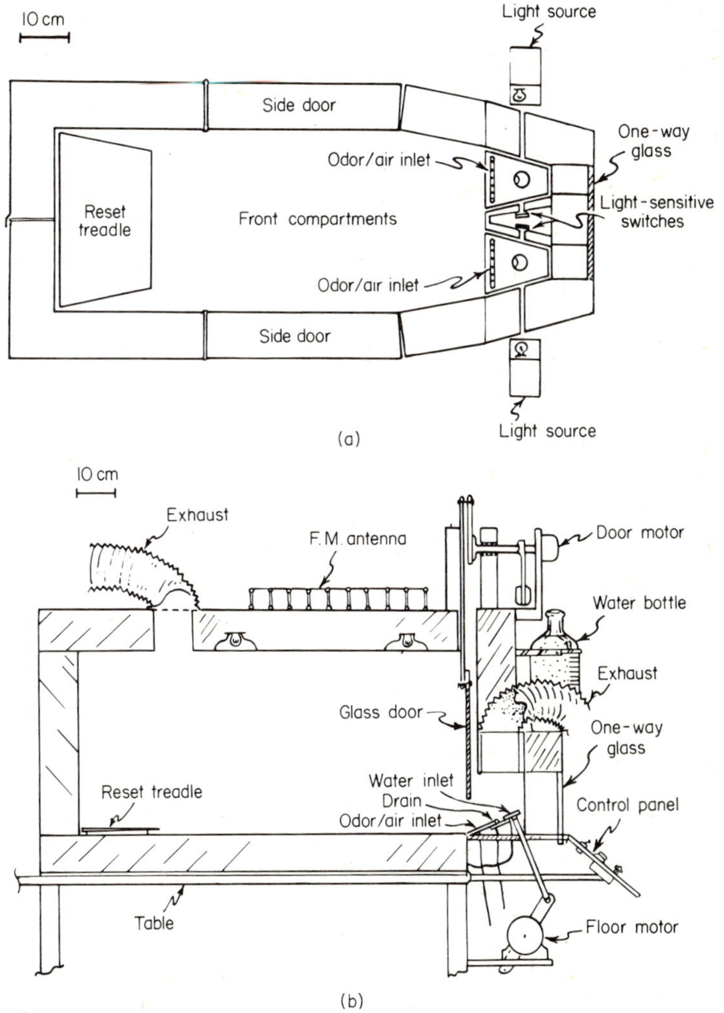

FIG. 5. (a) Plan of odour detection apparatus. (From Moulton, 1973.)
(b) Side view of odour detection apparatus. (From Moulton, 1973.)

One technique designed to accommodate the requirements for investigating simultaneously neural and behavioural responses to odour in rabbits is summarized by Moulton *et al.*, 1970 (see also Dravnieks, this volume, pp. 46–48 and Fig. 10; Moulton, 1973). The apparatus is illustrated in Fig. 5. The rabbits are trained to detect odours by a free-operant conditioning method. They indicate the presence of the odour by intercepting a photocell beam for 5 s for a water reward.

A thirsty rabbit, when placed in the air-swept main compartment of the apparatus, faces two bays. Odour flows vertically through one and air through the other. The two flows are delivered from an olfactometer and are interchangeable according to a randomly determined sequence. The animal can insert its head into either bay as

(c) Interior view of anterior section of odour detection apparatus showing rabbit with six telemetry units. The walls and doors are glass-lined to facilitate cleaning and the entire apparatus is electrically shielded.

frequently as it wishes until it sustains an interruption of the beam for a period of 5 seconds. When this occurs a choice is registered and the animal receives a water reward if correct. After a delay to allow the animal to drink—or immediately if the choice is incorrect—the floor of the bay rises up to eject the rabbit from the bay. Glass doors then descend to seal the bays from the main compartment between trials. Neural data, transmitted from telemetry units, are received by an omnidirectional antenna on the roof of the apparatus and by bipolar antennas (Moulton *et al.*, 1970; Çelebi, 1971).

In such a two-choice situation most rabbits will sample both bays before indicating a choice (at least when the odour concentrations are near threshold). This allows the investigator frequent opportunities to contrast neural activity associated with sniffing filtered air as against odorized air. The design of the apparatus also ensures that the animal's head is favourably oriented in relation to the receiving antennas during the period in which it is making a choice.

References

Adrian, E. D. (1950). The electrical activity of the mammalian olfactory bulb. *Electroencephalog. clin. Neurophysiol.* **2**, 377–388.

Annau, Z. and Weinstein, S. A. (1968). A miniature skull-mounted multiple connector for small animals. *J. Expt. Anal. Behav.* **11**, 187–198.

Barmack, N. H. and Hayes, D. F. (1970). A stepper motor controlled microdrive for recording from unanaesthetized animals. *Physiol. Behav.* **5**, 705–706.

Baru, A. V. (1954). Conditioned motor reflexes to chains of stimuli in rabbits. *Zh. vyssh. nervn. Deyatel.* **4**, 712–723.

Baust, N., Baumgarten, R. V. and Hellner, K. A. (1961). Drahtlos übertragene Mikroableitungen aus dem Gehirn der wachen, frei beweglichen Katze. *Pflüg Archiv. ges. Physiol.* **272**, 400–406.

Beechey, P. and Lincoln, D. W. (1969). A miniature FM transmitter for the radio-telemetry of unit activity. *J. Physiol.* **203**, 5P.

Békésy, G. von (1968). Problems relating psychological and electrophysiological observations in sensory perception. *Perspectives Biol. Med.* **11**, 179–194.

Black-Cleworth, P. and Verberne, G. (1971). Telemetered EEG and neuronal spike activity in olfactory bulb and amygdala in free moving rabbits. *In* "Biotelemetry" (Kimmick, N. P. and Vos, J. A., eds.), International Symposium 1971. Meander, N. V., Leiden, The Netherlands.

Blum, B. (1962). A method for chronic implantation of depth electrodes in the brains of experimental animals. *Electroenceph. clin. Neurophysiol.* **14**, 552–553.

Borg, G., Diamant, H., Ström, L. and Zotterman, Y. (1967). The relation between neural and perceptual intensity: A comparative study on the neural and psychophysical response to taste stimuli. *J. Physiol.* (*Lond.*) **192**, 13–20.

Borg, G., Diamant, H. and Zotterman, Y. (1970). Neural and perceptual responses to taste stimuli. *In* "Taste and Smell in Vertebrates" (Wolstenholme, G. E. W. and Knight, J., eds.), pp. 99–113. J. and A. Churchill, London.

Bridgeman, B. (1972). A simple system for extracellular micro-electrode recording from awake animals. *EEG Clin. Neurophysiol.* **33**, 116–118.

Brodie, D. A., Marshall, R. W. and Moreno, O. M. (1962). Effect of restraint on gastric acidity in the rat. *Am. J. Physiol.* **202**, 812–814.

Brown, M. W., Edge, G. M. and Horn, G. (1971). A miniature transmitter suitable for telemetry of a wide variety of biopotentials. *Electroenceph. clin. Neurophysiol.* **31**, 274–276.

Çelebi, G. A. (1971). Development and use of a multi-channel radio-telemetry system in a simultaneous behavioral and electrophysiological study of olfactory responses in the rabbit. Doctoral dissertation, Clark University.

Edelson, A. and Holdstock, T. L. (1969). A printed circuit swivel for recording from freely moving animals. *Electroenceph. clin. Neurophysiol.* **27**, 209–210.

Edge, G. M., Horn, G. and Stechler, G. (1969). Telemetry of single unit activity from a remotely controlled microelectrode. *J. Physiol.* (*Lond.*), **204**, 2P–4P.

Evarts, E. (1966). Methods for recording activity of individual neurons in moving animals. *In* "Methods in Medical Research" (Rushmer, R., ed.), Vol. II. Year Book, Chicago.

Frommer, G. (1970). A simple method for terminating chronically implanted electrodes in small animals. *Physiol. Behav.* **5**, 1501–1502.

Fuster, J. M. (1961). Excitation and inhibition of neuronal firing in visual cortex by reticular stimulation. *Science, N.Y.* **133**, 2011–2012.

Goldberg, S. J. (1970). Neural responses telemetered from the olfactory bulb of rats involved in an odor detection task. Doctoral dissertation, Clark University.

Goldberg, S. J. and Moulton, D. G. (1969). Multi- and single unit activity telemetered from the olfactory bulb of rats involved in an odor detection task. *Am. Zool.* **9**, 593.

Green, J. D. (1958). A simple microelectrode for recording from the central nervous system. *Nature, Lond.* **182**, 962.

Harrington, T. and Merzenich, M. M. (1970). Neural coding in the sense of touch: human sensations of skin indentation compared with the responses of slowly adapting mechanoreceptive afferents innervating the hairy skin of monkeys. *Expl. Brain Res.* **10**, 251–264.

Horovitz, A. P. and Chow, M. (1963). Technique of implanting permanent electrodes in cats for chronic stimulation and observation of EEG and behavioural effects. *J. Pharmac. Sci.* **52**, 198–200.

Hubel, D. H. (1960). Single unit activity in lateral geniculate body and optic tract of unrestrained cats. *J. Physiol.* **150**, 91–104.

Humphrey, D. R. (1970). A chronically implantable multiple micro-electrode system with independent control of electrode positions. *Electroenceph. clin. Neurophysiol.* **29**, 616–620.

John, E. R. and Morgades, P. P. (1969a). A technique for the chronic implantation of multiple movable micro-electrodes. *Electroenceph. clin. Neurophysiol.* **27**, 205–208.

John, E. R. and Morgades, P. P. (1969b). Neural correlates of conditioned response studies with multiple chronically implanted moving microelectrodes. *Expti. Neurol.* **23**, 412–425.

Johnson, J. H., Clemens, J. A., Terkel, J., Whitmoyer, D. I. and Sawyer, C. H. (1972). Technique for recording multiple-unit activity from the brain of the freely-moving rat. *Neuroendocrinol.* **9**, 90–99.

Jung, R. (1961). Neuronal integration in the visual cortex and its significance for visual information. *In* "Sensory Communication" (Rosenblith, W. A., ed.), pp. 627–674. Massachusetts Institute Technology Press, Cambridge, Mass.

Kiang, N. Y-S. (1965). "Discharge Patterns of Single Fibres in the Cat's Auditory Nerve" p. 154. Massachusetts Institute Technology Press, Cambridge, Massachusetts.

Knigge, K. M., Penrod, C. H. and Schindler, J. (1959). *In vitro* and *in vivo* adrenal corticosteroid secretion following stress. *Am. J. Physiol.* **196**, 579–582.

Korner, P. J., Uther, J. B. and White, S. W. (1968). Circulatory effects of chloralose-urethane and sodium pentobarbitone anaesthesia in the rabbit. *J. Physiol.* **199**, 253–265.

Löyning, Y., Oshima, T. and Yokota, T. (1964). Site of action of thiamyal sodium on the monographic spinal reflex pathway in cats. *J. Neurophysiol.* **27**, 408–428.

Mackay, R. S. (1970). "Bio-Medical Telemetry", 2nd edn., 533 pp. Wiley and Sons, Inc., New York.

Macrides, F. and Chorover, S. L. (1972). Olfactory bulb units: activity correlated with inhalation cycles and odor quality. *Science, N.Y.* **175**, 84–87.

Marg, E. and Adams, J. E. (1967). Indwelling multiple micro-electrodes in the brain. *Electroenceph. clin. Neurophysiol.* **23**, 277–280.

Marshall, D. A. and Çelebi, G. (1970). A tunable subminiature bio-telemetry transmitter. *Physiol. Behav.* **5**, 709–712.

Mathews, D. F. (1972). Response patterns of single neurons in the tortoise olfactory epithelium and olfactory bulb. *J. Gen. Physiol.* **60**, 166–180.

McElligott, J. G., Kado, R. T. and Zweizig, J. R. (1969). A miniaturized telemetry device for the transmission of the electrical activity of single nerve cells in the brain. *Proc. Nat. Telemetry Conf., Washington, D.C.,* pp. 207–210.

Merzenich, M. M. and Harrington, T. (1969). The sense of flutter-vibration evoked by stimulation of the hairy skin of primates: comparison of human sensory capacity with the responses of mechanoreceptive afferents innervating the hairy skin of monkeys. *Expl. Brain Res.* **9**, 236–260.

Milstein, V. (1969). Implantation of sterile electrodes in chronic animals. *Electroenceph. clin. Neurophysiol.* **27**, 442–443.

Moulton, D. G. (1973). The use of animals in olfactory research. *In* "Methods of Animal Experimentation" (Gay, W. I., Ed.), Vol. IV, pp. 143-223. Academic Press, New York and London.

Moulton, D. G. (1968). Electrophysiological and behavioural response to odour stimulation and their correlation. *Olfactologia* **1**, 69–75.

Moulton, D. G. (1960). Studies in olfactory acuity 3. Relative detectability of n-aliphatic acetates by the rat. *Quart. J. Exptl. Psychol.* **12**, 203–213.

Moulton, D. G., Çelebi, G. and Fink, R. (1970). Olfaction in mammals–two aspects: proliferation of cells in the olfactory epithelium and sensitivity to odours. *In* "Taste and Smell in Vertebrates" (Wolstenholme, G. E. W. and Knight, J., eds.), pp. 227–246. Churchill, London.

Moulton, D. G. and Tucker, D. (1964). Electrophysiology of the olfactory system. *Ann. N.Y. Acad. Sci.* **116**, 380–428.

Mountcastle, V. B. (1968). The problem of sensing and the neural coding of sensory events. *In* "The Neurosciences" (Quarton, G. C., Melnechuk, I. and Schmidt, F. O., eds.). Rockefeller Press, New York.

Mountcastle, V. B., LaMotte, R. H. and Carli, G. (1972). Detection thresholds for stimuli in humans and monkeys: comparison with threshold events in mechanoreceptive afferent nerve fibres innervating the monkey hand. *J. Neurophysiol.* **35**, 122–136.

Mozell, M. M. (1958). Electrophysiology of the olfactory bulb. *J. Neurophysiol.* **21**, 183–196.

Mundl, W. J. (1971). Preamplifier for recording of multiple-unit activity from moving animals. *Physiol. Behav.* **6**, 617–618.

Mykytowycz, R. (1965). Further observations on the territorial function and histology of the submandibular (chin) glands in the rabbit, *Oryctolagus cuniculus. Anim. Behav.* **13**, 400–412.

Naka, K. I. and Kido, R. (1967). Hypothalamic spike potentials recorded by chronically implanted tungsten microelectrodes. *Brian Res.* **5**, 422–424.

Oakley, D. A. (1971). Instrumental learning in neodecorticate rabbits. *Nature, New Biol.* **233**, 185–187.

Olds, J., Mink, W. D. and Best, P. J. (1969). Single unit patterns during anticipatory behaviour. *Electroenceph. clin. Neurophysiol.* **26**, 144–158.

Oomura, Y., Ooyama, H., Naka, F. and Yamamoto, T. (1967a). Microelectrode positioners for chronic animals. *Physiol. Behav.* **2**, 89–91.

Oomura, Y., Ooyama, H. and Yoneda, K. (1967b). Miniaturized high input impedance preamplifier. *Physiol. Behav.* **2**, 93–95.

Ottoson, D. (1959). Studies on the slow potentials in the rabbit's olfactory bulb and nasal mucosa. *Acta Physiol. Scand.* **47**, 136–148.

Ray, A. J. and Yates, J. T. (1970). Spot welding for connecting chronically implanted electrodes. *Electroenceph. clin. Neurophysiol.* **28**, 320–321.

Reeves, A. G. and Hagamen, W. D. (1968). A re-usable nylon socket holder for use with chronically embedded electrodes in the cat. *Electroenceph. clin. Neurophysiol.* **25**, 185–186.

Robertson, A. D. J. (1965). Anaesthesia and receptive fields. *Nature, Lond.* **205**, 80.

Rokushima, H., Naka, K. I. and Kido, R. (1968). A transistor amplifier for chronic micro-electrode recording. *Electroenceph. clin. Neurophysiol.* **25**, 183–184.

Rose, G. H. (1966). Permanently implanted recording electrodes for immature animals. *Electroenceph. clin. Neurophysiol.* **21**, 395–396.

Rosner, B. S. (1961). Neural factors limiting cutaneous spatiotemporal discriminations. *In* "Sensory Communication" (Rosenblith, W. A., ed.), pp. 725–737. Massachusetts Institute Technology Press, Cambridge, Massachusetts.

Shepherd, G. M. (1972). Synaptic organization of the mammalian olfactory bulb. *Physiol. Rev.* **52**, 864–917.

Shimamura, M., Yamuchi, T. and Aoki, M. (1968). Effects of chloralose anaesthesia on spinal reflexes. *Jap. J. Physiol.* **18**, 788–797.

Sia, J., MacNeil, D. A. and Sigg, E. B. (1971). A miniature three-channel preamplifier for unit recording in freely moving animals. *Physiol. Behav.* **7**, 121–122.

Skinner, J. F. (1971). "Neuroscience: a Laboratory Manual", 244 pp. W. B. Saunders Co., Philadelphia.

Skutt, H. R., Beschle, R. G., Moulton, D. G. and Koella, W. P. (1967). New subminiature amplifier-transmitters for telemetering biopotentials. *Electro-enceph. clin. Neurophysiol.* **22**, 275–277.

Spinelli, D. N., Bridgeman, B. and Owens, S. (1970). A simple single-unit micro-electrode recording system. *Med. Biol. Engng.* **8**, 599–602.

Stevens, S. S. and Davis, H. (1938). "Hearing". Wiley, New York.

Sutton, D. and Miller, J. M. (1963). Implanted electrodes: cable coupler for elimination of movement artefact. *Science, N.Y.* **140**, 988–989.

Swadlow, H. A. (1970). Operant conditioning of the restrained rabbit. *Physiol. Behav.* **5**, 627–630.

Talbot, W. H., Darian-Smith, I., Kornhuber, H. H. and Mountcastle, V. B. (1968). The sense of flutter-vibration: comparison of the human capacity with the response patterns of mechanoreceptive afferents from the monkey hand. *J. Neurophysiol.* **31**, 301–334.

Welker, W. I. (1967). A method for preparing brain casts. *Anat. Rec.* **158**, 239–244.

13 Psychophysical and Electrophysiological Experiments with Binary Mixtures of Acetophenone and Eugenol

E. P. KÖSTER
Utrecht University,
Utrecht, The Netherlands

P. MACLEOD
College de France,
Paris, France

I. Introduction

In 1970 Nato sponsored a summer school on "Odour Perception: Multidisciplinary Research Methods" at the State University of Utrecht. During the course of the meeting a number of electrophysiological and psychophysical experiments were conducted on the olfactory intensity of mixtures of two odorous compounds, acetophenone and eugenol.

It was the objective of these experiments to demonstrate the experimental techniques to the attendants of the summer school, to give them the possibility to experience the task of the subject in different psychophysical experiments and, last but not least, to investigate the correlation between psychophysical and electrophysiological data obtained with the same odorous stimuli. The substances to be used during the summer school had been selected on the basis of psychophysical experiments.

The main results of the experiments performed during the summer school and of the experiments carried out before the summer school will be presented in this paper.

II. Method

A. Odorous substances and concentrations

Highly purified acetophenone and eugenol were used in all experiments.*

The concentrations of the two substances to be used in the mixing experiments were selected in two matching experiments, one at approximate threshold level in humans and one at a fixed supra-threshold level which was judged to be equivalent in odorous intensity to a 0.5 Mol/l concentration of *m*-xylene in benzyl benzoate. All solutions of the two substances were also made in benzyl benzoate.

The concentrations selected for the series of experiments are given in Table I.

TABLE I. Concentrations of Acetophenone and Eugenol used in the mixing experiments (in Mol/L).

Threshold level		Supra-threshold level	
Acetophenone	Eugenol	Acetophenone	Eugenol
1.56×10^{-2}	1.24×10^{-1}	7.03×10^{-2}	3.71×10^{-1}
7.81×10^{-3}	6.19×10^{-2}	2.34×10^{-2}	1.24×10^{-1}
3.91×10^{-3}	3.10×10^{-2}	7.81×10^{-3}	4.13×10^{-2}
1.95×10^{-3}	1.55×10^{-2}	2.60×10^{-3}	1.38×10^{-2}
9.76×10^{-4}	7.73×10^{-3}	8.68×10^{-4}	4.59×10^{-3}
4.88×10^{-4}	3.86×10^{-3}		

Each pair of concentrations of eugenol and acetophenone mentioned in the same row of Table I (under threshold level or under supra-threshold level) were then mixed by volume in the following seven different mixing ratios

Acetophenone	100	85	70	50	30	15	0	%
Eugenol	0	15	30	50	70	85	100	%

* The substances were kindly provided and purified by I.F.F., Hilversum, the Netherlands.

Thus, for the threshold level 42 different stimuli (7 mixtures at 6 concentration levels) and for the supra-threshold level 35 different stimuli (7 mixtures at 5 concentration levels) were obtained. In the psychophysical experiments 5 ml of each of these stimuli was presented to the subjects in 25 ml Erlenmeyer bottles stoppered with an ordinary cork. In the electrophysiological experiments to be described here only the mixtures of the highest concentrations mentioned under supra-threshold level were used. These mixtures were presented to the rabbit by means of a blast injection olfactometer especially designed for this task. In these experiments pure butanol was used as a standard stimulus for matching purposes.

B. Experimental subjects

Psychophysical experiments

In the experiments carried out before the summer school 11 students (6 female and 5 male) took part. The age of the subjects ranged from 20–26 years. All of these students had had previous experiences with psychophysical experiments in olfaction. They were paid about 1 dollar (Dfl 3.50) per hour. In the experiments carried out during the summer school, the participants of the summer school served as subjects. Their number varied in the different experiments between 95 and 56. Their age varied from 24–65 years. Nearly all subjects were men and worked on an unpaid basis.

Electrophysiological experiments

In the electrophysiological experiments 4 male rabbits ("fauves de Bourgogne") aged about 3 months (weight between 2 and 2.5 kg) were used. The animals had been chronically implanted with electrodes in the olfactory bulb one week before the experiments.

C. Experimental procedures

Psychophysical experiments

Three types of psychophysical experiments will be described here: detection threshold measurements, supra-threshold measurements with a paired comparison method and supra-threshold measurements with category scaling.

1. Detection threshold measurement was carried out with a forced

choice method. Each of the 42 odorous stimuli was presented together with 3 blank stimuli which contained 5 ml of the diluent (benzyl benzoate) only. The subject was asked to indicate which of the four bottles presented to him smelled differently. The order of the 42 sets of 4 bottles and the position of the odorous stimulus within the set of 4 bottles were randomized. Between the presentations of the stimulus sets a one-minute interval was given to allow for recovery from adaptation.

2. Supra-threshold measurement with the paired comparison method consisted of paired concentrations of each of the 35 odorous stimuli with a fixed standard stimulus containing 0.5 Mol/l of *m*-xylene. The subject was asked to indicate whether he thought that the stimulus presented to him was stronger or weaker than the standard stimulus which remained in front of him during the whole experiment. The order of the 35 stimuli was again randomized and a one-minute interval was given between presentations.

Half of the subjects were asked to sniff first the standard stimulus and then the unknown stimulus at each presentation and the other half of the subjects was asked to work in the reverse order.

3. Supra-threshold measurement with category scaling was carried out with two different category scales, a 9 point scale without presentation of reference stimuli and a 5 point scale which was preceded by the presentation of the highest and the lowest concentrations in the set in order to give the subject a clear idea of the range of stimuli he was about to judge. The 35 stimuli were again presented in randomized order and a 30 second interval was allowed for recovery between presentations.

In all experiments the subjects gave their answers by making a mark on a "mark sensing card". For each presentation a separate card was used and as soon as the mark had been made the card was turned over. During the preparatory experiments each of the 11 subjects received each stimulus presentation 10 times, while during the summer school the subjects (56–95) received each stimulus presentation only once.

Electrophysiological experiments

The electrophysiological experiments consisted of recording the slow potentials evoked in both olfactory bulbs during bilateral synchronous olfactory stimulation. Each input of a differential amplifier was connected by a monopolar lead to each olfactory bulb.

Thus, the output signal reflects the difference in amplitude between the two bulbar responses. If one side of the nose is stimulated by a constant reference stimulus, it is always possible to find an intensity of stimulation at the opposite side which results in nulling the output signal by exactly matching the two evoked bulbar potentials. Such a matching is made very critical by the incidence of interbulbar reciprocal inhibition: the stronger stimulated side maintains the other one in a low responding state except at the matching point, where an unstable equilibrium is reached.

It is noteworthy that any centrifugal influence of telencephalic origin, being essentially bilateral, is cancelled by differential recording. Therefore, the paired olfactory organ acts as a null detector in which accuracy is increased by reciprocal interbulbar inhibition.

Although accurate, the intensity matching might be distorted by a dissymetry of the nasal passages, a bias which is avoided by resorting to the double weighting principle: the reference and the unknown stimuli are presented successively to the same side and matched to the same auxiliary reference presented to the other side. The following surgical procedures were used. The nasal and frontal bones were exposed under nembutal anaesthesia. Two holes were drilled above the central portion of the olfactory bulbs. A silver wire, 100 μm in diameter, insulated except at the very tip, was stereotaxically inserted 2 mm deep into each bulb and immediately secured by a drop of dental cement. A miniature socket bearing three additional electrodes (a reference and two EEG silver balls) was connected to the bulbar leads and attached to the skull by dental cement. All the electrodes were short-circuited before use to eliminate spurious potentials.

The animals were allowed 48 h to recover from surgery and then became available for at least one month of daily experiments. To immobilize the animal during the experiment the following procedure was used. The conscious animal has its body enclosed in a wooden box, with a small aperture encircling its neck. An iron piece attached to the cement block is held in a convenient position by a strong magnet. The magnetic attraction is adjusted to allow the animal to free itself by a voluntary movement without damaging the electrodes.

Since a strictly synchronous bilateral stimulation is required, a special dual injection olfactometer was developed. A loudspeaker motor drive drives a squeeze bottle which contains a sample of odour-saturated air. The bottle has a small opening, about 0.6 mm in diameter, located in front of the nostril along with a tiny thermistor bead. The thermistor monitors the animals' respiration and triggers at the onset of inspiration an electronic timing device which delivers a square pulse of adjustable delay, duration and intensity, to the moving coil. Electromechanical transduction of the electrical pulse results in injecting a "square puff" of odorized air into the inspired air stream.

Each injector is calibrated by direct measurement of the amount of air saturated with butanol delivered to a flame ionization detector. Calibration curves are plotted as functions of the duration and the intensity of the driving current pulse. The volumes delivered are linearly related to the electrical parameters with a 5% precision in a thousandfold range from 0.5 μl to 0.5 ml. Diffusion of odour during the inter-stimulus intervals is prevented by a constant aspiration at a flow rate of 2 ml/minute.

The inspiratory stroke of a rabbit being approximately 5 ml, the dilution factor may thus vary from $10-10^4$. The final stimulus concentration can be calculated when the vapour pressure of the odorous material inside the injector and the dilution factor are known. Each nostril has its own injector. The time parameters of the driving pulses are identical on both sides; only the amplitudes are altered in order to find a close intensity matching.

For recording, after the animal's head has been secured and the two injectors have been placed correctly in front of the nose, the short circuit inserted in the electrode assembly is replaced by a plug which is connected by a light, shielded cable to a highly sensitive, differential input two channel chart recorder. One channel is used for displaying the bulbar response, the other one for EEG monitoring of the animals' wakefulness.

The reference side, usually on the left, is first stimulated at a given intensity. A conspicuous monophasic response is evoked, a few tenths of a mV in amplitude and approximately one second in duration. Keeping the reference intensity constant, the test side, usually on the right, is now stimulated at varying intensities with a minimum recovery interval of three minutes between trials. If the test side intensity is lower than the reference, only slight changes in the initial unilateral response are observed. If it is stronger, amplitude and time-course of the response are similar, but with a reversed electrical sign. One finally finds in between these two possibilities a narrow range of intensity which evokes a very small diphasic response resulting from an almost perfect opposition of the two bulbar potentials of equal amplitudes and durations.

This point is taken as an indication of perfect intensity matching and is reproducible from day to day with an overall precision better than 10% in concentration.

As already mentioned, two successive matchings are made against the same auxiliary butanol reference according to the following schedule: a given butanol reference (BL) is first presented to the left nostril and matched with butanol on the right nostril (BR); keeping the same butanol reference BL, the different odour is then presented to the right nostril (OR) and again matched for intensity with BL. The final result is stated as BR = OR rather than BL = OR. BL, BR and OR are rated in microlitres of odorized air. The whole matching procedure usually takes less than one hour.

One of the main advantages of this method is that it gives reproducible results, with lower intra-individual than inter-individual variability, making it possible to measure and study intensity response curves of individual subjects.

D. Data treatment

Psychophysical experiments

In all psychophysical experiments only group data were used. This means that no deductions can be made from the data about the form of the underlying psychophysical functions, because intra-individual and interindividual sources of variability have been mixed. Since the authors were interested in comparing the strengths of the stimuli rather than in the form of the psychophysical functions, this is no serious problem.

In the threshold experiments the percentage of positive response to each of the 42 stimuli were corrected for chance performance, converted into *z*-scores and plotted against the concentration. For each mixing ratio the best fitting straight line through the six points found for the different concentrations was calculated and the concentration at which this line reached a *z*-score of zero was considered to be the threshold.

In the paired comparison experiments the percentages "stronger than standard" were calculated for each of the 35 stimuli, converted in *z*-scores and plotted against the concentration. Again, the best fitting straight line through the points for the 5 concentrations was calculated for each mixing ratio and the concentration at which this line reached the *z*-score of zero was considered to be the point of subjective equality with the standard for a given mixing ratio. These concentrations were then plotted against the mixing ratios to give seven isointensity points (Fig. 1).

In the category scaling experiment the average scale value for each of the 35 stimuli was calculated as well as their standard errors (ranging from 0.12 to 0.43 scale points with an average at 0.27). These average scale values were plotted against concentration and the best fitting curves were drawn through the points thus obtained for each mixing ratio at the different concentration levels. The concentrations at which these curves reached certain scale values were determined and these concentrations were plotted against the mixing ratios (Figs 3 and 4) as isointensity points.

III. Results

Psychophysical experiments

Three types of psychophysical experiments were performed. Their results will be discussed first.

A. Threshold experiments

The detection thresholds for each of the seven mixing ratios were determined with two groups of subjects, a small group of 11 subjects who received each stimulus ten times and a large group of 95 subjects who received each stimulus only once. The detection thresholds found for these two groups are given in Table II. Since it is the aim to compare these concentrations rather than to stress their absolute values, all concentrations are expressed as multiples of the lowest concentrations used (see Table I). Thus, 4.0 would mean that the concentration needed to reach the threshold for a certain mixing ratio was four times as high as the one at the lowest level in Table I.

TABLE II. Detection thresholds expressed in relative concentrations for 7 different mixing ratios of acetophenone and eugenol.

	Mixing ratio						
Acetophenone	100	85	70	50	30	15	0
Eugenol	0	15	30	50	70	85	100
11 subjects	4.0	5.0	4.0	5.0	3.4	2.9	3.8
95 subjects	2.9	3.0	4.0	3.0	3.5	2.0	2.9

As can be seen from Table II, the relative concentrations vary between 2.9 and 5.0 in the small group and between 2.0 and 4.0 in the large group. These differences are rather small and it is doubtful whether any significance may be attached to them. Moreover, the mixing ratios at which the highest threshold concentrations are found do not coincide for the two groups. Only at 15% acetophenone–85% eugenol they both seem to have relatively low thresholds.

In the case of the small group the individual thresholds for each mixing ratio were determined for each subject. According to a Wilcoxon matched pair signed rank test none of the thresholds of the intermediate mixing ratios was statistically different from the thresholds for the unmixed substances.

B. Paired comparison experiments

The supra-threshold experiments with the paired comparison method, in which the mixtures were matched against a standard of 0.5 Mol/l of *m*-xylene in benzyl benzoate, were also carried out with a small group of 11 subjects which received each set of stimuli ten times and a large group of 70 subjects which received each set of stimuli only once.

The results of these two experiments are given in Fig. 1.

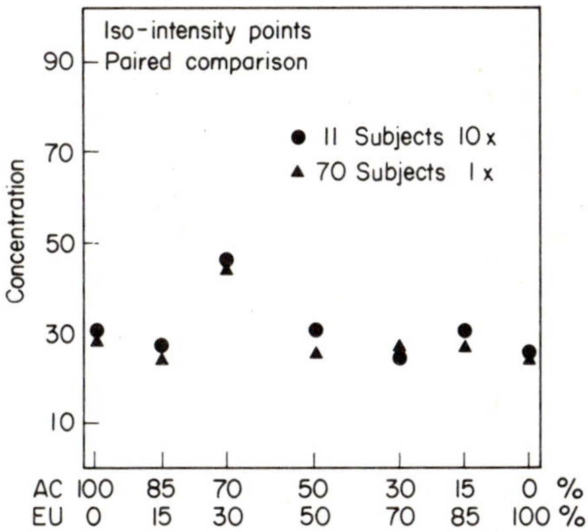

FIG. 1.

As can be seen from Figure 2 the data obtained from the two groups at supra-threshold level correspond very well. For both groups a much higher concentration is needed with the 70% acetophenone–30% eugenol mixture to make the matching than for either the unmixed substances or the other mixtures. According to the Wilcoxon matched pair signed rank test carried out on the individual data of the small group, the concentration differed significantly from both the unmixed acetophenone ($p < 0.02$; T = 4; N = 10 (one tie)) and the unmixed eugenol ($p < 0.01$; T = 5; N = 11 (no ties)). All differences with the other mixing ratios were at least significant at the 5% level.

None of the concentrations found for the other mixing ratios differ significantly from those of the unmixed substances or among themselves.

C. Category scaling experiments

Two types of category scales were used in these experiments. In the first one a 9-point rating scale was used without presentation of reference stimuli. The results of this experiment are given in Fig. 2. Seventy subjects took part in it. The isointensity concentrations at four different scale values are given.

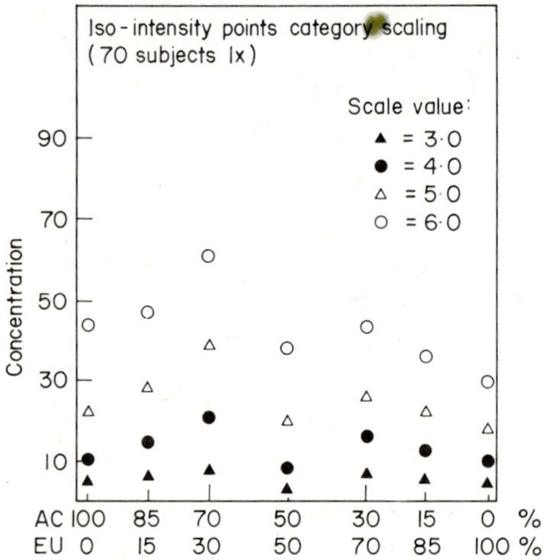

FIG. 2.

As can be seen from this figure the concentrations needed to reach a particular scale value were again highest for the 70% acetophenone–30% eugenol mixture at each scale level.

Although the significance of the difference between the concentration at this mixing ratio and the concentrations found for the unmixed substances can not easily be evaluated directly, it can be said that the averages of the scale values given by the subjects to the different stimuli (the five concentrations at which each mixing ratio was presented) were all lower than those found for the unmixed components at the same concentration levels. These differences were statistically significant at the 1% level for the three highest concentrations and at the 5% level for the one but lowest level.

Such significant differences were not found between the other mixing ratios and the unmixed substances at any concentration level.

The second experiment with the 5-point scale was carried out two days later with 56 subjects. Before the experiment, the subjects could smell the highest and the lowest concentrations that were to be presented to them. Thus they had a better idea of the range of concentrations to be judged. The results of this experiment are given in Fig. 3. The isointensity concentrations at four different scale values are given.

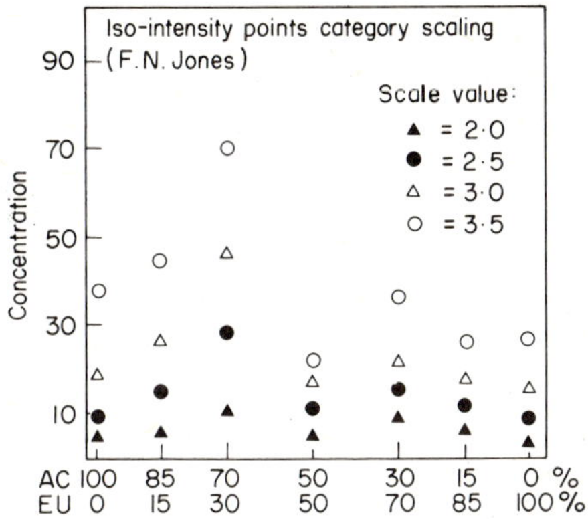

FIG. 3.

As can be seen from this figure, the results of this experiment were in agreement with those of the previous one and with those of the paired comparison experiment.

Here, the differences seem even more pronounced. Again the concentrations needed to reach a particular scale value are highest for the 70% acetophenone–30% eugenol mixture at each scale level.

Again the scale values found for this mixture at the five concentration steps were lower than the ones found for the unmixed substances at the same concentration level and again these differences were statistically significant (Wilcoxon) at the 1% level for the three highest concentration levels and at the 5% level for the one but lowest level. The average scale values of the other mixing ratios at the different concentration levels did not differ from those of the unmixed substances at the corresponding concentration levels.

Electrophysiological experiments

The raw data of the electrophysiological experiment are shown in Fig. 4.

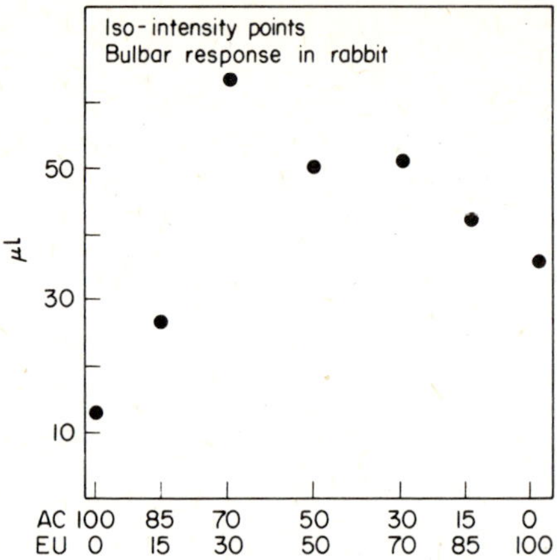

FIG. 4.

In this graph, on the abscissa the complementary percentages of acetophenone and eugenol in the binary odour mixtures are given. The ordinates give the volumes of the mixtures (OR) delivered at the right nostril, which are needed to match exactly an intensity of 90 μl of air saturated with *n*-butanol and delivered at the left nostril (BL). In this example the intensity of butanol needed at the right nostril (BR) to match the BL was 78 μl of air saturated with *n*-butanol.

It is therefore concluded that the points in this graph represent the mixture volumes which are exactly equivalent in odour intensity to 78 μl of *n*-butanol.

In order to make a direct comparison of these isointensity points with their human counterparts possible, the ordinates had to be standardized at both extremities to an arbitrary value of 1. Considering that 78 μm of butanol match 12 μm of acetophenone or 35 μm of eugenol, as well as a mixture of V_1 μm of acetophenone + V_2 μm eugenol, the ordinates are transformed into:

$$\frac{V_1}{12} + \frac{V_2}{35}.$$

The results of this transformation are given in Fig. 5, together with some data from the human experiments.

IV. Discussion

The most striking result of these experiments seems to be the correlation found between the odorous intensities perceived for the different mixing ratios at supra-threshold level in the psychophysical experiments and the intensities needed at these same mixing ratios to obtain a matching in the electrophysiological responses of the rabbit. More specifically, there seems to occur in both sets of data a reciprocal masking effect at one specific mixing ratio (70% acetophenone and 30% eugenol). Since it was demonstrated by Haring (1970) that those masking effects could not be attributed to any physicochemical intermolecular interaction, it seems that one can safely conclude that they are of biological origin. Similar reciprocal masking effects have been found earlier with mixtures of cyclohexanol and benzaldehyde in experiments in which psychophysical data (Köster, 1968) and electrophysiological data from the E.O.G. of the rabbit (MacLeod, 1968) were obtained.

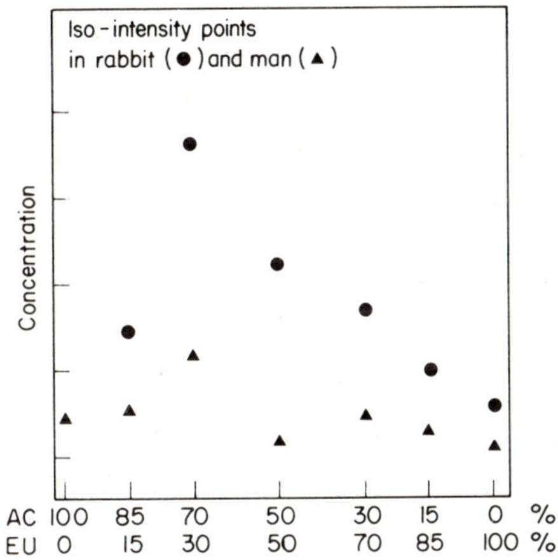

FIG. 5.

Had these previous experiments already shown that the masking effects were found at the most peripheral level of the olfactory system, because they were reflected in the EOG which is not subjected to any centrifugal influence, the results of the present experiment indicate that the secondary neurons from which the slow bulbar potential arises, follow the response patterns of the primary neurons in the same general way. Furthermore, if one allows for the speculative step from the animal to the human data and vice versa, it is clear from the results of the psychophysical experiments that the perceived intensity corresponds in a direct way with the bulbar response and hence with the EOG. It seems therefore that, whereas the quality coding in the olfactory system becomes increasingly specified as one proceeds from the peripheral to the bulbar level, but seems to remain almost unaltered between the bulbar and the perceptual level (Döving and Lange, 1967), the intensity coding remains essentially unaltered all the way.

The fact that, according to the psychophysical experiments, the masking effects are not found at the threshold level does suggest that they may be the result of competitive interaction of the molecules in occupying the receptor sites.

When few molecules are present, as in the case of threshold measurement, the competition is minimal since sufficient sites are available for all of them. If, however, the concentrations are higher the different molecules may block each others access to the receptor sites.

References

Döving, K. B. and Lange, A. L. (1967). Comparative studies of sensory relatedness of odours. *Scand. J. Psychol.* 8, 47–51.

Haring, H. G. (1970). Vapour pressure measurement. Paper given at the Nuffic Summer School: "Odour Perception: multidisciplinary research methods" held August–September 1970 at Utrecht.

Köster, E. P. (1968). Relative intensity of odour mixtures at supra-threshold level. *Olfactologia* 1, 29–41 (*suppl., Cahier d'Oto- Rhino- Laryngologie* 3, No. 5).

MacLeod, P. (1968). Interactions dans un mélange d'odeurs–Étude électrophysiologique. *Olfactologia* 1, 25–27 (*suppl., Cahier d'Oto- Rhino- Laryngologie,* 3, No. 5).

14 Pharmacological Aspects of Olfaction

M. G. J. BEETS
I.F.F. (Europe)
Hilversum, Holland

I. Introduction

The history of science shows a development through two distinct phases. In the first of these, which may be termed the philosophical phase, man attempted to understand his environment by logical thinking. Such thinking was in many cases based on traditional, religious and mystic interpretations rather than on objective observation of phenomena.

After a period of transition, the empirical phase takes over and objective observation and experimentation are gradually starting to assist and to guide logical thought.

We see in most disciplines a dramatic acceleration of progress after the beginning of the empirical phase.

The introduction of empiricism in science had three important effects; it provided a check on man's imperfect and often biased use of logic, it accelerated his supply of reliable information and it was the source of technology which enabled him not only to understand his environment but also to control it. The result of these three effects is the rapid expansion of our knowledge of nature and our control of matter and energy in little more than a century.

In the philosophical stages of science, the individual scientist was invariably a universalist who could discuss with authority almost any subject of human knowledge. The type survived during the early stages of empiricism but the avalanche of new scientific data, the increasing number of problems to be studied and finally, the appearance of industrial research gradually forced him to retreat in favour of a new type, the specialist.

We must admit that the principle of specialization in science has been responsible for many of the impressive achievements in the first half of our century. In recent decades however, increasing specialization has progressively narrowed the field of interest of the majority of scientists to an extent where serious interdisciplinary communication blocks have developed in a period in which the border lines between disciplines tend to become vaguer. Parts of different disciplines are merging to form new problem areas requiring novel approaches. It is becoming more and more apparent that future scientific progress will depend to an important degree on our ability to adapt our thinking to this new situation. New techniques of interdisciplinary communication and cooperation as well as some courageous reconsideration of the specialist concept will be needed.

The ideal scientist in the interdisciplinary phase of science seems to be a hybrid between yesterday's generalist and to-day's specialist. The latter must prepare himself to widen the scope of his interest and knowledge, to learn the language of his colleagues in other fields of research and to improve the accessibility of his own discipline by elimination of non-functional terminology. A major part of the present interdisciplinary communication problems is not due to the fact that the various disciplines are intellectually impenetrable to the uninitiated but to artificial barriers of professional jargon in research papers and discussions.

The study of olfaction is an ideal example of a multidisciplinary area of which the future development depends strongly on the removal of interdisciplinary barriers. The purpose of this book is to contribute towards this end by means of instructive chapters on the experimental approaches available for the study of olfaction.

In the process of olfaction, two partners are involved, a population of stimulant molecules and the olfactory system in the organism. Most of the preceding chapters are concerned with the function and structure of the latter. It seems justified to supplement these discussions by considering the role of the stimulant molecules and to suggest some possible experimental approaches on the pharmacological and on the closely related informational aspects of olfaction.

II. The pharmacology of chemoreception

Olfaction belongs to a group of closely related processes, connoted by the collective noun chemoreception.

The complexity of chemoreceptory systems varies widely between organisms but they may be described in general terms as a sequential arrangement of components in which chemoreceptory information generated in a peripheral receptor system is projected and processed and finally translated into whatever effect the function of chemoreception in the organism involved may require. In vertebrates, the process ends with a translation of information in the higher centres into behavioural effects, or, in man, into comparison with memory contents and verbal analysis of the odour or taste sensation. In monocellular organisms the information is translated into simple effects such as movement towards higher or lower stimulant concentration (Adler, 1972).

If we consider the chemoreceptory system in a given organism under a given set of conditions as an essentially invariable partner, the information generated in the interaction must be correlated directly with the structural parameters of the stimulant molecules.

This simple outline suggests that chemoreception is logically a subclass of the general category of processes initiated by interaction of alien molecules with components of living organisms. Consequently, we are justified in speaking of a pharmacology of chemoreception, from the study of which we can expect to learn about the relationship between stimulant structure and the number and nature of types of information generated in chemoreceptory interactions and also about the mechanisms involved in chemoreceptory processes.

In short, pharmacology ought to be considered as one of the important disciplines involved in the study of chemoreception.

That it has not yet attained this position has several reasons.

In the first place, pharmacological aspects of chemoreception have become associated in the minds of many with wild speculations on the mechanism of chemoreception because the theories proposed in the literature quite comprehensibly concentrate on the many things stimulant molecules can do with the receptor sites. There is, of course, no basis for such an association; pharmacology is an objective experimental discipline searching for relationships between phenomena.

Secondly, as we have indicated above, interdisciplinary communication in chemoreception research is still far from ideal. Many research projects could have yielded more or more valuable information if the author would have asked the advice of scientists representing other disciplines or if he had broadened his scientific field of view a little by familiarizing himself with the fundamentals of such disciplines.

Finally, a physiologist or a psychologist planning an experimental project may be quite willing to sit down with a pharmacologist to design a number of stimulant structures adapted to the requirements of his project but he will then have to find reliable and pure samples of the selected chemicals. Since the facilities for the production of these pure samples are frequently unavailable he will often be obliged to depend upon the less reliable alternatives which have been used in the past.

The initiative proposed during the 1966 Gordon Research Conference to make available a standard series of odorants of reliable purity to all scientists involved in chemoreception research could be an important step ahead since it would eliminate the coffee extract and tobacco smoke from the scene of chemoreception.

However, it will not suffice to provide a basis for the proper introduction of pharmacology in the orchestra of disciplines playing their parts in chemoreception research.

We shall firstly require the careful planning that is necessary if we are to meet the requirements of specific scientific projects, and the ready availability of pure samples. For most projects however, even this is not enough. Frequently, experimental results indicate certain trends which can only be explored if structural variants of the stimulants originally planned can be made available. This can only be realized by groups which are cooperating closely and on equal level with pharmacology-oriented synthetic chemists. Such ideal combinations are still extremely rare. However, several excellent examples which demonstrate practical feasibility are found in the literature of chemoreception.

The best thing to do for any group, planning a research project in chemoreception with possible pharmacological implications is to interest a nearby chemical group, to ask their advice and to recruit their active cooperation in the execution of the project.

Extensive discussions of the pharmacological approach to chemoreception in general (Beets, 1973) and to olfaction (Beets, 1970a, 1971) have been published recently. Rather than attempting to present a detailed review of the literature on this subject, I shall discuss in this short chapter a number of examples of experimental openings as well as the nature of the information we may expect.

III. Basic principles and objectives

In order to provide a practical foundation for research on the pharmacology of chemoreception, we need a framework of basic principles and a formulation of such principles which is sufficiently broad and devoid of preconceived details to be essentially non-speculative and to be generally acceptable.

A short discussion of such principles may facilitate the definition of objectives for pharmacological research in chemoreception. The fundamental principle of the pharmacological approach to chemoreception is that the molecular structure of a stimulant completely defines the nature of chemoreceptory response in a given organism and under a given set of conditions (Beets, 1957). In this formulation, the validity of the principle is obvious, since the molecular structure of a compound is necessarily the only source of all its characteristics, physical, chemical and physiological, as long as the partner (i.e. the instrument, the organism or the preparation) with which the compound interacts in the measurement of the characteristic remains invariable. Fortunately, however, the variance of chemoreceptory response between organisms or preparations from organisms belonging to the same species is limited over a reasonably wide range of conditions. This provides the practical basis for an experimental approach to structure-response relationships. Human subjects, for instance, can communicate about their observations of the same odour in a way which demonstrates that, although we have no way of knowing whether their experiences are the same, at least the structure and organization of the chemoreceptory information generated and processed in their olfactory systems and higher centres, are closely related. In cases where significant response differences between subjects or, for a single subject, under different conditions are observed, e.g. between normal and anosmic observers, before or after adaptation or at different stages of the oestrous cycle, such differences can be studied as an additional source of information on chemoreceptory processes. Examples will be mentioned in subsequent sections of this chapter.

The validity of inter-subject reproducibility of chemoreceptory response is tacitly assumed in several types of research, e.g. in the recording of single unit response patterns obtained by stimulation of preparations from different organisms of the same species, with a series of stimulants (Lacher, 1964; Schneider *et al.*, 1964; Døving, 1966b, 1968).

A. *Chemoreceptory interaction*

After the gradual rejection of all theories involving metabolization of stimulant molecules or teleactivation of the olfactory receptors by odorant molecules, the concept that olfactory as well as gustatory interaction take place in the framework of a reversible physical contact between a population of stimulant molecules and a population of receptor sites of necessarily molecular dimensions, has been generally accepted (Beets, 1971). This concept does not involve any speculation on the structure or the function of the receptor sites or of the mechanism of the interaction.

B. *Chemoreceptory information*

The product of the interaction is a pattern of primary chemoreceptory information consisting of a number of more or less discrete types of information.

Since the concept of chemoreceptory information plays an important part in this chapter, some explanation is required.

To say that the primary product of any of the senses is a pattern of information, having a specific organization and consisting of a number of types of sensory information each of which is represented in the pattern in a specific frequency or intensity, involves no speculation as long as we do not propose hypotheses on the nature of these aspects which are not supported by experimental fact.

A type of sensory information is never perfectly discrete in the sense that it has no extension in the space in which it is generated. A mono-chromatic colour such as yellow has no objective existence in nature. It is observed because the brain interprets the information generated by the interaction of a narrow range of wavelengths with the retina as yellow. The letter "A" may be pronounced by various persons in slightly different ways but it is still interpreted as "A". In the same way a type of chemoreceptory information such as "musk" or "sweet" has undoubtedly some extension in the space in which it originates. We have some reasons to interpret this by saying that chemoreceptory information patterns have topological characteristics and that a specific type of chemoreceptory information is generated in one or more specific areas of the receptor epithelium (Beets, 1970b, 1971).

C. Retention of information types

Processing of the primary information pattern in the chemoreceptory system may involve several operations such as transcoding, summation of identical bits and elimination of information below critical levels of intensity. Since the chemoreceptory system exists in order to make the primary information available to the organism and not to destroy or to blur it, we may assume that in no stage of projection identical bits will become different nor will different bits become identical. In short, the informational contents of the primary pattern are essentially retained in all stages of projection and processing in the chemoreceptory system. This means that we may learn about the informational structure of the pattern generated in the periphery by tapping information from any stage in the chemoreceptory system by means of electrophysiological techniques, e.g. from receptor units, from nerve bundles and from units in the olfactory bulb but also beyond this, by studying the translation products of chemoreceptory information such as behavioural effects and verbal analysis of odour and taste sensations as well as phenomena like anosmia, taste blindness and adaptation.

D. Receptor and receptor site

The literature of chemoreception shows much confusion regarding the concepts "receptor site" and "receptor". No clear distinction is made between the two and the receptor site is considered by many as a speculative concept.

However, the speculative nature of the receptor site is not in the concept itself but in experimentally unsupported ideas about the mechanism of its function in the chemoreceptory process. If we discard all mechanistic hypotheses until any of them will have found an experimental basis, we find that the concept of the receptor site is logically sufficiently clean to present a useful basis for thinking and experimentation.

If we accept that the chemoreceptory process starts with a reversible physical contact between a population of stimulant molecules and the epithelium, during which a pattern of chemoreceptory information is generated, we must also accept the existence of locations of necessarily molecular dimensions where stimulant molecules land, are reversibly absorbed, interact in some way and where elementary bits of information are generated. This defines the receptor site without involving any hypothesis about the mechanism of the interaction or about the nature of the information produced.

The word "receptor" is frequently used in the literature as an undefined conceptual hybrid of the receptor site and the membrane of the receptor cell. In this chapter we shall avoid the word "receptor".

What is the relationship between the receptor site and the receptor cell membrane?

The most neutral and consequently satisfactory concept of the receptor site seems to be that it is a normal structural feature of a proteinaceous membrane.

Although it must have a structure and configuration enabling it to accommodate specific types of stimulant molecules and not, or less effectively, other types, it may occur on any cell membrane. However, it is a receptor site because it happens to be located on the peripheral membrane of a receptor cell (Beets, 1973).

Even within the field of chemoreception, functional differentiation of sites may be entirely due to location rather than to structure. This means for instance, that the peripheral membrane of the olfactory receptor cell may very well contain structural features which can accommodate musk molecules and others which can interact with glucose molecules. However, only the former can have the function of receptor sites since glucose molecules, being non-volatile and water-soluble are never presented to the olfactory epithelium.

Although this reasoning contains speculative elements, it is not without logic and may present a basis for experimentation. It seems to be supported by the fact that the vapour of many volatile odorants generates some usually complex taste sensation on the tongue.

Since receptor sites must have molecular dimensions, the peripheral membrane of a single receptor cell must be assumed to contain many receptor sites of different types. There are reasons to assume that within a specific field of chemoreception such as olfaction, receptor cells differ because different types of receptor sites are represented on their membranes in different frequency ratios (Gesteland *et al.*, 1963).

The response specificity of each cell, derived from a varying distribution of a limited number of site types can only be fairly low. However, a topological arrangement of many millions of cells of low specificity can easily produce the complex information pattern required for the impressive discriminatory power of the olfactory system (MacLeod, 1963).

Topological characteristics of the response pattern of the chemoreceptory epithelium as a major source of chemoreceptory discrimination seem to have some experimental support and present promising openings for further experimentation. The subject has been discussed recently by the present author (Beets, 1970a, 1971, 1973). The impression of the organization of chemoreceptory information patterns as the source of discrimination we may derive from the heterogeneous collection of data in the literature, seems to suggest some analogy with geographical maps or paintings (Adrian, 1953; Beets, 1973).

E. Complexity of information patterns

Absorption processes are statistical phenomena. A population of identical molecules presented to a population of sites must be assumed to arrive in a variety of orientations and conformations with different steric requirements. The composition of these orientation, and conformation patterns will depend on many factors such as the polarity of the molecules, the steric accessibility of the polar group and the rigidity or flexibility of the molecules. Patterns can be expected to be more homogeneous for polar and rigid molecules than for non-polar and flexible molecules.

Since any population of identical stimulant molecules will present a variety of structural features to the receptor sites, the interaction pattern as well as the resulting information pattern can not be expected to be homogeneous and the latter will consist of a number of information types occurring in frequency ratios depending on molecular structure.

The conclusions are obvious. All chemoreceptory information patterns and consequently their derivatives resulting from projection and processing in the chemoreceptory system and in the higher centres, including the final odour and taste sensations, can be expected to consist of a variety of informational types.

In some cases, primarily for strongly polar and rigid molecules, a single type of information will strongly predominate while others occur in much lower frequencies. In other cases all informational types represented in the pattern will occur in comparable frequencies. A major consequence, which is supported by experimental evidence is that two different stimulants can never generate identical information patterns. Even stimulants generating patterns which share the same, strongly predominating component will have similar but not identical odours or tastes since the secondary components in the information patterns differ.

This concise explanation of some basic principles suffices for the purpose of this chapter. Detailed discussions on the theoretical background and on relevant experimental data, including literature references, have been published recently by the author (Beets, 1971, 1973).

The close relationship between chemoreceptory processes is stressed by the fact that the principles mentioned in this section are probably valid for all types of chemoreception. In view of the scope of this book we shall, from this stage on, concentrate our attention on olfaction.

What are the objectives of pharmacology-oriented research in olfaction and what can we expect such research to contribute to our knowledge of the olfactory system and its function?

Isolated from the more hypothetical suggestions, offered in the discussions of this section, the basic elements can be summarized in the following simple statements.

The olfactory process starts with a reversible physical contact between a population of stimulant molecules with a peripheral arrangement of receptor sites. This produces an information pattern consisting of a discrete number of informational types of which the frequency and the organization are entirely defined by the molecular structure of the odorant. The informational content of the primary pattern is essentially retained throughout all projecting and processing steps in the olfactory system and in the higher centres.

The objectives we are trying to define are fairly well indicated by these statements. They can be formulated in three major questions.

Which structural criteria of odorant molecules control olfactory response?

How many and which types of information are involved in olfactory response?

How are these types organized in the information pattern generated in olfactory interaction and what is the fate of this organization in the processing steps taking place in the olfactory system and in the higher centres?

A systematic study of the problems comprised in these questions can be expected to yield information on the olfactory process and the olfactory system not only supplementing the results of other disciplines but also suggesting new lines of approach. Although pharmacology has hardly appeared on the scene of olfaction research, it has already started to influence the direction of thought and experiment in other disciplines.

The most important reason however why the pharmacological approach should be fully included in olfaction research, is that it is a powerful catalyst for interdisciplinary cooperation. While other disciplines can continue working on their own, producing interesting information and being happily unconscious of the unnecessary limitations resulting from this attitude, the pharmacological approach can only be successful in multi-disciplinary team work.

After this, we may turn our attention, in subsequent sections, to the question; which experimental openings are available or can be conceived for the pharmacological approach to olfaction.

IV. Structural criteria: oriented profile

The major structural characteristics of a simple organic molecule with a single polar group and a rigid, saturated carbon-hydrogen skeleton are the molecular shape and the position, accessibility and nature of the functional group.

If a population of such molecules is presented to an absorbing interface, we may expect the absorption of most molecules to occur in the same orientation which is mainly determined by the parameters of the functional group. In such a case we can speak about a homogeneous orientation pattern. In cases where specific locations at the absorbing interface are equipped to interact with absorbed molecules according to some mechanism, identical suitable locations will interact with identically oriented molecules in identical ways, generating identical effects.

However, since absorption is a statistical phenomenon, orientational homogeneity can never be perfect. There will always be some molecules at the interface in different locations presenting different structural features to the locations where they land. Such molecules may or may not be able to interact more or less effectively with their landing sites. Also, molecules in the predominating orientation may find other types of locations with which they can interact, although perhaps less effectively.

The orientation pattern will become less homogeneous as the polarity of the molecules decreases and it can be expected to be almost random for non-polar molecules. In such cases the interaction pattern will produce a pattern of effects consisting of many different contributions. The structure is further influenced by structural flexibility of the absorbed molecules because in such cases conformational diversity and the possibility of each molecule adopting itself conformationally to the requirements of the locations where it happens to land, are superimposed over the factors mentioned above.

This is, very roughly, the picture we may expect when a population of structurally identical odorant molecules is presented to the olfactory epithelium.

Since, as we have discussed previously, molecular structure is necessarily correlated with olfactory response, we may ask which are the structural characteristics influencing or even determining the quality and intensity of the primary effect and consequently of the terminal product of the process, of the odour?

Systematic research on the pharmacological aspects of olfaction has hardly started but an abundance of data is available in the literature. Recent critical reviews of such data (Beets, 1971, 1973) strongly indicate that the major criterion determining the quality of olfactory response and of odour is the shape of the oriented molecule as it is presumably presented to the epithelium.

The importance of molecular shape for odour quality had been postulated in 1946 by Pauling and in 1952 by Amoore.

This principle seems to be useful as a starting point and as a working hypothesis for continued research.

Rather than repeating a discussion of data which have already been reviewed and analysed extensively, we shall consider in this and subsequent sections which types of experimental approach are available to future research.

We have seen that some classes of molecules, primarily those with rigid structures and accessible polar groups can be assumed to be absorbed at the olfactory epithelium in fairly homogeneous orientation, and conformation patterns. In the resulting olfactory information patterns a single component will strongly predominate.

By classification of odorants according to odour type we find groups with strongly similar odours such as musks, camphors, ambers, woody stimulants and many others.

If we consider one of these classes, we may conceive two different reasons for the odour similarity of its members. In the first case, odorants have similar odours because in their olfactory information patterns the same component predominates strongly. This has been termed fundamental odour similarity (Beets, 1970a). Classes of compounds with fundamental odour similarity can be indicated by the term "primary odorants" proposed by Amoore (1952).

Amoore's (1952, 1962) early attempt to assess the number of primary odours by semantic analysis of literature data, resulting in 7 primary classes, is well known. Later, on the basis of a preliminary study of anosmia, the same author (Amoore, 1969), postulated the existence of between 20 and 30 primary classes. This, translated in terms of the informational approach, may mean that olfaction is based on 20–30 information types.

In the second case the odour similarity is based on coincidence. Both information patterns may be composites of many components none of which predominates but their translation in the higher centres happens to result in similar odour impressions. On the other hand, in one of the patterns a single component may predominate strongly whereas the other pattern is complex. This again may result in similar odour impressions. This phenomenon, which has been termed pseudo similarity, is comparable with the similarity of two colours, one of which is monochromatic, the other being a spectral composite.

Before a meaningful study of structure-response relationship can be started we must know whether the odour similarity in the class to be studied is a case of fundamental odour similarity or of pseudo similarity. Only the former presents a basis for research.

The most practical technique which is available at this time is the study of specific anosmia. Other methods are cross-adaptation and correlation studies between different experimental methods of similarity assessment such as patterns of single unit recordings, odour similarity estimates and confusion estimates. These methods will be mentioned in subsequent sections.

Any well-established primary class of odorants can be used as an object for structure-response studies.

The questions to be answered are:

Which are the molecular profile criteria characteristic for the primary class we are studying?
Which is the scope of the class or, in other words, beyond which limits of the profile criteria do odour quality and intensity start to deteriorate in the sense of a dramatic decrease of odour intensity or a transition of the odour into a different type?

The first of these questions can only be studied if we have a reliable method to assess similarities of molecular profiles.

The simplest but also the crudest way to do this is to compare molecular models of pairs of odorants and to estimate similarities of profiles, oriented roughly with parallel dipole axes. Such estimates may be expressed in numerals between 1 and 10. A different way is to establish a sequence of profile similarities between a range of pairs. For an imaginary group of odorants A-H, the results will look as follows.

A/B = 8   B/D = 7   E/H = 4   etc.
A/H > B/D > F/C > G/J   etc.

Such numerical values or sequences can be compared with similar data on odour, or olfactory response similarities. Although this method is only a rough approximation it may be useful to obtain preliminary impressions of trends which can be checked by means of more accurate methods.

A second method, used by Theimer and Davies (1967), in the musk series is the selection of two profile parameters, e.g. the length and the breadth of the oriented molecule and the characterization of the molecular profile by means of the ratio between the two values.

Theimer and Davies (1967) demonstrated that this L/B ratio for a wide collection of musks lies in the range 2.8–3.3. They also showed that for the same collection the cross-sectional area of the oriented molecules is in the range 40–57 Å.

The most sophisticated method is the assessment of profile similarities by means of a computer (Amoore *et al.*, 1967).

The next target for research in this field is the assessment of profile criteria for the various primary classes. At this moment, the existence of three of these classes, the musks, the sweaty odorants and the urine odorants has been demonstrated. Only the profile criteria of the musks have been studied to some extent.

Continued research can be expected to define other primary classes. In each primary class the scope of the primary type in terms of profile parameters must be assessed by means of synthesis and evaluation of structural variants.

V. Structural criteria: functional group

In the previous section the importance of the oriented molecular profile for olfactory response has been discussed. Since many hydrocarbons are known to have odours and since, in polar molecules, the direction of the polar axis and consequently of the predominating orientation at an interface, is mainly determined by the functional group or groups, it seems that the position and steric environment of the latter are of more importance for odour quality than their nature.

Many data, analysed recently by the author (Beets, 1971, 1973) indicate that the presence of a functional group is an essential characteristic of primary classes of odorants. In its absence the orientation pattern can be expected to degenerate toward randomness and olfactory interaction will produce an information pattern consisting of many components on comparable levels of intensity. Similar effects are found in odorants with sterically hindered functional groups.

A number of characteristic examples is available in which the functional group in a primary odorant could be replaced by a different group without changing the odour quality. Such a replacement however, invariably influences the odour intensity.

Literature data also indicate that the presence of two functional groups in the same molecule may result, depending on their relative positions, in cooperation or in competition.

In the former case, the two groups are in neighbouring positions and act as a single polar group. This results in homogeneous orientation, and information patterns. Examples are macrocyclic lactones and macrocyclic esters of glycols and dicarboxylic acids. Beyond a certain distance, cooperation ceases and two different oriented profiles will dominate the orientation pattern to different extents, each controlled by one of the polar groups.

For an extensive discussion of these aspects, the reader is referred to recent reviews (Beets, 1971, 1973).

VI. Structural criteria: small molecules

Recently, Klopping (1971) has drawn the attention to the odour of small molecules such as hydrogen sulphide, methyl mercaptan and methyl isocyanide. Since molecular morphology can only play a minor role in the interaction of such molecules with the olfactory membrane, the efficiency of hydrogen bond formation between the functional group of the molecule and one or several functional groups in the membrane must be assumed to be the controlling factor in this case. This is of particular interest since it means that the same membrane can accommodate and interact with two entirely different populations of stimulant molecules according to different mechanisms.

This seems to agree well with the generalized concept of the receptor site indicated in one of the previous sections since it suggests that the receptor site is a normal feature of a proteinaceous membrane of which the functionality is derived from its location on the peripheral membrane of a receptor cell and from its structural relationship with a specific range of molecules.

Obviously the olfactory membrane can provide two different sets of receptor sites for two different kinds of molecules. "Normal" odorant molecules are accommodated by a set of profile-specific sites with low functional group specificity whereas small molecules find a set of hydrogen bonding sites with low profile-specificity (Beets, 1973).

The fundamental aspects of olfactory response by stimulation with small molecules have not yet been studied. The theoretical picture mentioned above, indicates that olfactory information patterns for small molecules can be expected to consist of many components on comparable levels of intensity. Although the types of sites interacting with very small molecules must be different from the set of receptor sites available to "normal" odorant molecules, their contribution to the olfactory information pattern may be assumed to consist of normal types of olfactory information. The latter are defined by the receptor cell on which the interaction takes place and not by the receptor site.

This would mean that no primary odorants will be found in this group and that, for instance, specific anosmia will not occur (see Section IX).

VII. Structural engineering: enantiomers

It has already been mentioned in one of the previous sections that the scope of primary types can be assessed by systematic structural variation of the odorant molecules.

However, an even more promising opportunity to obtain information regarding the influence of molecular morphology on olfactory response may be the application of structural engineering to pairs of enantiomers.

Enantiomers are identical in all scalar structural parameters and only differ in the shape of their oriented molecular profiles.

The question whether enantiomers have different odours has been studied by many authors. Most contributions however, involve pairs of enantiomers obtained from different sources or by different reaction paths. Very often purification methods are insufficiently defined. In short, in most publications claiming the discovery of odour differences between enantiomers, the possibility is not excluded that traces of impurities with low olfactory thresholds are responsible for the odour differences observed.

However, in a series of recent papers, Theimer and McDaniel (1970), Leitereg *et al.* (1971), Russell and Hills (1971) and Friedman and Miller (1971), have demonstrated in various ways the existence of significant differences between enantiomers.

Recently, Theimer and Klaiber (1972) found the *endo-* and *exo-*isomers of 4-camphyl-2-methyl-cyclohexanone to have strong urine odours, the odour of the *exo*-stereomer being 4–5 times stronger than that of the *endo*-stereomer. Resolution of the weaker *endo*-stereomer into its enantiomers showed the urine odour of the *d*-form to be 5–10 times stronger than that of the *l*-form.

In gustatory interaction two polar groups are functionally involved. This restricts the rotational freedom of the stimulant molecule to an extent depending on its structure. The odorant molecule retains its freedom to adapt itself rotationally to the structural characteristics of the receptor site, because its polar group is probably not involved in the interaction mechanism beyond its influence on molecular orientation.

Also, a huge number of odorants has to be accommodated by a limited number of site types and the assumption of a graded efficiency and a low stereospecificity of olfactory interaction seems inevitable.

This may explain the difficulties to demonstrate any difference in olfactory response for pairs of enantiomers as well as the weakness of the effect observed in most cases.

However, the question remains to be answered whether we can increase the response difference between antipodes by designing odorant molecules of which the enantiomers have oriented profiles with minimal superimposability in any rotational position around the polar axis. Although this can be done by means of space filling molecular models of fairly simple molecules, such as secondary or tertiary carbinols, of which the design and the synthesis would require relatively little effort, no systematic study in this direction has appeared in the literature.

Reliable data on the relationship between odour quality or odour intensity and the degree of superimposability of enantiomeric profiles in one or more series of odourants could very well give us a more accurate impression of the degree of sterospecificity involved in olfactory interaction.

Whenever such series of pure enantiomeric pairs become available, they should be used not only in psychophysical studies but also in electrophysiology for the recording of single unit response patterns.

A quantitative comparison of "enantiomeric" patterns from large numbers of units, e.g. on insect antennae (Lacher, 1964) or in the olfactory bulb (Moulton, 1967; Doving, 1965, 1966a, 1966b, 1968) may result in reproducible numerical values expressing response similarities.

In general we may expect the principle of structural engineering of stimulant molecules to be a rich source of information not only on structure-response relationships but also on functional and structural aspects of chemoreceptory systems. However, it will require the synthesis of large numbers of pure odorants as well as efficient cooperation between chemists, physiologists and psychologists.

VIII. Response similarity and structural similarity

In section IV of this chapter, correlation studies between various experimental methods of similarity assessment were mentioned as one of the techniques available to detect fundamental odour similarity.

Pseudo similarity of odour is necessarily an effect of coincidence; two different complex patterns of olfactory information happen to be translated into more or less similar odour impressions. This type of odour similarity must be assumed to have a superficial nature and it cannot be expected to survive an evaluation by means of psychophysical techniques. Also, the similarity is observed at the terminus of the olfactory process, in the odour sensation, and it cannot be expected to occur at other levels in the olfactory system. This means that significant correlations between response patterns on any level for odorants with similar odours can be interpreted safely in terms of fundamental similarity.

Døving (1966b) applied these principles to a group of 33 stimulants representing 5 of the 7 primary classes proposed by Amoore (1952). Recordings were made from a number of bulbar units in the frog and the response patterns were analysed statistically. In some classes such as the camphoraceous and the ethereal class, high chi-square values were obtained for some pairs but not for others. The only class for which chi-square values indicated a significant association between all response patterns was the musk class. It is interesting that the musks also produced response patterns with an unusually large number of unaffected units, suggesting a fairly selective interaction pattern. These results suggest that the descriptor "musk" can safely be used to label one of the basic types of olfactory information.

Døving and Lange (1967) continued the work published by Døving by studying correlations between chi-square values and two psychophysical techniques, similarity judgements and confusion measures. Rank correlation coefficients show that the agreement between the two psychological methods is better than between each of these and the chi-square method. This may mean that the psychophysical techniques are not entirely reliable to detect pseudo-similarity effects. This does not seem improbable since such techniques analyse the odour sensation itself, i.e. the source of the similarity effect.

Methods of this type should be further developed since they are useful tools in the study of structure-response relationships in olfaction.

IX. Specific anosmia

In section III the identification of the basic types of olfactory information was mentioned as one of the objectives of the pharmacological approach in olfaction research. Specific anosmia, i.e. the inability, in persons with otherwise normal olfactory acuity, to detect the odour of one or more odorants, appears to be a major tool in the study of this aspect.

The problem of characterization of informational types in olfaction is only meaningful if approximately identical sets of receptor sites are available to all mankind and perhaps even to all creatures possessing well-developed olfactory systems. Since sites are probably normal structural details of proteinaceous membranes, we may safely assume that this is the case.

However, structural details of proteinaceous matter in living tissues are genetically determined and may, to a limited extent, vary between individuals. For olfaction this may mean that a type of site which is available to most individuals may be lacking, or, perhaps for sterical reasons, not functioning properly, in the membranes of some. In such a deficient membrane, a specific informational component would not be generated and its derivative would be lacking in the final odour sensation. This would only become obvious for primary odorants in the odour of which the deficient component is the predominating one.

On this purely theoretical basis, observers can be classified for any primary group of odorants as belonging to one of the following categories.

Normal observers, in whose olfactory epithelium the receptor site type eliciting the predominating component is functioning normally and who, consequently, can smell and identify the latter.

Parosmic observers. The site type eliciting the predominating component is not functioning and its final derivative is not observed in the odour. Members of this group experience the minor components normally, usually at a much lower level of intensity and they are able to identify, to an extent depending on experience and acuity, some of these components.

Anosmic observers. These have the same deficiency in their epithelium as the parosmic observers but their acuity is too low to observe the minor components. Since in the information patterns elicited by odorants belonging to the same primary class the minor components may differ in quality and intensity, we may expect observers who are anosmics to some representatives of this class, to be parosmics for others.

These phenomena, specific anosmia and parosmia, actually occur in many subjects and since their characteristics seem to agree with those derived from the theoretical outline described above, we may assume that the explanation given is, within broad limits, a correct one (Beets and Theimer, 1970).

Guillot described various examples of specific anosmia (1948a, 1948b) and of parosmia (1953) and was the first to suggest that specific anosmia may be used for the detection of "odeurs fondamentales". Guillot (1948a, 1948b) was also the first to study anosmia for a primary class, the musks, and found some indications that anosmic patterns for different structural classes of musks do not coincide completely. Although this may mean that there exists more than one informational component of the musk type, other explanations are possible (Beets, 1971).

Also Amoore (1967, 1968) and Amoore *et al.* (1968) used specific anosmia to detect and to classify primary odorants or, according to the concept described in this chapter, to define and to label monoosmatic informational components.

Two experimental approaches to the problem of characterization of informational components have been described in the literature.

Amoore (1967, 1968) and Amoore *et al.* (1968) studied anosmia for the sweaty odour of isobutyric acid and related chemicals.

A series of buffered solutions of pure isobutyric acid, the concentration of each of which was half that of the next solution, was presented to a panel consisting of 97 normal subjects and 10 subjects with specific anosmia for the sweaty note. For each concentration the ability to distinguish from an odourless liquid was determined. The difference between the mean thresholds for isobutyric acid of the normal and the anosmic group was found to be reproducibly 5.6 ± 0.4 binary steps.

The experiment was repeated with a large series of structurally related odorants such as fatty acids with straight and branched chains, isobutyraldehyde, isobutyl alcohol and isobutyl isobutyrate. Isovaleric acid, which has by far the lowest threshold value of the series, showed the highest threshold difference. The obvious interpretation (Beets, 1970a) is that for fatty acids the ratio between the frequency of the predominating informational component, which may be labelled with the descriptor "sweaty", and the summated frequencies of the secondary components in the primary information pattern, shows a maximum for isovaleric acid.

Beets and Theimer (1970) reported a different method which, using specific anosmia as well as parosmia, does not require the measurement of threshold values. Three odorants, Δ^{16}-androstenone-3, $\Delta^{4,16}$-androstadienone-3 and *cis*-2-methyl-2-(4′-*tert.* butylcyclohexyl)-pentanone-4, having intensive and very similar urine-odours were presented in high dilution on a smelling strip to a panel of 100 subjects who were requested to estimate the odour intensity of each odorant and to describe the quality of the odour in terms of one or more associative key-words.

All normal observers described the odour of the three ketones as extremely intensive, using the descriptors urine and perspiration most frequently. 35 subjects were unable to smell the major note in any of the three ketones. Of these, 6 were anosmics and 29 were parosmics. The latter described the odours as weak and used descriptors which are not only totally unrelated to those proposed for the major note but which also varied widely in frequency of occurrence between odorants. However, approximately the same descriptors were used for all odorants.

The obvious conclusion is that there exists an informational component which can be labelled with the descriptor "urine", and that this component strongly predominates in the primary information patterns elicited by the three ketones. Using Amoore's terminology we may say that the three ketones belong to the same primary class.

Systematic application of the simple technique described by Beets and Theimer as well as of the more time-consuming method of Amoore brings the complete and definitive identification of all types of information involved in olfaction within reach. Amoore's method may have the added advantage that it yields some quantitative information on the intensity relationship between the major note and the secondary notes in the odours of primary odorants.

Decreasing molecular size and the disappearance of the steric requirements of the molecular profile can be expected to produce a drastic change in structure and complexity of the information pattern. As we have mentioned in section VI, very small molecules supposedly interact by hydrogen bonding with single functional groups in the membrane. Such molecules have a much higher chance of finding a suitable site than larger ones requiring a more complex profile-specific site for effective interaction. Also, this chance is probably more evenly distributed over the epithelium. The resulting information patterns can be expected to have a much more random character than those generated by a population of larger molecules. Since it is highly improbable that one of the informational types in such a complex pattern will predominate strongly, anosmia will probably not occur for very small molecules. This seems to offer an interesting opportunity for experimentation.

Systematic work on anosmia has hardly started. So far only three types of olfactory information have been identified with certainty, characterized by the descriptors "musk", "sweaty" and "urine", others, such as "camphoraceous" seem to be probable. Continued research on anosmia will probably enable us to define all or many of the others. In order to assist this type of research a simple informational code may be developed, using italic symbols for informational types which have been identified with certainty (e.g. *Mu* for musk, *Pe* for perspiration and sweaty and *Ur* for urine) and omitting the underlining for tentative labelling (e.g. Ca for camphoraceous). Greek letters may be used to indicate descriptors which are frequently used by subjects without having been identified as representing monoosmatic types of information.

Since the percentage contribution of a specific informational component to the total odour intensity of an odorant could be estimated by trained panels, consisting of normal and parosmic members, it seems that a complete and semi-quantitative definition of the odour of any odorant in terms of its basic components may, in the future, become a distinct possibility. Since such a definition would open the way to computer processing of odours, the operation of blending odorants in order to realize desirable odour effects may once become a task for the computer. Although this would have great practical value, it would only free the creative perfumer from the routine elements of his work, giving him more time for creative work and enabling him to produce results in much less time.

X. Cross adaptation

A second phenomenon which may find some use as a tool for the analysis of olfactory information is cross-adaptation. Adaptation is the temporary loss of sensitivity as a result of continuous or repeated stimulation. When the conditioning stimulus is not identical with the tested stimulus, the phenomenon is called cross-adaptation.

The process of adaptation seems to consist of two super-imposed components. The major one, characterized by fast recovery, is located in the olfactory bulb (Adrian, 1950; Ottoson, 1956), the minor one, characterized by slow recovery, in the epithelium (Stuiver, 1958).

Irrespective of the mechanism, it seems that adaptation causes a temporary loss or reduction of sensitivity of receptor site types or information-specific units in the bulb, involved in the generation or processing of the informational components present in the response pattern elicited by the conditioning stimulus. Consequently, stimulation with the tested stimulus immediately following conditioning can be expected to yield an odour sensation in which the odour components of the conditioning stimulus are lacking or weakened.

This suggests that adaptation is strongest when the conditioning and the tested stimuli are identical. Cross-adaptational effects must be strongest for odorants belonging to the same primary class and decrease with decreasing resemblance of the primary information pattern. In principle, Zwaardemaker has realized this as early as 1895 and the idea has been followed up by several modern authors.

Le Magnen (1948) studied cross-adaptation with odorants belonging to bitter almond, camphor and musk types. Stimulation with benzaldehyde after conditioning with benzonitrile leaves an odour sensation reminiscent of safrole. The residual odour after conditioning with a mixture of benzonitrile and safrole resembles indole whereas conditioning with a mixture of safrole and indole leaves an uncharacteristic benzene-like note. Musk-ketone and pentadecanolide, two structurally unrelated musks, showed no cross-adaptation. This seems to confirm Guillot's (1948a, 1948b) findings, mentioned earlier in this section, on anosmia with steroid and macrocyclic musks. The same explanations may be considered.

Cheesman (1960), Cheesman and Mayne (1953) and Cheesman and Townsend (1956) studied cross-adaptation for a group of 4 odorants (cyclopentanone, cyclopentanol, isopropanol and dioxan) by measuring the threshold concentration after exposure to a series of concentrations of the conditioning odorant. The results suggest a linear relationship between the logarithm of the threshold concentration of the tested odorant and the logarithm of the concentration of the conditioning odorant. The slope of this linear function can be used as a measure of the degree of similarity between the olfactory information patterns of both odorants. The highest values are obtained when the tested and the conditioning stimuli are identical. These values are approximately independent (0.7) of the nature of the odorant. The relationship between these results and molecular shape has been discussed by Beets (1971).

Extensive work on cross-adaptation has been published by Köster (1971).

XI. Informational complexity in olfaction

In previous sections we have seen that olfactory information patterns of widely varying complexities can be generated on the epithelium by stimulation of the latter with suitable odorants.

The simplest pattern is produced by a pure primary odorant such as a urine compound or a musk. Increasing informational complexity can be realized by stimulation with a mixture of two or more primary stimulants, with a non-primary stimulant and finally with a mixture of non-primary stimulants. It would be interesting to know, where, in terms of informational complexity, odorants with very low molecular weight (see section VI) fit into this series.

We do not know, whether and how the human mind experiences informational complexity of odour patterns. It is certainly not consciously experienced and consequently it cannot be assessed by means of verbal expression.

It seems that the measurement of duration estimates could be a suitable tool.

Ornstein (1969) on the basis of experiments with auditory and visual messages, formulated recently his storage size metaphor, which says that experience of duration lengthens with increasing complexity of the stimuli filling a given interval. The obvious explanation is that the more complex the information presented in a given period is, the larger the storage size in memory required.

Since this phenomenon seems to be experimentally rather solidly demonstrated, it seems possible to apply it to duration experience of intervals during which olfactory stimuli are presented. Such experiments could show whether the higher centres experience the increasing complexity of the presented information and whether they interpret it in the same way as has been observed for auditory and visual information, i.e. by lengthened duration estimates.

XII. Response intensity: indirect relations

Since all properties—chemical, physical and pharmacological—of a chemical compound, are direct consequences of its molecular structure, they are necessarily interrelated. In many cases the interrelationships can be expected to be too complex to be detected but in others we may discover simpler relations between two or three characteristics. Odour quality and odour, or olfactory response intensity may be found among the latter.

The work of Wright (1966, 1967) about the relationship between odour quality and specific frequencies in the far i.r. spectra as well as that of Theimer and Davies (1967) on the relation between musk odour intensity and the desorption rate from a water surface belong in this category. Attempts to define more complex relationships between olfactory thresholds (Laffort, 1969) and odour intensities (Dravnieks and Laffort, 1972) on the one hand and physical parameters on the other have been reported recently.

Tucker (1963) applied Beidler's (1954) taste equation to his electrophysiological recordings from small twigs of the olfactory nerve in the tortoise. He concluded that experience with quantitative olfactory data is parallel with the results of gustatory studies and that the two types of chemoreceptor sites may very well operate by a common mechanism.

These are examples illustrating that quantitative studies of relationships between odour intensity or olfactory response intensity and structural or physical characteristics may become a rewarding area of research.

XIII. Concluding remarks

In accordance with the objectives of this book, the purpose of this short chapter is mainly to suggest a number of openings for experimentation which may stimulate multi-disciplinary interest in the molecular approach to olfaction and which may help to establish the pharmacology of olfaction firmly as a new member of the group of disciplines involved in research on olfaction. Up to very recently, it has not been recognized as such although huge amounts of data on pharmacological aspects of olfaction have accumulated in the literature as by-products of other types of research. Since these data have been reviewed and analysed extensively in recent literature, I have only referred to them briefly where the text seemed to require this. Also the use of figures and structural formulae has been avoided.

Since the pharmacological approach is a new aspect of olfaction research, it seemed to be justified to include a theoretical outline of the subject which may serve as a starting point for experimentation and which may be either confirmed and developed or modified as the work proceeds.

References

Adler, J. (1972). Chemoreception in bacteria. *In* "Olfaction and Taste, IV" (Schneider, D., ed.), 70–80. Wissenschaftliche Verlags-Gesellschaft M.B.H., Stuttgart.

Adrian, E. D. (1950). The electrical activity of the mammalian olfactory bulb. *E.E.G. Clin. Neurophysiol.* **2**, 377–388.

Adrian, E. D. (1953). Sensory messages and sensation. *Acta Physiol. Scand.* **29**, 5–14.

Amoore, J. E. (1952). The stereochemical specificities of human olfactory receptors. *Perf. Ess. Oil Rec.* **43**, 321–330.

Amoore, J. E. (1962). The stereochemical theory of olfaction. 1. Identification of the seven primary odours. *Proc. Sci. Sect. Toilet Goods Ass.* **37** (*suppl.*), 1–12.

Amoore, J. E. (1967). Specific anosmia: a clue to the olfactory code. *Nature, Lond.* **214**, 1095–1098.

Amoore, J. E. (1968). Specific anosmias and primary odours. *In Proc. of the Nato Summer School,* Istanbul, 71–85.

Amoore, J. E. (1969). A plan to identify most of the primary odors. *In* "Olfaction and Taste", Vol. III (Pfaffmann, C., ed.), 158–171. Rockefeller University Press.

Amoore, J. E., Palmieri, G. and Wanke, E. (1967). Molecular shape and odour: Pattern analysis by PAPA. *Nature* (*Lond.*) **216**, 1084–1087.

Amoore, J. E., Venström, D. and Davis, A. R. (1968). Measurement of specific anosmia. *Percept. Mot. Skills* **26**, 143–164.

Beets, M. G. J. (1957). Structure and odour. *In* "Molecular Structure and Organoleptic Quality", 54–90. S.C.I. Monograph No. 1. Society of Chemical Industry, London.

Beets, M. G. J. (1970a). The molecular parameters of olfactory response. *Pharmacol. Rev.* **22**, 1–34.

Beets, M. G. J. (1970b). Relationship of chemical structure to odor and taste. *In Proceedings 3d International Congress of Food Science and Technology,* Washington, 379–394.

Beets, M. G. J. (1971). Olfactory response and moleçular structure. *In* "Handbook of Sensory Physiology", Vol. IV, 257–322 (Beidler, L. M., ed.). Springer-Verlag, Berlin, Heidelberg, New York.

Beets, M. G. J. (1973). Structure-response relationships in chemoreception. *In* "International Encyclopaedia of Pharmacology and Therapeutics", Section 5, pp. 225–295 (Cavallito, C. J., ed.). Pergamon Press Ltd., Oxford.

Beets, M. G. J. and Theimer, E. T. (1970). Odour similarity between structurally unrelated odorants. *In* "Taste and Smell in Vertebrates. A Ciba Foundation Symposium", 313–321 (Wolstenholme, G. E. W. and Knight, J., eds.). H. and A. Churchill, London.

Beidler, L. M. (1954). A theory of taste stimulation. *J. Gen. Physiol.* **38**, 133–139.

Cheesman, G. H. (1960). Odour and chemical constitution—a new approach. *Proc. R. Aust. Chem. Inst.* **27**, 70–73.

Cheesman, G. H. and Mayne, S. (1953). The influence of adaptation on absolute threshold measurements of olfactory stimuli. *Q. J. Exp. Psychol.* **5**, 22–30.

Cheesman, G. H. and Townsend, M. J. (1956). Further experiments on the olfactory thresholds of pure chemical substances, using the "sniff-bottle method". *Q. J. Exp. Psychol.* **8**, 8–14.

Døving, K. B. (1965). Studies on the responses of bulbar neurons of the frog to different odour stimuli. *Rev. Laryng.* (suppl.), 845–854.

Døving, K. B. (1966a). An electrophysiological study of odour similarities of homologous substances. *J. Physiol.* (*Lond.*) **186**, 97–109.

Døving, K. B. (1966b). Analysis of odour similarities from electrophysiological data. *Acta Physiol. Scand.* **68**, 404–418.

Døving, K. B. (1968). Electrophysiological studies of sensory relatedness. *Proc. Nato Summer School.* Istanbul, 493–508.

Døving, K. B. and Lange, A. L. (1967). Comparative studies of sensory relatedness of odours. *Scand. J. Psychol.* **8**, 47–51.

Dravnieks, A. and Laffort, P. (1972). Physico-chemical basis of quantitative and qualitative odour discrimination in humans. *In* "Olfaction and Taste", Vol. IV, 142–148 (Schneider, D., ed.). Wissenschaftliche Verlags Gesellschaft, Stuttgart.

Friedman, L. and Miller, J. G. (1971). Odor incongruity and chirality. *Science, N.Y.* **172**, 1044–1046.

Gesteland, R. C., Lettvin, J. Y., Pitts, W. H. and Rojas, A. (1963). Odour specificities of the frog's olfactory receptors. *In* "Olfaction and Taste", Vol. I (Zotterman, Y., ed.). Pergamon Press Ltd., Oxford.

Guillot, M. (1948a). Anosmies partielles et odeurs fondamentales. *C.R. Hebd. Séanc. Acad. Sci. (Paris)* **226**, 1307–1309.

Guillot, M. (1948b). Sur quelques caractères des phénomènes d'anosmie partielle. *C.R. Séanc. Soc. Biol.* **142**, 161–162.

Guillot, M. (1953). Anosmies et parosmies individuelles spécifiques. *Recherches* (Oct. issue), 26–31.

Klopping, H. L. (1971). Olfactory theories and the odors of small molecules. *Agricult..Food Chem.* **19**, 999–1004.

Köster, E. P. (1971). "Adaptation and cross-adaptation in olfaction". Thesis, University of Utrecht.

Lacher, V. (1964). Elektrophysiologische Untersuchungen an einzelnen Rezeptoren für Geruch, Kohlendioxyd, Luftfeuchtigkeit und Temperatur auf den Antennen der Arbeitsbiene und der Drohne (Apis Mellifica L.). *Z. Vergl. Physiol.* **48**, 587–623.

Laffort, P. (1969). A linear relationship between olfactory effectiveness and identified molecular characteristics, extended to fifty pure substances. *In* "Olfaction and Taste", Vol. III, 150–157 (Pfaffmann, C., ed.). Rockefeller University Press, New York.

Leitereg, T. J., Guadagni, D. G., Harris, J., Mon, T. R. and Teranishi, R. (1971). Evidence for the difference between the odours of the optical isomers (+)- and (–)-carvone. *Nature, Lond.* **230**, 455–456.

Le Magnen, (1948). Analyse d'odeurs complexes et homologues par fatigue. *C.R. Hebd. Séanc. Acad. Sci. (Paris)* **126**, 753–754.

MacLeod, P. (1963). Histologie et physiologie du récepteur olfactif. *La France et ses Parfums* **6**, 233–237.

Moulton, D. G. (1967). Spatio-temporal Patterning of response in the olfactory system. *In* "Olfaction and Taste, Vol. II", 109–116 (Hayashi, T., ed.). Pergamon Press, Oxford.

Ornstein, R. (1969). "On the Experience of Time". Science of bchaviour, Penguin, Harmondsworth.

Ottoson, D. (1956). Analysis of the electrical activity of the olfactory epithelium. *Acta Physiol. Scand.* 1–83. **35** (suppl. 122) 1–83.

Pauling, L. (1946). Analogies between antibodies and simpler chemical substances. *Chem. Eng. News* **24**, 1064–1065.

Russell, G. F. and Hills, J. I. (1971). Odor differences between enantiomeric isomers. *Science, N.Y.* **172**, 1043–1044.

Schneider, D., Lacher, V., Kaissling, K. E. (1964). Die Reaktions-Weise und das Reaktionsspektrum von Riechzellen bei Antheraea pernyi (Lepidoptera, Saturniidae). *Z. Vergl. Physiol.* **48**, 632–662.

Stuiver, M. (1958). "The Biophysics of Smell". Thesis, University of Groningen.

Theimer, E. T. and Davies, J. T. (1967). Olfaction, musk odour and molecular properties. *Agricult. Food Chem.* **15**, 6–14.

Theimer, E. T. and McDaniel, M. R. (1970). Odour and optical activity. *In, Preprints Joint symposium on Perfumery. British Society of Perfumers; Society of Cosmetic Chemists Great Britain,* Eastbourne.

Theimer, E. T. and Klaiber, E. (1972). Personal communication.

Tucker, D. (1963). Physical variables in the olfactory stimulation process. *J. Gen. Physiol.* **46**, 453–489.

Wright, R. H. (1966). Odour and molecular vibration. *Nature, Lond.* **209**, 571–573.

Wright, R. H. (1967). The musk odour. *Perf. Ess. Oil Rec.* **58**, 648–650.

Zwaardemaker, H. (1895). "Die Physiologie des Geruchs". Wilhelm Engelmann, Leipzig.

Author Index

Numbers in italics refer to pages where references are listed at the end of each article

Subject Index